T0133392

# FIELD TECHNIQUES FOR SEA ICE RESEARCH

# FIELD TECHNIQUES FOR SEA ICE RESEARCH

Edited by Hajo Eicken, Rolf Gradinger,
Maya Salganek, Kunio Shirasawa, Don Perovich,
and Matti Leppäranta

UNIVERSITY OF ALASKA PRESS
FAIRBANKS

UNIVERSITY OF ALASKA PRESS
P.O. Box 756240
Fairbanks, AK 99775-6240

ISBN 978-1-60223-059-0

Library of Congress Cataloging-in-Publication Data
Field techniques for sea ice research / edited by Hajo Eicken ... [et al.].
p. cm.
ISBN 978-1-60223-059-0 (lithocase : alk. paper)
1. Oceanography—Fieldwork. 2. Sea ice. 3. Sea ice—Measurement. I. Eicken, Hajo.
GC57.F54 2009
551.34'3—dc22
2009012544

This publication was printed on acid-free paper that meets the minimum requirements for ANSI / NISO Z39.48–1992 (R2002) (Permanence of Paper for Printed Library Materials).

We gratefully acknowledge support and sponsorship by the following organizations:

Cover illustration *Camden Bay, Beaufort Sea,* by David Mollett, 36 × 48 inches, oil on canvas, © 1990.
Cover and interior layout by Alcorn Publication Design.

Printed in China

# Contents

# Preface

This book originates in part from a series of graduate and undergraduate sea ice field courses taught by several of the editors in Japan, Finland, and the United States over the past two decades. In the absence of dedicated textbooks providing an introduction to and overview of sea ice field measurements, students in these classes were provided with a range of handouts, copied pages from manuals, or other documentation. However, with growing interest in the subject area of sea ice studies, it became increasingly clear that a handbook that would lay out all this information in an organized fashion would be a useful resource for students and others with an interest in the field. Given the breadth of research methods and disciplines represented in field courses and modern studies of sea ice, an international collaborative effort appeared to be a promising approach in compiling material for such a resource. We felt that we would also want to try and preserve some of the diversity, interdisciplinarity, and breadth of perspectives offered by the international community of sea ice researchers and experts. In pursuing this approach, we recognized that conformity and homogeneity in the layout, scope, and depth of the individual chapters had to be balanced with a commitment towards diversity, and, occasionally, idiosyncrasy. The latter allowed the different contributors to present their perspective and insight into the subject matter with a didactic approach appropriate to the subject matter.

The Fourth International Polar Year (IPY) from March 2007 to March 2009 provided a unique opportunity to benefit from the flurry of ongoing field measurements and collaborative projects, drawing researchers and students into the polar regions. Thus, in May of 2008 we were able to assemble a highly qualified, international group of students and practitioners for a field course in Barrow, Alaska, that provided the foundation for this publication. In addition to those who contributed to this graduate-level class, we also solicited contributions from a number of colleagues on relevant topics not covered in the field course. The resulting product hence still retains some of the flavor of a course textbook, with a few missing ingredients that were lacking because of budgetary or other constraints, and with some of the chapters more narrowly focused because of their origin in a specific set of measurements made during the class. However, the topics considered key to field studies of sea ice—such as ice thickness measurements, ice biology, or ice sampling—are covered in depth. Also included are two broader chapters on the role of remote-sensing data and the use of models in field research, as well as a contribution on the principal aspects of data management relevant to field studies. While these topics are covered in other textbooks in much more depth, we felt it important to discuss these subject areas from the perspective of field research in the context of this volume. Despite a trend toward specialization in scientific research and training, we hope that this book may help spawn exchange and collaboration among these different approaches to studying sea ice. Moreover, in planning field measurements, remote sensing and model simulations have become increasingly valuable tools with which field researchers should be familiar.

The format of the field course allowed the inclusion of a class on visual anthropology and documentary filmmaking. The instructor Maya Salganek and her students compiled and edited video footage (accessible on the multimedia DVD that comes with this book) illustrating applications of the different field methods and providing a visual almanac of the techniques introduced in the book. At the same time, we also see the footage on this DVD as a survey of sea ice field research as practiced in the Arctic during the Fourth IPY, and hope that future historians of science may gain some insight (and entertainment?) in comparing what may well appear as crude efforts by the time the Fifth IPY comes around.

The aim of this publication is to serve as both an introduction and a reference for techniques and methodologies employed in sea ice field research. It is addressed primarily at students as well as practicing scientists, technicians, or engineers who find themselves facing a question or problem that requires measurements or data collection on sea ice. At the same time, the book may also serve as a resource to scientists from other fields, arctic communities, and members of the public who are interested in learning more about how knowledge about sea ice is derived from measurements in the environment. For the latter, the DVD may serve as an introduction and provide a glimpse of the different research activities and approaches. More in-depth knowledge can then be obtained from the different chapters in the book. For those new to the field but interested in a comprehensive introduction to the design and execution of sea ice field research programs across a range of relevant disciplines, the different chapters provide a solid foundation, with further reading material, resources, and video demonstrations of specific approaches compiled on the accompanying DVD.

As pointed out by one of the reviewers of an earlier draft of this book, this is an arctic-centric publication. We see this as both a strength and a weakness—that is, a result of the origins of this book in a place-based field course. While many of the chapters discuss methodologies specific to arctic sea ice research in detail, much of the book does apply to sea ice regardless of its location, even if many of the examples we draw on come from the northern hemisphere. At the same time, the reviewer was right in implying that much is to be gained from a perspective on sea ice that includes both poles (and lower latitudes) as vantage points. We may well have failed in that regard, but see this as a challenge and motivation for future work. The same may be said about the varying approaches taken in the different chapters with respect to depth and breadth of coverage. While we have strived for some uniformity, the remaining diversity does reflect more than simply chapter authors' preferences and is indicative of different approaches to the field by different disciplines. In that regard, we see this book as the first, modest contribution to a broader exchange between the different disciplines and knowledge systems with an interest in sea ice. Much of this book may be obsolete already by the year that arctic summer sea ice is predicted to disappear completely by the most dramatic forecasts (2013 as of this writing). It is our sincere hope, however, that this volume may spark broader collaboration among sea ice researchers to document and refine the best-practice approaches to sea ice field studies. The use of modern tools such as collaborative online documentation—for example, in the form of wikis—has much to offer and can both grow out of and supercede the chapters that follow.

Sea ice research is a highly collaborative enterprise and we have been fortunate to have great support from a range of different people and organizations. Hence, we would like to gratefully acknowledge the constructive criticism of the reviewers of this book, colleagues and students who commented and helped us improve upon various aspects of this work. Support in the field is crucial and requires a substantial network of experts who provide a range of logistics support and we are thankful for their help. The course this book builds on has received particularly valuable support from the Barrow Arctic Science Consortium, a science-support organization that is built on the century-long partnership between the Iñupiat Eskimo of northern Alaska and scientific researchers. Funding agencies play a key role in enabling both research and instruction and we are more than grateful for the financial support of the U.S. National Science Foundation, the Geophysical Institute, the International Arctic Research Center, the Center for Global Change and Arctic Systems Research and the School of Fisheries and Ocean Sciences at the University of Alaska Fairbanks, the North Pacific Research Board, the Institute of Low Temperature Science, and the International Antarctic Institute of the Graduate School of Environmental Science of Hokkaido University, as well as the Climate and the Cryosphere Program of the World Climate Research Program and the Census of Marine Life, Arctic Ocean Diversity Program. Finally, we acknowledge the help of Jane Brackney, Mette Kaufman, and staff at the University of Alaska Press, in particular Sue Mitchell and Elisabeth Dabney, in helping us put this book together.

H. E., R. G., M. S., K. S., D. K. P., M. L.

Fairbanks, Alaska, March 2009

**Foreword**

# Encounters with Northern Sea Ice

*Ned Rozell*

## INTRODUCTION

While many people have heard of sea ice, relatively few have seen it. Even fewer have stepped on it. In Alaska, villagers on the western and northern coasts have some interaction with the enigmatic ice that forms on top of frigid saltwater, as do some scientists and ship captains who venture into the Arctic. But most people don't know much about sea ice.

Northern sea ice is an ever-moving jigsaw puzzle of large and small pieces of ice, some formed by the cold air of the present year and some thicker ice that has survived at least one summer. Collectively, this ice forms a disc surrounding the North Pole that waxes and wanes with the seasons, pushed by currents and winds.

Geologists have found at the bottom of the Arctic Ocean the fossils of plants that now grow off Southeast Asia, suggesting that millions of years ago the northern ocean was not only ice-free but warm. With our short memories and life spans, we tend to put more weight on current trends. The recent shrinkage of the northern ice pack—revealed by satellites that have given us an overhead view since the late 1970s—now makes headlines each fall, when the pack reaches its minimum in mid-September. A new record minimum was reached in 2007, with 2008 not far off the mark.

Through this period of transition from what we know as the northern ice extent to what it will become, people are out there every day interacting with this ephemeral substance.

These people depend on measurements that make sense of the ice, be they a hammer tied to a rope and dropped through a hole, or a view from 500 miles above captured by a hulk of orbiting metal and plastic. Because of sea ice's ability to modify the world around it, measurements of its properties, no matter how slight, are some of the most important gathered today. That information may help us quantify and understand what is perhaps the most far-reaching physical change of the Earth in our lifetimes.

## OPENING THE WAY

One way to learn about sea ice is to drive a few thousand miles through it. Dan Oliver lives in the fishing town of Seward, Alaska, where he's the project manager for a new university research ship, the Alaska Region Research Vessel. Before his new career, he spent much of the last three decades cutting paths through sea ice on U.S. Coast Guard icebreakers.

"I find all aspects on the icebreakers pretty fascinating—how you drive them, the ability to hit stuff, where you [can] go, their machinery plants," he says in his office, on the waterfront in Seward.

Oliver, a boyish 51, first saw sea ice when he was in the Coast Guard Academy back in 1978. Back then, the native of San Luis Obispo, California, served on the Coast Guard icebreaker *Westwind*, which cut a trail north through the ice for other ships headed to Thule Air Force Base in Greenland. During his career, he served on several icebreakers, in positions from engineer to ship commander.

He has a special relationship with the 420-foot, 16,000-ton *Healy*, the largest ship in the Coast Guard's fleet. Oliver was the first executive officer on the *Healy*, riding with her out of the shipyard in Louisiana to the ship's home port in Seattle, and thousands of miles beyond.

"I've been all over the world," he says. "Four trips to Antarctica, to the east and west Arctic. The only continent I haven't been to is Africa. I've seen quite a bit of the world, all because of icebreakers."

It takes years of experience and observation of sea ice to take the helm of one of these 400-foot behemoths with a million-gallon gas tank. Over the years, Oliver learned that deep blue ice is multiyear ice, thick stuff that "beats you up pretty good" when you drive into it. And, when the pole nods toward the sun in midsummer, the freshwater of melt ponds on sea ice can ease the ship's passage by acting as a hull lubricant. Conversely, cold, fresh snow on top of sea ice makes an eerie squeaking sound as it robs the ship of momentum. And a captain eventually develops an eye for where to attack a pressure ridge, where two large ice floes are straining against each other (or maybe not to attack it at all). And he always knows which way the wind is blowing.

Because its bridge doesn't allow a captain to see very far ahead, the *Healy* is an interesting ship to drive, Oliver says. Whenever heavy ice is ahead, the captain heads to the "aloft con," a crow's nest about 80 feet above the waterline.

"It's a lot of fun to drive from up there," Oliver says. "In aloft con, you've got your helm and your throttles up there. It's like you're driving from a flying bridge of a small boat."

Way up there above the sea ice, a captain has to look out and think ahead to choose a path for the ship, using clues like a "water sky," an area of darkened clouds that promises open water.

"The open water may be beyond what you can see in the aloft con, but it's a good indicator of what direction you want to take ahead," Oliver says. "At your broadest viewpoint, you're looking out over next day with helicopter, satellite, looking for areas of heaviest ice to avoid, and large leads. You may have a particular spot you want to go to, particularly [to deploy a] science station, but you're never going to drive right to it. . . . For the most part, you're never going in a straight line; you just don't want to go backwards."

When dealing with massive ice floes that dwarf even the largest nuclear icebreaker of the Russian fleet, there are times when the ice wins. Oliver was on the *Healy* when it got stuck west of Barrow for four days.

"We were trying to get to a coring spot in early June," Oliver says. "The person driving the ship tried to cut through a floe rather than going around, and the one floe was really two big ones."

The ship got stuck hard between the two floes, and there was nothing the crew could do but keep the screws moving to keep ice from the propulsion system. Ice from a pressure ridge rose above the level of the main deck, and the ship lost four days of science time.

"If you get stuck in the early part of summer, it gets frustrating, but ice is so dynamic," Oliver says. "The wind's going to shift, the pressure's going to get off the hull.

"When you're stuck, you're stuck, and best thing you can do is shut down and wait it out, because sooner or later, in a half day or four days, it's going to change."

Sea ice can be a difficult substance even when it's not trapping your ship. Returning from a North Pole trip in 2005, the *Healy* and the Swedish icebreaker *Oden* resorted to the "backing and ramming" technique to make it home through thick, multiyear ice.

"Every ship length we made was a success in that one," Oliver says. "We backed up one-and-a-half or two ship lengths, then went full ahead at seven or eight knots."

Oliver and the Swedish captain used their momentum to drive the ships up on ice and bend it to the breaking point, time and time again.

"Twenty miles was a good day on that trip."

In early summer, icebreaker captains can be more aggressive, because the heat shrinks ice floes and opens leads. Late in the season, with ice growing each day, it's different.

"In fall, you've got to be more conservative," Oliver says.

Much sea ice has formed and disappeared since 1978, when Oliver, then 21, rode the *Westwind* up to Greenland. He has noticed long-term changes in the northern ice pack since his first trip, about the same time satellites began showing us the ice from above.

"In 1978, it took a number of days to work ourselves through the ice to get to Thule," he says. "The last time I was at Thule, in 2003, at the same time in late July, there was open water there."

As for why there's less ice, "I'm pretty convinced it's the effect of global warming," Oliver says. But he knows the ice well enough to predict that it isn't going anywhere for a while.

"Certainly, I would tell you the ice breaks out earlier," he says. "But I don't think you'll ever see the Northwest Passage open year-round."

## WALRUS CHOOSE THEIR FLOES

First-year sea ice may look the same to most people, but marine mammals might prefer specific types of ice, varieties that might be on the wane in a warmer climate.

Gary Hufford, the regional scientist with the National Weather Service in Anchorage, has, along with his colleagues, found that walruses seem to gravitate towards broken, angular ice floes—rectangles, triangles, and other sharp-edged cakes of ice. He compared years of data about walrus location to ice-zone classifications and found that walrus were most often riding jagged ice floes, rather than rounded ones.

"It fit like a glove," Hufford says of what he saw when he lay maps of walrus concentration over images of angular ice floes.

Over the years, Hufford has verified the notion by watching walrus from the bridge of icebreakers, such as the U.S. Coast Guard's *Healy*. The huge animals—the largest weighing more than a pickup truck—most often lounge on sharp-edged ice floes, possibly because more open leads exist between angular floes, which don't fit together as snugly as rounded ones. Angular ice floes provide more open water, through which walrus can dive to the sea floor, where they use their tusks and muzzles to stir up invertebrates. A favorite food of the walrus is clams, which they suck from shells with powerful, pistonlike tongues.

Hufford, a 66-year-old physical scientist who has worked with the National Weather Service since 1978, first looked at possible correlations between marine mammals and types of sea ice years ago. When climate warming came into the public consciousness in the 1990s, he revisited the subject, wondering if walrus habitat in the western Bering Sea, "the home of the broken ice pack," might be dwindling.

"If we look at the images from 1999 to the present, it seems there's more meltout in the western Bering than in the eastern Bering," Hufford says.

He's also using high-resolution images from several satellites, including MODIS (250 m resolution) and SAR (10 m resolution) to see if jagged floes preferred by walrus are melting faster than rounded floes, which have less surface area related to their volume that is exposed to open water.

Walrus use sea ice as a platform for breeding, birthing, resting, and migration. They climb aboard sea ice when it forms in the Bering Sea slightly north of St. Paul Island and ride it northward into the Chukchi Sea. Less angular ice floes would especially affect female walrus, which use the ice as a platform to raise their pups in summer. Hufford wonders if they can adapt.

"[There are] two problems," Hufford says. "Floes they're using are disappearing quickly now, and what happens with total reduction [of ice] over time? Will they go to a rounded floe and feel comfortable?"

Another problem walrus face is riding the floating ice farther north into deeper water, which would prevent them from diving to the bottom. And the moving ice spreads the walrus harvest of clams in a random pattern, which is important when a single animal can eat hundreds of pounds of clams in a day.

"If the ice totally goes away, just how far can they forage without eating themselves out of house and home?" Hufford says.

In the worst-case scenario, extreme loss of sea ice might doom all the walrus in the Bering and Chukchi seas. Hufford and his coworkers are also looking at that possibility. The loss of thousands of walruses churning the bottom of the cold oceans might mean less productivity in those famously rich waters.

"The loss of walrus could affect the entire Bering Sea," Hufford says.

Along with walrus and the well-documented polar bear, ribbon seals are also at risk in this time of change, Hufford says. The seals seem to prefer marginal ice that supports their weight but is too flimsy to hold a polar bear. But a warmer world also endangers their favorite habitat.

"If diminished sea ice conditions persist, and if ribbon seals are not able to adapt to land as a substrate for reproduction, the prospect for them is bleak," Hufford's colleague G. Carleton Ray wrote in a recent paper.

Hufford has for three decades cruised the cold waters of the Bering and Chukchi seas, as well as the ocean surrounding Antarctica. With his experience riding icebreakers, Hufford doesn't need satellite images to tell him that things have changed in the last 30 years.

"I'm used to seeing meter-thick ice, and now I'm seeing half-meter ice," he says.

## GENERATIONS ON ICE

At the end of March, Joe Leavitt straddles his snowmachine, fires it up, and drives away from his hometown of Barrow, Alaska, and onto the snow-covered sea ice. Stopping his machine a mile from shore, he squints at the blinding-white horizon and smiles, knowing that this year will be a good one for the whalers. The winds have favored them. The ice is thick, more than one year old, which makes him feel nostalgic for days gone by. And it makes him feel safe.

Leavitt, 50, is a member of a crew that hunts bowhead whales offshore of the farthest-north U.S. community. Since he was old enough to make coffee for his father's whaling crew, he has been on the sea ice in springtime. Years ago, as a small boy, he played with his friends on the ice in front of town no matter what the season.

"The ice used to hug Barrow all year round, even in the summer—barges had a hard time beaching in Barrow," he says. "Now, we're getting cruise ships into Barrow every summer."

Despite the recent changes, Leavitt and others are able to continue their tradition of harvesting saltwater mammals that weigh 100,000 pounds and spend their entire lives—which can be as long as two centuries—in the vicinity of sea ice. Near Barrow, sea ice still grows offshore starting in December (rather than October, as it did when Leavitt was a boy).

By March, a few dozen whaling teams venture out on the ice, making trails toward open water that end at flat white slabs, nice platforms for hunters to winch a whale using a block-and-tackle pulley system (sometimes two, when the whale is extra large).

The whalers follow trails out to open water, often about four miles off Barrow, dragging skiffs with their snowmachines. During the best years, they find ice that has survived several years in the Arctic Ocean, growing each winter and shrinking but not disappearing each summer. But sometimes multiyear ice doesn't blow in, and they hunt on ice that formed during the current winter. They operate on ice one foot thick rather than older ice that can be two stories thick.

"Some of the first-year ice won't hold a whale," Leavitt says. "We have to find a pressure ridge to pull up a whale. Multiyear ice right off Barrow, that's the best you can ask for. When we have multiyear ice, we don't have to run [grab all their stuff and go when currents and wind bring in new masses of ice]. We know it's stronger than the ice that's coming in. It's a lot safer. Sometimes we don't have to move away from the lead all spring.

"When ice is too young, we have to get away from the lead when more ice comes, because the ice will crack behind you, and you'll get taken out. And once you get drifted out, you don't want to get drifted out again."

In May 1997, Leavitt and 146 other people were hunting on an ice floe when the tide lifted up a young, "glued-together" sheet ice during a bad ice year. A fracture developed between the hunters and shore, and they were adrift.

"We tried to make it to the beach but the ridging had already started there and we had to head back out toward sea to be on the safe side. But when we were out there, it started ridging right beside us."

A helicopter rescued the whalers, but they had to leave boats and snowmachines behind.

"We left all our gear out on the ocean," he says. "The only way we could get back into town was a chopper. That was the baddest feeling in the world. We even left my father's flag in the boat—that was the biggest mistake. But the elders went out to the boats and collected all the flags. I'll never leave that flag again. . . . We eventually got our gear back, about 29 miles east of the point."

Leavitt says that first-year ice is now the more common form that collects in front of Barrow. But, once in a great while, strong winds will push in the strong, thick, multiyear ice the whale hunters prefer.

"Once it comes back, it's like everything's back to normal," Leavitt says.

## ICE-FREE SUMMERS AND WEATHER

Northern sea ice has made headlines in recent years for shrinking to its smallest extent in the era of satellites (2007), and for having made a slight recovery (2008). Some researchers wonder if the ice has dwindled to a "tipping point," which could result in an ice-free Arctic Ocean during summer.

Though summer seems an obvious period to examine the effects of an ice-free Arctic Ocean, John Walsh thinks autumn might cause more changes in northern weather.

"If we lose the summer ice, we're going from a situation where we had an oceanwide ice bath [sea ice with water ponded on top and open leads] to one where we'd have an open ocean within a few degrees of freezing," says Walsh, chief scientist with the International Arctic Research Center at the University of Alaska Fairbanks. "In both cases you provide moisture to the atmosphere and provide heat to the atmosphere.

"I think the real difference comes in fall and early winter," he says. "If you open up the ocean in summer, you can absorb more energy in the upper 50 meters or so, and that really delays freeze-up. As you get into October and November, you then have an open ocean surface that's a lot warmer than what we had in prior decades, when we'd get freeze-up in September or October, all the way down to the Bering Strait."

With open ocean comes more moisture at a time when the sea was formerly locked up in ice, and the air was colder and dryer.

"In the autumn and early winter, the drier atmosphere is ripe for taking up the additional moisture, which feeds storms," Walsh says. "We're setting ourselves up for a double whammy—you're not only having the ice coming back later, but you're favoring the fall storms . . . so the coastal communities are more vulnerable."

An ice-free summer would affect more than just villages touching the ocean in northern and western Alaska, Walsh says. For example, more moisture available to the atmosphere above Alaska's North Slope, combined with the seasonally cooling air, sets the stage for more snow.

"You could potentially have both warmer and snowier conditions for the first half of winter, compared to now," Walsh said. "Both of those favor warmer ground."

Permafrost researchers like Tom Osterkamp have written that heavy snows do their share of permafrost thawing because of snow's insulating effect. Less permafrost could be a direct effect of an open Arctic Ocean.

"That would apply to Alaska as far south as the Brooks Range and perhaps into the Interior . . . and large parts of northern Europe and Asia," Walsh says.

Walsh says more snow and warmer autumns due to the lack of summer sea ice could also be in store for Europe and Asia to about 50° N latitude, which includes most of Scandinavia and northern Russia.

But, at the time of this writing in 2009, an open Arctic Ocean in summer is not a certainty. Though several scientists have intimated that northern sea ice has crossed a threshold, Walsh says there are too many unknown factors out there for him to make such a prediction.

"I'm a little more cautious than some people as to whether we've passed the tipping point," he says. "When you get down to it, the Arctic Ocean has not been ice-free for tens of thousands of years. . . . If these '07 and '08 events were influenced to a large extent by natural variability—if there happened to be a few years when a lot of ice was transported out of the Arctic by wind forcing—history would say the Arctic Ocean can recover.

"What's different now is the extra couple of watts per square meter of greenhouse forcing, and the fact that we're seeing warmer water coming in from the North Atlantic. Those two factors suggest why we could be in a different ball game from the past 10,000 years."

Walsh says that northern-ocean researchers found signs of a reversal of the recent warming of the Atlantic water in the Arctic Ocean during a 2008 cruise, and that's another factor that makes him less convinced the ocean will be ice-free very soon.

"[There were people at the 2008 American Geophysical Union Fall Meeting] saying we could be ice-free within the next decade," Walsh says. "It's conceivable. I certainly wouldn't bet the ranch on that scenario, but it could happen."

But Walsh says he will err on the side of caution for the time being.

"We've got a research area that's become a lot more prominent in the public eye, which is good," Walsh says. "But we want to make sure we don't oversell or oversimplify or exaggerate what may be about to happen. The odds are high that [an ice-free ocean in summer] will eventually happen over the 50-to-100-year time frame, but the road to an open Arctic Ocean may not be a straight line."

## THE FUTURE OF SEA ICE

Sitting in his sunny office above Cook Inlet in Anchorage amid stuffed manila envelopes that weighed down every tabletop, Lawson Brigham needed to make two PowerPoint presentations before he left for London one autumn night. Brigham spends a good deal of time thinking of the future of the Arctic Ocean, and people have invited him overseas to share his views of a world with less sea ice.

In London, Brigham would meet with marine insurers from all over the world to discuss the risks associated with ships operating for longer periods of time in oceans with ice in them. A few days later, in Berlin, he would attend a meeting on how the United States and European nations can work together to improve safety in the Arctic in a new era with less ice and more ship traffic.

Brigham, 60, chairs the Arctic Marine Shipping Assessment of the Arctic Council, but he learned about sea ice by bashing into it. During a career that spanned three decades with the U.S. Coast Guard, Brigham spent a few years as captain of the icebreaker *Polar Sea*, which has a home port of Seattle. In summer of 1994, he piloted a ship from Bering Strait to the North Pole and back down through Fram Strait. It was the first crossing of the Arctic Ocean by a ship that wasn't a submarine. He has also opened channels through ice off Antarctica, in the Great Lakes, and in the Baltic Sea.

"I've sailed just about everywhere on the planet that has ice," he says.

He is now working on an assessment for the Arctic Council on shipping in the Arctic, because he and others believe that big changes are coming to the region.

"From 1977 to 2008, there have been seventy-seven voyages to the North Pole, and thirty-three out of those seventy-seven were in the last four years," he says. "That's a lot. Who would have guessed?"

That increased ship travel in an area usually choked with ice includes scientific missions and three or four voyages each summer of Russian nuclear icebreakers on which tourists can reach the North Pole for fees that range around $40,000. These trips occur during the polar summer, from about June to September, and they may be on the increase if sea ice continues to decline.

Increasing values of oil, gas, coal, nickel, and zinc are also fueling the pull northward into regions that were until recently too icebound to consider for development. The largest zinc mine in the world—Red Dog Mine northeast of Kotzebue—is north of the Arctic Circle, and Canadians will soon develop a giant iron mine on Baffin Island.

"There are plenty of these resources in the Arctic and the question over the decades has been, 'is it economical?'" Brigham says. "And now it is."

With the potential rush to the north in mind, Brigham has written in the magazine *The Futurist* about four different scenarios that could develop in a warmer Arctic Ocean region. These could be the northern reality in 2040:

A "globalized frontier," a free-for-all in which rising prices make it feasible to develop nickel, copper, coal, oil, and freshwater sources in the high Arctic. Indigenous people along the arctic coasts benefit from more jobs relatively close to home. Booming development puts the Arctic at risk for severe environmental problems, and industrialization trumps environmental issues.

An "adaptive frontier," in which countries have come together to put restrictions on shipping and air travel for environmental protection and safety. Instead of an all-out blitz on resources, leaders have first agreed on treaties to protect fisheries and encourage sustainable development. Native leaders have a real say in economic and environmental issues. The Arctic becomes a model for habitat protection.

A "fortress frontier," in which leaders of different countries lock up their sections of the Arctic, restricting passageway through their airspace and waters. Arctic states also bar other countries from fishing in their waters. Indigenous people gain wealth but also move from their homelands in some cases due to extreme environmental events. The arctic states have great economic and military strength.

An "equitable frontier," in which transportation is one of the most important arctic industries. Fresh, clean water from the Arctic becomes one of the more valuable commodities. Arctic states create new national parks, enhancing both the environment and tourism. Poverty is low among indigenous people because of revenue sharing from the transportation, tourism, and mining industries. Military presence is low but security is high. Tensions among the arctic countries are almost nonexistent.

Brigham said he has no idea which, if any, of these scenarios will play out, and that there are wild cards in the form of land claims and the unpredictability of climate change. The only sure bet seems to be that things won't remain the same in the Arctic, largely because of changes in one substance, sea ice.

"There can be little doubt that extraordinary change is coming to the entire region and its people," Brigham wrote.

But that change may not be as helter-skelter as some imagine, Brigham says. The future Arctic will still be a pretty cold place, with only a few months of ice-free passage.

"There has never been a commercial trip crossing the Arctic Ocean, but reading the newspapers you'd think there's a lot coming soon," he says. "We have to remind the rest of the world that because of the polar night and the cold temperatures in winter, there's always going to be arctic sea ice cover."

# FIELD TECHNIQUES FOR SEA ICE RESEARCH

# Chapter 1

# Introduction

*H. Eicken, R. Gradinger, K. Shirasawa, D. K. Perovich, M. Leppäranta, M. Salganek*

## 1.1 THE NEED FOR SEA ICE FIELD MEASUREMENTS

As much as one-tenth of the world's oceans are covered with sea ice at some point during the year. Where present, sea ice plays an important, often defining, role in the natural environment. It controls heat exchange between ocean and atmosphere and is a key component of the global climate system. Polar and subpolar ecosystems are both dependent on and constrained by sea ice. Human activities are intimately tied to and affected by sea ice, be it indigenous hunters' access to marine mammals from ice platforms or the hazards presented by drifting ice to shipping and industrial development. The importance of sea ice in these different contexts is well recognized and has been the subject of a substantial number of publications and monographs dedicated to a specific aspect of the role of sea ice in the world (e.g., Untersteiner 1986; Melnikov 1997; Leppäranta 1998; Wadhams 2000; Krupnik and Jolly 2002; Thomas and Dieckmann 2003; Leppäranta 2005).

Over the past few years interest in sea ice has grown substantially. In part, this increased attention appears to be linked directly to the realization that some of the substantial reductions observed in arctic summer minimum ice extent in recent years figure prominently in a broad suite of environmental, socioeconomic, and geopolitical changes. In 2007 and 2008, arctic summer minimum ice extent was reduced by roughly one-fifth below the previous record minimum observed during the satellite era (1979 to the present) (Figure 1.1) (Stroeve et al. 2008). These reductions appear to be in line with enhanced warming forecast for the Arctic due to ice-albedo feedback, amplifying surface heating by reducing the highly reflective ice cover and exposing the dark oceans underneath (Holland and Bitz 2003). Currently, a number of arctic nations are reevaluating their territorial claims under the United Nations Convention of the Law of the Seas (UNCLOS). Part of this reevaluation appears to be spurred by renewed interest in natural resources in or adjacent to ice-covered waters, such as those of the Chukchi and Beaufort seas (Brigham

2007). Industrial development and shipping, in turn, are paying close attention to changes in the ice cover.

At the same time, arctic ecosystems and the communities and people who depend on these ecosystems are experiencing significant, in some areas rapid, change (Krupnik and Jolly 2002). In arctic (coastal) seas, much of this change is intimately tied to a thinner, less extensive sea ice cover. Such alterations in the ice cover in turn greatly affect key species such as fish, walrus, and other marine mammals (Grebmeier et al. 2006).

While not necessarily expressing causal relationships, Figure 1.1 provides two different perspectives on this suite of changes. On the one hand, the top panel indicates the magnitude of interannual variability, cyclic variations, and a superimposed trend in the summer arctic ice extent. On the other hand, the bottom panel provides an indication of the increasing public attention, as expressed in media coverage, given to sea ice. More detailed analysis indicates that this media attention was both a response to (geo) political and socioeconomic developments (e.g., listing of the polar bear as a "threatened" species under the provisions of the U.S. Endangered Species Act; increasing oil and gas exploration activities in the Arctic; revision of territorial claims under UNCLOS); growing concern over climate change; and a reflection of the media covering scientific research findings associated with a reduction in arctic sea ice extent and in particular the record minimum of 2007.

At present, we find that a diminishing arctic sea ice cover is focusing our attention on the different functions that sea ice performs in a global and regional context. The uses of the ice cover and the services it delivers, such as its role as an important platform for marine mammals, indigenous people, and industry, are not only being impacted by reductions in ice extent and the length of the ice season. Activities on or among sea ice are also increasingly occurring in parallel and in potential conflict as their overall scope grows while the sea ice area shrinks. Here, the concept of sea ice system services—explained in more detail in subsequent chapters—as the tangible and intangible benefits that ecosystems and humans derive from sea ice can be of significant value in furthering our understanding of how these different processes act in concert.

However, an important aspect that is key to our understanding of sea ice as it receives increasing attention from stakeholders and decision makers, as well as the public, relates to the question of how scientific knowledge and insight into the processes driving variability and change of the global ice cover are actually generated or obtained. The current focus is mostly on information derived from satellite remote sensing and large-scale model simulations. The accuracy and utility of such information hinges on the generation and acquisition of more fundamental knowledge obtained through qualitative and quantitative observation of processes and changes over time in the natural environment. In fact, it is typically only after we have gained some insight into sea ice as a process and a phenomenon that remote

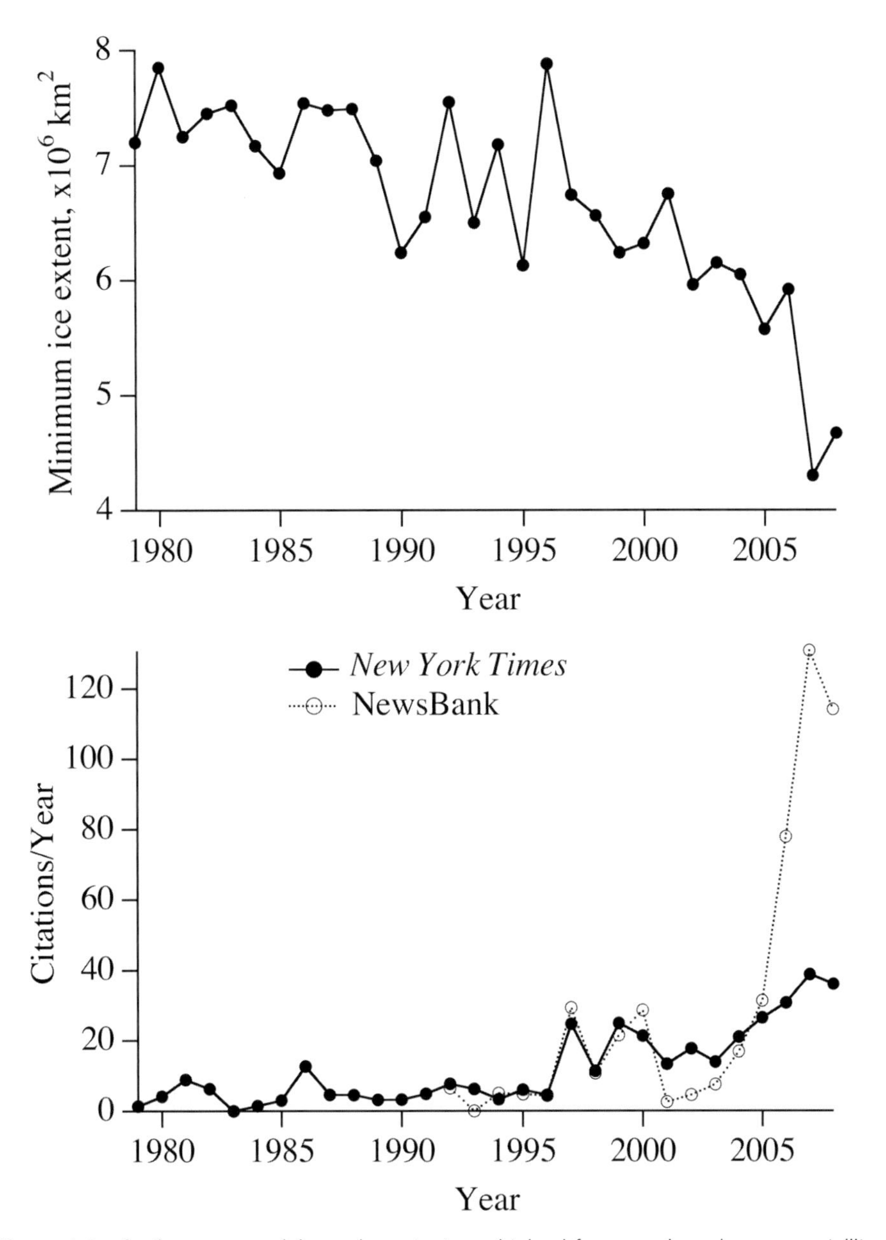


**Figure 1.1.** Arctic summer minimum ice extent as obtained from passive-microwave satellite data (top; data provided by the National Snow and Ice Data Center) and number of articles published in the *New York Times* and headlines of United States Newspapers (as referenced in NewsBank database) containing the phrase "sea ice" (bottom). The number of citations per year has been normalized to the keyword "Saturday" to account for the increase in number of publications and articles.

sensing and modeling can come into play. The present book focuses on the different approaches and methodologies that allow us to extract information from sea ice field measurements and that hence shape our current understanding of the role of sea ice in local and global contexts. While exhaustive in neither breadth nor depth, it does also provide a glimpse of the status of an interdisciplinary field of study at the end of the Fourth International Polar Year.

## 1.2 OVERVIEW OF BOOK CONTENTS

Preceding this introduction is a Foreword, "Encounters with Northern Sea Ice," by Ned Rozell, a science writer, who portrays a number of sea ice experts and sea ice users. His contribution provides a more tangible perspective on what sea ice means to different people and the potential value of information obtained from field measurements.

Chapter 2, "The Sea Ice System Services Framework: Development and Application," is an attempt to provide something of a link between sea ice policy and sea ice research by examining the services and benefits (or costs) associated with the use of sea ice by different groups from within and outside of the polar regions. Even though some of the concepts introduced in the chapter are firmly rooted within political science, it may help shed light on the origins of stakeholders' interests in sea ice and the potential contributions that field-based research can offer.

The next section of the book consists of eighteen chapters that constitute the main body of reference material in this volume. Sea ice field research often starts with the removal of a snow layer to gain access to the ice underneath. Chapter 3.1, "Field Techniques for Snow Observations on Sea Ice," provides a broad introduction to the principal methods employed in characterizing snow to gain more insight into its important role in affecting a number of key ice properties and processes.

Chapter 3.2, "Ice Thickness and Roughness Measurements," is a comprehensive survey of the different approaches employed in measuring one of the most important properties of sea ice, its thickness.

Sampling methods and basic ice-core analysis are discussed in Chapter 3.3 ("Ice Sampling and Basic Sea Ice Core Analysis"), serving in some ways also as a foundation for some of the subsequent chapters. Thus, Chapter 3.4 introduces measurements of the "Thermal, Electrical, and Hydraulic Properties of Sea Ice," covering a broad range of different approaches, including ever-more-important data acquisition through in situ sensor systems.

The mechanical properties of sea ice and measurements of strength are discussed in an applied context based on structural engineering in Chapter 3.5, "Ice Strength: In Situ Measurement." This chapter also illustrates how a range of key sea ice uses that depend on the structural integrity of the ice as well as the role of ice as a hazard are tied to fundamental properties discussed in some of the earlier chapters.

Chapter 3.6, "Sea Ice Optics Measurements," lays the foundation for some of the discussion of biooptics in later chapters, but more important, provides guidance on the different approaches to conduct one of the most fundamental measurements over sea ice, that of ice albedo.

Chapter 3.7 introduces the important field of study of ice-ocean interaction by covering the topic of "Measurements and Modeling of the Ice-Ocean Heat Interaction." With a focus on heat exchange, the chapter also provides a perspective on how these measurements can be employed to estimate exchange of nutrients and other compounds.

Chapter 3.8, "Biogeochemical Properties of Sea Ice," introduces the suite of ice-ecology contributions and focuses on sampling and fundamental measurements in biogeochemical studies of sea ice. It is followed by Chapter 3.9, which introduces a number of approaches for the "Assessment of the Abundance and Diversity of Sea Ice Biota" and demonstrates how both physical and biological processes have to be taken into consideration as the distribution of biota throughout the ice cover is determined.

The study and in particular capturing of ice-associated seals is the topic of Chapter 3.10, "Studying Seals in their Sea Ice Habitat: Application of Traditional and Scientific Methods." An important component of the sea ice field course as a demonstration of sea ice use by marine mammals, this chapter also provides a good perspective on the successful marriage of local, indigenous (northern Alaska Iñupiaq) knowledge and Western scientific methodology. Discussion of studies of all ice-associated mammals and birds is beyond the scope of this book; however, this chapter provides some insight into the measurements and methodologies of such studies using the example of ice seals.

Chapter 3.11, "Community-Based Observation Programs and Indigenous and Local Sea Ice Knowledge," provides an overview and guidance on how to work at the interface between academia and the vast body of knowledge and understanding accumulated by indigenous arctic residents.

Chapter 3.12 focuses on "Ship-Based Ice Observation Programs." With numerous icebreaker expeditions into ice-covered waters each year and more frequent "ship-of-opportunity" cruises, ship-based observations provide an important link between remote sensing and highly localized on-ice measurements.

The use of drifting sensors and longer-term installations of sensor systems is discussed in Chapter 3.13, "Automatic Measurement Stations." It provides insight into the autonomous sensor platforms that have resulted in increasing deployment of ever-more-sophisticated automated systems in recent years.

Chapter 3.14, "Data Management in Sea Ice Research," is a brief introduction to key principles of data management as pertinent to the other topics covered in this book.

Chapter 3.15, "Principal Uses of Remote Sensing in Sea Ice Field Research," is a comprehensive overview of the relevant remote-sensing resources. While remote sensing of ice surfaces depends greatly on validation through ground-based observations, modern field studies in turn increasingly rely on remote-sensing data for planning and interpretation of measurements. Along the same lines, Chapter 3.16, "The Use of Models in the Design and Interpretation of Field

Measurements," discusses how a similar mutual dependence between field studies and model simulations.

Chapter 3.17, "Integrated Sea Ice Observation Programs," is something of a synthesis of the preceding chapters and provides two examples or case studies (one of a coastal arctic sea ice observatory, the other of a highly interdisciplinary ship expedition in antarctic waters) of how a suite of different measurements can be tied together in an integrated fashion to help address more complicated questions that transcend disciplinary boundaries.

Alice Orlich, a participant in the 2008 sea ice field course with prior fieldwork experience, presents an overview of "Personal Field Logistics" in Chapter 3.18, touching on a range of important topics that are prerequisite to safe, comfortable, and successful fieldwork. The accompanying DVD contains additional material on field safety as pertinent to the aforementioned field course that may be of interest to readers.

Chapter 4 concludes the main body of the book with a set of observations and a brief outlook on present-day and future relevance of sea ice field measurements.

The book is accompanied by a multimedia DVD, produced by Maya Salganek and a class of documentary film students, with computer animations by Miho Aoki. The DVD features video footage accompanying most of the main book chapters and demonstrating the application of different measurement approaches in the field and laboratory. It also provides a perspective on sea ice and the animals and people that depend on it in a coastal Alaska setting. In addition to these video documentaries, the DVD also contains a range of other resources that link back to specific chapters, including programs and handbooks for ice observations or derivation of basic ice properties, and other electronic materials. The DVD also contains the data and analyses of the different field course components, edited by Sinead Farrell, one of the course participants.

## REFERENCES

Brigham, L. W. (2007), Thinking about the Arctic's future: Scenarios for 2040, *Futurist*, *41*, 27–34.

Grebmeier, J. M., J. E. Overland, S. E. Moore, E. V. Farley, E. C. Carmack, L. W. Cooper, K. E. Frey, J. H. Helle, F. A. McLaughlin, and S. L. McNutt (2006), A major ecosystem shift in the northern Bering Sea, *Science*, *311*, 1461–1464.

Holland, M. M., and C. M. Bitz (2003), Polar amplification of climate change in coupled models, *Climate Dyn.*, *21*, 221–232.

Krupnik, I., and D. Jolly (2002), *The Earth Is Faster Now: Indigenous Observations of Arctic Environmental Change*, Arctic Research Consortium of the United States, Fairbanks, AK.

Leppäranta, M. (2005), *The Drift of Sea Ice*, Springer, Berlin.

Leppäranta, M. (1998), *Physics of Ice-Covered Seas* (2 vols.), Helsinki University Printing House, Helsinki.

Melnikov, I. A. (1997), *The Arctic Sea Ice Ecosystem*, Gordon and Breach Science Publishers, Amsterdam.

Stroeve, J., M. Serreze, S. Drobot, S. Gearheard, M. M. Holland, J. Maslanik, W. Meier, and T. Scambos (2008), Arctic sea ice extent plummets in 2007, *Eos, Trans. Am. Geophys. Un.*, *89*, 13–20.

Thomas, D. N., and G. S. Dieckmann (2003), *Sea Ice: An Introduction to Its Physics, Biology, Chemistry and Geology*, Blackwell Science, London.

Untersteiner, N. (1986), *The Geophysics of Sea Ice*, Martinus Nijhoff Publ., Dordrecht (NATO ASI B146).

Wadhams, P. (2000), *Ice in the Ocean*, Gordon and Breach Science Publishers, London.

# Chapter 2

# The Sea Ice System Services Framework: Development and Application

*Amy Lauren Lovecraft and Hajo Eicken*

## 2.1 INTRODUCTION

This chapter outlines how a sea ice system services (SISS) approach towards the problems posed by diminishing and potentially more variable sea ice coverage, in particular in the Circumpolar North, may help guide sea ice measurements and address relevant information needs. It does so in two ways. First, the chapter is designed to introduce the reader to a conceptualization of sea ice that is broader than a mere location of data collection. Second, this breadth should enable the researcher to situate his or her own sea ice projects within temporally and spatially bounded information needs of the citizens, industries and their requirements, and governmental objectives. In other words, by the end of this chapter you should be able to consider what your measurements may be used for as well as what new measurements may be needed as the sea ice system changes in coming decades. This can help you plan your research more effectively, write better grant proposals, form more inclusive collaborations, and present your data to stakeholders, decision makers, and colleagues.

As noted in the Foreword, sea ice has been and continues to be used by many different social groups, animals, and other living organisms. While the ice has long served as both a resource and hazard for a wide variety of users, it was not until the documented shrinking of its coverage and increasing socioeconomic and geopolitical interests in the Arctic that any comprehensive approach towards analyzing sea ice from a social science perspective arose (e.g., Eicken et al. 2009). In light of the focus on climate change and its amplified effects at the poles (ACIA 2005), it is timely to consider sea ice from a systems perspective. More specifically, in order to develop policy options that can be effective in addressing arctic (and to a lesser extent antarctic) sea ice variability and change as a whole and over time, rather than as a reaction to a particular crisis or development from narrow disciplinary, commercial, or national agendas, a systems approach is essential. This chapter briefly reviews our methodology. It then provides a comprehensive approach to sea ice studies and proposes a form of policy analysis that can be used to evaluate plausible future states of this system in flux. Individuals and societies study what they believe to be of value. The research measurements explained in the following chapters are directly tied to how people, organizations, and governments define aspects of the sea ice system in terms of importance (e.g., as habitat for marine mammals, as a hazard to marine transport). Specifically, this chapter will provide readers with the necessary information to

- characterize social-ecological systems;
- explain how ecosystem services are derived by societies;
- understand the public policy concept of problem definition and
- use it to approach analysis of social-ecological systems; and
- consider stakeholders' shared concerns and points of conflict related to the jurisdictional and political consequences of expected sea ice change.

## 2.2 FRAMEWORK DESIGN

In order to begin to review a society's policies towards its natural resources and its own peoples one must be able to set system boundaries. Good social and natural science can identify and measure variables and their interactions (King et al. 1994). This requires appropriate, carefully designed frameworks for conceptualizing a system of study in order to capture the key variables and their interaction to produce, ultimately, a less distorted depiction of reality. In the case of sea ice, we have found the correct combination of analytical tools in four different disciplinary perspectives. First, and perhaps most important, we argue that one must understand the subsystem in which one's research is situated from multiple perspectives. The newly developed and growing field of social-ecological systems analysis justifies selecting "interactive subsystems" of environmental management in coupled human-natural systems (Anderies et al. 2004). For example, initial research of a wildfire management regime is likely to analyze specific aspects of forestry related to when and to what extent an area may burn. Likewise a review of municipal water management measures pumping capacity and flow in order to provide a service to water users. However, if we specifically want to understand how water scarcity affects wildfire abatement capacity, we must select analytical boundaries to engage this question rather than examine all water and forest policies for a given region, or neglect the interrelated dynamics of the two systems. The interactive subsystem researched will include both water and forest policies as they interact. Thus we look at the institutions, which one can think of as rule sets (e.g., laws, decision guidelines, budgets, jurisdictions) jointly affecting water use and wildland fire.

Our SISS framework selects sea ice as the key geophysical variable around which many formal and informal institutions exist. So, to do research on sea ice one should have the correct subsystem in mind. For example, measurements of ice thickness alone cannot explain the safety of travel, one must also measure the kinds of traffic (e.g., bears, people on foot, snowmobiles) and the extent of such traffic (e.g., daily, monthly) to accurately depict what the ice thickness measurement means to travel. This traffic is subject to sets of rules and knowing these rules can enhance the accuracy of data collection and presentation. Furthermore, knowing when the rules may conflict with the reality of reported travel is essential, otherwise correct data collection will produce less-than-useful results.

Second, the use of the ecosystem services perspective permits a researcher to examine the interplay among people dependent on natural systems at multiple scales, the impacts of their decision making on these systems, and the direction of research that may most benefit society (Millennium Ecosystem Assessment 2005). As people evaluate what services (benefits gained from) and hazards (negative impacts of) sea ice present over time different parts of society will find different services relevant. For example, the necessity of sea ice for indigenous hunters can be a concurrent hazard to the oil and gas industry. Which measurements might each group value more? How might each direct data collection? What research is most likely to be funded and by whom?

Third, problem definition literature from the field of policy studies can assist those interested in collecting and explaining scientific data for public and private decisions. By understanding how a society "problematizes" aspects of the natural world sea ice researchers can perform more useful research and learn to more effectively communicate the meaning of their research to inform, for example, potential legislation or agency action (Seidman and Rappaport 1986; Rochefort and Cobb 1994; Fischer 2003; Layzer 2006). As sea ice diminishes it causes social problems, such as removing the buffer zone for coastal communities, but it also reduces the hazards associated with arctic shipping—in both cases one needs to consider how the sea ice "problem" will become a part of social decision making. For example, how will funding agencies direct their efforts in data collection? Those readers who regularly apply for grants and other forms of research support from public and private sources already have intimate knowledge of how social values, what we define as a problem or a solution, directly impact the measurements taken.

Last, anthropological, biological, and geophysical data must be considered in how one actually constructs the elements of the social and ecological systems to be analyzed. These aspects are highlighted in greater detail in other chapters (for example, Chapters 3.3, 3.6, and 3.14). It should be noted that Lovecraft (2007; 2008), Chapin et al. (2006), and most recently in relation to sea ice, Eicken et al. (2009) have previously written about the importance of connecting social and ecological systems through an examination of (eco)system services in order to evaluate or propose governance techniques or institutions to address human-nature interactions. This chapter draws directly from these earlier works and places their conclusions in an educational context. It demonstrates how this sort of research is performed and why it is of potential value to sea ice researchers.

### 2.2.1 SOCIAL-ECOLOGICAL SYSTEMS

The feedbacks that can be observed between human actions and changes in natural systems are so tightly linked that they are often described as a coupled social-ecological system (Chapin et al. 2006; Berkes and Folke 1998). Examination of these coupled systems has produced scholarship across natural and social sciences characterizing the attributes of social-ecological systems (SESs) and researching the effects of different institutional arrangements and management practices (Ostrom 1990; Costanza et al. 1993; Berkes and Folke 1998; Gunderson and Holling 2002; Anderies et al. 2004). Anderies et al. (2004) define a social-ecological system as "an ecological system intricately linked with and affected by one or more social systems." In other words, both the social and ecological components of any given system (e.g., a village, such as Barrow, on the North Coast of Alaska) will have self-organizing independent relationships contained within them (e.g., election cycles are independent of the period of bowhead whale migration off the coast of Alaska) as well as having some interactive subsystems (e.g., the strength and stability of landfast sea ice will determine the location and timing of camps set up on the ice by those hunting bowhead whales). In these subsystems interactions are "the subset of social systems in which some of the interdependent relationships among humans are mediated through interactions with biophysical and non-human biological agents" (Anderies et al. 2004). In other words, the socioeconomic and political natures of human populations are in part determined by how these people and their systems of governance set rules for interaction with their natural surroundings.

In the case of sea ice as an SES, we also need to consider that in addition to its ecosystem functions (those functions which depend on, or facilitate, biological attributes of a system) much of the function of sea ice is in fact tied to the geophysical processes and properties associated with the ice cover. In other words, our framework opens up the ecosystem services concept to include the purely geophysical processes, which may not be technically included as ecosystem services, but are of fundamental importance to human and all other life. Thus, the role of sea ice in the climate system is almost entirely controlled by physical processes that determine ice albedo or its impact on atmosphere-ocean heat exchange. It is these processes in turn that are modified by human activities such as release of greenhouse gases into the atmosphere, which in turn may impact the state of the ice cover itself. Throughout this chapter, we are including such sociogeophysical subsystems in the overarching definition of SES, and hence refer more generally to sea ice system services.

The importance of SES theory to evaluating institutions, or the potential for rule sets, stems from the intertwining of social worlds and ecological processes.

This chapter defines institutions using the fairly standard working definition of political and policy studies:

> Institutions are enduring regularities of human action in situations structured by rules, norms, and shared strategies, as well as by the physical world. The rules, norms, and shared strategies are constituted and reconstituted by human interaction in frequently occurring or repetitive situations. Where one draws the boundaries of an institution depends on the theoretical question of interest. (Crawford and Ostrom 1995)

In the case of sea ice the intertwining is particularly complex for two reasons. First, there are multiple and often competing uses of sea ice by different stakeholders within the system: for example, between those seeking to remove or alter ice for passage by surface vessels and those wanting the ice to remain for animal habitat. It should be noted that this point includes those researchers doing work on the sea ice itself, as most of such work only exists due to government funding, a clear expression of social values directing data collection. Second, there is no comprehensive set of institutions to govern the use of sea ice from place to place or year to year. In other words, sea ice has been viewed only as a constant feature providing a service or hazard, not as a key variable in a social-ecological system. Our framework places the ecosystem attribute in flux, in this case sea ice, at the center of analysis. It treats the sea ice system as an independent variable in relation to human society as well as a dependent variable affected by human activity. We ask, what services have been provided by this feature and how will they change as climate changes? This perspective forces the examination of a broader set of social, biological, and geophysical variables in order to understand not only how the arctic system's properties will change but how human societies will adapt through the removal or creation of new institutions. Rather than only examining the social or geophysical properties or processes, such holistic research can advance understanding of how societies perceive, use, and govern the interactive relationships that depend on sea ice not merely as a material but as a system. Studying sea ice as a system—that is, an assemblage of interdependent features and processes involving physical, chemical, biological, and social factors—allows us to examine the ways in which ecological/geophysical and social subsystems are linked through human actions (Young 2002). In this chapter we explain our work to transverse multiple disciplines in order to create the most effective depiction of the sea ice problem.

### 2.2.2 ECOSYSTEM SERVICES

The recently completed Millennium Ecosystem Assessment (MA) is a multiyear synthesis project explaining global landscape change and assessing the relationships between ecosystems and human well-being through ecosystem services. The project defines *human well-being* as having five components derived from services provided by natural systems: security, basic material needs for good life, health, and good social relations. These four support the fifth: freedom of choice and action, meaning "opportunity to be able to achieve what an individual values doing and being" (MA 2005).

The MA classifies ecosystem services as four types, which, at different geographic and time scales, provide for human well-being: the capacity for one to achieve what one values. *Provisioning* services are the tangible goods people obtain from their natural environment such as food, genetic resources, fuels, and freshwater. *Regulating* services are the benefits derived from the regulation of biochemical processes in the ecosystem itself such as climate regulation from ice coverage, pollination through plant and animal interactions, and the quality of water as filtered by soils, or coastal marshes. Third, ecosystems provide societies with *cultural* services by offering a location (and the characteristics of a place) for spiritual enrichment, knowledge transmission, social relationships, and other nonmaterial benefits. Last, the foundations of societies rely on *supporting* services, which are those necessary long-term processes that produce all the services noted above. Soil formation, regulation of atmospheric radiation, and nutrient cycling on land and in oceans are fundamental processes key to life in ecosystems.

It is the first service we most commonly associate with daily human life as people act upon different forms of sea ice at different times throughout the year as a platform for hunting or social activities, to obtain freshwater, or as a hazard to be circumnavigated in order to travel arctic waters. Indirectly, as the ice provides a platform for many marine mammals to flourish (e.g., polar bears and seals) this also serves the human population in terms of sustenance or hunting. The third service is also significant and other chapters in this text document the importance of the ice for the peoples who have lived with it for millennia. The second and fourth services often fall to the natural sciences while the first and third tend to be the purview of policy makers and social scientists focused on providing open water access or species preservation along jurisdictional lines of governance. However, the layers of institutions designed to conserve, harvest, or distribute resources related to sea ice are currently neither comprehensive nor well designed to handle rapid change. While sea ice provides across the four services, human understanding of the services has been piecemeal and policy to manage across these services for multiple scales of social needs has been generally uncoordinated. If, as researchers, we ask the question "how can my work enhance human well-being?" we have to consider where our measurements and data fit into the social understanding of ecosystem services.

### 2.2.3 POLICY ANALYSIS AND PROBLEM DEFINITION

Governance of natural resources across the globe has, more than ever before, accepted science as the backbone of policy making. Scientific results address questions about the natural and social world that help answer routine, but fundamental, questions of modern governance. Which lands need to be preserved? Where have the fish gone? What are the economic, environmental, and social costs of producing energy? Why are there changes in sea ice coverage? Answers to questions such as these encourage agency officials, politicians, corporations, or the public to make certain decisions or support certain actions, including the prioritization of fundamental and applied research related to social-ecological systems. However, the relationship between science and policy making has become subject to political interpretation by a wider audience and in a more public manner than ever before. Recognition of this trend in policy development and analysis is crucial as more varied stakeholders (those with a recognized interest in a region or issue), particularly in the Arctic, are able to affect government policy, though to varying degrees. This chapter proposes that the use of the policy concept of "problem definition" can not only help justify the boundaries of a given research project but also help the researcher derive information out of measured variables that has meaning to the people the research affects. Furthermore, as science changes how people perceive the world around them by presenting them new facts, policy problems will also change. Researchers are often most familiar with this through grant calls by funding agencies such as the U.S. National Oceanic and Atmospheric Administration, or other federal organizations seeking information related to current "problems." *Where* do these problems come from? *Why* are they problems at all? A person, community, or government had to perceive them as such.

Until the last decade or so there has been comparatively little interest in arctic shipping routes and offshore oil and gas development was greatly limited. New information suggesting that the Arctic may have ice-free summers by 2040, if not sooner, in combination with rising demand and fluctuating prices for fossil fuels coupled with technological developments have triggered, for example, the U.S. Minerals Management Service to conduct highly successful Outer Continental Shelf lease sales in the Chukchi Sea (Brigham 2007). Agencies and corporations, as well as activist groups for indigenous peoples and others concerned with environmental impacts have been scrambling to develop rules for arctic marine shipping and other activities, such as fisheries. In other words, there was not something requiring action, a "problem," until recent scientific predictions and observations began to demonstrate that sea ice is diminishing and will continue to do so. However, the results of such a change in sea ice will create different opportunities as well as hardships among different peoples, industries, and governments who have operated with sea ice as a dominant feature of their existence, whether to enculturate children, pilot an icebreaker, or rely on jurisdictional or regulatory boundaries now in flux.

In summary, any student or researcher of the sea ice regime may want to consider who will use the information generated and to what ends. For example, it has been demonstrated that the use of scientific arguments can serve to legitimize some policy options by converting contested social issues into technical issues to be handled by experts. In other words, the institution of science acts as a legitimizing tool for policy decisions because it can translate issues deeply controversial in society (e.g., when and where should oil drilling proceed off the northern coasts of Alaska) into technical issues that appear to be value neutral (Weeks 1995). This textbook and other publications (e.g., Krupnik and Jolly 2002) have highlighted the importance and accuracy of local observation and necessity of local knowledge (see, for example, the Foreword and Chapters 3.11 and 3.17). This chapter further argues the need to include local and minority standpoints on problems in the context of research design, since these may reflect relevant expertise that has not yet found its way into the scientific literature and since scientific research informing policy options needs to cover the unexplored but relevant territory as much as the explored to be truly useful. In order to minimize the distortion of reality and effectively address a policy problem, one must maximize the knowledge and information of all who are asserting that a problem exists.

"Listening carefully to what marginalized people say—with fairness, honesty, and detachment—and trying to understand their life worlds are crucial first steps in gaining less partial and distorted accounts of the entire social order." This serves as a starting point to provide "a causal, critical account of the regularities of the natural and social worlds and their underlying causal tendencies" and rejects, with careful analysis of data, epistemological relativism (Harding 1992, p. 583).

### 2.2.4 APPLYING SOCIAL-ECOLOGICAL SYSTEMS AND SEA ICE SYSTEM SERVICES CONCEPTS TO THE STUDY OF SEA ICE

How do the concepts discussed above actually enter into the design and completion of a sea ice research program or project? As argued in a number of publications, such as Holton and Sonnert (1999), Stokes (1997), in a world of limited resources and increasing complexity of social-ecological systems, the merging of fundamental (or basic) and applied research into science that is located within a broad problem area of recognized societal relevance and urgency is one way to prioritize research activities. Casting sea ice research problems in terms of the services provided by the ice cover (as listed in Table 2.1) underscores that sea ice studies appear to meet the criterion of broad societal relevance, even for regions of the globe that are far removed from the poles where the regulatory services of the ice cover may still be functioning. At the same time, the concept of sea ice system services (SISS) can help maximize the benefit of the scientific work to stakeholders and the broader public. Understanding how science research is needed to create applied knowledge to solve problems can generate more accurate, applicable, and widely usable hypotheses, measurement techniques, and data interpretation.

The example of climate-regulation services may illustrate how implicitly even fundamental research that seemingly does not have a particular application in mind does contribute to the body of knowledge that is relied upon in policy and decision making. Thus, assessing the role of sea ice as a regulating factor in the global climate system requires an understanding of the role of sea ice for the earth's radiation budget, which in turn strongly depends on the ice optical properties and specifically albedo (Table 2.1), discussed in more depth in Chapter 3.6. At the same time, areal extent of the ice is crucial in determining the amount of reflected radiation in relation to the low-albedo ocean, requiring remote sensing studies of the polar regions (Chapter 3.15). The thickness distribution of the ice cover and its velocity are key in determining the transfer of heat between atmosphere and ocean as well as the heat exchange between high and low latitudes (Chapter 3.2). Measurements of the key variables associated with the climate-regulation service can serve as a diagnostic of past and potential future change. However, in conjunction with climate model simulations, which in turn depend on such observations both for model development and validation, we are able to quantitatively assess the magnitude of climate regulation provided by sea ice. Such an analysis suggests that sea ice may contribute as much to changes in radiative forcing and hence climate change as a doubling of atmospheric carbon dioxide concentrations, that is, the "greenhouse effect" (Hall 2004; Eicken et al. 2009). Since research also shows that this impact is largely through ice-albedo feedback, casting fundamental studies of the sea ice heat and mass budget in terms of SISS establishes a direct link to policy and decision making, such as in the context of international agreements to curb emission of greenhouse gases. It also provides guidance on further research that is required.

In fact, the different categories in Table 2.1 provide guidance on how the different methods discussed in this textbook interrelate and are part of a larger scientific framework. Table 2.1 also illustrates just how closely sea ice research is tied to fundamental questions of policy, jurisdiction, and regulation, whether pertaining to the fate of ice-associated species such as the polar bear or the risk associated with shipping and industrial development in ice-covered waters. Ideally, curiosity-driven research, which is seemingly detached from the overarching framework of the sea ice social-ecological system, and research that specifically seeks to minimize uncertainty and maximize benefits for decision makers and stakeholders should actually converge. The target variables and methodologies shown in Table 2.1 are in fact often common to both of the two approaches. The challenge in generating scientifically accurate and relevant research results that are also relevant in a broader context is twofold. First, a measurement design that does in fact generate information about variables or processes that are relevant to different sea ice users may require substantial work and involvement by those users at different stages of the work. Second, in contrast to fundamental studies of, for example, the mass budget

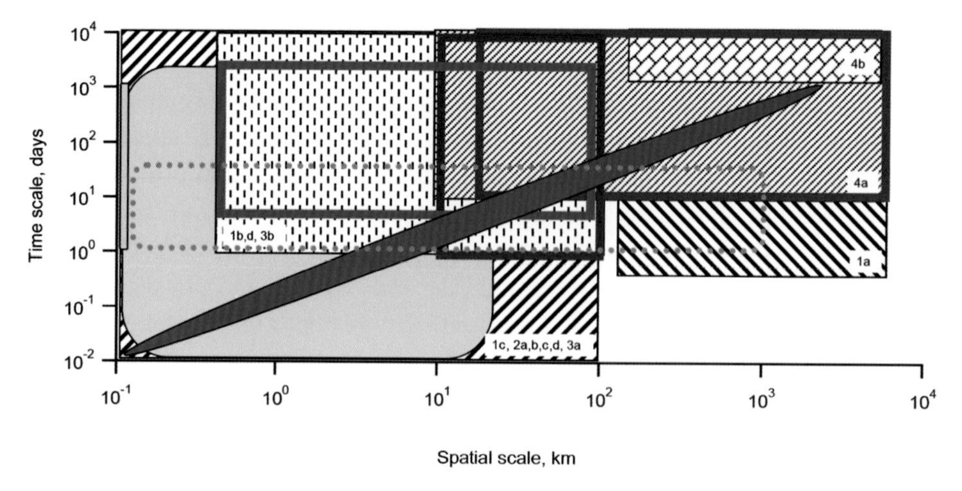


**Figure 2.1.** Schematic depiction of different spatio temporal scales relevant for sea ice system services and design of field measurement campaigns. The numbers indicated in the different boxes correspond to the categorization of services listed in Table 2.1. The colored, outlined boxes indicate the approximate coverage of different satellite products discussed in Chapter 3.15 (blue: passive microwave satellite data on ice concentration and extent; red: ice charts; green: synthetic aperture radar). The orange, dotted box corresponds to aerial and ship surveys of ice thickness and morphology (Chapters 3.2 and 3.12). The yellow box covers most of the methods discussed in this book that involve coring of sea ice and analysis of these core samples. The green, stretched oval indicates the sampling rate and sampling area of drifting sensors, while the solid, light blue box indicates the spatial and temporal scales covered by community-based observations.

of the polar sea ice covers, much of the use-relevant information has to be acquired at spatial and temporal scales that are finer than those commonly employed in large-scale studies of sea ice, though the results may be applied on the global scale in terms of international policy making.

To illustrate the importance of scale and put the subsequent chapters in this book into context, consider Figure 2.1, which shows the time and spatial scales relevant for the different services listed in Table 2.1. Note that most of the sea ice field methods discussed in this book, such as any that rely on the extraction of ice cores or measurements at a particular site, are covered by the yellow box at the very left-hand side of the diagram. Drifting sensors (Chapter 3.13) sample the ice locally but can collect data on ice-ocean and air-ice interaction along a drift path represented in rough outlines by the green, stretched oval. Also shown are the approximate coverage of typical airborne or ship surveys of ice thickness and ice morphology (dotted outline). Remote sensing methods generally cover the upper half of the diagram, mostly toward the right because of relatively poor resolution of key sensor systems introduced in Chapter 3.15. Finally, the solid blue box represents the spatiotemporal scale covered by community-based observations. Since the observations are typically closely tied to sea ice use (Chapter 3.11), they are made by default at the relevant scale. Use of the SISS concept in the context of the design phase of

**Table 2.1.** Sea ice system services categorization* (from Eicken et al. 2009)

| Service category | Type of service | Target variables | Measurement approach |
|---|---|---|---|
| **1. Regulating** | a. Sea ice as regulator of arctic and global climate | Albedo, extent, thickness, mass flux | Satellite, aerial, and submarine surveys, moorings |
| | b. Hazard for marine shipping and coastal infrastructure | Concentration and extent, multiyear (MY) ice fraction, thickness | Satellite, aerial surveys |
| | c. Stabilizing element for coastal infrastructure and activities | Thickness and morphology, ice season duration, MY ice presence | High-resolution satellite imagery, LK†, aerial and ground-based surveys |
| | d. Ice as a geologic agent: Erosion control through damping impacts of storms, buttressing permafrost coastline (bottom-fast ice), enhancing erosion through ice rafting of sediments | Duration of ice season, sediment entrainment | Satellite surveys, ground-based measurements and sampling |
| **2. Provisioning** | a. Transportation corridor | Stability, morphology, thickness/strength | Satellite, ground-based surveys, LK, coastal radar, in situ instrumentation |
| | b. Platform for a range of activities (subsistence hunting and fishing, oil and gas development, etc.) | Stability, morphology, thickness/strength, MY ice fraction | Satellite, ground-based surveys, LK, coastal radar, in situ instrumentation |
| | c. MY ice as source of freshwater | MY ice fraction and age | LK, satellite, ground-based surveys |
| | d. Access to food sources | Morphology, ice biota | LK, aerial, ground-based surveys, sampling |
| **3. Cultural** | a. Subsistence activities on and among ice | Extent, morphology, duration of ice season, ice ecosystems | LK, satellite, ground-based surveys, sampling |
| | b. Ice as part of cultural and spiritual landscape (incl. tourism, e.g., Hokkaido) | Extent, morphology, duration of ice season, ice ecosystems | LK, satellite, aerial, ground-based surveys |
| **4. Supporting** | a. Sea ice based food webs and ice as a habitat (ice algae, under-ice fauna, seals, walrus, polar bears, etc.) | Extent, stability, morphology, duration of ice season, ice ecosystems | LK, ground-based surveys, sampling |
| | b. Reservoir and driver of biological diversity (e.g., extremophiles) | Extent, stability, morphology, duration of ice season, ice ecosystems | Sampling, LK |

*Categories follow those identified by Millennium Ecosystem Assessment (2005).
†LK refers to the local knowledge used to observe and understand the natural environment and may include indigenous and traditional components.

field measurement programs can help identify key gaps or important spatial or temporal scales that need to be covered. For example, as illustrated in Figure 2.1, few sensors or measurement methods provide information at the spatial and time scale most relevant to safety and hazard mitigation by sea ice users, occupying the lower-most left of the diagram.

## 2.3 DISCUSSION OF ARCTIC FUTURES

Let us consider this information about social-ecological systems, ecosystem services, and problem definition and apply it to what may arise in the Arctic as sea ice diminishes. In light of directional climate change, how might clusters of people, groups, industries, and governments develop into advocacy coalitions pressing claims on national governments or international organizations? There is a growing suite of issues problematic across levels of governance tied to changing climates and corresponding changes in social systems:

- Increased marine and air access require new transportation systems and routes that will pass through areas currently used for subsistence harvesting or otherwise impact natural systems people depend upon for provisioning services.
- Resource development of oil and gas, fisheries, and freshwater will be less expensive now that these resources are more accessible.
- Indigenous and rural arctic peoples will experience disruption to cultural patterns and practices of well-being, cultural transmission, and access to land and sea. These disruptions will be perceived differently by different people.
- Conservation measures may be harder to enforce based on increased and cheaper access across the Arctic.
- Different levels of governance, from local councils to national actors, will have to address constituencies divided on these issues. These levels of government also have their own interests to look after such as financial stability, capacity to exercise power, and popular support through electoral cycles.

Brigham (2007) has proposed a suite of scenarios for the future of the Arctic based on what may happen with the different interests noted above. These futures are only possible due to the shrinking of sea ice cover; sea ice presented a barrier during the Cold War years, but this feature of the top of the world is now fading.

**Globalized:** Increased access and economic focus with a free-market approach to resources; private dominates public.

**Adaptive:** Increased access tied to carefully structured development with strict protections; incremental changes to government (e.g., enhanced reliance on existing entities); public-private partnerships.

**Fortress:** Restricted access and development only by direct stakeholders (arctic nations) tied to movement of indigenous peoples in favor of development; heavy public-security focus.

**Equitable:** New forms of governance focused on sustainability; public and indigenous dominance partnered with restricted private development.

As this text goes to press, there has been an effort to stem what appears to be a rush for arctic territory in the Ilulissat Declaration from May 2008. During the Arctic Ocean Conference in Greenland, which was to review impacts of climate change, including new maritime shipping routes, a political meeting took place among the five arctic circumpolar nations. The declaration is only two pages in length, but some key language related to the future(s) of the Arctic is evident:

> By virtue of their sovereignty, sovereign rights and jurisdiction in large areas of the Arctic Ocean the five coastal states are in a unique position to address these possibilities and challenges. In this regard, we recall that an extensive international legal framework applies to the Arctic Ocean.... Notably, the law of the sea provides for important rights and obligations concerning the delineation of the outer limits of the continental shelf, the protection of the marine environment, including ice-covered areas, freedom of navigation, marine scientific research, and other uses of the sea. We remain committed to this legal framework and to the orderly settlement of any possible overlapping claims. This framework provides a solid foundation for responsible management by the five coastal States and other users of this Ocean through *national implementation and application of relevant provisions.* We therefore see *no need to develop a new comprehensive international legal regime to govern the Arctic Ocean.* We will keep abreast of the developments in the Arctic Ocean and continue to implement appropriate measures. (Ilulissat Declaration 2008; italics added)

This declaration does seem to indicate a fortress approach to deciding the fate of the Arctic by highlighting the "unique position" of the five arctic nations. In fact, while noting that the Law of the Sea is the correct mechanism to resolve Arctic Ocean interest conflicts, the declaration also notes that "national implementation" is the means by which to address the mechanism. This clearly means national political institutions and processes are the predominant force shaping arctic policy among these five nations and, conversely, it implies that subnational or regional activities of indigenous groups or nongovernmental organizations are not (outside of their nations) a part of the "orderly settlement of any possible overlapping claims." Although it could be argued that the document may leave room for incremental adaptation to the changes in the Arctic, as it does not seek any new "international legal regime," it is more likely the sentence indicates the commitment to the current national-level arctic actors to engage in nation-nation debates over the future of the Arctic.

## 2.4 CONCLUDING THOUGHTS

This chapter began by addressing the important role sea ice plays in human activities, from the most local activities permitting the feeding of one's family, and ends on a simmering global conflict over international rights to exploit hydrocarbons and transit routes because sea ice no longer will pose the same hazard it has over the past hundreds of years. The framework we propose is meant to encourage the creation of linkages across human management of ecosystem services that reveal the interconnectedness between geophysical, biological, and social attributes of a social-ecological system. The current competitive nation focus towards the Arctic also indicates that sea ice has played a role as a barrier whose growing absence reveals an Arctic not yet prepared to handle the multiple human relationships tied to sea ice. It seems unlikely that new forms of governance necessary to create an Equitable Arctic according to Brigham have yet been seriously considered at the national level. However, there have been different accords regionally between subnational actors to provide ecosystem services in the past (Meek et al. 2008), and these may hold the institutional capacity to provide for an arctic social-ecological system that adapts rather than collapses.

### REFERENCES

ACIA (2005), *Arctic Climate Impact Assessment—Scientific Report*, Cambridge University Press, Cambridge.

Abel, T. (1998), Complex adaptive systems, evolutionism, and ecology within anthropology: Interdisciplinary research for understanding cultural and ecological dynamics, *Georgia Journal of Ecological Anthropology, 2*, 6–29.

Agrawal, A. (2005), Environmentality: Technologies of Government and the Making of Subjects, Duke University Press, Durham, NC.

Anderies, J. M., M. A. Janssen, and E. Ostrom (2004), A framework to analyze the robustness of social-ecological systems from an institutional perspective, *Ecology and Society*, *9*(1), 18.

Berkes, F., and C. Folke (1998), Linking social and ecological systems for resilience and stability, in *Linking Social and Ecological Systems: Management Practices and Social Mechanisms for Building Resilience*, edited by F. Berkes and C. Folke, pp. 1–25, Cambridge University Press, Cambridge.

Bosso, C. (1994), The contextual bases of problem definition, in *The Politics of Problem Definition: Shaping the Policy Agenda*, edited by D. Rochefort and R. Cobb, pp. 182–204, University Press of Kansas, Lawrence.

Brigham, L. W. (2007), Thinking about the Arctic's future: Scenarios for 2040, *The Futurist, 41*, 27–34.

Chapin, F. S., A. L. Lovecraft, E. S. Zavaleta, J. Nelson, M. D. Robards, G. P. Kofinas, S. F. Trainor, G. D. Peterson, H. P. Huntington, and R. L. Naylor (2006), Policy strategies to address sustainability of Alaskan boreal forests in response to a directionally changing climate, *Proceedings of the National Academy of Sciences of the United States of America, 103*, 16,637–16,643.

Constanza, R. L., L. Wainger, C. Folke, and K. Maler (1993), Modeling complex ecological economic systems: Toward an evolutionary, dynamic understanding of people and nature, *BioScience, 43*(8), 545–555.

Crawford, S., and E. Ostrom (1995), A grammar of institutions, *American Political Science Review, 89*(3), 582–600.

Edelman, M. (1988), *Constructing the Political Spectacle*, University of Chicago Press, Chicago.

Eicken, H., A. L. Lovecraft, and M. L. Druckenmiller (2009), Sea ice system services and Arctic observing systems: A model to reconcile scientific and stakeholder information needs, *Arctic*, 62(2), 119–136.

Fischer, F. (2003), *Reframing Public Policy: Discursive Politics and Deliberative Practice,* Oxford University Press, Oxford.

Folke, C., T. Hahn, P. Olsson, and J. Norberg (2005), Adaptive governance of social-ecological systems, *Annual Review of Environmental Resources, 30*, 441–473.

Gunderson, L. H., and C. S. Holling (eds.) (2002), *Panarchy: Understanding Transformations in Human and Natural Systems*, Island Press, Washington, DC.

Hall, A. (2004), The role of surface albedo feedback in climate, *J. Climate, 17*, 1550–1568.

Harding, S. (1992), After the neutrality ideal: Science, politics, and strong objectivity. *Social Research, 59*(3), 567–587.

Heclo, H. (1978), Issue networks and the executive establishment, in *The New American System,* edited by A. King, pp. 87–124, The American Enterprise Institute for Public Policy Research, Washington, DC.

The Ilulissat Declaration (2008), Arctic Ocean Conference, Ilulissat, Greenland, 27–29 May 2008; electronic document retrieved at http://www.oceanlaw.org/index.php?name=News&file=article&sid=77, on December 4, 2008.

Holton, G., and G. Sonnert (1999), A vision of Jeffersonian science, *Issues Sci. Technol., 16*, 61–65.

King, G., R. Keohane, and S. Verba (1994), *Designing Social Inquiry: Scientific Inference in Qualitative Research*, Princeton University Press, Princeton, NJ.

Krupnik, I., and D. Jolly (2002), *The Earth Is Faster Now: Indigenous Observations of Arctic Environmental Change*, Arctic Research Consortium of the United States, Fairbanks, AK.

Layzer, J. (2006), Fish stories: Science, advocacy, and policy change in New England fishery management, *The Policy Studies Journal, 34*(1), 59–80.

Lovecraft, A. L. (2007), Bridging the biophysical and social in transboundary water governance: Quebec and its neighbors, *Quebec Studies, 42*, 133–140.

Lovecraft, A. L. (2008), Climate change and Arctic cases: A normative exploration of social-ecological system analysis, in *Political Theory and Global Climate Change,* edited by S. Vanderheiden, pp. 91–120, The MIT Press, Cambridge, MA.

Meek C. L., A. L. Lovecraft, M. T. Robards, and G. P. Kofinas (2008), Building resilience through interlocal relations: Case studies of walrus and polar bear management in the Bering Straits, *Marine Policy, 32*(6), 1080–1089.

Millennium Ecosystem Assessment (2005), *Ecosystems and Human Well-Being: Synthesis*, Island Press, Washington, DC.

Ostrom, E. (1990), *Governing the Commons: The Evolution of Institutions for Collective Action,* Cambridge University Press, New York.

Rochefort, D. A., and R. W. Cobb (eds.) (1994), *The Politics of Problem Definition: Shaping the Policy Agenda,* University Press of Kansas, Lawrence.

Seidman E., and J. Rappaport (eds.) (1986), *Redefining Social Problems,* Plenum Press, New York.

Stokes, D. E. (1997), *Pasteur's Quadrant: Basic Science and Technological Innovation,* Brookings Institution Press, Washington, DC.

Weeks, P. (1995), Fisher scientists: The reconstruction of scientific discourse, *Human Organization, 54*(4), 429–436.

Young, O. R. (2002), *The Institutional Dimensions of Environmental Change: Fit Interplay and Scale,* MIT Press, Cambridge, MA.

## Chapter 3.1

# Field Techniques for Snow Observations on Sea Ice

*Matthew Sturm*

### 3.1.1 INTRODUCTION

From September through June, snow covers the arctic sea ice. Even newly formed ice (as in leads) is rapidly covered by snowfall or by snow blown in from adjacent snow-covered floes. Consequently, the most common surface of the sea ice—for travelers, biota, surface energy exchange, and remote sensing—is snow, not ice. The snow covering the ice is a complex layered material (Colbeck 1991), the layering a consequence of the fact that the snow is laid down by a sequence of discrete winter weather events: snowfall, rain-on-snow, wind drifting, and melting. The snow layers are generally unstable in their initial depositional form, so once deposited, they undergo metamorphism that changes their characteristics and physical properties. The metamorphism can range from sintering (German 1996), leading to cohesive, rock-hard layers of snow, to temperature gradient (TG) driven kinetic growth (Colbeck 1982), which produces coarse-grained friable snow structures. Liquid water can infiltrate the snow cover and refreeze, producing icy layers. The end result of these primary and secondary processes is a blanket of snow covering the ice that varies both vertically and laterally, in some cases dramatically. The challenge in making snow observations on the sea ice is to capture the properties of key interest, and some measure of their spatial variation, without becoming overwhelmed by the heterogeneity.

For this reason, it is helpful to think of making measurements on the *snow cover* rather than the *snow*. The word *cover* has equal importance as the word *snow*: It reminds us that both the sequence and the character of the snow layers at a point, as well as the spatial variation of these layers, are what produce the impact of the snow on the sea ice system. For example, a snow-cover property that is of prime importance in determining the services the sea ice can deliver is the thermal conductivity ($k$) (Maykut and Untersteiner 1971). It determines, to first order, how much congelation ice forms during the winter. In order to compute the bulk thermal insulation of the snow cover, we have to measure $k$ on the full vertical sequence

**Table 3.1.1.** Snow on sea ice services arranged by stakeholder groups

| Stakeholders | Why do they care about snow on sea ice? | Knowledge needs |
|---|---|---|
| Coastal communities | subsistence: seals | snow drift size for dens |
| | subsistence: food chain | snow depth and transmitted PAR; freshwater lakes |
| | subsistence: hunting/whaling | snow depth and density for roadways & water supply |
| | travel safety | blowing snow, hidden thin ice, flooded ice |
| | cultural value | integral part of the ice; the ice is integral to life |
| Marine transportation | weather/safety/access | blowing snow, snowfall impedes visual navigation |
| | operating costs: ice thickness | more snow less ice; less snow more ice |
| | operating costs: ice-breaking efficiency | snow on ice increases hull friction, costs of ice breaking |
| Oil and gas | regulations | avoiding seal dens, polar bears, other wildlife |
| | weather/safety/access | blowing snow; snowfall impedes visual navigation |
| | operating costs: ice thickness | more snow less ice; less snow more ice |
| | operating costs: ice-breaking efficiency | snow on ice increases hull friction, costs of ice breaking |
| Tourism | intact cultures | snow critical but somewhat intangible |
| | pristine environments & wildlife | snowmobile trips and skiing |
| Public interests | climate change | extent and thickness of snow in snow-ice albedo feedback |
| | species diversity (charismatic megafauna) | control of light penetration through ice; dens; habitat |
| | romantic notions of wilderness | the "look" of the ice |

of snow layers. However, the spatial distribution of areas of deep and shallow snow can also have a large impact on ice heat losses (Sturm et al. 2002). In short, it is the integrated properties of the snow that matter and that need to be captured in a measurement program.

In practice, because making direct measurements of many critical snow properties is difficult and time-consuming, we are often forced to tie the properties that are of interest to more easily observed snow properties like depth or density that we can measure more quickly and in more places. The actual property we measure, and the reasons for measuring it, vary with the stakeholder group for whom the measurements are made. As an integral part of the sea ice, it is impossible to understand the ice, its development, and its thermal and mechanical properties without some reference to the snow. Nonetheless, some of the more salient services provided by the snow alone are given in Table 3.1.1, arranged by stakeholder groups.

## 3.1.2 THE DEVELOPMENT OF THE SEA ICE SNOW COVER

The snow cover starts to accumulate on the tundra in northern Alaska and Siberia in September, a time of year when the sea ice extent is near minimum and there is considerable open water (Meier et al. 2007). Snow that falls in the ocean does not end on the ice, so the sea ice snow cover lacks the early season layers found on land. In both cases the snow covers consist of relatively few snow layers, the product of a limited number of snowfall and wind events (five to nine events might take place in a typical winter) and a relatively dry winter climate. As Figures 3.1.1 and 3.1.2 illustrate, most of the snow cover build-up takes place between October and December (see also Untersteiner 1961; Hanson 1980). Recent climate-driven changes favoring later ice build-up could have a large impact on the nature of the winter snow cover because of this asymmetrical seasonal pattern of snowfall.

After layers are deposited, they undergo (a) compaction, (b) metamorphism, and possibly (c) wind erosion. Compaction (Kojima 1966), the process by which individual grains of snow are forced closer together under the action of gravity and the weight of overlying layers of snow, tends to be limited in tundra and sea ice snow covers because overburden pressure is limited: The snowpack is very thin. Metamorphism (Colbeck 1982), on the other hand, can be extreme, with most layers being altered significantly. The two main metamorphic pathways are wind transport (leading to cohesive wind slabs; see Benson and Sturm 1993; Pomeroy and Gray 1995) and temperature gradient-driven metamorphism leading to depth

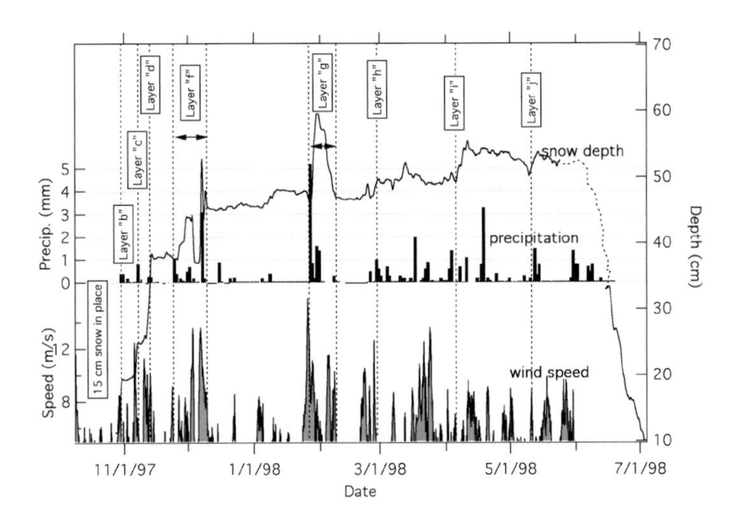


**Figure 3.1.1.** The build-up of the snow cover at an ice camp on multiyear ice in the northern Chukchi Sea showing that most of the accumulation was between October and December. The events (wind and snowfall) that control the build-up are relatively limited in number.

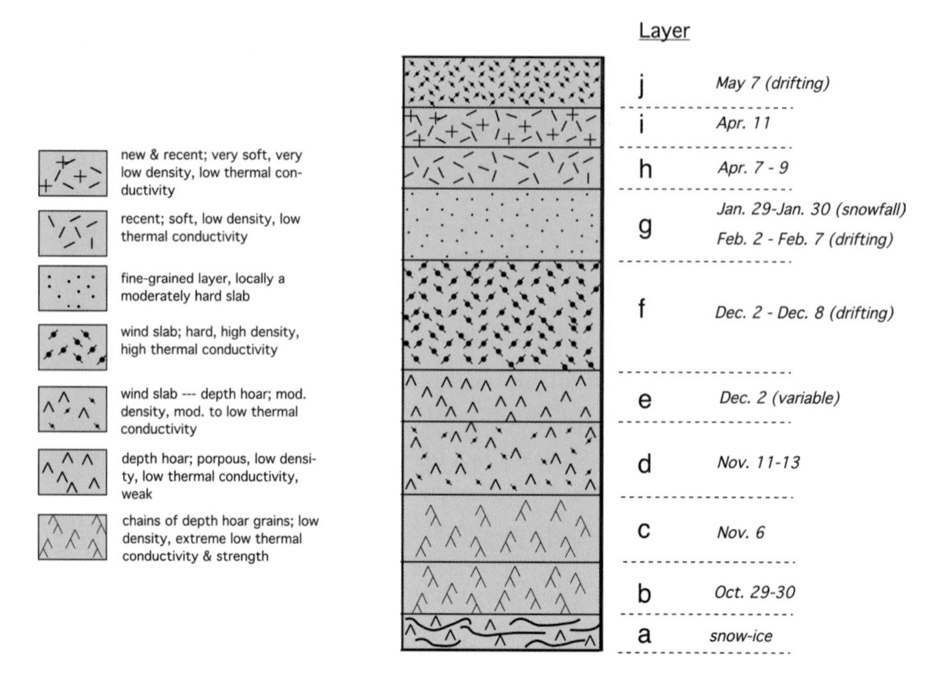


**Figure 3.1.2.** The layer stratigraphy for Figure 3.1.1 showing the date each layer was deposited. The symbols are from the *International Classification for Snow on the Ground* (Colbeck et al. 1992), which is available as part of the research tools on the multimedia DVD.

hoar formation (Akitaya 1972; Sturm 1991). These two types of layers are about as dissimilar as seasonal snow layers can be. One (wind slab) has the smallest of all snow grains (0.1–0.5 mm), is well bonded and has small pores and low permeability. The other (depth hoar) has the largest grains (5–15 mm), is poorly bonded and has large interconnected pores that result in permeability values thirty times higher than wind slabs (Shimizu 1970). Similarly, one (wind slab) has the highest thermal conductivity (Sturm et al. 1997), while the other (depth hoar) has the lowest.

As the snow cover thickens and develops through the winter, the snow surface is constantly changing in response to the wind, new snow accumulation, surface hoar formation, and surface melt. Erosional and depositional forms (dunes, barchans, erosion pits, and drifts in the lee of ice blocks) form and change in each windstorm (Doumani 1966; Watanabe 1978; Goodwin 1990) (Figure 3.1.3). Mist, fog, or rain on snow can glaze the surface, producing firnspiegel, a shiny ice coating. In the lee of ice ridges and rafted blocks, snowdrifts in excess of 2 m deep can form. In late May or early June, the snow cover melts and the resultant runoff forms melt ponds on the ice.

**Figure 3.1.3.** Typical surface features of the sea ice snow cover, including both depositional (the triangular-shaped snow apron in the foreground under the orange notebook) and erosional features (the sastrugi shown in the inset).

## 3.1.3 BASIC SNOW COVER OBSERVATIONS

Basic snow observations are made by (a) digging a snow pit and recording information about the layers, (b) conducting a traverse along which spot measurements of snow properties are recorded, or (c) using aerial photography or remote-sensing products to infer areal snow properties. A comprehensive description of standard snow layer measurements is given in the International Classification for Seasonal

Snow on the Ground (Colbeck et al. 1992): A copy is included as part of the research tools on the accompanying DVD. The Classification is due to be updated in 2009.

The current layer observation suite includes (symbol and units listed):

- Density ($\rho$) (kg m$^{-3}$ or g cm$^{-3}$)
- Grain shape ($F$)
- Grain size ($E$) (mm)
- Snow strength ($\Sigma$) (Pa)
- Snow hardness ($R$) (Pa)
- Snow temperature ($T$) (°C)
- Layer thickness ($L$) (cm)
- Salinity (not in the International Classification but important on sea ice) (psu)

More difficult-to-measure properties, like thermal conductivity and dielectric constant, can sometimes be inferred from simpler-to-measure layer, grain, and density observations.

Spot measurements along traverses include the following:

- Depth ($H$) (cm)
- Snow water equivalent ($SWE$) (mm or cm)
- Snow-covered area ($SCA$) (% or fraction)
- Snow surface roughness

Traverse measurements can sometimes allow more detailed pit measurements to be extrapolated spatially and also provide a statistical base from which mean properties (like average snow depth) can be computed. Snow measurements from aerial photography or remote sensing products are covered in Chapter 3.15.

### 3.1.3.1 SNOW PIT MEASUREMENTS

A typical snow pit kit is shown in Figure 3.1.4. It contains the tools needed to make the snow layer measurements listed above. Because the sea ice snow cover can be quite hard, it is useful to have both a steel shovel and an aluminum grain scoop for excavating the pit (see Figure 3.1.5). The idea is to create a smooth pit wall in which layers will be easy to delineate. A saw, snow knife, or straight shovel should be used to slice a clean, vertical face, keeping the original surface snow intact as a reference for measurements. A pit about 1 m wide works well. A steel spatula or trowel is useful for cleaning up the face after shoveling. A whisk broom can be used to brush the clean face. This will accentuate the snow layering. Snow layers have different hardness and the brush will remove snow from weaker layers but not from stronger ones. In addition to the sharp boundaries revealed by

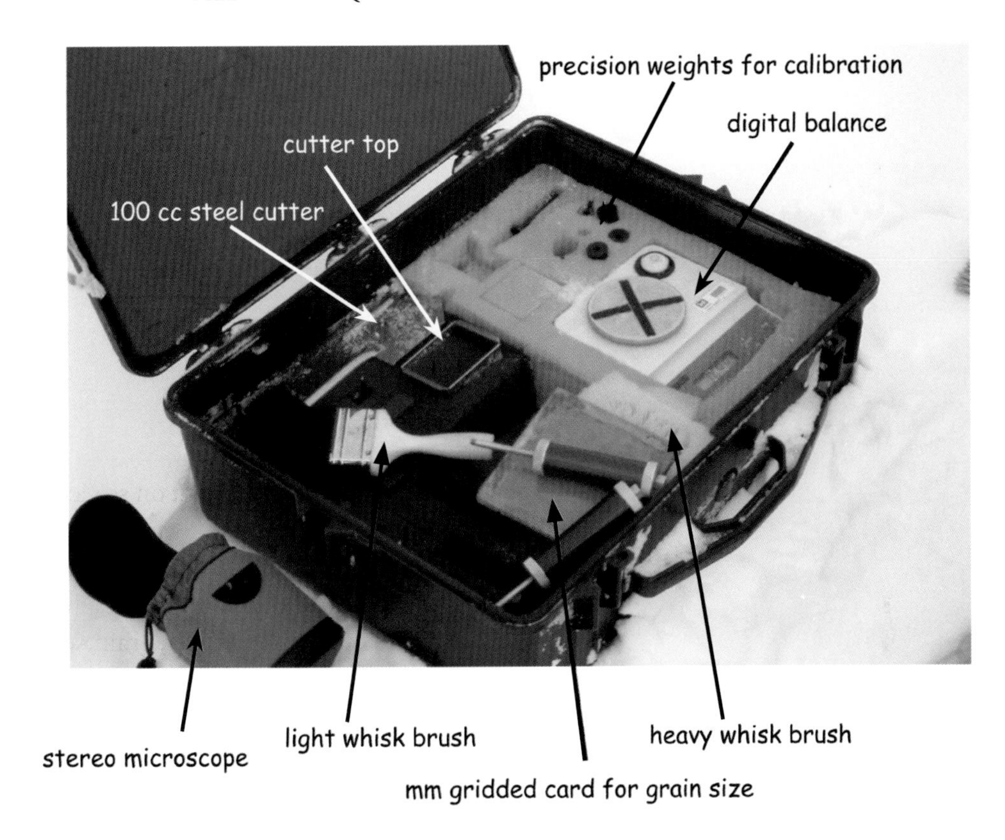


**Figure 3.1.4.** A typical snow pit kit. A folding rule and spatula is also included but is not shown here.

**Figure 3.1.5.** Digging a snow pit on the sea ice

**Figure 3.1.6.** A brushed snow pit wall with a folding rule in place, zero (0 cm) at the ice surface

brushing, subtle differences in grain size, overall texture, and gray-white color will differentiate layers.

Layer measurements are referenced to the sea ice surface (zero height) (Figure 3.1.6). It is useful to place a ruler against the pit wall so that the top and bottom heights of each layer can be read off the ruler. There are many ways of recording layer observations from the pit, one of which is shown in Figure 3.1.7.

| Height (H) | Grain Shape (F) | Layer | Density (ρ) | Grain Size (E) | Hardness (R) | Temperature (T) |
|---|---|---|---|---|---|---|
| 34 | | j | 0.186 | 4 | fist | -23 (34 cm) |
| 31 | | h & i | 0.316 | 1 | pencil | |
| 27 | | g | 0.318 | 2 | 4 fingers | |
| 24 | | f | 0.401 | 2 | knife | -16 (19 cm) |
| 16 | | e | 0.291 | 4 | fist | |
| 15 | | d | 0.339 | 3 | 1 finger | -11 (15 cm) |
| 11 | | c | 0.265 | 7 | fist | -7 (9 cm) |
| 6 | | b | 0.343 | 9 | fist | |
| 2 | | a | 0.492 | | ice | -4 (0 cm) |
| 0 | | | | | | |

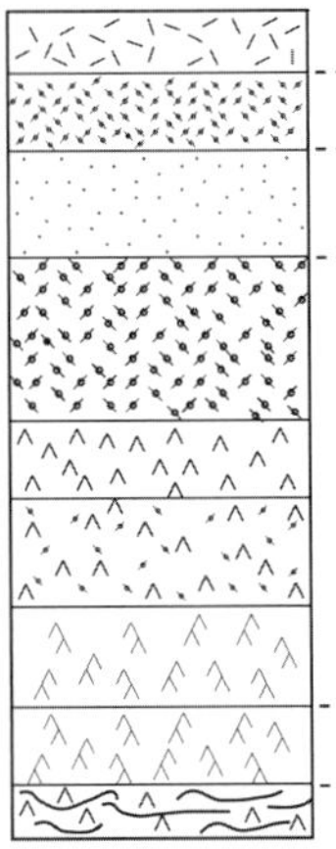

**Figure 3.1.7.** A tabular-graphical method of recording snow pit information. Grain shape (F) is indicated by the International Classification Symbols (see Figure 3.1.2).

*Density*: Density is measured by removing a known volume of snow from a layer and weighing the volume. Various cutter designs can be used (Figure 3.1.8). The advent of digital balances that run on batteries and work in the cold have made these measurements faster and more accurate. It is customary to tare out the balance using the empty cutter so that the balance reads the net weight of the snow. For a 100 cm$^3$ cutter, the density is then the observed reading divided by 100 (i.e., 36.7 g equals a density of 0.367 g cm$^{-3}$). Virtually all digital balances utilize a strain meter for weighing. While balance output is linearized for temperature, rapid changes in temperature can lead to errors. Frequent checks of the balance performance using a known weight is advisable. Cutters range in height from 3 to 10 cm, so they

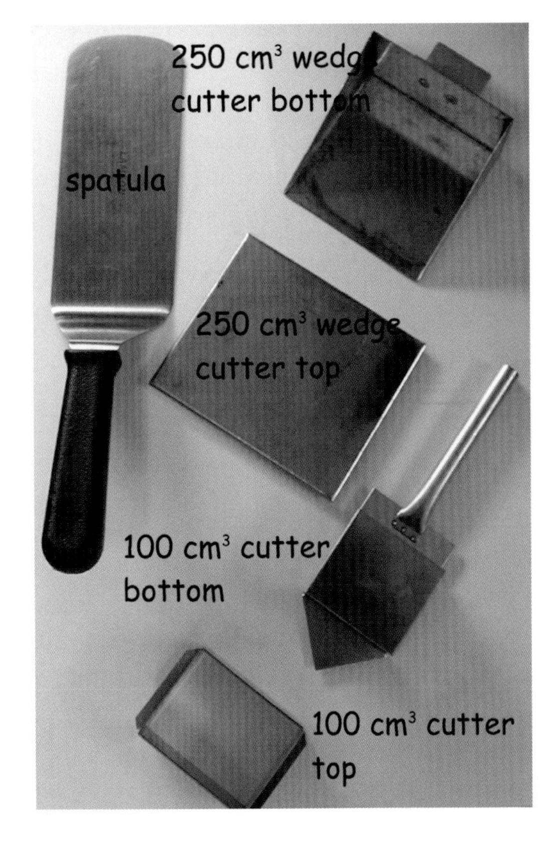


**Figure 3.1.8.** Density cutters

may not span the thicker snow layers. In that case, multiple samples in a vertical line should be taken so that they span the layer. This will capture any vertical density gradient present within the layer. The bulk density ($\rho_{bulk}$) for the pit can be computed from the individual measurements ($\rho_i$):

$$\rho_{bulk} = \frac{1}{h_{total}} \sum_{i=1}^{n} \rho_i h_i \qquad \text{(Equation 3.1.1)},$$

where $h_{total}$ is the total snow depth and $h_i$ is thickness of the $i^{th}$ layer.

*Grain Shape*: Field identification of grain shape requires experience, a decent hand lens or microscope, and good lighting. A simple guide is given here, but readers should refer to the International Classification for Seasonal Snow on the Ground, which has many photomicrographs of snow grains, as well as the work of Seligman (1936), LaChapelle (1969), Pahaut and Marbouty (1981), and Libbrecht (2006). The goal in making grain shape measurements is to make an *identification of snow type* because the type can be used to infer a number of physical characteristics.

There are four basic types of snow present on arctic sea ice (Figure 3.1.9):

- depth hoar
- wind slab
- new or recent snow (stellar dendrites, capped columns, needles)
- thin, often hard-to-detect ice crusts or melt grain layers

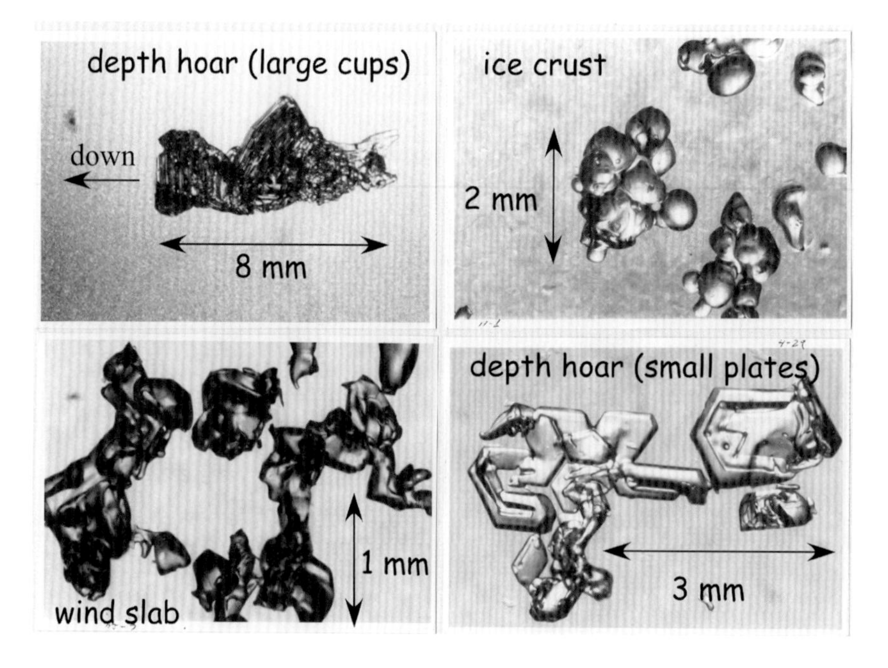


**Figure 3.1.9.** Three of the four typical types of snow grains found on arctic sea ice (see text)

These four can be roughly subdivided into either faceted grains (depth hoar and new snow) or rounded grains (wind slab and melt grains). Faceted grains are indicative of rapid kinetic growth under a strong temperature gradient. They are also the product of what is known as TG (temperature gradient) metamorphism. Rounded forms are indicative of slow growth, or no growth (i.e., equilibrium grain forms), a type of metamorphism that is known as ET (equi-temperature) metamorphism (Sommerfeld and LaChapelle 1970; Colbeck 1982).

In addition to facets, the hardness of a layer can assist in grain type identification. Wind slabs are hard (it takes a pencil or a knife to penetrate them), new and recent snow is soft (penetrated easily by a fist), and depth hoar can be soft (fist or three fingers) or moderately hard (pencil) if it was originally a wind slab that has metamorphosed over time into depth hoar. The latter type of depth hoar, common on the sea ice, is called *indurated* or *hard* depth hoar. It is also called *wind slab-to-depth hoar*. One final type of snow grain common on sea ice is a type of depth hoar called *chains-of-grains* (Trabant and Benson 1972). It is made up of linked, downward-facing pyramidal cups (Figure 3.1.9, upper-left picture, where down in the snowpack is to the left in the picture) that show a distinctive vertical structuring and are the product of a long period of kinetic or TG metamorphism.

*Grain size* is estimated by placing a number of snow grains on a gridded card, examining the card with a hand lens or microscope, and making a visual estimate of the mean size. Most snow layers will actually exhibit a grain size distribution (Figure 3.1.10), so estimating the mean takes practice. There is a tendency for novice

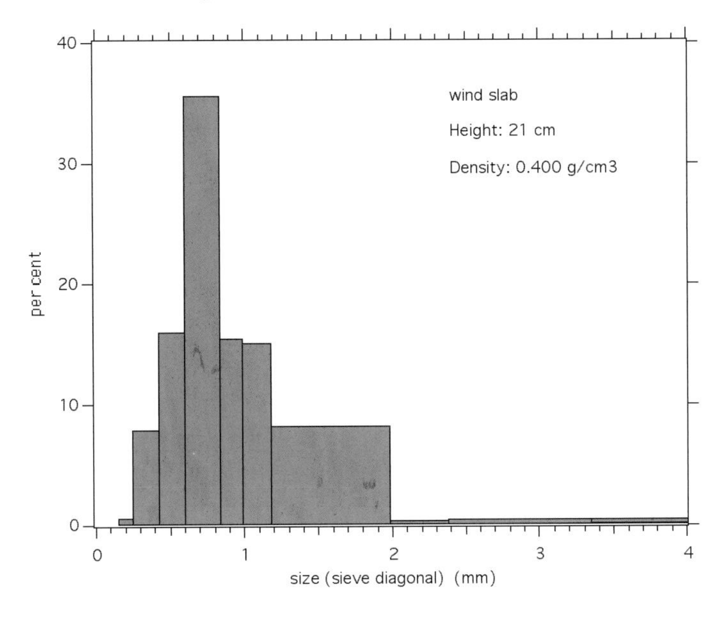


**Figure 3.1.10.** Grain size distribution (%) for a wind slab layer on the North Slope of Alaska. Note that the grain size (as measured by sieving) ranges from 0.2 to 2 mm.

observers to discount the smaller grains, leading to an overestimate. Grain size is often measured in order to model the passive or active microwave signal from the sea ice snow cover for remote-sensing studies. Mätzler (2002) suggests that for that application, what is needed is an estimate of the optical grain size. He indicates that the thinnest dimension of the grains is the best estimator for this purpose. For depth hoar, which can form thin plates or cups, this would be the plate or cup wall thickness.

*Snow strength* is rarely measured and requires special equipment, but *snow hardness* is an easy measurement to make. Up until about ten years ago, a Rammsonde (a slide-hammer penetration rod with a conical tip) (Roche 1949) was the standard for measuring an index of snow strength, though it was difficult to convert readings to true strength measurements. Recently a micropenetrometer housing a stress sensor (Schneebeli and Johnson 1998) has been developed that reports values related to bond strength, but the device is expensive and requires considerable effort to operate: there are only about twenty such devices worldwide. Snow hardness, though a subjective measurement (Table 3.1.2), requires no equipment other than a hand, a pencil, and a knife. Tests with multiple observers indicate reasonable consistency in values, and the measurements are useful in differentiating one wind slab from another on the sea ice.

**Table 3.1.2.** Snow hardness

| Term | Rammsonde (N) | Strength (Pa) | Hand test |
|---|---|---|---|
| Very low | 0–20 | $0–10^3$ | Fist |
| Low | 20–150 | $10^3–10^4$ | 4 fingers |
| Medium | 150–500 | $10^4–10^5$ | 1 finger |
| High | 500–1000 | $10^5–10^6$ | Pencil |
| Very high | >1000 | $>10^6$ | Knife blade |
| Ice | | | |

*Snow temperature, layer thickness,* and *salinity* measurements are always made as a function of height in the snow. A digital temperature reader and probe or dial thermometers can be used to make the temperature measurements. The probes are inserted into the pit wall and allowed to equilibrate. Then the temperature and height recorded. It is necessary to shade the probes or thermometers if it is a sunny day; otherwise solar heating can take place. Vertical profiles of temperature and salinity can indicate the thermal state of the snow and how much seawater flooding has taken place. One particularly useful measurement is the snow-ice interface temperature. Collecting a number of these measurements along traverse lies and then plotting the value as a function of snow depth (Figure 3.1.11) is a good way to explore the impact of the snow cover on the thermal balance of the ice.

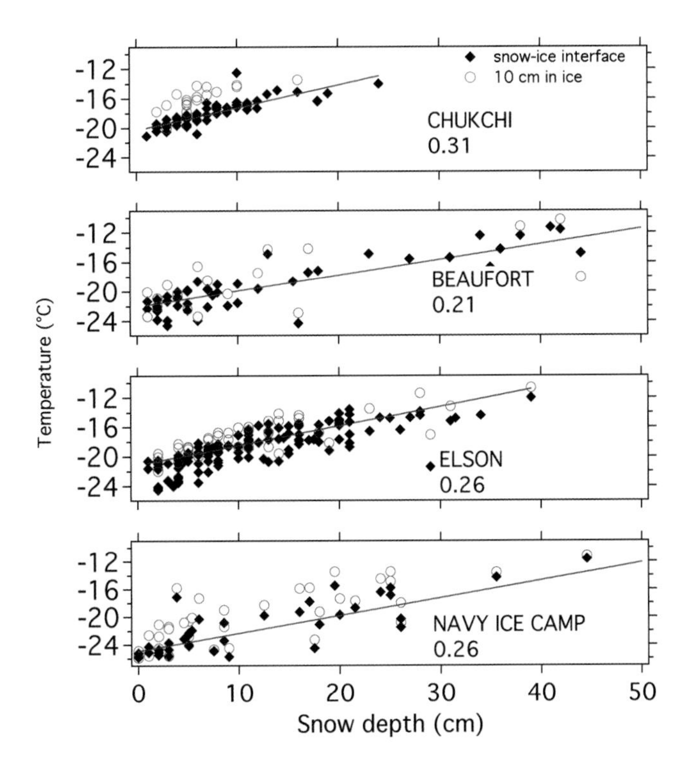


**Figure 3.1.11.** The snow-ice interface temperature as a function of snow depth for several areas near Barrow, Alaska. The number beneath each location is the slope of the best fit line to the data. Differences suggest variations in snow cover thermal properties between the four areas.

## 3.1.3.2 TRAVERSE MEASUREMENTS

Snow traverse observations consist of spot measurements made along lines or routes. These can be straight or circuitous, and of any length. For example, the USDA National Resource Conservation Service operates thousands of snow courses in the western United States that consist of routes 100 to 300 m long along which five to ten measurements of depth and snow-water equivalent are made monthly (Davis et al. 1970). On the sea ice, traverses have ranged from 100 m to 10 km in length (Sturm et al. 2006). Choice of route length, orientation, type of measurement, and frequency depend on the time available for the work, the tools available, crew mobility, the weather, and the study design.

*Snow depth* is the primary measurement made on traverses. In principle it is an easy measurement to make on ice because there is little ambiguity as to where the base of the snow is located (on the tundra this can be uncertain to ±4 cm). Any graduated rod or ruler can be pushed through the snow to refusal and the depth read off

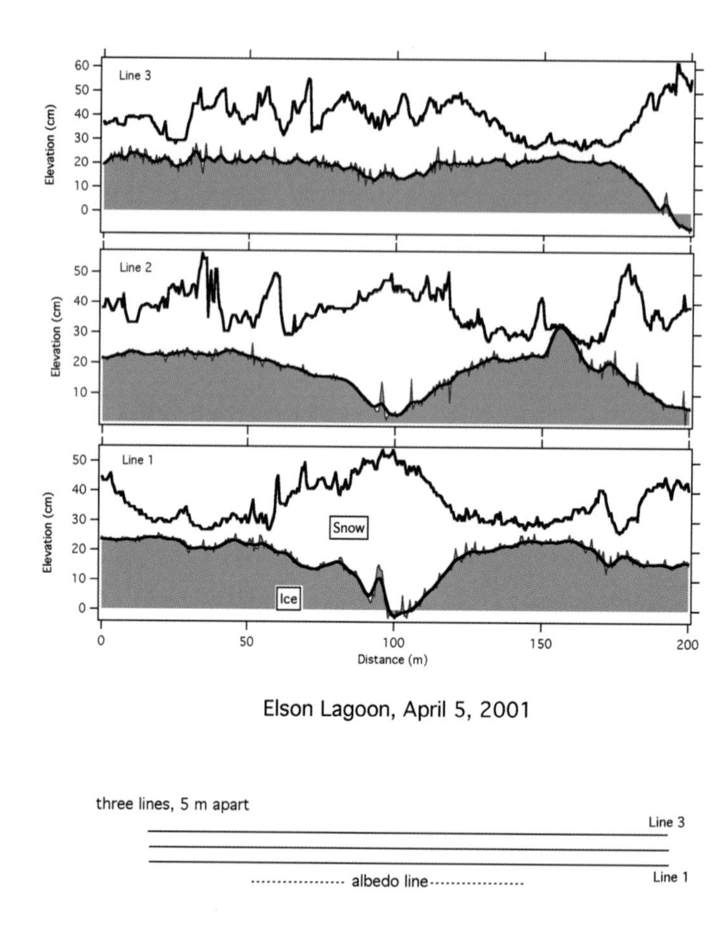


**Figure 3.1.12.** Three typical snow depth cross-sections from arctic sea ice showing depths that vary from 6 to 54 cm over less than 50 m (data from Elson Lagoon, April 5, 2001). The dark shaded area is sea ice, which shows some evidence of deforming under the weight of the snow (see Chapter 3.2).

the side of the rod. However, a single measurement is usually of little value because the wind-drifted snow cover exhibits a wide range of depths over a small area (Figure 3.1.12). There is no hard-and-fast rule for deciding how many measurements to make and over what length of line to make them, but Figure 3.1.13 provides some guidance for lines of several hundred meters. The data in the figure were computed using several long, oversampled lines of snow depth data from Barrow. Each line was 600 m long, sampled every 0.5 m. Line length ($L$) and number of sample points on that line ($n$) were selected (sample spacing was $L/n$) and then a line of length $L$ was randomly located within the 600 m line. This was repeated one thousand times, each time selecting the line in a random location along the measured line. The root mean squared error for all those $L/n$ values was computed, another set of $L$ and $n$ values chosen, and the process repeated. The results suggest that lines shorter than 100 m with $n$-values of less than 40 may have substantial bias.

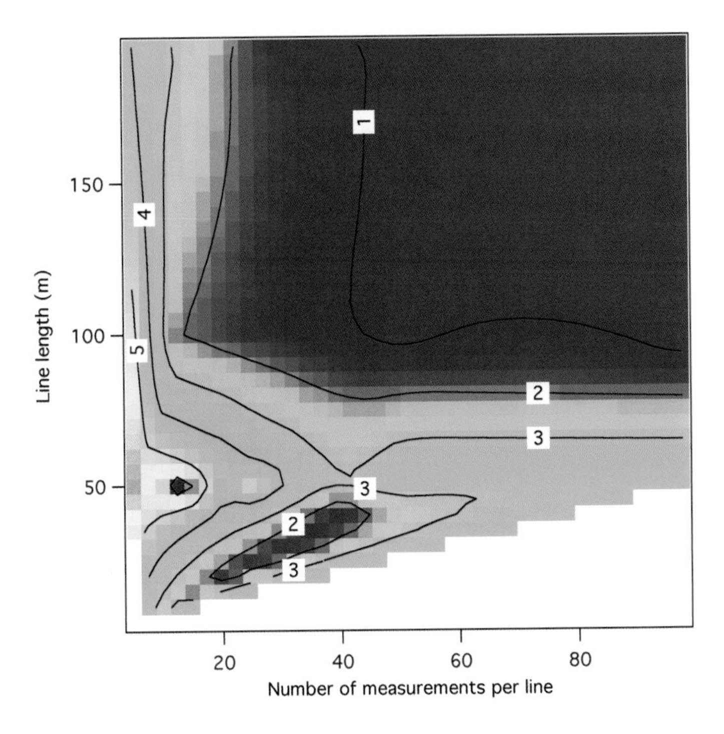


**Figure 3.1.13.** Monte Carlo simulations of the probable error in mean snow depth as a function of traverse line length (*L*) and number of measurements per line (*n*). The approximate error in mean value is indicated by the isolines, which have units of standard deviation ($\sigma$) in cm. See text for how these were computed. Errors in the mean are reduced as lines exceed 100 m and 50 measurements per line (Sturm, unpublished data, 2008).

In recent years a new device (GPS-MagnaProbe) (Figure 3.1.14) that measures and records snow depth and GPS position automatically allows much more rapid collection of depth data than with a ruler or graduated rod. The device utilizes a floating basket that rests on the snow surface. The operator pushes the steel rod down through the basket to the ice surface, at which time he/she pushes the trigger switch to record the depth. At the end of the day, all data can be collected from the probe using a laptop computer. A single operator can collect several thousand depths in a day, 10 to 20 times more than can be collected by hand.

One depth-sampling strategy that has been used effectively for arctic sea ice is to measure depth along short (100 to 500 m) traverses in various types of ice: smooth first-year ice, rubble fields, ridge areas, multiyear ice, etc. (Sturm et al. 2002). Distinctively different distribution patterns emerge (Figure 3.1.15).

*SWE* is the second most common traverse measurement after depth. A coring tool (Figure 3.1.16) is used to obtain a snow core of known volume (depth times cross-sectional area of the corer), which is weighed to determine the mass. Three types of corers are in use in the United States and Canada: the aluminum Federal Sampler,

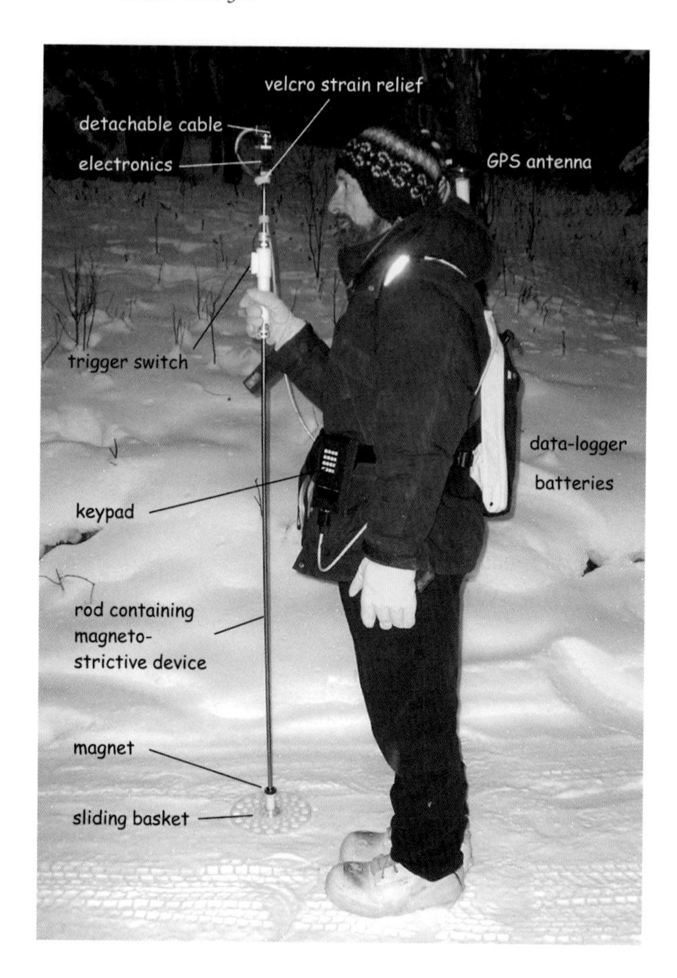


**Figure 3.1.14.**
A GPS-MagnaProbe for measuring snow depth and GPS position

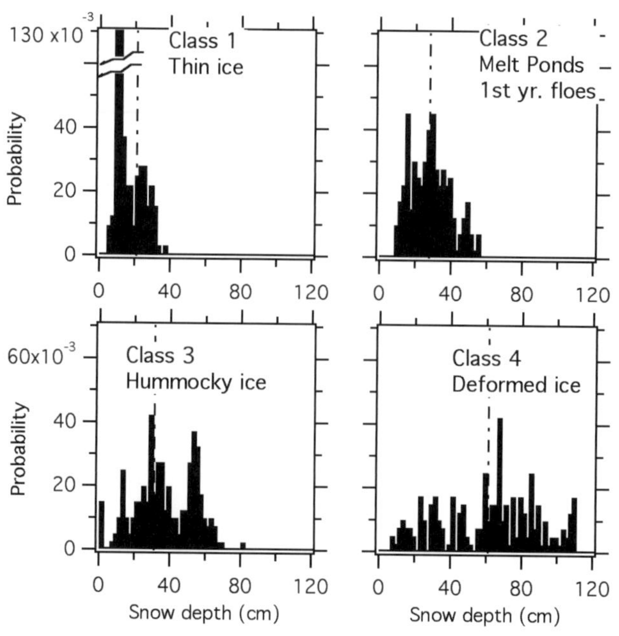


**Figure 3.1.15.**
Snow depth probability distribution functions (pdfs) by ice type from an ice camp in the northern Chukchi Sea (Sturm et al. 2002)

the lexan Snow-Hydro sampler, and the steel ESC Canadian corer. All three have a sharp circular cutting head that penetrates through wind slab well if rotated aggressively. But taking a snow core on the sea ice differs from taking one on the tundra because the corer will not core the ice and therefore will not take a plug at the end of the core. A plug is necessary to keep the snow from falling out of the corer as it is removed, so on the ice it is necessary to dig down to the base of the snow and insert a spatula or trowel under the corer to prevent this loss. The bulk density ($\rho_{bulk}$) is computed from:

$$\rho_{bulk} = m/(Ah_{total}) \text{ (Equation 3.1.2),}$$

where $m$ is the mass of the snow core, $A$ is the cross-sectional area of the corer, and $h_{total}$ is the snow depth, usually read from a scale engraved on the side of the corer. If mass is measured in g, the area in $cm^2$, and the depth in cm, the density will be in g $cm^{-3}$. The SWE (cm) will be:

$$SWE = \rho_{bulk} \cdot h_{total} \quad \text{(Equation 3.1.3).}$$

Equation 3.1.3 indicates that bulk density and depth are equally important in determining the SWE, but density is a conservative variable in comparison to the depth (the dynamic range of the latter is several times larger than the former) (Dickinson and Wheatly 1972; Steppuhn 1976). In practice, depth controls SWE to first order. It is generally possible to develop a robust regression of SWE vs. depth from about 40 measurements, though a larger sample is better (Figure 3.1.17). Drift snow (snow > 60 cm deep) is denser than nondrift snow (0.38 vs. 0.34 g $cm^{-3}$), a fact that should be taken into account if the drifts are extensive in the area that is being measured and across which one wishes to extrapolate SWE based on depth.

**Figure 3.1.16.** A typical SWE corer

*Snow-covered area* (SCA) is also an important measurement. It can be estimated from aerial photographs, but care must be taken to differentiate snow from weathered sea ice. It can also be determined from snow depth probe lines from the percentage of

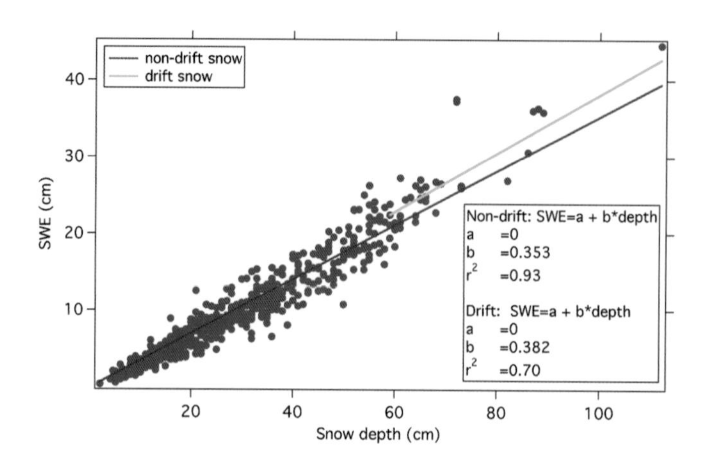


**Figure 3.1.17.** Snow water equivalent (SWE) as a function of depth. Data ($n$=592) come from several sea ice campaigns near Barrow, Alaska. The slope of the regression line (b) is one estimate of the bulk snow density. Note that the drift snow regression line has a steeper slope than nondrift snow.

zero-depths recorded along a traverse. It is a particularly important measurement during the spring melt period, when the ice cover transitions from near 100 percent snow cover to little or no snow.

Last, the nature and orientation of the surface drift and erosion features are often recorded along traverses. Watanabe (1978) recognized four types of features: barchans, dunes, sastrugi, and erosion pits. Some of these features are elongated parallel to the wind (dunes and sastrugi), while others are elongated perpendicular to the wind (barchans). In general, large drift aprons in the lee of ice blocks can help determine the primary drift direction. Lanceolate sastrugi and anvil-head drifts, often common on the ice (Figure 3.1.18) also provide unambiguous transport

**Figure 3.1.18.** Lanceolate sastrugi on the Chukchi Sea near Barrow. The wind was from the lower left corner of the photograph.

directions. Doumani (1966), Watanabe (1978), and Goodwin (1990) provide useful pictures and descriptions of these surface drift features.

### 3.1.4 ANALYSIS OF SNOW PIT AND SNOW TRAVERSE DATA

In some cases, the measurements from a snow pit on the sea ice will have a specific use, but more often, they are part of a program to characterize the snow in a general way. One method of analyzing the snow pit data is to tabulate it and from the individual snow pits derive a set of summary statistics. Essentially, through this process, one describes or defines an "average" layered snow cover for the area. For example, based on 68 snow pits dug in 2000 and again in 2002, Sturm and Liston (2003) defined an average snow cover for the lakes of the Arctic Coastal Plain of Alaska. This snow cover had between five and six layers, averaged 32.5 cm deep, and had the textural composition shown in Figure 3.1.19. The value in producing this sort of "average" is that it is then possible to derive for the "average" snow a set of secondary properties that might not otherwise have been measured, such as the thermal conductivity. For example, from published values (Sturm et al. 1997), each type of snow layer can be assigned a thermal conductivity, and the bulk thermal conductivity ($k_{bulk}$) of the snow cover computed from:

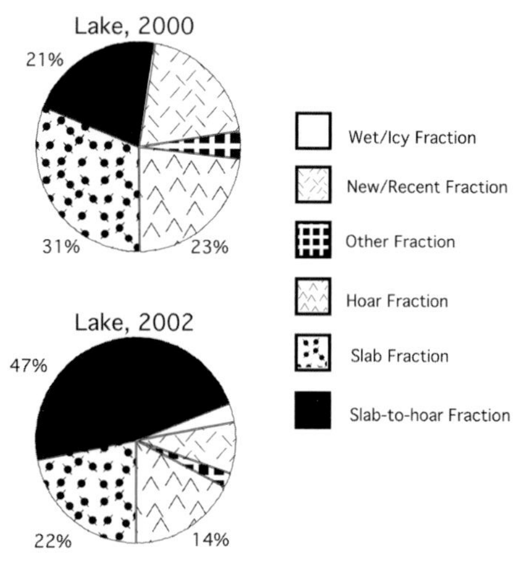


**Figure 3.1.19.** The textural composition of the snow cover on lake ice of the Arctic Coastal Plain (Sturm and Liston 2003).

$$\frac{1}{k_{bulk}} = \frac{1}{\overline{H}} \sum_{i=1}^{\overline{n}} \frac{\overline{h}_i}{k_i} \quad \text{(Equation 3.1.4)},$$

where $\overline{h}_i$ is the average thickness of the $i^{th}$ layer; $k_i$ is the thermal conductivity of that layer based on the texture, density, and grain characteristics; $\overline{n}$ is the average number of layers; and $\overline{H}$ is the average snow cover depth.

Traverse measurements can be plotted in cross-section (Figure 3.1.12), analyzed using normal statistics to produce pdfs (Figure 3.1.15), or examined using spatial statistics (Figure 3.1.20) to infer information about the structure of the snow depth distribution. Alternately, the measurements can be used to establish functional relationships, such as the one between snow depth and SWE (Figure 3.1.17), that are useful for extrapolation. In some cases, it may be possible to use depth or SWE as a proxy measure for inferring snow textural data from one location to

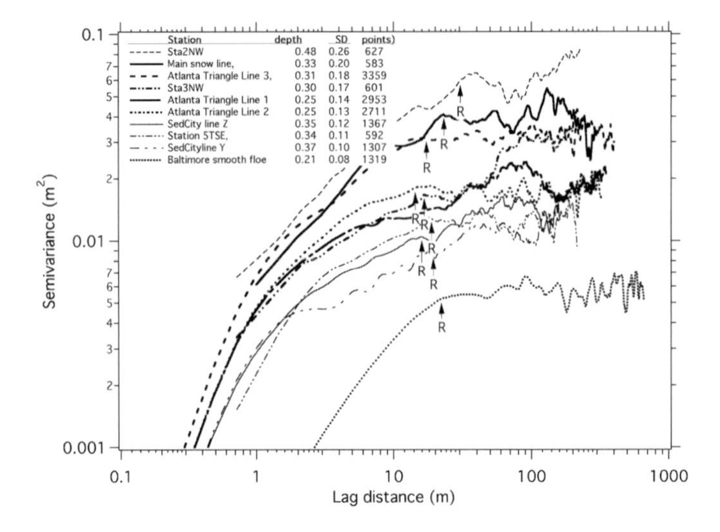


**Figure 3.1.20.** Semivariograms (see Isaaks and Srivasta 1989, for an explanation of these) for snow depth traverses measured at an ice camp in the northern Chukchi Sea showing that the range (*R*=structural length) across many different types of ice was about 20 m. One implication of this constant range is that snowdrifts independent of the ice topography were the cause.

another, for example in the case where the snow pits establish a clear relationship between layering and depth or SWE.

## 3.1.5 CONCLUSIONS

The snow cover on arctic sea ice is thin, has five to nine layers, is heavily wind affected, and is always heterogeneous. Observations of this snow cover are usually collected in order to characterize both the local and areal distribution of the snow properties. Frequently, thermal, optical, or dielectric properties are the main interest, but because these are difficult to measure and require specialized equipment, simpler properties like the depth, density, grain shape, and layer stratigraphy are measured. From the simpler properties, the more complex properties can be inferred (with acceptable accuracy) using look-up tables or published textural-property relationships.

Field measurements consist of either layer measurements made in snow pits or spot measurements made along traverse lines. The former include layer thickness, density, grain shape and size, and temperature. The latter include depth, snow water equivalent, and surface conditions. A wide range of sampling strategies have been employed for both types of measurements, though in most cases, given the strong heterogeneity of the material, undersampling is the main problem. As there is almost never enough time and manpower to collect a comprehensive suite of measurements, careful thought should be given to the purpose of the snow measurements so that the measurements that are made will be appropriate to the problem at hand.

In the future, it may be possible to measure many of the snow properties discussed here much more quickly and extensively using remote sensing methods. For example, in principle, snow depth can be (and has been) measured using airborne radar (e.g., FMCW radar), while surface roughness can and has been measured using ground-based and airborne lidar. While these advanced techniques show promise, there is a fundamental problem that will remain difficult to overcome: The sea ice snow cover rests on sea ice. The two are composed of essentially the same material with differing amounts of air, and they have a complex functional relationship with one another. As a consequence, electromagnetic and stand-off (remote) measurements often have to deal with at best subtle differences that differentiate the two materials. These advanced technologies will improve and may eventually supplant direct field measurements, but that is likely to take many years to happen, and in the interim, the field methods described here will continue to be used.

## REFERENCES

Akitaya, E. (1972), Some experiments on the growth of depth hoar, in *Physics of Ice and Snow*, pp. 713–723.

Benson, C. S., and M. Sturm (1993), Structure and wind transport of seasonal snow on the Arctic slope of Alaska, *Annals of Glaciology*, *18*, 261–267.

Colbeck, S. C. (1982), An overview of seasonal snow metamorphism, *Reviews of Geophysics and Space Physics, 20,* 45–61.

Colbeck, S. C. (1991), The layered character of snow covers, *Reviews of Geophysics*, *29*, 81–96.

Colbeck, S., E. Akitaya, R. Armstrong, H. Gubler, J. Lafeuille, K. Lied, D. McClung, and E. Morris (1992), *The International Classification for Seasonal Snow on the Ground, 23*, The International Commission on Snow and Ice of the International Association of Scientific Hydrology/International Glaciological Society/U.S. Army CRREL, Hanover, NH.

Davis, R. T., W. T. Frost, T. A. George, B. L. Whaley, R. E. Melsor, R. W. Enz, J. N. Washichek, D. W. McAndrew, P. E. Farnes, G. W. Peak, A. G. Crook, G. L. Pearson, P. Keil, M. W. Nelson, and others (1970), *SCS National Engineering Handbook: Section 22, Snow Survey and Water Supply Forecasting*, Soil Conservation Service, U.S. Dept. of Agriculture.

Dickinson, W. T., and H. R. Whitely (1972), A sampling scheme for shallow snowpacks, *IASH Bulletin*, *17*, 247–258.

Doumani, G. A. (1966), Surface structures in snow, in *International Conference on Low Temperature Science: I. Physics of Snow and Ice*, pp. 1119–1136, Sapporo, Japan.

German, R. M. (1996), *Sintering Theory and Practice*, John Wiley & Sons, Inc., New York.

Goodwin, I. D. (1990), Snow accumulation and surface topography in the katabatic zone of eastern Wilkes Land, Antarctica, *Antarctic Science, 2*, 235–242.

Hanson, A. M. (1980), The snow cover of sea ice during the Arctic Ice Dynamics Joint Experiment, 1975–1976, *Arctic and Alpine Res., 12*, 215–226.

Isaaks, E. H., and R. M. Srivastava (1989), *An Introduction to Applied Geostatistics*, Oxford University Press, New York.

Kojima, K. (1966), Densification of seasonal snow cover, in *Physics of Snow and Ice*, pp. 929–951, Institute of Low Temperature Science, Sapporo, Japan.

LaChapelle, E. R. (1969), *Field Guide to Snow Crystals*, University of Washington Press, Seattle and London.

Libbrecht, K. (2006), *Ken Libbrecht's Field Guide to Snowflakes*, Voyageur Press, St. Paul, MN.

Mätzler, C. (2002), Relation between grain-size and correlation length of snow, *J. Glaciol., 48*, 461–466.

Maykut, G. A., and N. Untersteiner (1971), Some results from a time-dependent thermodynamic model of sea ice, *Journal of Geophysical Research, 76*, 1550–1575.

Meier, W. N., J. Stroeve, and F. Fetterer (2007), Whither Arctic sea ice? A clear signal of decline regionally, seasonally and extending beyond the satellite record, *Annals of Glaciology, 46*, 428–434.

Pahaut, E., and D. Marbouty (1981), Les cristaux de neige, *Neige et Avalanches, 25*, 3–42.

Pomeroy, J. W., and D. M. Gray (1995), *Snowcover Accumulation, Relocation, and Management*, National Hydrology Research Institute, Science Report Number 7.

Roch, A. (1949), *Report on Snow and Avalanches Conditions in the U.S.A. Western Ski Resorts from January 1–April 24, 1949, 39*, Federal Institute for Research on Snow and Avalanches—Weissfluhjoch-Davos, Davos, Switzerland.

Schneebeli, M., and J. B. Johnson (1998), A constant-speed penetrometer for high-resolution snow stratigraphy, *Annals of Glaciol., 26*, 107–111.

Seligman, G. (1936), *Snow Structure and Ski Fields*, International Glaciological Society, Cambridge.

Shimizu, H. (1970), Air permeability of deposited snow, *Low Temperature Science, Series A, 22*, 1–32.

Sommerfeld, R. A., and E. R. LaChapelle (1970), The classification of snow metamorphism, *J. Glaciol., 9*, 3–17.

Steppuhn, H. (1976), Areal water equivalents for prairie snowcovers by centralized sampling, in *Proceedings, 44th Annual Western Snow Conference*, pp. 63–68.

Sturm, M. (1991), *The Role of Thermal Convection in Heat and Mass Transport in the Subarctic Snow Cover*, Report 91-19, USA-CRREL.

Sturm, M., J. Holmgren, M. König, and K. Morris (1997), The thermal conductivity of seasonal snow, *J. Glaciol., 43*, 26–41.

Sturm, M., J. Holmgren, and D. K. Perovich (2002), The winter snow cover of the sea ice of the Arctic Ocean at SHEBA: Temporal evolution and spatial variability, *J. Geophys. Res.-Oce.*, *107*, doi:10.1029/2000JC0004000.

Sturm, M., and G. Liston (2003), The snow cover on lakes of the Arctic Coastal Plain of Alaska, U.S.A., *J. Glaciol.*, *49*, 370–380.

Sturm, M., J. A. Maslanik, D. K. Perovich, J. C. Stroeve, J. Richter-Menge, T. Markus, J. Holmgren, J. F. Heinrichs, and K. Tape (2006), Snow depth and ice thickness measurements from the Beaufort and Chukchi Seas collected during the AMSR-Ice 03 Campaign, *IEEE Transactions on Geoscience and Remote Sensing*, *44*, 3009–3020.

Sturm, M., D. K. Perovich, and J. Holmgren (2002), Thermal conductivity and heat transfer through the snow and ice of the Beaufort Sea, *J. Geophys. Res.-Oce.*, *107,* doi:10.1029/2000JC000409.

Trabant, D., and C. S. Benson (1972), Field experiments on the development of depth hoar, *Geological Society of America Memoir*, *135*, 309–322.

Untersteiner, N. (1961), On the mass and heat budget of Arctic sea ice, *Arch. Met. Geoph. Biokl.*, *A12*, 151–182.

Watanabe, O. (1978), Distribution of surface features of snow cover in Mizuho Plateau, *Memoirs of National Institute of Polar Research*, 44–62.

# Chapter 3.2

# Ice Thickness and Roughness Measurements

*Christian Haas and Matthew Druckenmiller*

## 3.2.1 INTRODUCTION

Apart from ice concentration and ice extent, which are related to the presence or absence of ice, thickness is probably the most important sea ice property, defining its quality and suitability for providing the services discussed in Chapter 2. In this chapter, basic aspects of the ice thickness distribution will be discussed, measurement methods will be presented, and applications of the methods for various users of sea ice services will be demonstrated.

Throughout the chapter, the term *ice thickness* will be used to describe the distance between the ice surface and the ice underside. This term is more objective than *ice depth*, which is sometimes used instead and seems more obvious for observers standing on the ice and wondering about what is below them. Similarly, snow thickness is sometimes referred to as *snow depth*. Definitions of ice thickness often include the thickness of snow, in which case it should rather be referred to as *total ice thickness*. Definitions for all these terms are given below (Section 3.2.2), so that one may properly define each variable and observation and avoid notation errors that may significantly miscommunicate data.

### 3.2.1.1 The Ice Thickness Distribution

Figure 3.2.1 shows aerial photographs of typical sea ice covers, both during the winter (left) and summer (right). It can be seen that the ice surface is covered by miniature mountain ranges, so-called ridges and rubble, which result in a considerably rough surface. As sea ice floats on the water and is generally in isostatic equilibrium, it is clear that ridge sails at the ice surface must be accompanied by ridge keels below the ice, and that the ice is considerably thicker at those locations than at the adjacent level ice. In the Arctic and even at the North Pole, the snow and upper ice layers typically melt during the summer, resulting in meltwater that collects in so-called melt ponds (Figure 3.2.1, right). Because of their low albedo, melt ponds

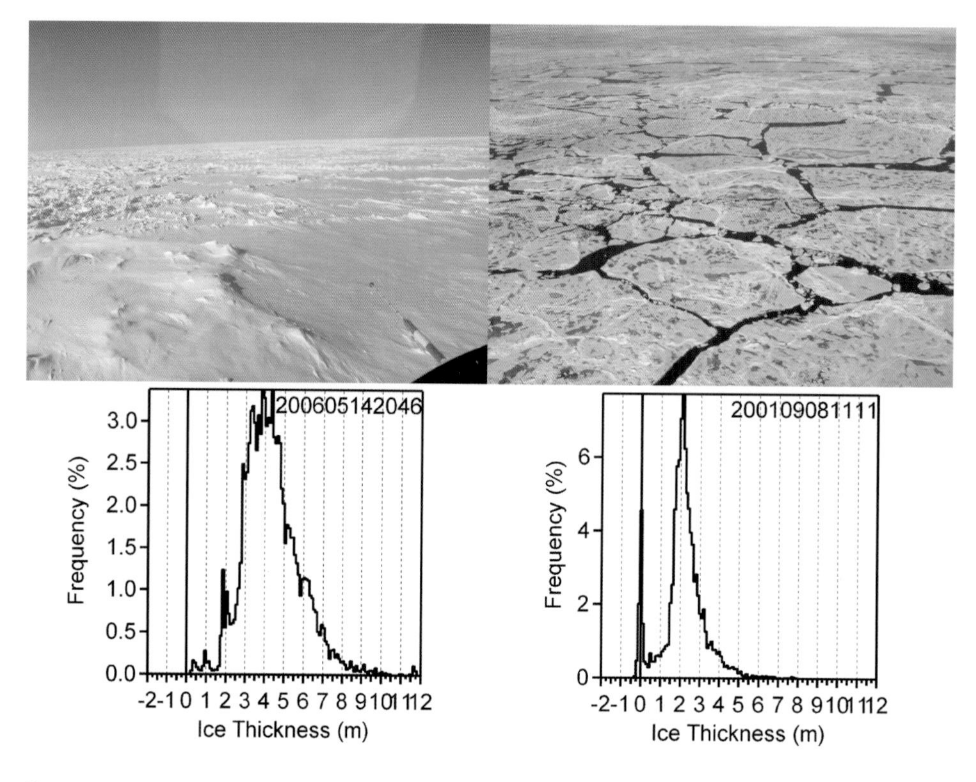


**Figure 3.2.1.** Aerial photographs and thickness distributions from airborne electromagnetic sounding typical of arctic sea ice in winter (left) and summer (right), from old multiyear ice in the Lincoln Sea (left) and second- and multiyear ice in the region of the North Pole (right). See text below on aspects of airborne electromagnetic sounding for deriving these thickness distributions.

enhance local melt (Section 3.2.9.5). Due to this preferential melting, the ice is typically thinnest at melt ponds, which additionally increases the roughness of the ice.

From this discussion of small-scale thickness variability it already becomes clear that a single ice thickness measurement may not be sufficient to characterize the thickness of an ice floe. Instead, a larger number of measurements is required, and should extend across a representative section of ice, comprising both level and rough ice, and possibly melt ponds. This so-called thickness profile can best be represented by means of a histogram or thickness distribution, as illustrated in Figure 3.2.1.

The thickness distribution is defined as a probability density function $g(h)$ of the areal fraction of ice with a certain ice thickness (Thorndike et al. 1975). The probability density function (pdf) of ice thickness $g(h)$ is given in Equation 3.2.1 by

$$g(h)\, dh = dA(h,h+dh) \,/\, R \qquad \text{(Equation 3.2.1)},$$

where $dA(h,h+dh)$ is the areal fraction of a region $R$ covered with ice of thickness between $h$ and $(h+dh)$. In practice, the thickness distribution is mostly obtained along linear profiles, and d$A$ and $R$ are one-dimensional, with $R$ as the total length of the profile. g($h$) is derived by dividing a frequency histogram of ice thickness data by the bin width (d$h$). Thus, its dimension is $m^{-1}$. Note that with a pdf the numerical value of each thickness bin is independent from the bin width used in calculating the histogram. This may be required if numerical values of thickness histograms are to be compared with other distributions, or are used to parameterize the thickness distribution in computer models. For most practical applications, it is sufficient to calculate the frequency distribution and to give results in fractions or in percentages.

Figure 3.2.1 shows typical thickness distributions representing the winter and summer conditions seen in the aerial photographs. The left histogram is from old multiyear ice in the Lincoln Sea north of Ellesmere Island, Canada. It possesses multiple local maxima, so-called modes, and a long tail towards thick ice. Note that there is almost no open water, as the fraction of ice with thickness 0 m is zero. In contrast, there is > 4 percent open water in the summer thickness distribution, which was obtained over second- and multiyear ice in the region of the North Pole in summer. This thickness distribution possesses only one clear mode (at 2.1 m), and its tail drops off towards thick ice considerably faster than in the example on the left.

The thickness distributions shown in Figure 3.2.1 give an accurate representation of ice thicknesses present along the surveyed profiles. They demonstrate in particular that ice thickness is mostly nonuniform, but very variable on small scales of meters to tens of meters. This small-scale variability is caused by the various and interacting processes of freezing, melting, and deformation. In fact, the thickness distribution bears information on the history and relative importance of these processes. Sea ice is only a relatively thin layer on the water, and thus rapidly responds by motion or drift to external forces exerted by winds and currents. The resulting forces are often nonuniform due to the divergence of winds and currents, and due to internal forces of the ice or the presence of obstacles like islands or coasts. Therefore, the ice cover frequently opens in divergent regions to form leads and polynyas, or ice floes collide with one another in convergent regions. If resulting forces exceed the fracture toughness of the ice, the ice breaks and ice blocks and fragments are piled above and below the adjacent ice to form pressure ridges and rubble fields (Figure 3.2.2 and animation on accompanying DVD). In contrast, new ice growth commences in open water once it is exposed to the cold air, adding regions of thin ice of variable thickness to the ice cover. This thermodynamically grown ice is mostly undeformed and level, while dynamically formed, deformed ice is typically very rough.

Most ice covers consist of larger regions of thermodynamically grown level ice, intersected by smaller regions of dynamically formed, deformed ice. This can be seen in the thickness distribution from the region of the North Pole, for example (Figure 3.2.1). The strong mode of 2.1 m indicates the thickness of the majority of

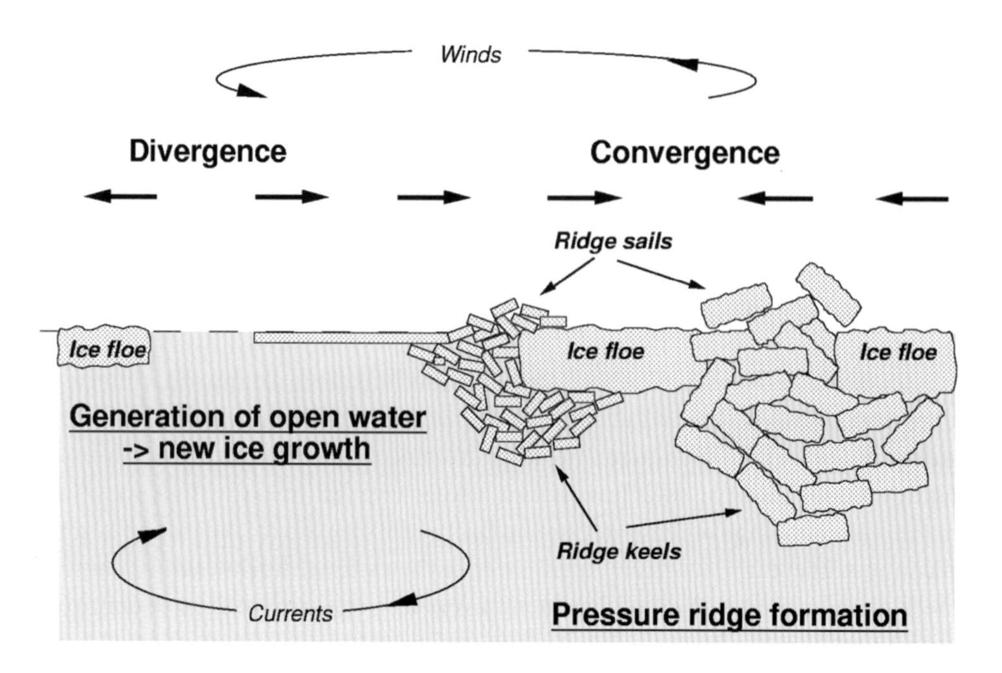


**Figure 3.2.2.** Illustration of the different dynamic and thermodynamic processes contributing to the development of the ice thickness distribution of a sea ice cover

the ice, which in this case was level second-year ice. As such, it is an indication of thermodynamic growth conditions (freezing and melting) throughout the history of that ice cover. Changes in thermodynamic growth, for example due to the seasonal cycle or long-term climate changes, would leave the shape of the distribution almost unchanged, but would result in shifts of modal thickness towards thinner or thicker ice. In contrast, the tail of the distribution represents the thickness and amount of deformed ice, and is therefore a measure of the intensity of deformation throughout the evolution of the ice cover. The fraction and thickness of deformed ice is affected by changes in ice motion, which can result from changes in atmospheric circulation patterns or ocean currents. It is also affected by the age of the ice cover, since more and more ice is typically added to it the longer it evolves. However, it is important to note that the thick ice of ridges is also subject to preferential ablation during the summer (Perovich et al. 2003).

These contrasts are clearly seen when comparing the second-year thickness distribution in Figure 3.2.1 (right) with the old multiyear ice thickness distribution in Figure 3.2.1 (left). The latter is characterized by multiple modes representing various classes of young ice and multiyear ice, as well as by a well-developed tail with significant amounts of deformed ice thicker than 10 m.

The discussion of Figure 3.2.1 has shown that only a description of the complete thickness distribution can reveal the different aspects of an ice cover's developmental history and the importance of the underlying dynamic and thermodynamic growth processes. For example, the mean thicknesses of 4.33±1.48 m (± one standard deviation) and 2.22±0.80 m resulting from the two distributions in Figure 3.2.1 bear no information about the occurrence and fractions of individual ice types (with modal thicknesses of 0.5 m, 1.0 m, 1.9 m, 3.9 m, 4.4 m in the Lincoln Sea, and 0.0, and 2.1 m at the North Pole, respectively), open water, or deformed ice. However, the standard deviation of a mean thickness indicates the range of occurring thicknesses and can therefore serve as a measure for sea ice roughness as well.

While this discussion was mostly focused on regional scales from meters to tens of kilometers, it should be noted that the same dynamic and thermodynamic processes also act on basinwide or hemispheric scales. For example, the long-term mean drift systems of the Transpolar Drift and Beaufort Gyre in the Arctic Ocean remove ice quasi-permanently from the Siberian Arctic and move it across the North Pole towards the coasts of Canada and Greenland. As a result, polynyas and thin ice prevail along the coasts of Siberia, while the thickest ice is found off the coasts of North America. Similar conditions are observed in the Southern Ocean, where for example the Weddell Gyre pushes ice against the Antarctic Peninsula, resulting in ice almost as thick as observed in the Arctic.

## 3.2.2 INTRODUCTION TO MEASUREMENT TECHNIQUES IN RELATION TO VARIOUS SEA ICE SERVICES

It is clear that a full description of the ice thickness distribution is far beyond the aims, capabilities, or requirements of many activities related to the various sea ice services. Table 3.2.1 provides an overview of the services, and of the different relevant temporal and spatial scales on which thickness information is required or relevant. From these, it becomes already clear that not every method discussed below is suitable to provide the required data.

Table 3.2.1 is contrasted by an overview of the available and most frequently used thickness measurement methods in Table 3.2.2. These methods will be described in detail in the sections below. Suitable thickness measurement methods can be identified by matching the relevant and achievable spatial and temporal scales and resolutions. Some examples for the application of certain methods in using various sea ice services are given in Section 3.2.9. The values in Table 3.2.2 are only rough estimates. Explanations for the accuracy are given in the sections below. Here, only the accuracy in the actually observed variable is given, for example surface elevation in the case of altimeter measurements, or total thickness in the case of electromagnetic (EM) sounding. It should be noted that with any of the methods described below, accuracy varies with spatial scale, ice type and roughness, and even ice thickness itself. For example, the accuracy of drill-hole measurements degrades for ice thicker than 10 m or so, and EM sounding strongly underestimates maximum ridge thicknesses, while possessing high accuracy over level ice. Spatial resolution and temporal resolution are stated for the most common applications.

**Table 3.2.1.** Sea ice system services and related activities that may benefit from ice morphology and thickness measurements *(from Eicken et al. 2009)*

| Sea ice services and related activities[a] | Examples of the role of ice morphology and thickness | Scales of relevance [b] | |
|---|---|---|---|
| | | Spatial (m) | Temporal |
| Regulator of arctic and global climate | Partially control albedo; related to the probability of first-year ice surviving the summer melt season | 0.01–100,000 | days–years |
| Hazard for marine shipping and coastal infrastructure | Key variables of importance when deciding whether to navigate through icy waters; determines the load imparted by an ice floe impacting an offshore structure | 1–100 | minutes–years |
| Stabilizing element for near-shore infrastructure<br>Subsistence activities on or from the ice | Contribute to whether or not the ice is stably grounded in the landfast ice zone or able to withstand pressures from the adjacent ice pack | 1–100 | hours–months |
| On-ice travel and a platform for industrial activities | Central to determining load-bearing capacity; surface roughness relates to trafficability (see Section 3.2.9) | 0.1–100 | hours–months |
| Sea ice–based food webs and ice as a habitat | Related to the ability of marine mammals, such as polar bears and ice seals, to find suitable conditions for denning | 0.1–1000 | months–years |
| Reservoir and driver of biological diversity | Control the amount of light available to biota within and under the ice | 0.1–1 | months–years |
| Oil spill response | Partially determine pathways and reservoirs for oil spilled beneath sea ice | 0.1–100 | hours–days |

[a] *Categories adopted from Section 2, Table 1, of Eicken et al. (2009).*

[b] *These spatial and temporal scales relate to the importance of ice morphology and thickness to the listed sea ice services. Other variables may be of importance on different scales. For example, permeability on subcentimeter scales is important to responding to oil spilled beneath ice.*

Clearly, drill-hole measurements could be performed with higher resolution, for example, but there is a limit of feasibility, especially as the destructive nature of this technique is considered. Similarly, the spatial coverage is given as that most commonly achieved. Of course, drill-hole measurements could be extended to cover larger regions, as more efficient modes of transportation are explored, or airborne surveys could be extended by using larger planes or fuel caches on the ice. Real-time capability is very important for many sea ice services, as can be seen by small temporal scales listed in Table 3.2.1. Also, the numbers of individuals or institutions using methods are only rough estimates. They are meant to represent the number of researchers who can actively perform or process measurements rather than only

**Table 3.2.2.** Overview of ice thickness and roughness measurement techniques and their various characteristics. Variables represent most common applications and present technical feasibility, and only rough quantitative estimates.

| Method | Accuracy of observed variable (m) | Spatial resolution (m) | Temporal resolution | Spatial coverage (km) | Real-time capability | Applied by N individuals/ institutions |
|---|---|---|---|---|---|---|
| Drilling | 0.02 | 0.5–5 | weeks–years | 0.1–10 | Yes | >100 |
| EM sounding (ground-based) | 0.1 | 5 | weeks–years | 0.1–10 | Yes | 20 |
| Laser surveying | 0.02 | 0.5–5 | weeks–years | 0.1–10 | Yes | 10 |
| DGPS surveying | 0.05 | 0.5–5 | weeks–years | 0.1–10 | Yes | 10 |
| IMBs[a] | 0.1 | n/a | hours–days | local | Yes | 10 |
| ULS[b] submarine | 0.1 | 1–5 | years–decades | 500–5000 | No | 2 |
| ULS[b] moored | 0.1 | 5–50 | minutes–hours | Local, or depending on ice drift speed (tens to hundreds of kilometers) | Not yet | 20 |
| AEM[c] | 0.1 | 3–5 | weeks–years | 10–100 | Yes | 6 |
| Airborne laser profiling | 0.1 | 0.2–5 | weeks–years | 10–1,000 | No | 10 |
| Satellite laser altimetry | 0.07[d] | 170–25,000[d] | hours–days–half-yearly[d] | 1,000–10,000 | No | 5 |
| Satellite radar altimetry | 0.07[d] | 330–100,000[d] | hours–days–weeks–months[d] | 1,000–10,000 | No | 5 |

*a Ice mass balance buoy (IMB)*
*b Upward-looking sonar (ULS)*
*c Airborne electromagnetic sounding (AEM)*
*d Depending on spatial and temporal averaging*

use the final data. For example, many scientists have used ice thickness data from submarine upward-looking sonar (ULS) which are freely available through the Internet. However, only two researchers or institutions, namely in the United States and United Kingdom, can at present contribute to future mission planning and performance, if at all. Similarly, the acquisition and processing of satellite altimetry data is dependent on the availability and orbits of satellites, and processing often requires close insights into satellite system parameters and access to auxiliary data. These, as well as the planning of future satellite missions, are only available to very few research groups.

The next sections provide overviews of the most commonly used methods of ice thickness and roughness surveying. These are ordered from the most simple towards the most advanced techniques, with less emphasis on the latter as they

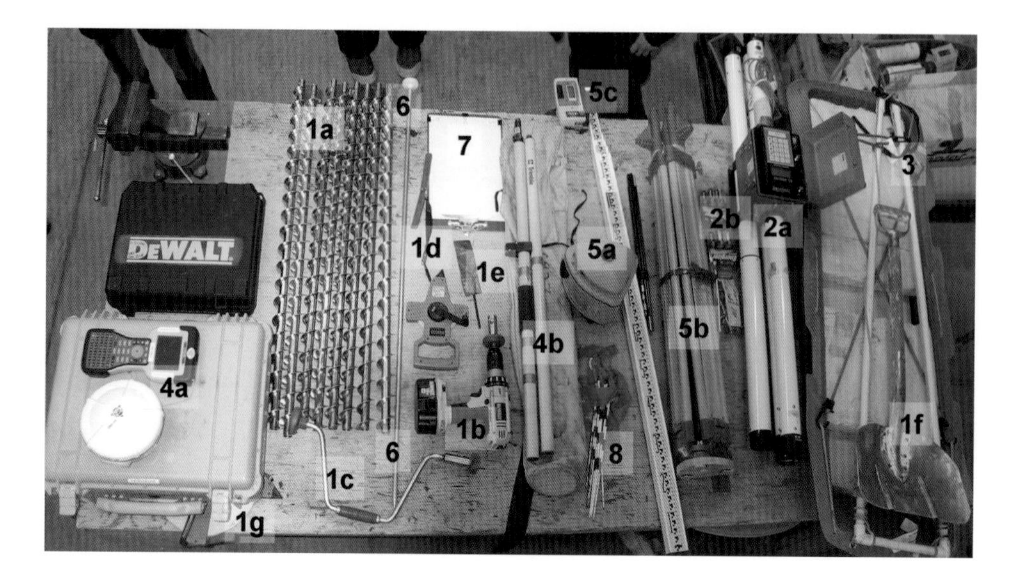


**Figure 3.2.3.** Assemblage of different instruments for simple ground-based thickness measurements (one or several of those may be used):

1. Thickness auger with numerous drill extensions (a; drill bit not visible), with cordless power drill (b), emergency hand brace (c), thickness gauge (d), button release tool (e), shovel (f), and 60 m ruler tape (g).
2. EM31-MK2 ground conductivity meter with onboard data logger (a), with spare batteries (b).
3. Pulka sledge for transport of equipment or dragging of EM31.
4. Differential GPS antenna and data logger (a), and range pole (b).
5. Rotating laser (a) with tripod (b) and telescopic range pole with laser detector (c).
6. Snow thickness meter stick.
7. Clipboard for paper sheets.
8. Pegs for marking field sites and fixing ruler tapes.

typically require significant expert knowledge and heavy logistics. There are several trade-offs between those methods. The most simple measurements like drilling are often the most accurate, but progress is slow, and measurements require hard work. Therefore, they are mostly also spatially limited. In contrast, advanced methods like airborne or satellite altimetry provide repeat data over large regions. However, their accuracy can be questionable, and logistical support and data access is often limited to a few individuals.

Figure 3.2.3 shows all the equipment required to perform drill-hole and ground-based EM thickness measurements, as well as laser and DGPS surveying—four techniques discussed in this section. This equipment can easily be assembled and taken by almost anyone to the ice to gather a thorough thickness and roughness data set.

More information on details of measurements can also be found in the *Handbook for Community-Based Sea Ice Monitoring* (Mahoney and Gearheard 2008).

That book focuses on drill-hole and hotwire measurements that can easily be performed by nonexperts. This chapter will not discuss visual observations of ice thickness, which, for example, can be performed from icebreakers when ice floes turn and are pushed up along the hull, revealing their cross-sectional profile, which can then be compared with a scale. However, it should be noted that this method is quite important, and has contributed much knowledge, particularly of the large-scale thickness distribution in the Southern Ocean, more than is available from any other method to date (Worby et al. 2008). It is also further discussed in Chapter 3.12. The chapter will not address satellite remote-sensing methods other than altimetry, although the thickness of thin ice can sometimes be successfully retrieved from thermal infrared imaging. Imaging remote sensing methods are further discussed in Chapter 3.15.

## 3.2.3 DRILLING

Figure 3.2.4 defines the most important variables commonly referred to with respect to ice thickness. Ice thickness is the distance between the ice underside (or ice-water interface) and the ice surface (or snow-ice interface), while snow thickness is the distance between the snow-ice interface and the snow surface. Their sum is referred to as *total thickness*. *Freeboard* is the height of the ice surface above the water level, while *surface elevation* or *snow freeboard* are commonly used to describe the height of the snow surface above the water level. *Draft* is the depth of the ice underside below the water level. These variables do not only yield information on the overall thickness or mass of ice and snow, but ratios of freeboard and thickness, for example, can also be used to study the isostasy of the ice, and to calculate the densities of ice and snow. This principle is utilized by recent satellite altimetry missions like ICESat and CryoSat to derive ice thickness from measurements of sea ice surface elevation or freeboard (see below).

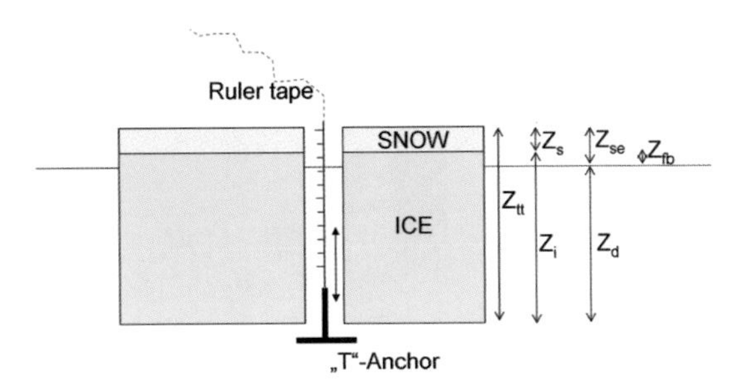


**Figure 3.2.4.** Measurement of total thickness ($Z_{tt}$), ice thickness ($Z_i$), snow thickness ($Z_s$), surface elevation ($Z_{se}$), draft ($Z_d$), and freeboard ($Z_{fb}$) by means of a thickness gauge (ruler tape with T-anchor) in a drill hole.

All these variables can be measured with drill holes through the ice. In the drill hole, the water level provides a reference datum for observations of draft, freeboard, and surface elevation. Note that only three variables have to be measured, and that all other variables can be calculated by subtraction or addition from those measurements.

Two different means are commonly used to drill holes though the ice: either mechanically by means of a motor-driven ice auger or thermally by means of a steam or hot-water drill.

Motor-driven ice augers are available with various metal flight diameters between 5 and 25 cm. With increasing diameter, drilling becomes increasingly difficult, and more engine and man power are required to drill through thicker ice. An ice corer can also be used to drill a hole (see Chapter 3.3). For extensive measurements, 5 cm diameter auger flights are most widely used, as for example manufactured by Kovacs Enterprises Drilling Equipment Inc. (see photos in Figures 3.2.3 and 3.2.5). These stainless-steel flights are 1 m long and join one to another via a push-button connector, which allows for quick connection of one auger section to another. This method of assembly means that there are no pins or connector bolts to lose or care for and no bolts on which clothing can snag. At the lower end of the lowest flight, a 5.1 cm wide ice-cutting bit is used for the actual drilling.

**Figure 3.2.5.** Equipment and procedures for mechanical ice thickness drilling. (a) Auger flights penetrating into the ice through a snow pit; (b) a two-stroke gas engine with custom-made handle bars; (c) Thick ice (>5 m) drilled with a battery powered drill; and (d) Kovacs Enterprise flight and gas power head.

Augers can be powered by two-stroke gas engines or electric power drills. Recent high-end 18 VDC cordless electric hand drills are powerful enough for most applications, and batteries may last for 30 to 50 drill meters. Drilling rates of 1 m in 15 to 20 seconds are achievable with most power drills. If one wishes to turn the augers by hand, a hand brace can be used (Figure 3.2.3).

For ice thicknesses between 1 and 2 m, it is convenient to start the drilling off with 2 m of flights, if the driller is tall enough to hold the system and safely reach the power head at > 2 m elevation. If the ice is thicker than 2 m, after every meter of drilling the drill has to be removed from the upper flight with the flights still remaining in the hole, and another extension can be inserted between the uppermost flight and the drill. By this, multiyear pressure ridges up to 24 m thick and grounded ice islands up to 23 m thick have been drilled. However, note that one flight weighs approximately 1.5 kg, and therefore the equipment becomes successively heavier and harder to manage.

When using a drill with a chuck, instead of a pin, the weight of numerous auger flights may become too heavy for the drill's chuck to hold. For this reason it is important for the bit that connects the uppermost auger flight to the drill's chuck to have a disk (rubber or metal) that is of a greater diameter than the hole (see Figure 3.2.3) in order to prevent the flights from being lost under the ice if the bit happens to slip out of the chuck.

Therefore, over thick and deformed ice thermal drills are sometimes used instead. With these, hot water or steam is generated by boiling water in a reservoir, and pumping it under high pressure through a hose into a metal rod with a typical diameter of 3 cm (Figure 3.2.6). The hot rod tip as well as the steam melt the ice at and below the tip, allowing the rod and hose to easily enter vertically into the slowly forming and deepening hole. Drilling progress is comparable to mechanical drilling, but the hose is lighter and much easier to use. The boiler and pump are usually powered by fuel or kerosene. A water reservoir, typically filled with water obtained from under the ice through a drill hole, is required for steam generation.

**Figure 3.2.6.** Operation of hot-water drill comprising of a generator, pump, boiler, and hose set up on a sledge (left). A stainless-steel drill rod is used to direct the hot steam under pressure vertically into the ice (right; Photo courtesy Pekka Kosloff).

Custom-tailored hot-water drills are manufactured by Kovacs Enterprise Inc. as well. Their geometrical dimensions of $40 \times 50 \times 50$ cm$^3$ and typical weights of > 70 kg require heavier logistics for transport and operation. Therefore, they are often set up at only one place (e.g., close to a pressure ridge) and then a wide area is reached by means of a 50 m long hose.

With both drilling methods, complete penetration of the ice is easily felt as the drill falls freely into the water underneath. Before making any measurements, shavings around the drill hole should be carefully removed such that the original ice or snow surface is well visible. Now, the depth of the water level can be easily determined in the drill hole with respect to the ice surface, and all parameters can be measured very accurately within 2–5 cm with a thickness gauge (Figures 3.2.3 and 3.2.4, and video on accompanying DVD). The gauge consists of a ruler tape and a foldable T-anchor, which is lowered through the hole and then pulled up until the T-anchor catches the ice underside (Figures 3.2.3 and 3.2.4). Note that the narrow holes of thermal drills require a slim thickness gauge. If none is available, thickness has to be measured according to the known length of hose inserted into the ice before the instant when it melts through the ice-water interface indicated by a sudden fall. This may cause significant measurement errors.

Depending on the character of the snow (thickness, hardness), one can either drill through the snow or remove it before drilling. Snow thickness is usually measured with a meter stick or ski pole with a glued measuring tape by ramming it vertically into the snow until it encounters the underlying snow-ice interface. With metamorphous snow, the stick has to be rammed firmly several times to confirm the penetration to the hard ice surface (see Chapter 3.1). In any case, care should be taken not to disturb the original snow surface for measurements of total thickness or surface elevation.

Figure 3.2.7 shows a typical ice thickness profile obtained by mechanical drilling as described above (black lines). The profile is 400 m long and extends over both level and deformed ice. A point spacing of 5 m was chosen to properly sample the roughness of the deformed ice. Less spatial resolution would have been required to sample the very uniform level ice sections. However, the chosen point spacing should always be equidistant to allow calculation of representative statistics. It can also be seen that in this case 400 m was long enough to sample at least two zones of deformed ice, and to verify that the adjacent level ice was of the same thickness throughout, indicating the same age and origin of the ice along the whole profile. Note that snow thickness is generally larger over the deformed ice, as the roughness of ice blocks and ridge sails retains more snow during wind-redistribution events (see Chapter 3.1). In the example of Figure 3.2.7, drilling was performed by two people while a third did the actual measurements and wrote them into a notebook. The whole drill-hole survey of eighty-one holes was completed in approximately six hours. Overall, a total thickness of 194 m of ice was drilled, and 20 m of snow measured. Mean ice and snow thicknesses along the profile were $2.15 \pm 0.84$ m ($\pm$ one

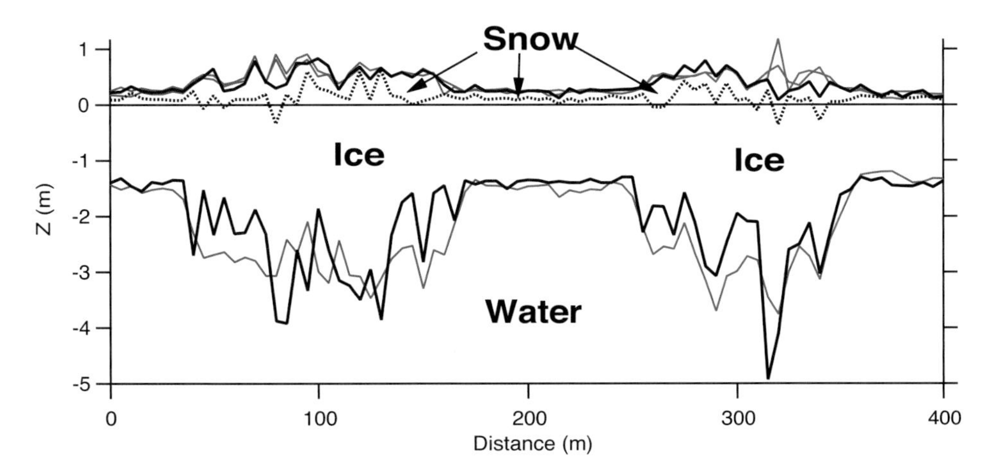


**Figure 3.2.7.** 400 m long ice thickness profile obtained by drilling with a point spacing of 5 m, on first-year ice at Barrow, Alaska. Z = 0 m represents the water level, and black lines indicate ice draft and surface elevation (solid), and freeboard (stippled). Blue lines show draft from ground-based EM sounding and surface elevation from laser and DGPS surveying (see below).

standard deviation) and 0.25±0.17 m, respectively, and their modal thicknesses were 1.5 and 0.15 m, respectively.

It should be noted that these results are only valid for this specific 400 m long thickness profile. Due to the small-scale roughness and variability, particularly in the deformed ice zone, the mean values will change when more data are added. However, the modes of the thickness distribution are very narrow and significant, and will not change as long as the profile extends over ice of the same origin and age. In general, the length of a planned thickness profile and the measurement point spacing will depend on the actual purpose of the measurement, and if regional or local data are required. Two-to-five-meter spacing is required to fully resolve the small-scale roughness due to rafting and ridging, although one would need an even finer resolution to detect very small-scale roughness. Note however that drilling is a destructive method, and the original ice underside could be easily disturbed if a drill-hole spacing of 0.1 m would be chosen, for example. A wider spacing might be chosen if regional results are of more interest, or if the ice is very level and uniform. Additional notes or photos should be taken to characterize the ice in general, and should be taken into account when interpreting the data later.

As drilling can generally be applied by anybody and is most accurate, much of our information about ice thicknesses worldwide stem from this method. Apart from data from upward-looking sonar (and recently from satellite altimetry; see Sections 3.2.8.1 and 3.2.8.4), data sets from drill-hole measurements are probably still the most extensive data source today. Almost all knowledge about antarctic sea ice thickness comes from drill-hole data (e.g., Wadhams et al. 1987; Lange and Eicken 1991; Worby et al. 1996), and there has been a synthesis of more than 123,000 drill-hole measurements to study sea ice variability in the Canadian Archipelago (Melling 2002).

*Tips and Tricks for Mechanical Drill-Hole Measurements*

- Don't forget to take a shovel if you want to remove the snow before drilling; with thick, hard snow it might be easier to measure total ice thickness without removing the snow.
- Keep drill holes clean from snow or shavings (e.g., by moving the flights up and down repeatedly to flush the hole and then wiping the ice surface clear with your shoes) as they may clog the drill hole and flights can get stuck, particularly when it is very cold.
- In thick ice with high freeboard, seeing the water level can be difficult, particularly if the hole is not clean. You can better see it with a flashlight, a little float lowered into the hole, or by lowering a metal or wooden stick into the hole—you will see the water level from where the stick has become wet.
- Have a tool for release of auger flights' push-button connectors as these can be difficult to operate in the cold when wearing gloves. A small screwdriver or pin will do, and you could tie it around your wrist to never lose it.
- Make sure couplings are tight, as well as clutches if an electric drill engine is used; many flights have been lost to the sea floor.
- Avoid bending of flights by ideally disassembling them as they are retrieved from the borehole. The guys in Figure 3.2.5c must have been insane.
- Watch your hands when touching flights and couplings with the motors attached: Serious injuries ranging from cuts to dislocated fingers have been reported. Also, watch clothing (e.g., scarfs), which can get wound up and strangle you.
- Stand on the windward side to avoid engine exhaust, and watch those drops of engine oil that can spatter you.

## 3.2.3 EM SOUNDING

Drill-hole measurements are so tedious and slow that very often it is desirable to use a simpler method with a better performance. In addition, the accuracy of drill-hole measurements and additional information about isostasy are often not required, or freeboard or surface elevation could be obtained from laser or DGPS surveying (see below). For these cases, the classical geophysical method of electromagnetic (EM) induction sounding provides a perfect alternative to drilling.

EM sounding has been used by geophysicists for many generations on land to study the conductivity structure of the underground. This is important for mapping of objects or geological features with distinct conductivities contrasting with the conductivity of the background rock, for example for ore bodies, groundwater, waste deposits, or unexploded ordnance. EM instruments consist of transmitting and receiving coils of wire to generate and detect low-frequency EM fields with frequencies typically ranging between a few hundred to 100,00 or 200,000 Hz. The primary field emitted by the transmitter coil penetrates through the underground, where it induces eddy currents whose strength and phase depends on the depth and conductivities of the underground materials. These eddy currents in turn generate secondary EM fields, whose strength and phase are measured by the receiving coil. From these measurements the conductivity structure or layering of the underground can be derived.

Since the mid-1980s, this method has been applied to sea ice (Kovacs et al. 1987a; Kovacs and Morey 1991). The sea ice environment provides an ideal, approximately two-layer case of highly resistive ice over infinitely deep, conductive seawater. Therefore, the primary EM field penetrates the ice almost unaffectedly into the water, where induction takes place only in a relatively thin layer under the ice, because the saline, conductive water prevents deeper penetration of the fields (Figure 3.2.8; see also animation on accompanying DVD). Strength and phase of the resulting secondary field are therefore closely related to the distance between the instrument containing the transmitting and receiving coils, and the ice-water

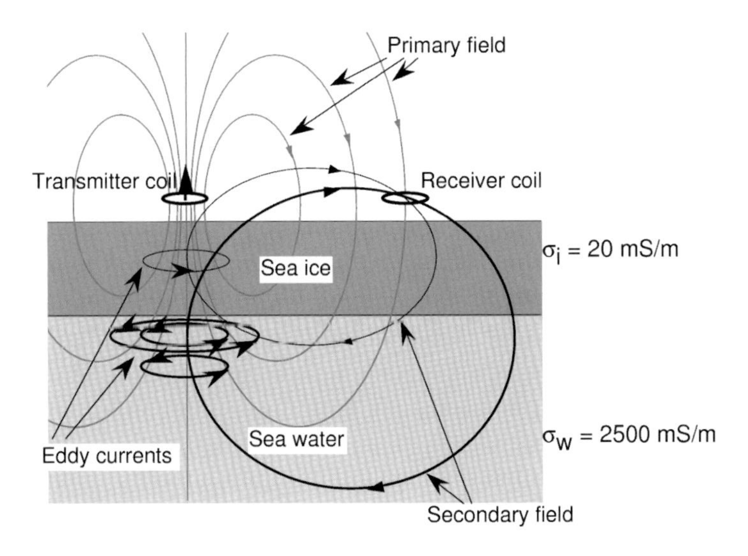


**Figure 3.2.8.** Principle of EM induction measurement of sea ice thickness. A primary field generated by a transmitter coil induces eddy currents primarily in the conductive water under the ice, which results in the generation of a secondary EM field, whose strength and phase are measured by a receiving coil. Strength and phase of the secondary field depend on the distance between the coils and water, which relates to ice thickness. Note that the sketch shows coils in vertical dipole configuration, which is typically not used for ground-based surveys.

interface. If the instrument rests on the ice surface, this distance corresponds to the ice thickness. However, if the instrument rests on the snow surface, the measured thickness represents total, snow-plus-ice thickness.

For many applications, total thickness is a sufficient observable, as the snow is generally much thinner than the ice. In summer in the Arctic, snow normally melts completely, such that the measured EM thickness corresponds to the ice-only thickness. However, given the importance of the snow cover as an independent climate variable and for ice thermodynamics, it is often desirable to perform additional snow thickness measurements along the same profiles, or statistically in the same region. This is also required if comparing measurements from different years to separate changes of ice thickness from changes of snow thickness.

Strength and phase of the secondary field depend not only on ice thickness and water and ice conductivity, but also on the instrument configuration, that is, the spacing between transmitting and receiving coils and the frequency of the transmitted EM field. The most commonly used EM instrument today for sea ice thickness measurements is a Geonics EM31, which has a coil spacing of 3.66 m and operates with a signal frequency of 9.8 kHz. All equations and figures following in this section refer to this instrument configuration. The instrument yields a reading of apparent conductivity $\sigma_a$ in units of millisiemens per meter (mS/m), which is computed from the imaginary or quadrature component of the measured secondary EM field (McNeill 1980).

Transmitting and receiving coils of an EM instrument form magnetic dipoles. The geometry of the intersection of the primary and secondary field with the ice-water interface and the resulting field strengths differ for vertical and horizontal dipoles. The relationship between the measured secondary EM field and ice thickness differs accordingly for surveys in vertical or horizontal dipole modes (VDM or HDM), which are performed with horizontal or vertical coplanar coils (HCP or VCP), respectively. The sketch in Figure 3.2.8 illustrates horizontal coplanar coils operated in VDM. Figure 3.2.9 shows the relationship between the EM signal in VDM and HDM (expressed as apparent conductivity $\sigma_a$) and ice thickness. Note that the VDM response is arbitrary, as for typical seawater conductivities it drops off for both thinner and thicker ice from a maximum of 412 mS/m at a thickness of 0.9 m. It should only be used if it is clear that only thicknesses smaller or larger than 0.9 m are present along the profile. Therefore, the instrument is mostly operated in HDM mode, which also has a slightly smaller footprint (Kovacs and Morey 1991), the area in which most of the eddy currents are induced and over which the thickness retrieval is averaged (see below).

The relationship between the EM signal (expressed either as a relative field strength Z, in parts-per-million [ppm] of the primary field, or as apparent underground conductivity $\sigma_a$ [McNeill 1980]) and ice thickness can be derived theoretically for given conductivities of the ice and water, according to:

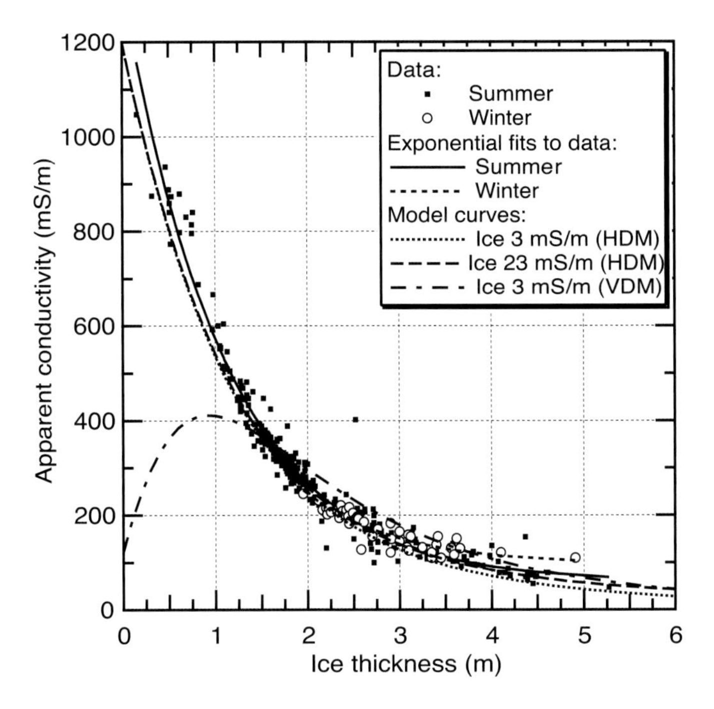


**Figure 3.2.9.** Measured apparent conductivity versus ice thickness for two winter and summer data sets and their exponential fits (Equations 3.2.3a,b; from Haas et al. 1997). Also plotted are three two-layer 1-D model curves for ice floating on water with a conductivity of 2600 mS/m. Ice conductivities of 3 and 23 mS/m have been assumed, and curves are shown for both, HDM and VDM.

$$Z = -r^3 \int_0^{\infty} r_{TE} e^{-2\lambda h} \lambda^2 J_0(\lambda r) d\lambda \quad \text{(Equation 3.2.2)},$$

with $r$ being the coil separation, $h$ the receiver and transmitter height above ground, $\lambda$ the wave number, and $r_{TE}$ a recursively determined transverse electric (TE) mode reflection coefficient resulting from the electromagnetic properties of the underground. The underground is assumed as a one-dimensional model of horizontal layers with infinite lateral extent. Equation 3.2.2 is a so-called Hankel transform with a Bessel function of the first kind of order zero ($J_0$), which can only be solved numerically.

However, relationships between EM signal and ice thickness can also be determined empirically by comparisons between EM and drill-hole measurements as shown in Figure 3.2.9 for the EM31. It can be seen that in HDM $\sigma_a$ decreases negative-exponentially with increasing ice thickness, and agrees very well with model results. Least-square fitting of a negative exponential equation can yield the desired transformation equation for deriving ice thickness $z_i$ from measurements of $\sigma_a$. In the example of Figure 3.2.9, the fitted equations were:

$$\sigma_{a_w} = 95.8 + 1095.5 \times \exp(-0.995 \times z_{i_w}) \quad \text{(Equation 3.2.3a)},$$

and

$$\sigma_{a_s} = 57.2 + 1270.9 \times \exp(-0.900 \times z_{i_s}) \quad \text{(Equation 3.2.3b)},$$

for measurements under winter and summer conditions (sub-indices w and s, respectively; Haas et al. 1997). Inversion yields

$$z_{i_w} = 7.03 - \ln(\sigma_{a_w} - 95.8)/0.995 \quad \text{(Equation 3.2.4a)},$$

$$z_{i_s} = 7.94 - \ln(\sigma_{a_s} - 57.2)/0.900 \quad \text{(Equation 3.2.4b)},$$

for ice thickness.

Figure 3.2.9 and Equations 3.2.3 and 3.2.4 show that the performance of EM measurements under summer and winter conditions is equally good. In fact, under summer conditions the ice is warmer and more porous than in winter, but the salinity of the brine is strongly reduced. Therefore, the overall conductivity of the ice changes only slightly. From numerous ice core analyses, Haas et al. (1997) have shown that the ice conductivity varies only between 3 and 23 mS/m between winter and summer. Model curves in Figure 3.2.9 are in very good agreement. Similarly, melt ponds are mostly composed of fresh meltwater, and have a minor effect on the validity of the transformation Equations 3.2.3 and 3.2.4 (Haas et al. 1997; Eicken et al. 2001).

However, it should be noted that the equations given above are only valid for the ranges of water and ice conductivities of 2600 mS/m and 3–23 mS/m given in the caption of Figure 3.2.9. The EM response is particularly sensitive to changes in water conductivity of a few 100 mS/m, which implies that different equations have to be derived, for example, for brackish water in the Baltic or Caspian seas.

The agreement between ground-based EM and drill-hole measurements lies generally within 5 to 10 percent both in winter and summer (Haas et al. 1997). The exponential fits in Figure 3.2.9 explain 91 percent (winter) and 98 percent (summer) of the observed variability indicated by the scatter of individual measurements. The negative-exponential decline of the EM response with ice thickness implies also that with greater ice thickness small thickness variations do not cause strong signal changes, and therefore cannot be detected. In HDM, thickness changes of 0.1 m can typically be detected with the EM31 with ice thicknesses of up to 5 m. The sensitivity is slightly better in VDM, which could be well used over thick ice.

The accuracy of EM measurements and agreement with drill-hole measurements is reduced by the footprint of the EM method, which is the area under the instrument over which ice thickness is averaged. Due to the lateral extent of the eddy currents, the resulting secondary field is induced over the area of the eddy currents. A general definition of the footprint size is the area within which 90 percent of the secondary field is induced (Liu and Becker 1990; Kovacs et al. 1995; Reid et al. 2006). The footprint implies that in general no perfect agreement between drill-hole measurements, which only yield ice thickness within a few centimeters of the drill hole, and EM sounding can be achieved if ice thickness varies laterally. While this variation is naturally small over level ice, it is large over rough, deformed ice and pressure ridges. Consequently, the maximum thickness of pressure ridges is generally underestimated because the water adjacent to the keels contributes to the EM signal and increases it (Figure 3.2.7). In contrast, on ridge flanks, ice thickness can be overestimated because the adjacent keel can lead to reduced induction of eddy currents. Overall, experience shows that maximum ridge thicknesses can be underestimated by as much as 50 percent, while the mean cross-sectional ice thickness across ridges agrees within 10 percent of drill-hole measurements (Haas et al. 1997; Haas and Jochmann 2003).

Apart from its efficiency and accuracy, the advantage of EM ice thickness measurements is that they are nondestructive and do not require any mechanical contact with the ice or snow. Nondestructive measurements are required in situations where drill holes could disturb the hydrostatic equilibrium and therefore the thermodynamic balance in longer-term studies of ice thickness change. This is relevant where negative freeboard due to heavy snow load could cause flooding after drilling (Haas et al. 2008c), or when drill holes could form artificial drainage channels during the melt season (Eicken et al. 2001). The contact-free nature of EM measurements means that an EM instrument could be deployed in a sledge or kayak to allow easy towing (e.g., by a snowmachine) and protection over various surfaces including melt ponds. However, it also allows for the deployment of EM instruments from platforms above the ice, for example from icebreakers, helicopters, or lighthouses (see Section 3.2.9 for examples), or from more advanced platforms like fixed-wing airplanes, airships, or hovercrafts.

*Tips and Tricks for Ground-Based EM Measurements*

- Some drill-hole measurements should always be performed at EM measurement sites to confirm the validity and accuracy of the transformation equation (according to Figure 3.2.9 and Equations 3.2.3 and 3.2.4) and to obtain a seawater sample for measurements of water conductivity, if unknown.
- Most EM instruments should be used in horizontal dipole mode to avoid ambiguities in the EM response. The EM31 operates in VDM by default. However, it can be easily operated in HDM when turned 90 degrees around its long axis, such that the instrument lies on its side and the display shows sideways.

- Geonics also provides an EM31 ICE instrument, which is supposed to be configured in HDM by default, and includes an ice thickness module that displays readings of ice thickness based on similar transformations as those in Equation 3.2.4. However, the module can only be calibrated for water conductivities between 2,000 and 3,000 mS/m.
- Use a sledge for easy pulling of instruments by snowmachines or while walking. Use a kayak for better protection and profiling over melt ponds.
- Use a data logger for continuous (analogue or digital) recording along extended profiles. Ideally, with every EM measurement the data logger should also record synchronous GPS positions to allow for later geocoding of measurements and derivation of an equidistant data set, which is required for calculation of ice thickness statistics independent of variations of ice drift or survey speed.
- Snow thickness is important! The only current operational method is using a meter stick (see Figure 3.2.3, Section 3.2.2, and Chapter 3.1). In this case, care should be taken to accurately coregister snow thickness and EM measurements, which is difficult with a data logger. Ideally, two people can efficiently read snow thickness and EM response at every measurement location and note them by writing into a notebook.
- Freeboard or surface elevation can be measured along the EM profile by means of laser or differential GPS surveying (Sections 3.2.5 and 3.2.6) to allow studies of isostasy or ice density. Obtaining a coincident data set at the same measurement locations of the EM measurements requires careful logging and/or documentation of every individual measurement.
- Use an external battery for the EM instrument to extend its longevity under cold conditions.
- Make sure batteries are equally charged! In most EM instruments, the accuracy and stability of the EM response depends critically on two equal voltages, typically +6 VDC and –6 VDC. With the EM31, voltages down to ± 4.8 VDC are acceptable.

### 3.2.4 GROUND-PENETRATING RADAR (GPR)

In a classical, geophysical sense, another type of EM measurement for snow and ice thickness is radio echo sounding (RES), also called ground-penetrating radar (GPR). GPR measurements employ high-frequency EM waves ranging between a few hundred megahertz to a few gigahertz. On land, GPR is successfully used in environmental and engineering geophysics to determine the thickness of thin layers (e.g., of soils, pavements, landfills, or aquifers). The method is based on measurements of the travel time of radar pulses, which travel from the transmitting antenna to an interface, where they are reflected and detected by a receiving antenna. The distance $d$ to the reflecting interface is obtained from $d = c\ t/2$, with radar wave propagation speed $c$ ($2.998 \times 10^8$ m/s in air) and ("two-way") travel time $t$ between antennae and the reflecting interface and back.

According to general principles, sea ice and snow thickness measurements could provide ideal applications for GPR measurements. However, there are only very few examples published in the literature where this has been successfully achieved (e.g., Kovacs and Morey 1986; Sun et al. 2003; Otto 2004).

The problems of applying GPR to sea ice thickness measurements are manifold. Compared to other thin layers commonly surveyed in engineering geophysics, ice and snow thicknesses—typically ranging from 1 to 3 m, and 0.1 to 0.8 m, respectively—are often too thin to be accurately measured. Resolving these thicknesses requires minimum radar wavelengths between 0.4 and 12 m, corresponding to four times the layer thickness. For radar propagation speeds of approximately 0.22 m/ns in snow and 0.16 m/ns in sea ice, this results in minimum radar frequencies of 550 MHz for snow and 13 MHz for ice. For improved resolution of and distinction between the reflection from internal layers and the ice-water interface, higher frequencies are used, typically between 250 MHz and 1 GHz.

Radar reflections are caused primarily at interfaces with contrasting dielectric properties, like at the snow-ice and ice-water interfaces. The radar wave propagation velocity $v = c/\sqrt{\varepsilon_r}$ is also dependent on the dielectric constant $\varepsilon_r$ of ice, snow, and brine. The dielectric constant of sea ice is strongly dependent on the amount and distribution of brine within the ice matrix (Stogryn and Desargant 1985; Kovacs et al. 1987b). Therefore, the propagation speed varies in dependence of ice salinity and temperature. In addition, brine inclusions themselves form small scatterers for the radar waves, leading to internal reflections and low signal-to-noise ratios. Internal scattering and absorption ("loss") are particularly strong in saline first-year ice and in warm and wet ice during the ablation season.

Figure 3.2.10 shows an example of a multiyear profile, where a good reflection from the ice underside was received (Otto 2004). The bottom panel of the figure shows a comparison of ice and snow thicknesses derived from the radar reflections with results from EM induction and drill-hole measurements. In general, a nice agreement between methods to within 0.1 to 0.2 m is visible.

The example shows that radar measurements of snow and ice thickness are generally possible, at least over cold multiyear ice. However, the radargram also reveals some general problems with varying bottom reflection amplitudes, blurring of reflections, and loss of trace-to-trace correlation, particularly in zones of rougher ice. These signal characteristics prevent easy and automatic extraction of radar two-way travel times to the reflecting interfaces, which makes processing of the radar data very labor-intensive and difficult, and which degrades their accuracy. In addition, knowledge of radar propagation speed is required to convert travel times into ice and snow thickness. These can vary by more than 0.03 m/ns, resulting in a thickness uncertainty of a few decimeters.

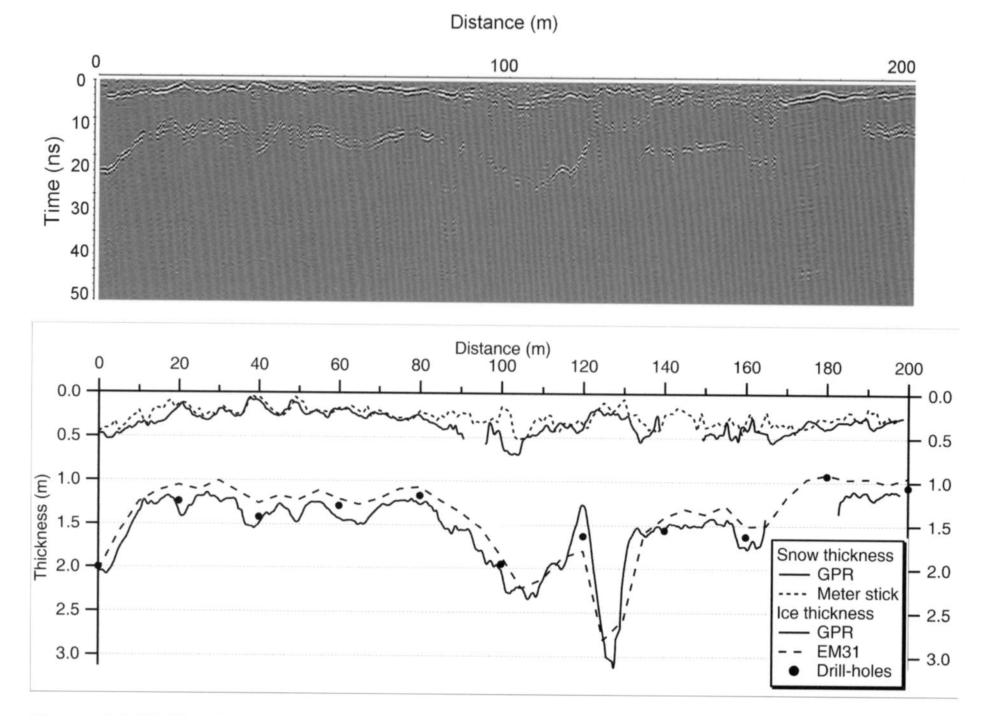


**Figure 3.2.10. Top:** Radargram of a 200 m long profile in the Barents Sea, obtained with a 800 MHz antenna on 1.2 m thick, cold multiyear ice, covered by 0.2 to 0.3 m of snow. **Bottom:** Comparison of radar-derived ice and snow thicknesses with results from EM sounding and drill-hole/meter-stick measurements. Ice and snow thickness was calculated using radar propagation velocities of $v_{ice}$ = 0.158 m/ns and $v_{snow}$ = 0.218 m/ns, respectively (figures modified from Otto 2004).

The blurred reflections from the undersides of both snow and ice over rougher zones are partially a result of reduced coherence of the reflected signal due to the presence of many scatterers in the radar footprint with variable distances to the antennae. This problem, and issues related to energy dispersion, become more important if radar measurements were performed from above the ice (e.g., from an icebreaker or helicopter).

However, it could be expected that many of the problems can be overcome in the future with improved instrumentation. One such example is the use of wide-band, continuous-wave, frequency-modulated (CWFM) radars, which have better penetration, resolution, and signal-to-noise characteristics than conventional systems. The potential of this technology for snow and ice thickness measurements has been demonstrated by Kanagaratnam et al. (2007) and Holt et al. (2008).

## 3.2.5 LASER SURVEYING OF THE GEOMETRIC SEA ICE ROUGHNESS

The small-scale variability of the thickness of sea ice implies a considerable surface roughness. Sea ice is rough on many scales from millimeters to tens of meters.

Centimeter-scale roughness is important for sea ice microwave properties and is sometimes referred to as "radar roughness." Here and in the subsequent sections, we focus on the "geometric roughness" of the ice, which relates to features with dimensions of decimeters and greater. Geometric roughness is generally caused by features like pressure ridges, rubble, rafts, snowdrifts, and sastrugis, or melt ponds. Geometric roughness is important because it can possibly be used as a proxy for ice thickness; it strongly modifies the surface drag of the ice, and therefore the interaction between ice and atmosphere; it acts as obstacle for the redistribution of snow; and it determines the hydrological drainage network during the melt season. A measure of surface roughness is the root-mean-square (rms) roughness, which is the standard deviation of a roughness or surface elevation profile. Pressure ridge distributions—that is, the number, height, and spacing of ridges in short intervals along a profile or within a given area—are also used to quantify surface roughness. The derivation of ice thickness from measurements of surface elevation is discussed further in Section 3.2.8.4.

Many geodetic surveying methods are available for observations of the surface morphology and roughness. A fast and accurate alternative to standard theodolite measurements is surveying with a rotating, self-leveling laser (Figure 3.2.11). This is a horizontally rotating laser whose laser beam forms a horizontal reference plane.

**Figure 3.2.11.** Rotating laser on a tripod placed on a small ridge. Two people with the laser detector attached to a range pole can be seen in the far back.

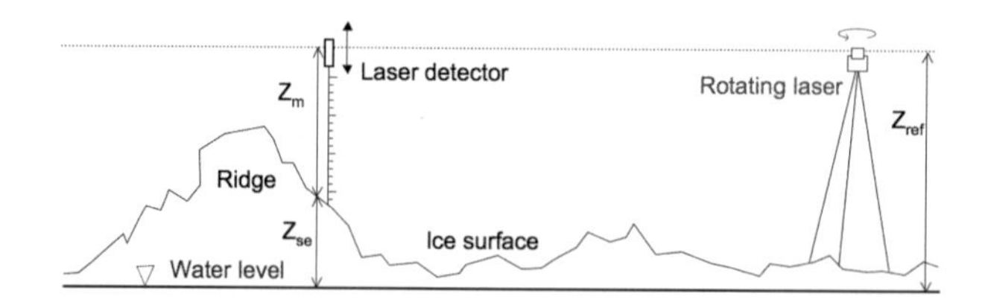


**Figure 3.2.12.** Illustration of laser surveying of ice surface elevation $Z_{se}$ by means of a rotating laser forming a reference plane at height $Z_{ref}$ above the water level. The height $Z_m$ of the reference plane above the snow surface is measured with a laser detector, usually mounted to a telescopic range pole.

The height $Z_m$ of this reference plane above the ice surface can be measured with a laser detector (Figures 3.2.3 and 3.2.12), which acoustically or optically indicates when it is detecting laser light and is therefore exactly within the reference plane. The laser detector can be mounted onto a telescopic range pole for easy height measurements. The height $Z_{ref}$ of the reference plane above the water level has to be measured once at the beginning of the survey (e.g., over a drill hole or crack in the ice). Surface elevation results as $Z_{se} = Z_{ref} - Z_m$ (Figure 3.2.12).

A typical surface elevation profile surveyed by laser leveling is included in Figure 3.2.7, where it is also compared with the surface elevation from drilling. A good agreement between both measurements can be seen. The accuracy of laser surveying, which is mainly determined by the narrowness, stability, and horizontal alignment of the laser beam and by the sensitivity of the detector, ranges between 0.5 and 1.0 cm. Significant disagreement with drill-hole measurements results mainly from slight variations in the actual sampling sites in rough ice or over a rough snow surface. This can be seen in a few locations in Figure 3.2.7 where the drill hole has been drilled in the lowest locations (e.g., next to a rafted ice block), while the laser measurement has been performed on the crest of the ice block, with only a lateral distance of 0.1 to 0.2 m between them.

Note that either the snow surface elevation or ice surface elevation (freeboard) can be measured, and that both can be derived from each other if snow thickness is known. However, measuring the snow surface elevation is generally easier, although over soft snow precautions have to be taken to avoid the laser detector range pole from penetrating into and below the snow surface.

Efficient surveying requires at least two people, one to perform the measurement and one to take notes. However, there are also range poles available with attached data loggers, which allow one person to perform the measurements alone.

*Tips and Tricks for Laser Surveying of the Geometric Surface Roughness*

- Place the laser tripod firmly onto the ice surface, as later settling of the snow or movements of individual legs due to melt could result in small laser movements with consequent interruption of its rotation and the possible loss of the original geometry and reference level.
- Place the laser in the center of the profile, such that its range (approximately 200 m for most lasers under clear weather conditions) can be fully utilized on both sides of the laser location.
- Place the laser tripod on a high point along the profile and raise the laser just high enough so that it is above the height of the highest ice feature along the profile plus the minimum length of the range pole.
- Take careful notes of measurement location for later merging with other data sets, for example from EM sounding.
- Fog, rain, or snow can scatter the laser beam and result in a loss of the signal at the detector. The detector can be irritated by direct incident sunlight and might have to be shadowed.
- Avoid penetration of the range pole into the snow if snow surface elevation is to be measured. Over soft snow this can be tedious, but the range pole can be placed on a foot, for example, which is stamped into the snow with its upper edge located at the same height as the snow surface.

### 3.2.6 DGPS SURVEYING OF SURFACE ELEVATION

The collection of differential global positioning system (DPGS) data is another method for obtaining measurements of freeboard, surface morphology, and roughness, with accuracies achieved at ±0.01 m. Eventually, these measurements can also be used to derive ice thickness (see Section 3.2.8.4 for a detailed discussion of potentials and constraints). DGPS methods improve upon the accuracy of standard GPS measurements through the use of phase information of the GPS signals received by two GPS receivers. One receiver is used along the profile to collect the height measurement (the roving receiver) and the other is used as a stationary reference (or base) receiver positioned at a known point. Figure 3.2.13 shows an example of what this setup might look like in the field. Using simultaneously collected data from the base receiver, the solution obtained for the position of the roving receiver is corrected for various errors, including those associated with satellite clock and orbital drift and regional ionospheric conditions. The accuracy of DGPS is improved the smaller the baseline distance is between the two receivers. This is because the corrections applied to the roving receiver's solution generally rely on the two receivers using the same satellite constellation and also because a more accurate calculation is made when these shared satellites are located high above the horizon (i.e., the signal path length between satellite and receiver is minimized).

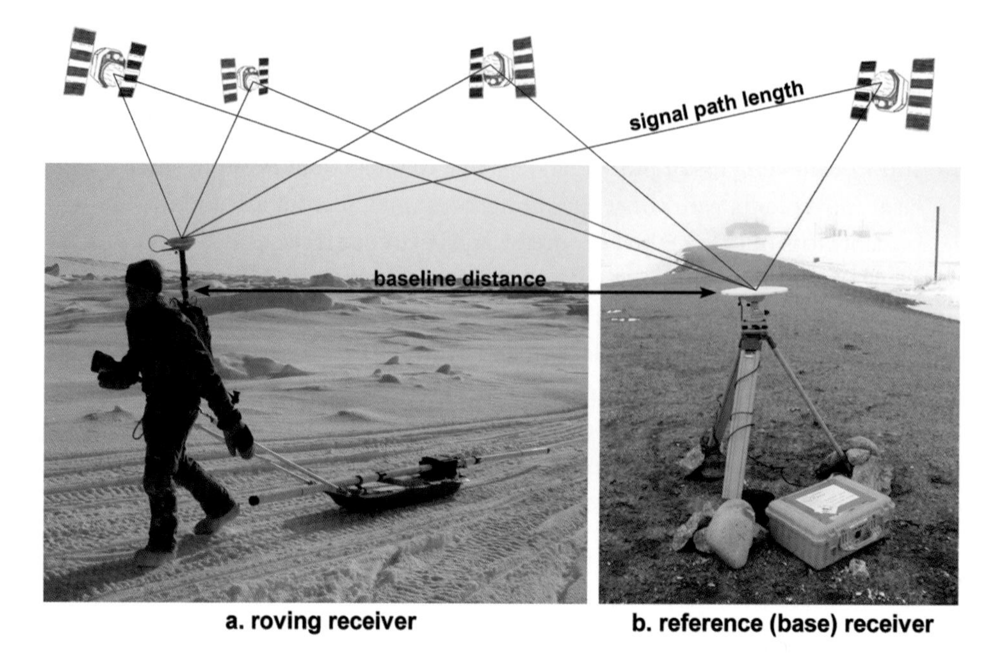


**Figure 3.2.13.** Example DGPS setup for measuring the surface roughness of landfast sea ice. The roving receiver in Figure 3.2.13a is mounted on a backpack. An EM-31 is in the sled being pulled. The reference, or base, receiver is shown in Figure 3.2.b. The straight line distance between the two receivers is referred to as the baseline distance.

DGPS positioning can be obtained either through postprocessing of data or in real time using a radio-transmitted correction from the base to the roving receiver. In the latter case, the survey depends on the two receivers maintaining radio contact, which can be difficult when surface topography is highly variable.

The survey style should be chosen based on the desired spatial coverage, sampling frequency, time available for the survey, and desired level of accuracy. Continuous surveys, set to make observations at either selectable time or distance intervals, are typically employed when sampling over large distances, field time is limited, or when the roving receiver setup is not conducive to point-based measurements, such as when attached to a moving vehicle (e.g., helicopter, snowmachine, sled). The time stamp in continuous DGPS data is typically used to coregister the measurements with other forms of simultaneously collected data sets, such as EM-derived measurements of ice thickness.

In contrast to continuous surveys, point measurement surveys rely on the occupation of specific points of interest for a set period of time in order to obtain greater accuracy. An additional advantage to point measurements is that the data is easily labeled and assigned to features of interest. For example, if one wished to measure surface elevation along a 100 m transect at 5 m intervals and merge this data set with ice thickness measurements made at the same spacing, a point survey would be preferred (cf. Figure 3.2.7). A disadvantage to this method is that it is time-intensive compared to continuous surveys.

Major advantages of DGPS surveying compared with laser surveying are its principally unlimited range, allowing one to measure long profiles without the requirement to relocate any instruments like the laser, and its additional provision of very accurate location information, which is required to geolocate the height measurements and eventual additional observations. Figure 3.2.7 shows a typical surface elevation profile obtained by DGPS surveying, and compares it with the surface elevation from drilling and laser surveying. Again, a good agreement between both measurements can be seen. As discussed above with respect to laser surveying, disagreement with drill-hole measurements results mainly from slight variations in the actual sampling sites in rough ice or over a rough snow surface.

DGPS surveys for ice morphology and roughness can be performed in a number of different setups. The most common are with the receiving unit on a pole, tripod, backpack (see Figure 3.2.13), or vehicle. In any case, the following should always be considered when performing a survey:

*Tips and Tricks for DGPS Surveying of the Geometric Surface Roughness*

- Plan the survey for a time of day when the number of the satellites in the constellation overhead is relatively great.
- Ensure setup of the receivers allows for unobstructed views of the sky at all times.
- Keep notes of the type of GPS receivers and antennas used for the survey, as this information is very important during data processing.
- Make detailed notes on receiver settings, such as observation frequency and those settings specific to how the receiver chooses which satellites to use (e.g., antenna mask, position dilution of precision [PDOP], etc.).
- Always remember to measure antenna height above the surface of interest (e.g., the surface of the ice) before the survey, and also afterwards to ensure the antenna height was not altered during the survey.
- Note that DGPS-derived surface elevation measurements refer to a global ellipsoidal geodetic model. This can locally deviate from the geoid (i.e., the true height of the water surface) by several tens of meters. Therefore, the difference between the geoid height and ellipsoidal height has to be compensated before "true" surface heights above the local water level can be obtained. If possible, a hole should be drilled into the ice, either immediately before or after the survey, to make a point observation of the surface elevation of the water. This measurement allows for the observations of ellipsoidal ice surface elevation to be easily converted to measurements of freeboard, which will equal the surface elevation minus the water elevation. Note that ocean tides cause a slow drift in the surface elevation measurements, which can be corrected by two reference measurements

over water before and after the survey. Otherwise, all DGPS measurements can be referred relative to some local surface elevation of unknown absolute height.

- Surveying an ice surface with a snow cover requires special consideration. For point surveys, it may be easiest to shovel the snow and measure directly from the ice surface. When continuously surveying from a moving vehicle, such as a sled or snowmachine, one may only be concerned with average snow thickness. In this case, estimating the vehicle's depression into the snow cover may help correct for a more representative survey of the ice surface. However, DGPS can also be very useful in measuring the surface roughness of the snow cover as well.

## 3.2.7 ICE MASS BALANCE BUOYS AND HOT-WIRE THICKNESS GAUGES

All methods presented so far can be used to obtain long profiles of ice thickness data within a relatively short time. However, it is difficult to perform measurements over a longer time period to study temporal changes, as field personnel has to be present throughout. In contrast, automatic measurement stations can be deployed on the ice to provide long-term data of ice thickness changes at one location, either on fast ice or on a drifting ice floe, in which case ice thickness changes are observed along the drift track of the floe. These automatic stations are commonly called ice mass balance buoys (IMBs) (Richter-Menge et al. 2006). A more comprehensive discussion on how to operate these buoys automatically and transmit their data via satellite is given in Chapter 3.13.

A typical IMB consists of acoustic range finder sounders to measure snow accumulation and ablation and ice bottom growth and melt, and of a thermistor string to observe internal ice temperature profiles (Richter-Menge et al. 2006). The latter normally extents into the water and therefore provides additional measurements of water temperature, while an additional air temperature sensor provides above-surface temperature data. Relative changes of surface elevation and draft can be easily calculated from changes of the acoustically derived distance to these surfaces. Reductions in measured distances indicate snow accumulation and bottom growth, while increased distances are due to surface and bottom melt. The accuracy of these measurements is typically better than 1 cm. Note that absolute changes of freeboard, surface elevation, and draft with respect to the water level cannot easily be determined. Freeboard changes can only be determined if an additional pressure sensor is used in the water. All measurements together not only allow thickness change observations, but can also be used to address the role of various atmospheric and oceanic boundary conditions for these changes.

The combination of surface and bottom sounding with air, ice, and water temperature measurements is a powerful means to study the thermodynamic development of the ice. However, as IMBs obtain data only at one location, they cannot yield any information on dynamic ice thickness changes due to rafting and ridging, and therefore cannot access the complete ice mass balance on a floe or at the regional scale. In addition, results are strongly dependent on local conditions, for example, whether the sensors are located on ponded or unponded ice, or if a snowdrift develops under the sounders. The buoy itself might modify the sea ice mass balance through preferential melting by absorption of radiation during the summer, or by disturbing airflow and therefore the wind-induced redistribution of snow.

Longer-term changes of draft and surface elevation can also be measured by means of so-called thickness gauges. These consist of an ablation stake and a hot-wire thickness gauge, and their design and operation is described in great detail by Mahoney and Gearheard (2008). Extensive measurements were made, for example by Perovich et al. (2003) during the Arctic Surface Heat Budget of the Arctic (SHEBA) drifting station. A long, wooden stake painted white and marked with metric tape can serve as an ablation stake, frozen into a deep drill hole that does not extend through the ice. The hole should be deep enough to prevent melting out of the stake during the summer. Surface elevation can be measured off the stake with an accuracy of 0.5 cm by noting the intersection of the stake with the snow surface. A hot-wire thickness gauge consists of a stainless-steel wire with a steel rod attached as a crossbar on the bottom end and a wooden handle on the top end. It is installed adjacent to the ablation stake. To make a measurement, the stainless-steel wire is hooked to a generator that is also connected to a copper wire grounded in the ocean. The electrical resistance of the stainless-steel wire melts it free, and the handle can be pulled upward until the steel rod hits the bottom of the ice. Then, the handle position is read off the ablation stake to give the position of the ice bottom. Instead of a stainless steel wire and generator, electric heating cables as used for example in pipe antifreeze applications can be frozen into the ice. These cables heat when connected to a 12V or 24V battery, for example. Uncertainties of stake and gauge readings are typically less than 0.5 cm.

Operation of thickness gauges is almost similar to performing repeated drill-hole measurements at the same location. However, the mechanical and thermal destruction and disturbance of the original ice and snow layers are reduced to a minimum. The holes resulting from melting the wires is very narrow, and refreezes quickly after the power source is disconnected.

### 3.2.8 ADVANCED METHODS

This section presents more advanced techniques for sea ice thickness measurements. Although the methodological principles are generally simple, major challenges exist in the technical realization of these concepts and to meet the required accuracies. In addition, all sensors over or under the ice have to be deployed by some means of transport and logistics, usually ships, aircraft, submarines, or satellites. This involves significant cost, prohibiting broad and flexible application. Even though today the operation of satellites is taken for granted by many, it should be noted that the ground segment (i.e., satellite control and the reception, processing, and distribution of data) requires significant computational resources and expertise. In addition, mission planning and extension is often politically driven, and may prevent the acquisition of long-term, systematic data sets. However, there are many efforts underway to improve this situation. In addition, initiatives are progressing to at least make the high-level, final thickness data sets available to a broad public in uniform formats, such that data sets from individual methods can be merged to yield long and spatially dense records of general ice thickness conditions in the polar oceans.

### 3.2.8.1 Upward-Looking Sonar (ULS) Profiling of Ice Draft

Due to the density of sea ice of approximately 920 kg/m$^3$ (Chapter 3.3), the draft of freely floating sea ice is approximately nine-tenths of its thickness if the snow cover is neglected. This suggests that draft can be used as an accurate proxy for ice thickness. Indeed, most of the ice thickness data gathered to date have been derived from measurements of ice draft performed by upward-looking, range-finding sonar (ULS) or ice-profiling sonar (IPS). These have been operated on both submarines (e.g., Bourke and Garret 1987; Rothrock et al. 1999; Wadhams and Davis 2000; Rothrock et al. 2008) and oceanographic subsea moorings (e.g., Vinje et al. 1998; Strass and Fahrbach 1998; Melling et al. 2005), respectively.

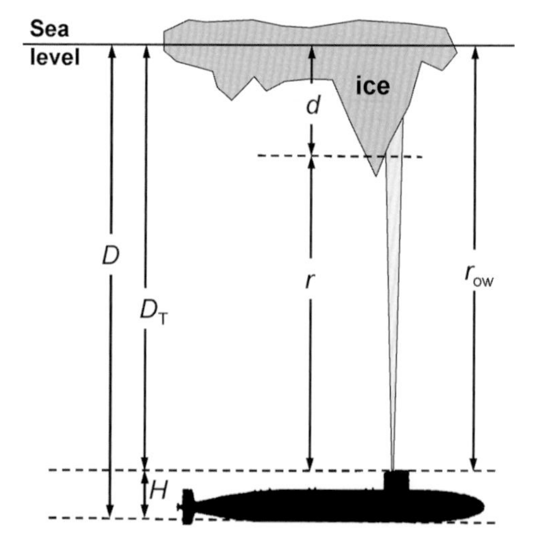


**Figure 3.2.14.** Schematic drawing showing how ice draft d measurements are obtained by ULS from a submarine (from Rothrock and Wensnahan 2007). The depth of the submarine $D$ is determined by a sensor that measures pressure $p$ and is calculated as $D = (p - pa)/((\text{water density}) \times g)$, where $g$ is the acceleration of gravity and pa is sea level pressure. The height $H$ is the vertical distance from the pressure sensor to the sonar transducer, which is mounted on top of the submarine at a depth $D_T$. The range $r$ is the distance to the ice, while $r_{ow}$ is the distance to open water, determined when the sonar passes under open water.

The measurement principle is illustrated in Figure 3.2.14 for the case of a ULS operated on a submarine. Ice draft is derived from measurements of the two-way travel time of short acoustic pulses between the sonar and the ice underside, from where they are reflected or scattered back to the sonar. The distance $r$ between ice and sonar transducer results from $r = v\ t/2$, with two-way travel time $t$ and sound velocity in water $v \approx 1436$ m/s. The depth $D_T$ of the transducer below the water level is determined with a pressure sensor, and by taking into account the vertical distance $H$ between pressure sensor and transducer. Draft $d$ results from the difference of the depth and distance measurement: $d = D_T - r$.

Sonars used for ice draft profiling typically operate with acoustic signal frequencies of a few hundred kilohertz, corresponding to subcentimeter wavelengths. They have focused beams of 1 to 3 degrees width, limiting the sonar footprint at the ice underside to less than 10 m even for ULS depths as low as 150 m below the water level. Many sonars do not only measure travel time, but also yield an estimate of the echo amplitude, which allows retrieval of the roughness and type of the reflecting interface (e.g., open water; thin, new ice; or thick ice) (Melling 1998).

The accuracy of ULS draft measurements is affected by system precision, uncertainties in the sound velocity profiles between the transducer and ice, and weather-related variations of air pressure, which directly enter the depth measurement. For correction of sound velocity and air pressure effects it is essential that open water is repeatedly present along profiles to provide a zero-draft reference range $r_{ow}$ to which measurements under ice can be related (Melling et al. 1995). Unfortunately, detection of open water can be very difficult when leads are narrow or when waves or swells alter the echo characteristics of the sonar signal. Some of these problems can be better addressed if high pulse repetition rates are chosen to obtain a good spatial resolution of leads, and if echo amplitude information is utilized. High pulse repetition frequencies pose a challenge for data logging and battery capacities of moored ULS operated for long periods of at least a year.

Other uncertainties of the draft measurements result from the sonar footprint (Vinje et al. 1998), which leads to systematic overestimates of ice thickness in regions of deformed ice, and from instrument tilt due to current drag on moorings or variations in submarine trim. However, detailed studies of ULS measurements show accuracies of around 0.05 m for moored instruments (Melling et al. 1995; Strass and Fahrbach 1998) and biases of +0.29 m with a standard deviation of 0.25 m for former U.S. Navy submarines (Rothrock and Wensnahan 2007).

With ULSs mounted on submarines or autonomous underwater vehicles (AUVs), long profiles can be surveyed in a very short time to provide snapshots of regional and basinwide thickness distributions. In contrast, ULSs operated on moorings can provide long time series of ice thickness change at one location. If ice drifts over the mooring, as is mostly the case, a ULS provides Eulerian information about the spatial ice thickness distribution along the trajectory of the ice. One of the longest records has been obtained in the Beaufort Sea, and now provides a more than twelve-year-long observational data set of ice thickness variability and trends for analyses of arctic climate variability (Melling et al. 2005).

Despite the great potential of submarine measurements, it should be noted that these can only be performed onboard military nuclear submarines, to which only few scientists from the United States and United Kingdom have access. Military submarine operations are usually classified, and data is only released after many years and with some geographic blurring, which does not allow any near-real-time studies. However, data from some forty cruises between 1975 and 2000 is now publicly available at the National Snow and Ice Data Center in Boulder, Colorado, providing comprehensive data of ice thickness decline in the Arctic (http://nsidc.org/data/g01360.html). Although there were dedicated science missions onboard U.S. submarines within the Scientific Ice Experiments (SCICEX) program between 1995 and 1999, the future of that program and of the release of new data is presently uncertain. In the Southern Ocean, military and nuclear submarine operations are prohibited by the antarctic treaty. However, AUVs could potentially gather as extensive and accurate data once navigational issues are resolved to facilitate more reliable operations under ice (Dowdeswell et al. 2008).

Many difficulties exist also with the operation of moorings. Their deployment, operation, and recovery in deep or ice-covered water is very challenging and expensive, as it requires logistics support by ships, airplanes, or helicopters and potentially divers. In shallow water, there is always a risk of damage or loss of equipment due to ice scouring by ridge keels or icebergs. Iceberg collisions are common in the Southern Ocean, where tabular icebergs with drafts of several hundred meters sometimes collide with moorings. There, ULSs are normally operated in greater depths to avoid collisions with even more abundant smaller icebergs, despite consequences for the accuracy of the measurements due to greater beam divergence and sound travel distance. In almost every seasonal ice zone, moorings are sometimes damaged or removed by commercial trawl fishing.

An alternative to single-beam sonars are Acoustic Doppler Current Profilers (ADCPs) (Shcherbina et al. 2005; Bjørk et al. 2008). ADCPs are normally used to measure profiles of water current speeds by means of the Doppler shift of acoustic signals emitted at various wide angles by the ADCP and scattered back from individual water layers. However, ice draft can be estimated using the distance between the ADCP and the range cell with maximum echo intensity. The bin size of the ADCP resolution cells is typically 1 m, but it is possible to increase the resolution to about 0.1 m by fitting a Gaussian model curve to the vertical echo intensity distribution. The maximum point of the Gaussian curve can then be used to determine the distance from the instrument to the lower ice surface.

Apart from single-beam sonar profiling, multibeam sonars have also been used to obtain quantitative two-dimensional information about draft and roughness of the ice underside (e.g., Wadhams et al. 2006). These can be used from both submarines and AUVs. Like with laser scanning (Section 3.2.8.3), two-dimensional mapping allows for observing the across-track spatial distribution of under-ice features. This reduces sampling biases in the analysis of thickness data, and can contribute to improved detection of open water. However, for distinguishing between thin ice and open water it is still required to carefully analyze echo amplitudes or the textural characteristics of the target area. Two-dimensional data is valuable for the validation of measurements with larger footprints, for example, from EM sounding and satellite remote sensing.

Due to the large draft-to-thickness ratio *R* of approximately 0.9, uncertainties in the draft profiles result only in relatively small errors in calculated ice thickness. Consequently, ice thickness is normally simply calculated from draft using a single *R* value, e.g., of 0.89 (Rothrock et al. 1999) or 0.93 (Rothrock et al. 2008), depending on whether the effect of snow is included or what snow thickness and density are being assumed. Although the uncertainty of *R* introduces uncertainties of some centimeters to a few decimeters in calculation of ice thickness, depending on actual ice and snow thicknesses, the problem is much more serious with the calculation of ice thickness from measurements of freeboard or surface elevation (Section 3.2.8.4).

### 3.2.8.2 Airborne EM

As mentioned in Section 3.2.3, one advantage of EM measurements is that they do not require contact with the ice, and can therefore be performed from airborne platforms. Consequently, helicopter-borne EM (HEM) surveys of sea ice thickness have been performed since the 1980s (Kovacs et al. 1987a). Similar to EM sounding in general (Section 3.2.3; see also animation on accompanying DVD), this method is principally based on the generation of a primary field and measurement of a relative secondary field strength from which the distance between the EM instrument and ice-water interface $h_w$ is derived (Figure 3.2.15).

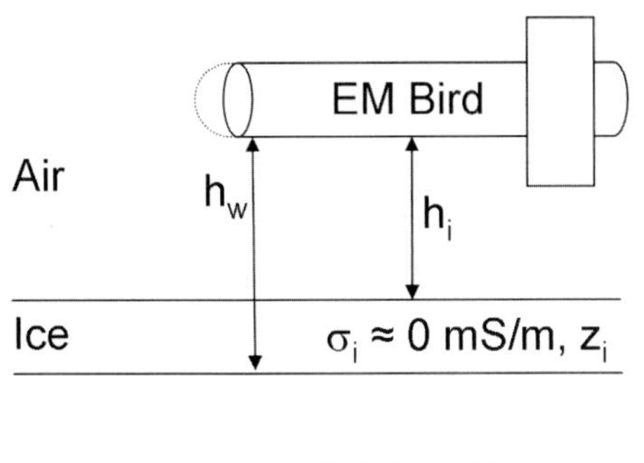


**Figure 3.2.15.** Principle of EM thickness sounding, using a bird with transmitter and receiver coils and a laser altimeter. Ice thickness $Z_i$ is obtained from the difference of measurements of the bird's height above the water and ice surface, $h_w$ and $h_i$, respectively; $h_w$ is obtained with the assumption of a negligible ice conductivity $\sigma_i$, known water conductivity $\sigma_w$, and horizontal layering (Haas et al. 2008b).

**Figure 3.2.16.** EM bird in operation

Normally, the transmitting and receiving coils are housed in a cylindrical shell towed by a helicopter and called an "EM-Bird" (Figures 3.2.15 and 3.2.16) (Kovacs et al. 1987a; Prinsenberg and Holladay 1993; Haas et al. 2008b). However, there are also systems where the shell is hard mounted at the nose of the helicopter (the so-called IcePic) (Prinsenberg et al. 2002), or where the coils are at the wing tips of a fixed-wing aircraft (Multala et al. 1996). In any case, as the EM system is operated at some altitude, it is required to also measure its height $h_i$ above the ice surface, which is normally achieved with a laser altimeter. Then, ice thickness $z_i$ results from the difference of the two distance measurements (Figure 3.2.15):

$$z_i = h_w - h_i \quad \text{(Equation 3.2.5)}.$$

Here, it is important to note that this approach and equation imply that ice thickness is actually total ice thickness (i.e., ice-plus-snow thickness; cf. Figure 3.2.15). While this is not a problem during the arctic summer, when ice is usually snow-free, snow thickness at other times of the year has to be obtained by other means if ice and snow thickness are to be distinguished. A helicopter system has the advantage that snow thickness can be measured manually in situ at a few locations where the helicopter lands.

A key issue in the retrieval of ice thickness is the inversion of the electromagnetically determined field strength to ice thickness. As discussed in Section 3.2.3, this can be done by numerically inverting the solution of a forward model, either for one-dimensional (Equation 3.2.2) (Rossiter and Holladay 1994) or two-dimensional

(Liu and Becker 1990) underground models. However, those procedures require very accurate sensor calibration and the use of several frequencies to invert not only for ice thickness but also for the conductivity of the underlying seawater.

However, the inversion can also be performed more empirically by fitting an exponential function to the open-water-only response of the EM signal (Haas et al. 2008b), similar to the procedure described in Section 3.2.3 with respect to Figure 3.2.9 and Equations 3.2.3a,b. Figure 3.2.17 shows the relationship between bird height above the ice surface and measured and modeled EM responses for a flight over a typical sea ice profile. The model results have been computed using Equation 3.2.2 for open water (ice thickness of 0 m) with a seawater conductivity of 2,500 mS/m, representative of in situ salinity measurements. The model curve provides the general means of computing the height of the bird above the water surface $h_w$ or ice from a measurement of the EM field strength at a certain height above the water. Measurements at different heights are obtained because the altitude of the helicopter and bird vary between 10 and 25 m during the flight. The data can be separated into two classes: (1) where open water measurements at different bird heights agree well with the model curves; and (2) where the presence of sea ice leads to a reduction of the measured EM signal at a given laser height (Figure 3.2.17). Therefore, the

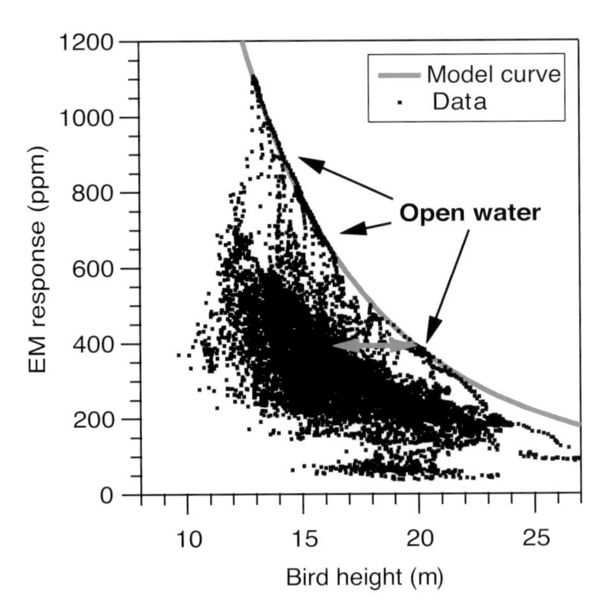


**Figure 3.2.17.** EM signal (Inphase component) versus bird height $h_i$. A model curve for open water with a conductivity of 2500 mS/m and data over a typical ice cover with some leads are shown. The horizontal arrow illustrates how ice thickness (4 m) is obtained for a single data point from the difference between $h_i$ and the model curve $h_w$ for a given EM field strength (Equation 3.2.5). Reprinted from the Annals of Glaciology with permission of the International Glaciological Society.

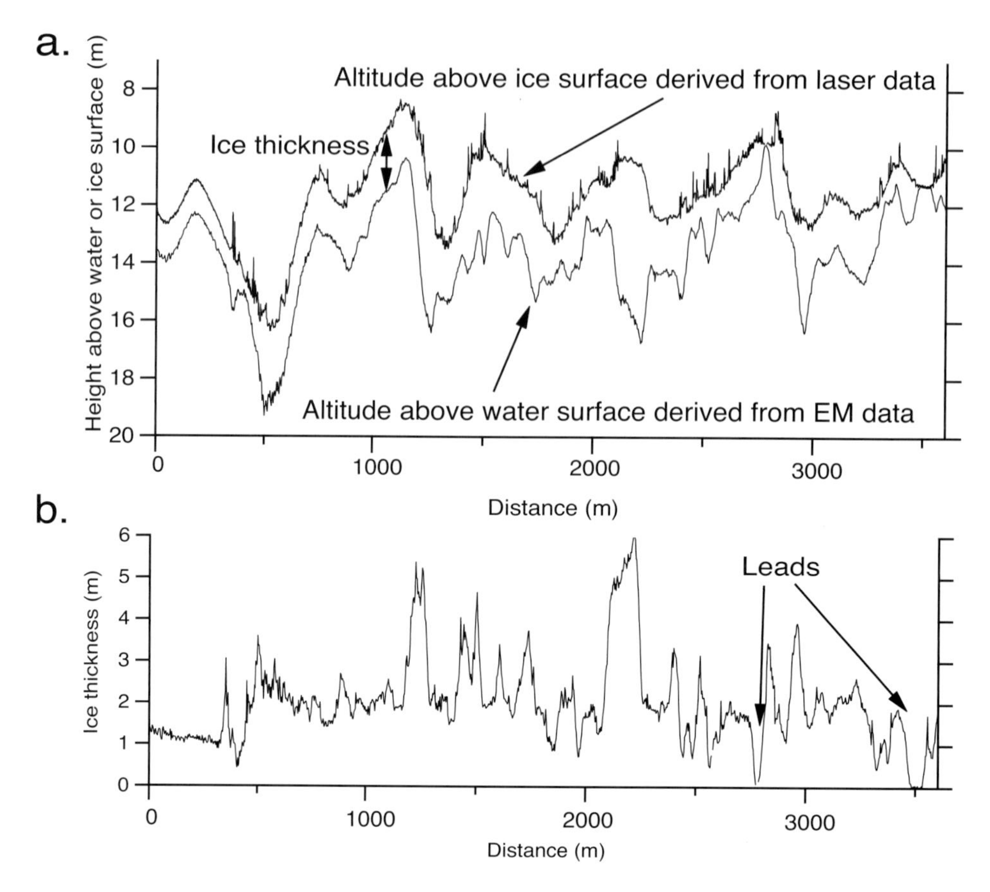


**Figure 3.2.18.** (a) EM- and laser-derived bird height above the water $h_w$ and ice surface $h_i$, respectively, and (b) ice thickness profile resulting from subtraction of the latter from the former (Haas et al. 2008b).

scattered cloud of data points below the model curve represents measurements over ice. Ice thickness is computed by subtracting the laser height measurement over sea ice from the model curve. It can also be visually estimated from the horizontal distance between each EM measurement and the model curve (Figure 3.2.17). The thickness computation assumes a negligible sea ice conductivity of < 20 mS/m—a reasonable assumption in most cases (Haas et al. 1997; Pfaffling et al. 2007).

The advantage of this approach is that it can be performed with only one channel of EM data, and is less sensitive to inaccuracies of the instrument calibration and uncertainties in the conductivity of ice and water. Figure 3.2.18 illustrates the two steps of determining the height above the ice and water surfaces $h_i$ and $h_w$, and obtaining ice thickness from the difference of these measurements. Individual ice floes with ridges, as well as thin ice on leads, can be clearly seen.

The occurrence of open water provides a convenient way to verify the calibration of the instrument. With this respect, airborne EM measurements take as much advantage of open water as all other advanced methods discussed in this chapter (i.e., ULS and airborne laser profiling, and satellite altimetry). However, once the conductivity of the seawater is known, which can be assumed with sufficient accuracy for most regions of the Arctic and Southern oceans, the calibration of the measurements depends only on the electronic stability of the instrument components, and not on external factors like tides, currents, or sea-level pressure as with the other methods. As was shown by Haas et al. (2008b), the calibration of typical airborne EM systems is stable to within 0.1 m over a long time. Only thermal drift can be significant, but can be compensated during high-altitude flight periods.

As with ground-based EM measurements, it is important to realize that the EM response is strongly dependent on the conductivity of the underlying water. Therefore, the sensitivity and accuracy of EM ice thickness measurements decreases with decreasing salinity. However, even with salinities as low as 3 ppt as typical, for example, for the northern Baltic Sea, successful measurements have been performed (Multala et al. 1996; Haas 2004). The EM response is also strongly dependent on operating altitude: Better results are obtained for lower flying heights. Very low altitudes can be assumed with a fixed-mounted system; however, the flying speed and range are then limited (Prinsenberg et al. 2002).

The accuracy of airborne EM measurements is typically 0.1 m over level ice (Pfaffling et al. 2007; Prinsenberg et al. 2002). However, the footprint of the method is much larger than for ground measurements, and is estimated to range between 3.7 and 10 times the flying altitude for horizontal coplanar coil configurations (Kovacs et al. 1995; Reid et al. 2006). This results in stronger underestimation of maximum keel depths, although a proper validation of the accuracy of EM measurements over three-dimensionally varying natural ridges is still pending. However, the method is well capable to provide reasonable mean ice thicknesses across ridges and to compare the relative difference in abundance and thickness of ridges in different regions. This is demonstrated in Figure 3.2.19, showing a comparison of drill-hole and HEM thicknesses along a 500 m long profile. The bird altitude was 20 m, resulting in a footprint size of ~60 to 80 m. The underestimation of the ridges is very obvious. However, two parallel drill-hole profiles obtained 20 m to both sides of the plotted line, and aerial photography also

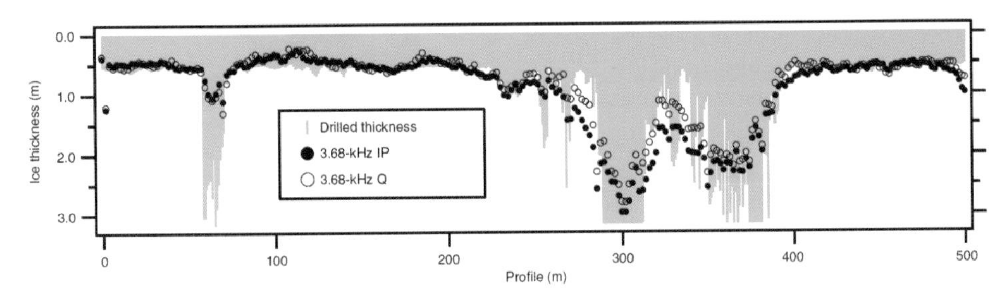


**Figure 3.2.19.** Comparison of drill-hole and HEM-derived (In-phase IP and quadrature Q channels) ice thickness profile (Pfaffling et al. 2007). The graph does not display the maximum 5.8 m drilled ridge thickness at 305 m.

showed strong lateral inhomogeneities in the main ridge structure over the footprint area, which contributes to the disagreement visible in the one-dimensional profile in the figure.

Due to the requirements imposed by arctic logistics and by operations from small helicopters and from landing pads on icebreakers, EM birds for sea ice thickness are much smaller than instruments normally used in exploration geophysics, which provides a challenge for their signal-to-noise characteristics. Typical instruments are 3 to 5 m long and weigh 100 to 150 kg. Single and multifrequency instruments are used with typical signal frequencies between 1 and 100 kHz. Sampling rates are up to 10 Hz, corresponding to a typical point spacing of 3 to 5 m between individual measurements, depending on flying speed. The laser data are normally acquired with a higher sampling rate (e.g., 100 Hz), resulting in a relatively high-resolution profile of surface roughness (cf. Figures 3.2.19 and 3.2.21). The laser data require separate processing as described in the next section.

### 3.2.8.3 Airborne Laser Profiling

Due to the general isostasy of sea ice, and following the principles outlined with respect to measurements of ice draft in Section 3.2.8.1, sea ice surface elevation and freeboard can be used as a proxy for ice thickness. The thicker the ice, the higher its surface rises above the water level. In fact, with proper knowledge of the thickness of the snow cover and of the densities of ice and snow, ice thickness can be accurately calculated from surface elevation. A detailed discussion of the associated uncertainties is presented in Section 3.2.8.4. In addition, the surface profile includes information about the height and frequency of ridges and ice morphology, therefore providing information on the deformational history of the ice which can also characterize different ice regimes.

Long profiles of surface roughness and elevation can be obtained by means of airborne laser altimetry, with a vertically downward-looking laser distance meter

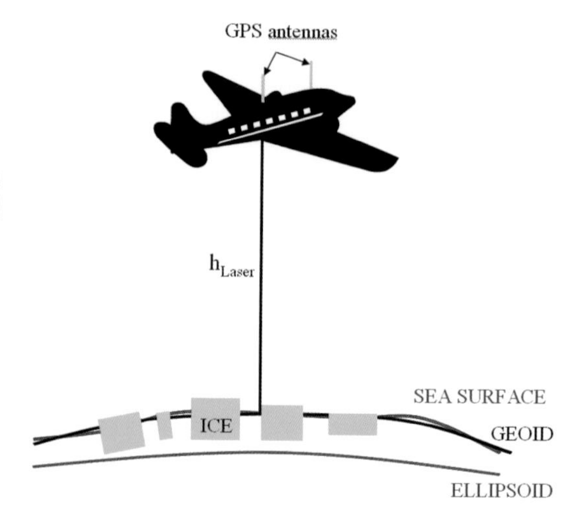


**Figure 3.2.20.** Principle of airborne laser altimetry (courtesy R. Forsberg). The height $h_{Laser}$ of the aircraft above the ice and water surface is measured by means of a laser altimeter. The actual sea surface height above a reference ellipsoid is dependent on tides, geostrophic currents, and sea level pressure (dynamic sea surface topography), and varies additionally with geoid undulations due to the mass distribution within the Earth's crust and upper mantle.

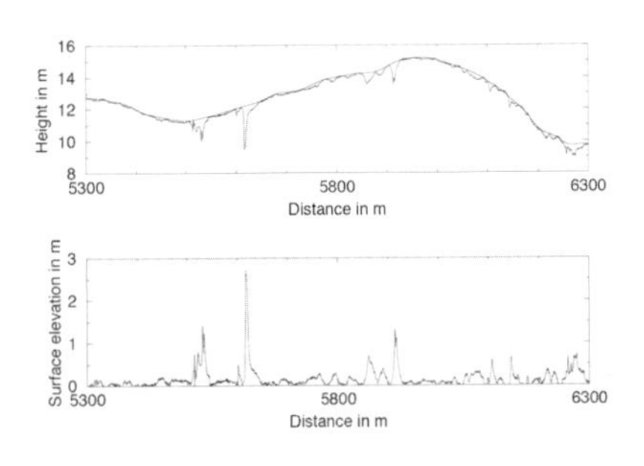


**Figure 3.2.21.** One-kilometer long section of a helicopter-borne laser profile. The raw data and the derived helicopter motion are shown (top) as well as the profile of surface height resulting from subtraction of the latter from the former (bottom).

operated from a helicopter or airplane (Figure 3.2.20). Such studies were performed as early as the 1960s (e.g., Mock et al. 1972). Figure 3.2.21 shows an example of a short profile thus obtained. The data are characterized by a high-frequency signal resulting from the small-scale changes of ice surface elevation including ridges, superimposed on a slowly varying signal resulting from altitude variations of the aircraft. The challenge for data processing of any kind of altimetry data is the removal of this slowly varying altitude variation, to obtain a profile of surface elevation and ice morphology only. Different methods exist to perform altitude removal, and their application depends on the availability of additional, auxiliary data, for example, from differential GPS and inertial navigation systems. Standard high-pass filtering procedures are generally not adequate because there is some spectral overlap between ice and aircraft signals, and because the surface elevation is not normally distributed.

However, Hibler (1972) introduced a three-step filtering method to bypass this difficulty, which is followed by most studies in the absence of auxiliary data. First, a conventional high-pass filter is applied to identify minimum points along the high-pass-filtered time series, which represent locations with the lowest surface elevation along the profile, including open water. Then, a polygon connecting these low points in the original data set is constructed. The polygon is then low-pass filtered to represent the smoothly varying aircraft motion. Surface elevation results from subtraction of the measured distances from the reconstructed aircraft altitude (Figure 3.2.21).

It is important to note that due to the filtering process the obtained surface elevations do not represent the true surface elevations above the water level, but are relative to some datum, which typically follows the surface of the level ice. This implies that the obtained elevations are normally too inaccurate to calculate ice thickness from. However, surface roughness (i.e., the variability of the elevation profile) and in particular ridge distributions can be well determined. The latter can be identified from local maxima along the profile by some kind of Rayleigh criterion, which requires that the local elevation maxima are at least twice as high as the deepest points of the neighboring troughs (e.g., Hibler 1975). Ridges are only considered if their height exceeds a certain cut-off height of typically 0.8 m, for example, to avoid bias from smaller roughness features like snowdrifts or processing artifacts from floe edges.

From the height of the extracted ridges and their spacing, various statistical parameters can be derived, including mean ridge height and spacing, as well as ridging intensity, which is the ratio of mean height or squared mean height over mean spacing. From these, significant regional differences can be well determined (e.g., Dierking 1995; Granberg and Leppäranta 1999), which also provide important information for the validation of satellite data, particularly from synthetic aperture radar (SAR) (see also Chapter 3.15).

Most single-beam laser altimeters operate in the near infrared with a wavelength of 905 nm. Sampling frequencies range between 100 and 2,000 Hz, resulting in a spacing of 0.02 to 0.5 m between individual measurements, depending on flying speed. Distance can be measured with an accuracy of 0.01 to 0.05 m, and therefore even the small-scale roughness of the ice surface can be well resolved. Inaccuracies result mainly from pitch and roll of the aircraft, leading to slanted viewing angles, and from uncertainties in the aircraft altitude compensation. Laser altimeters are typically flown at altitudes of 30 to 300 m.

*Improved Removal of Aircraft Altitude Variations*

The problems due to aircraft motion are more severe for helicopters than for fixed-wing aircraft, which are much more stable in flight and whose altitude varies more slowly. However, altitude variations provide a serious problem for fixed-wing airplanes as well. Today, they can be measured independently with high precision using a combination of differential GPS and inertial navigation systems. These measure the position, altitude, and attitude of the aircraft (Hvidegaard and Forsberg 2002). Sea ice surface elevation $Z_{se}$ is then obtained from $Z_{se} = h_{GPS} - h_{Laser} - N - \Delta h$ where $h_{Laser}$ is the laser range, $h_{GPS}$ the GPS-measured airplane height in a GPS reference system relative to an ellipsoid (e.g., WGS84), N is the geoid height, and $\Delta h$ is a residual term describing the dynamic sea-surface topography due to tides, currents, and sea-level pressure variations (Figure 3.2.20). These terms are required because GPS heights are determined relative to a worldwide ellipsoid that does not represent the actual geoid at a given location. The geoid elevation N is obtained from a high-resolution geoid model, for example, from the Arctic Gravity Project (Forsberg and Skourup 2005).

The residual sea-surface topography term has magnitudes of less than 1 or 2 m and is of a long-wave nature (see also Figure 3.2.23). As the actual sea surface corresponds to the lowest points in the obtained surface elevation profiles, either from open water or newly frozen leads, similar filtering procedures as described above for stand-alone lasers can be used to remove the remaining bias. Local minima at leads or cracks can be identified as open-water tie points and can be connected to represent the actual sea level along the flight track (Hvidegaard and Forsberg 2002). This procedure also removes any residual bias due to possible laser offsets and misalignments, GPS errors, geoid errors, and the dynamic sea-surface topography. As all these residual biases vary only very slowly, surface elevation can be obtained with an accuracy of between 0.05 and 0.13 m (Hvidegaard and Forsberg 2002; Forsberg and Skourup 2005). The uncertainty in derived ice thickness is approximately tenfold, and depends on assumptions about ice and snow density, and snow thickness (see Section 3.2.8.4). Accurate retrieval of ice thickness from measurements of surface elevations remains a major challenge (Hvidegaard et al. 2006).

As with multibeam, upward-looking sonar measurements (Section 3.2.8.1), two-dimensional information about the ice surface elevation can be obtained from airborne across-track laser scanning (Figure 3.2.22). Most laser scanners have across-track viewing angles of 45 degrees, resulting in a swath width approximately equal to the flying height. Measurements are performed at angular steps of, for example, 1.5 degrees, resulting in a spatial sampling rate of 1 m for a flying altitude of 300 m. The accuracy of individual measurements is comparable to those of single-beam lasers (i.e., ranges between 0.02 and 0.05 m). However, accurate correction of aircraft attitude variations by means of DGPS and inertial navigation system data is particularly important, as slight roll angles lead to lateral shifts of the laser swath and will result in wrong projections of each individual beam on a horizontal reference plane. Laser returns from open water are often spurious, depending on the wave-related roughness of the water surface and the amount of impurities in the water. However, with off-nadir beams this problem is even larger, as most energy is reflected away from the laser detector. This can be seen in Figure 3.2.22, where data gaps indicated by white filling exist in the regions of open water and thin, level new ice. Laser scanner data is particularly useful for the validation of lower-resolution airborne or satellite measurements which include information

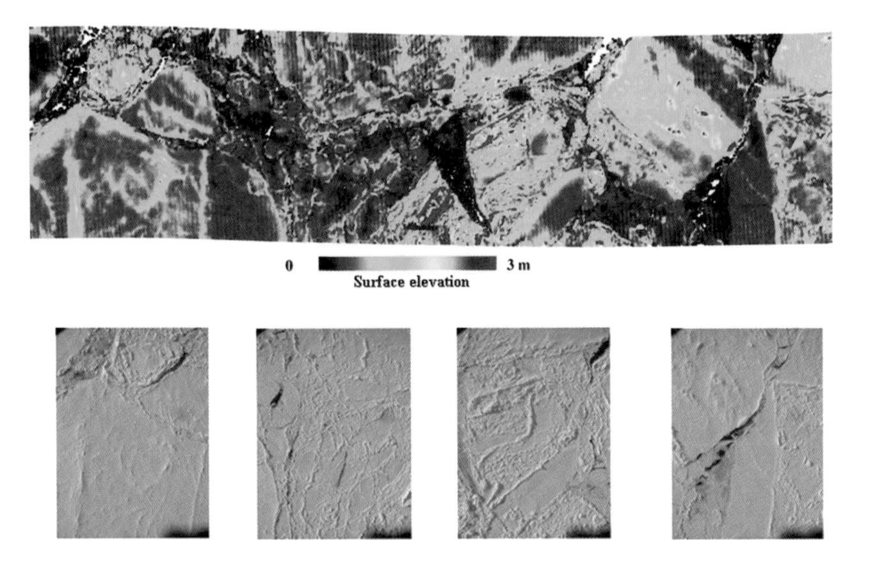


**Figure 3.2.22.** Typical laser scanner swath of the two-dimensional distribution of sea ice surface elevation, and comparison with aerial, nadir-looking photographs. The width of the swath is approximately 300 m (courtesy R. Forsberg).

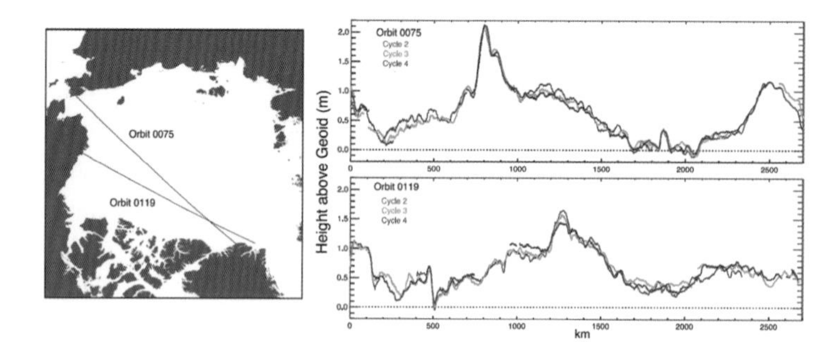


**Figure 3.2.23.** Sample elevation profiles across the Arctic Ocean from two exact repeat ground tracks of ICESat (see below), separated eight days apart, and after removal of the best available arctic geoid and tidal effects (Kwok et al. 2006). Large variations along orbits and between repeat-tracks can still be seen.

from the two-dimensional thickness distribution and roughness, as was shown with respect to airborne EM surveys (Hvidegaard et al. 2006) and ICESat surface elevation measurements (Skourup and Forsberg 2006).

To overcome some of the operational constraints, like limited operating time and range and high cost of manned aircraft, unmanned aerial vehicles (UAVs) have been used for airborne laser surveying as well, both with single-beam and scanning laser altimeters. Dedicated laser instruments are small enough to be carried by these vehicles, which generally have a limited payload. However, costs of these relatively new platforms are still very high, and it is not yet proven that their operational constraints are less than those of manned aircraft. There are also serious problems with air-worthiness and flying permissions.

### 3.2.8.4 Satellite Altimetry

Altimeters can also be operated from satellites to observe sea ice surface elevation or freeboard, which can be transformed into estimates of ice thickness. At first sight, satellite measurements seem to be superior over all other methods in their temporal and spatial coverage of the polar oceans. And this is certainly true if large-scale climate aspects of sea ice thickness are considered. However, for most other sea ice service applications, satellite altimetry is probably the least suitable method, due to constraints imposed by orbit geometry and survey repeat times. Altimetric measurements are essentially one-dimensional along the satellite track, and therefore there are large regions of the Earth's surface which are not covered by satellite altimeters at all. There are trade-offs between orbit inclination, repeat orbit intervals, and ground coverage. For example, higher across-track coverage can be achieved with longer repeat intervals, but then temporal changes cannot be so well resolved. A typical orbit repeat period is thirty days. More frequent measurements are then only performed at cross-over locations of descending and ascending orbits. In addition, the uncertainty of individual point measurements can be large, and sufficient accuracy is only obtained with significant spatial and temporal averaging.

Satellite altimeters measure the distance between the satellite and the surface of the Earth. Relative surface height differences between the ice and water are observed to estimate sea ice freeboard or surface elevation. In the absence of frequent open-water regions, the processing again has to rely on assumptions of linear changes of the local sea-surface topography between open-water tie points. Although orbit variations of satellites are much smaller and happen on longer spatial scales than altitude variations of airplanes, reconstruction of the local sea surface height as a datum for surface height measurements of the snow or ice surface is still a major challenge. Small-scale sea-surface height variations occur due to tides and currents, unknown geoid undulations, and temporal variations due to weather-related surface pressure changes (Figure 3.2.23). In addition, calculation of ice thickness from measurements of surface elevation or freeboard relies on several assumptions about snow thickness and density, as well as the densities of ice and snow.

Two different kinds of instruments exist: radar and laser altimeters. These are generally different in their penetration characteristics for snow and sea ice, and in their spatial resolution. Their different penetration characteristics are particularly important for sea ice measurements. While the near-infrared wavelengths of lasers do not penetrate into snow and ice and are scattered at the upper surface, radar altimeter wavelengths, typically at Ku-band (e.g., 13.8 GHz or 2.2 cm), penetrate the snow to some degree, and the reflections are generally believed to originate from the snow-ice interface (Laxon et al. 2003). Therefore, with laser altimeters the elevation of the snow surface $Z_{se}$ is obtained, while with radar altimeters the freeboard $Z_{fb}$ of the ice is retrieved (cf. Figure 3.2.4). Accordingly, different equations for the calculation of ice thickness $Z_i$ are applied, which result from Archimedes' law and the general isostasy of the ice:

For laser altimetry: $$z_i = \left(\frac{\rho_w}{\rho_w - \rho_i}\right) z_{se} - \left(\frac{\rho_w - \rho_s}{\rho_w - \rho_i}\right) z_s$$ (Equation 3.2.6a),

For radar altimetry: $$z_i = \left(\frac{\rho_w}{\rho_w - \rho_i}\right) z_{fb} - \left(\frac{\rho_s}{\rho_w - \rho_i}\right) z_s$$ (Equation 3.2.6b),

with the densities $\rho_w$, $\rho_i$, and $\rho_s$ of water, ice, and snow, respectively, and snow thickness $Z_s$.

With typical densities of $\rho_w = 1{,}024$ kg/m$^3$, $\rho_i = 925$ kg/m$^3$, and $\rho_s = 300$ kg/m$^3$, the first term in these equations implies an approximately tenfold amplification of freeboard uncertainties for the calculation of ice thickness for both methods. However, it is also important to note that the second terms are different, resulting in a stark difference in the sensitivity of thickness retrievals to uncertainties in snow thickness. The term is approximately 7 for laser altimetry and approximately 3 for radar altimetry. Therefore, snow thickness uncertainties in laser altimeter data contribute to more than twice as large uncertainties in retrieved ice thicknesses than in radar altimeter data.

A comprehensive analysis of the sensitivity of ice thickness calculations according to Equations 3.2.6a,b to uncertainties in snow and ice properties has been performed by Kwok and Cunningham (2008). While snow thickness on arctic sea ice is generally derived from a climatology compiled by Warren et al. (1999), Kwok and Cunningham (2008) show that improved estimates of snow thickness can be obtained from reanalysis models from the European Center for Medium Range Weather Forecast (ECMWF). In addition, the authors show improved analyses of seasonal changes of snow density. With these, the overall uncertainty of thickness retrievals from altimetry can be reduced to less than 0.7 to 0.5 m. More discussions of the comparison of surface elevation and ice draft can also be found with Wadhams et al. (1992), who compared draft and surface elevation data obtained along coincident tracks of a submarine and an airplane.

*Radar Altimetry*

Radar altimeter observations of sea ice were first performed with Seasat in 1978, and more routine observations are available from ERS-1 and ERS-2 since 1991. Since 2002, radar altimetric measurements over sea ice are obtained by Envisat. These measurements are generally limited to sea ice regions south of 81.5° N and north of 81.5° S due to the inclination of the satellite orbits of 98.5° at best. In late 2009, the European Space Agency will launch CryoSat, a dedicated ice radar altimetry mission, which will obtain measurements up to 88° N and S. The ERS altimeter had a pulse repetition frequency of 1,020 Hz. Fifty pulses were averaged into one waveform (see below) to yield an effective sampling interval of 0.49 s, corresponding to a point spacing of approximately 330 m on the ground. Note that the point spacing is much smaller than the footprint (see below).

General aspects and details of radar altimetry and sea ice measurements are described by Fetterer et al. (1992). Satellite radar altimetry is routinely employed to measure sea surface heights and wave heights over open oceans. Satellite altimeters use pulse-width-limited geometry, where a pulse of microsecond-scale length is emitted within a < 1-degree-wide beam. The pulse echo returned from the surface of the Earth is received with temporal resolution of the energy arriving back at the satellite. The area on the surface illuminated by the pulse grows with time, until the trailing edge of the pulse leaves the lowest reflecting points at nadir. The maximum area simultaneously illuminated is usually referred to as pulse-limited footprint, and depends on the roughness of the surface. It can vary between one and several kilometers in diameter. Returns received at later times represent energy scattered back from off-nadir locations. The data set of received energy versus time is generally referred to as waveform, and its shape and instance are used to assess the distance to and characteristics of the Earth's surface.

Over sea ice regions, most waveforms are very narrow with high power. This is a result of specular reflection from small fractions of smooth open water or nilas in cracks or leads with high backscatter within the large footprint (Peacock and Laxon 2004), which reflect rather than scatter the radar energy back to the satellite. The resulting "peakiness" of waveforms can be quantified and is a powerful separator between ice-covered and open ocean (Fetterer et al. 1992). However, when the footprint is entirely covered by ice, diffuse waveforms occur within the sea ice zone as well (Laxon et al. 2003; Peacock and Laxon 2004). This is usually observed in less than 5 to 10 percent of all measurements, depending on season and ice conditions. The different height retrievals from specular and diffuse waveforms within the sea ice zone can then be used to determine the surface heights of the water and ice surfaces, whose difference is freeboard.

The construction of the sea surface from specular waveforms is not straightforward, as it is generally unclear which time instance of the waveform represents the distance to the pulse-limited footprint region at nadir. This instance is normally determined by the satellite electronics based on diffuse waveforms and the time history of previous waveforms, and can vary on short and longer time scales along the orbit (Peacock and Laxon 2004). Therefore, extensive, sea ice–specific reprocessing ("retracking") of waveforms is required to reduce the noise. Additional low-pass filtering is applied, and then the resulting profiles of sea-surface height have noise levels of less than 0.25 m (Peacock and Laxon 2004). Diffuse waveforms are treated similarly, and freeboard is obtained from the height differences obtained from specular and diffuse waveforms (Laxon et al. 2003). Ice thicknesses of less than 1 m are normally excluded from the analysis as their freeboard is too small to be discerned with high confidence. The details of the processing vary with sensor characteristics and ice conditions, and also have to include corrections for radial orbit errors and tropospheric and ionospheric effects. They generally require some fine-tuning. For example, in the Arctic freeboard can often only be retrieved between October and March, as broken-up ice fields and a wet, melt pond–covered ice surface significantly reduce the number of diffuse reflections and radar backscatter. Statistical accuracies are further improved by spatial and temporal averaging, e.g., over 0.25 × 0.25 degree grid areas and monthly fields. With this, Laxon et al. (2003) estimate their ice thickness accuracy to be within 0.11 m to 0.2 m, which implies an accuracy of their freeboard retrieval of 0.01 to 0.02 m, according to Equation 3.2.6b.

A thorough validation of satellite radar altimetry thickness retrievals is pending. However, Figure 3.2.24 shows a comparison of radar-altimeter-derived thicknesses with results from ULS profiling. Large uncertainties are visible both with individual radar altimeter and ULS data points, pointing to scale-related problems due to the differing footprints of radar altimeter and ULS measurements. On a large-scale, long-term average, however, radar altimeter retrievals of ice thickness seem to have a high accuracy. This is also indicated by the linear least-square fit in Figure 3.2.24, which yields a slope of 0.98 and intercept of 0.12 m (Laxon et al. 2003).

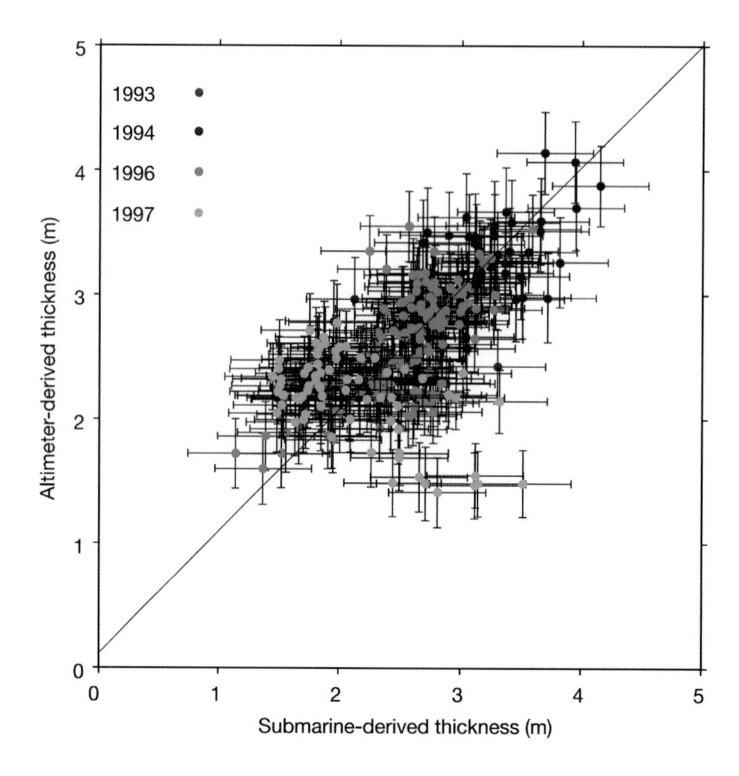


**Figure 3.2.24.** Comparison between satellite altimeter and submarine ULS-derived ice thickness in the Beaufort Sea (Laxon et al. 2003). Submarine thicknesses are shown for 50 km profile segments gathered during four missions. Altimeter thickness estimates were generated from observations within fifteen days and 100 km of the submarine tracks. Submarine thicknesses exclude ice thinner than 0.5 m and open water, because (owing to difficulties in discriminating thin ice from open water) progressively less altimeter data can be retrieved once thickness falls below 1 m. Error bars show uncertainties in altimeter thickness due to measurement errors and snow depth variability and an error in submarine thickness of 0.4 m. Diagonal line shows result of a least-squares fit (see text).

*Laser Altimetry*

A satellite laser altimeter has so far only been launched once, and is still operating (as of April 2009) onboard the ICESat satellite launched in 2003. Its Geoscience Laser Altimeter System (GLAS) operates with a laser wavelength of 1,064 nm to profile the Earth's surface (Zwally et al. 2002). With a beam width of approximately 110 mrad and a pulse rate of 40 Hz, it obtains a measurement approximately every 170 m with a footprint diameter of 70 m. The orbit inclination is 94 degrees, allowing surveys up to 86° N and S. Due to the narrowness of the beam, accurate attitude determination is very critical, in addition to the general necessity of precise orbit determinations. Unfortunately, there were technical problems with the laser systems of ICESat, which have limited its operation to February-to-April and October-to-November seasons; however, this is still sufficient for the observation of seasonal ice thickness variations.

As with radar altimeters, not only the travel time of the emitted laser pulse between satellite and the Earth's surface is recorded, but also its time-dependent return at the laser receiver. The resulting waveforms are analyzed for width and amplitude, allowing conclusions on surface roughness and reflectivity (Zwally et al. 2002). The wider the received waveform, the rougher the surface.

The ability to determine surface reflectivity can be ideally used to delineate regions of snow-covered ice and new ice or open water, as they are characterized by high and low albedos, respectively, even in the 1,064 wavelength range (Figure 3.2.25). Therefore, open water can be detected from regions of low reflectivity, and can be used as tie points for the reconstruction of sea level (Kwok et al. 2006). Figure 3.2.25 shows that low reflectivity of a newly opened lead coincides with a region of the lowest surface elevation as expected. The figure also shows that there are additional regions of thinner ice identified by their low surface elevation, which do not have lower reflectivities. These represent younger ice, which is already snow covered to some extent. The figure illustrates that these low-level locations could be mistakenly identified as open water or new ice if no other,

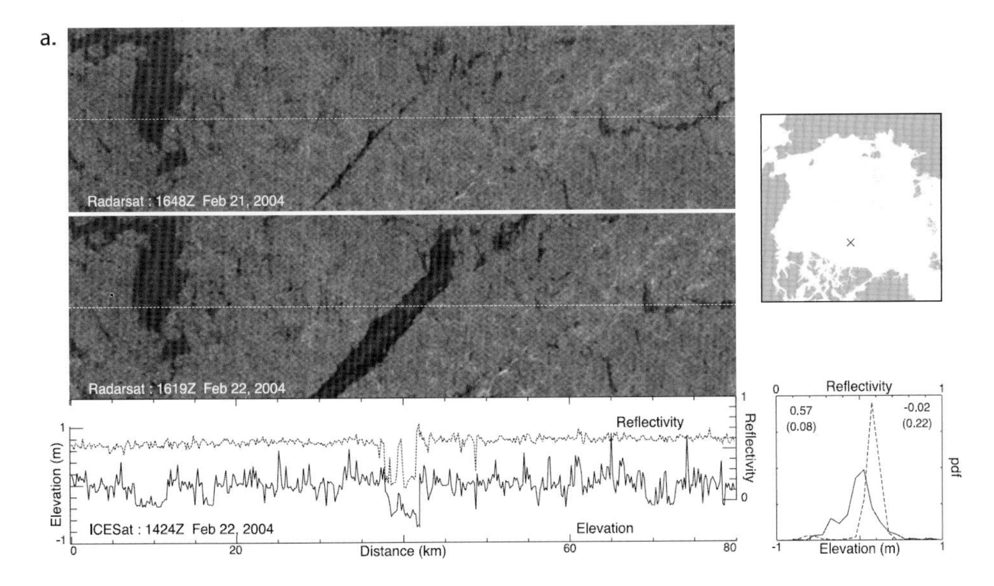


**Figure 3.2.25.** Example of 80 km long ICESat profile from the Arctic Ocean (cross on map) and comparison with near-coincident Radarsat SAR images (from Kwok et al. 2006). The SAR images were obtained one day apart, and show the opening of a new lead in the center of the image, visible by its low backscatter (dark color). The ICESat track is shown as white dashed line. The overflight took place only two hours before the acquisition of the lower image. Lower panel shows the elevation (solid, centered around mean) and reflectivity profiles (dashed). The bottom right panel shows the distributions of elevation and reflectivity, and their mean and standard deviations (numbers).

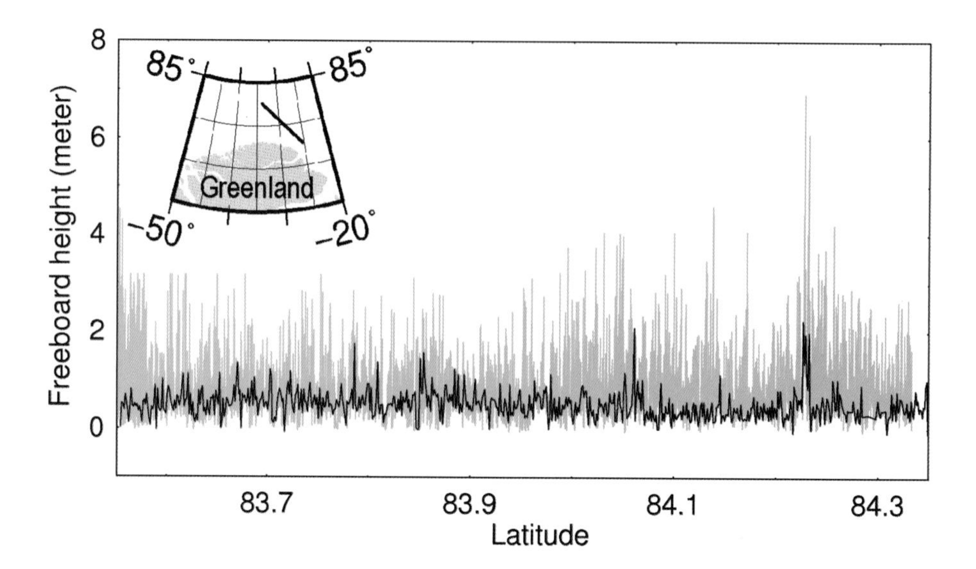


**Figure 3.2.26.** Comparison of freeboard heights from airborne (blue) and satellite (ICESat; black) laser altimetry. The inset map shows the track location north of Greenland (from Forsberg and Skourup 2005).

even lower locations would be observed. Only with proper reflectivity information can open-water locations be clearly identified. Experience shows that usually there is some occurrence of open water along profiles of 25 km length, which is normally taken as the minimum distance over which at least one tie point should be observed (Hvidegaard and Forsberg 2002; Forsberg and Skourup 2005; Kwok et al. 2007).

Overall, Kwok et al. (2007) estimate that the uncertainty of ICESat freeboard retrievals along 25 km segments is better than 0.07 m on average. Again, validation of these results is difficult due to the spatial and temporal scales and the different footprints. Figure 3.2.26 shows two 80 km long, coincident ICESat and aircraft laser profiles (Forsberg and Skourup 2005). Sea level has been derived from an along-track lowest-level filtering scheme with a resolution of ~15 km, neglecting reflectivity information in the ICESat data. Both profiles are in good qualitative agreement. However, the height of lowest levels differs by ~25 cm. This difference is due to the large footprint of ICESat of 70 m, which prevents the detection of narrow leads or cracks. Similarly, individual pressure ridges cannot be profiled, as they are averaged over the footprint area.

Figure 3.2.27 shows another example for the validation of ICESat thickness retrievals (Kwok and Cunningham 2008). ICESat freeboard has been retrieved with more sophisticated methods than for the example in Figure 3.2.26, using SAR images and reflectivity information for the determination of open-water tie points (see above). Ice draft has been calculated with the best available estimates for snow thickness and snow and ice density (Kwok and Cunningham 2008). Results are

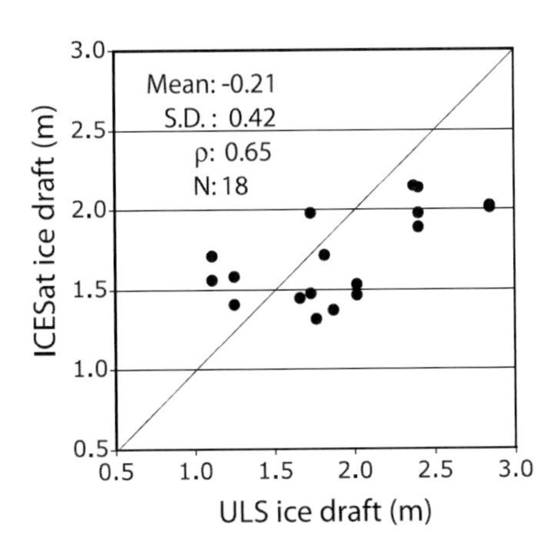


**Figure 3.2.27.** Comparison of mean ICESat and ULS draft retrievals for overlapping and coincident, 25 km long profiles (Kwok and Cunningham 2008).

compared with moored ULS measurements, which have been temporally averaged to be comparable with 25 km long ICESat profiles. Taken together, the overall difference between ICESat and ULS draft estimates is –0.21±0.42 m, with a correlation of 0.65 between the two ice draft populations (Figure 3.2.27).

With both satellite radar and laser measurements, the question remains to be answered whether the surface elevation retrievals represent mean ice thickness, modal ice thickness, or some other measure of the thickness and morphology of the ice. As the modal ice thickness represents the thickness class with the largest areal coverage (Section 3.2.1), it would be plausible that the instance of maximum energy return at the satellite receiver agrees with the two-way travel time to the level ice surface. In contrast, ridges contribute to the roughness of the ice surface, and their high elevations should lead to earlier arrivals and higher energy of the leading edge of the received waveforms.

## 3.2.9 ICE THICKNESS MEASUREMENTS FOR VARIOUS SEA ICE SERVICES

### 3.2.9.1 Ice Thickness Measurements for the Detection of Climate Variations

Sea ice is a regulator of the global climate, and its thickness is an indicator for the state of the climate in the polar regions and its variability. Ice thickness measurements have to be performed in a systematic manner and over long times to detect climate changes, given the large seasonal and interannual variability of sea ice and the redistribution of thick and thin ice to different regions. There are many examples of observations of ice thickness changes on various scales, ranging from long-term point measurements on fast ice using drill holes, to long time series of moored ULS measurements and submarine ULS profiles. The arguably most spectacular changes have been reported from a comparison of U.S. submarine cruises performed in the period 1958–1976 with data from the 1990s (Rothrock et al. 1999) (Figure 3.2.28). Overall, ice thickness has decreased by 1.8 m between

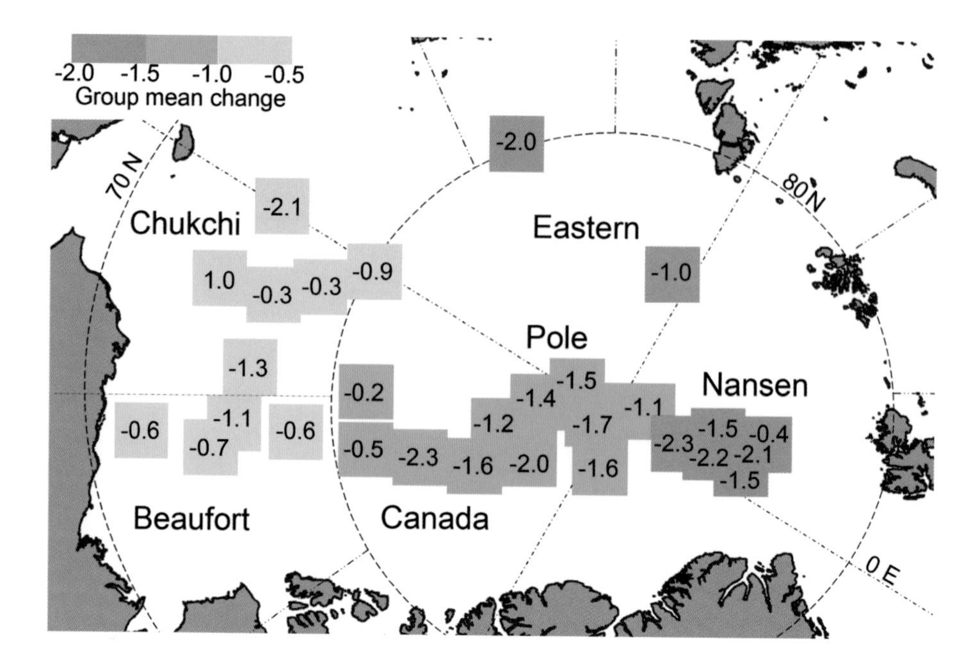


**Figure 3.2.28.** Changes in mean ice draft between the periods 1958–1976 and 1993–1997 (Rothrock et al. 1999). Boxes show regions of 150 km width, over which overlapping data between both periods have been compared.

periods—a thinning of 42 percent from a mean of 3.1 m. Apart from the average decrease, all regions with overlapping data showed thinning, which was strongest in the eastern Arctic and least in the Beaufort Sea (Figure 3.2.28). However, model studies showed later that the interpretation of the changes is difficult because data were obtained only sporadically within the compared periods, and because only the central region of the Arctic Ocean over which data were declassified could be analyzed. This did not allow the detection of redistribution of thick ice within the whole Arctic, and particularly towards the coasts of Greenland and Canada.

Another example of observations of most recent, climate-related thickness changes in the region of the North Pole is given in Figure 3.2.29 (Haas et al. 2008a). Data was obtained by means of ground-based and helicopter-borne EM sounding, using an icebreaker as a moving base to provide a large, regional observational range. Results show another rapid reduction of modal and mean thicknesses of 53 and 44 percent between 2001 and 2007, respectively. Again, data have only been obtained in six summers over a seventeen-year period, because icebreakers are not regularly available for this kind of research. Therefore, the interannual variability cannot be derived from the measurements to better confine the trend. Also, data from April 2007 had to be seasonally adjusted for comparison with the summer measurements in order to account for seasonal changes due to ablation during the melting season. However, with additional data now available from buoys, satellites, and meteorological reanalyses, the causes of the thinning can be much better identified. Figure

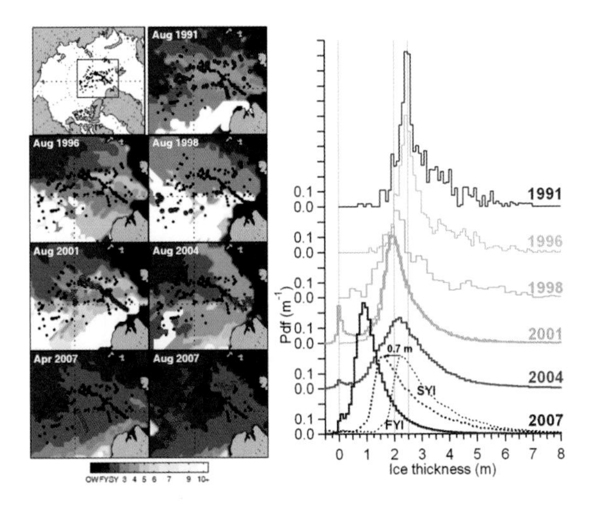


**Figure 3.2.29.** Late-summer ice age (left) and thickness (right) change in the region of the North Pole between 1991 and 2007 (Haas et al. 2008a). Thicknesses were obtained by means of ground-based (thin lines, red circles) and HEM sounding (thick lines, magenta triangles). The second-year ice (SYI) distribution obtained in April 2007 was seasonally adjusted by 0.7 m to represent summer conditions (Haas et al. 2008a).

3.2.29 shows that the observed thinning is accompanied by a reduction of the age of multiyear ice, and even by a replacement by first-year ice in the study region. Presentation of thickness distributions like in Figure 3.2.29 and analyses of decreasing modal thicknesses also allow for conclusions that much of the observed thinning is due to changes in thermodynamic boundary conditions. However, these observations alone do not allow us to judge whether the thinning is due to changes in atmospheric or oceanic boundary conditions, or both. Nor can conclusions be drawn on the importance of reduced winter growth or summer ablation for the observed thinning. Ideally, these measurements should be accompanied by observations from ice mass balance buoys (Section 3.2.6), or they should be performed at least twice a year, at the end of the freezing season (April/May) and ablation season (September).

### 3.2.9.2 Thickness Measurements in Support of On-Ice Travel, Subsistence, Ice-Runway Preparation, and Biological-Physical Process Studies

Travelers on sea ice, whether they be local hunters on foot or snowmachine, engineers designing ice roads, or scientists conducting field campaigns, are often very concerned with both ice thickness and roughness. Ice thickness is directly related to ice's load-bearing capacity and integrity; ice thickness and surface roughness determine whether ice is stably grounded in shallow waters; and surface roughness has clear implications for trafficability—the ability of a given traveler to traverse the ice cover. Although sea ice users may have differing understandings of how ice properties relate to these on-ice travel and operational concerns, general relationships do exist and can often be informed by the measurement techniques discussed in this chapter.

When determining whether or not ice of a given thickness will support an activity on the ice, one must consider a number of variables. Given the viscoelastic response of ice to an applied load, both the weight of the load and the rate of loading are important. Also when transporting a heavy load across level ice, one must consider how speed relates to wave dynamic responses in the water beneath the ice that may contribute to failure. The load-bearing capacity of ice depends on ice type (first year or multiyear), temperature, salinity, porosity, stress history, and cracks and nonhomogeneities (Potter and Walden 1981). These variables also have implications for how a given piece of ice will withstand longitudinal pressures imparted by adjacent ice. While knowledge and formulas used to evaluate strength and load-bearing capacities vary in sophistication and are typically derived empirically, ice thickness is perhaps the most important variable and also the easiest and most straightforward to measure.

Grounded sea ice ridges, which may be thought of as roughness on the scale of $10^0$ to $10^1$ m, play a fundamental role in whether or not landfast sea ice may be considered stable from an ice user's perspective. Of importance to travelers on ice are break-out events in which large sections of ice are destabilized and detach from the coast or from other remaining sections of landfast ice. For example, throughout history and continuing into the present day, such events have presented a significant threat to hunters traveling on sea ice in the coastal regions of the Arctic (George et al. 2004). Grounded ridges may be either formed in place through deformation or may be advected in from elsewhere and deposited. Assessing the stability of landfast sea ice is very complex as it is influenced by a range of interacting variables—sea-level fluctuations, water temperature, currents, winds, ice morphology, salinity, pack ice interaction, and "weak spots" or cracks within the ice. In general, the presence of grounded ridges within the landfast ice cover, most of which is typically floating, contributes to its ability to resist destabilizing forces. However, in some cases, the keels of weakly grounded ridges may in fact contribute more to a potentially destabilizing drag by the current than to serving as anchoring points (Mahoney et al. 2007).

Roughness on scales ranging from $10^{-2}$ to $10^1$ m typically has implications for trafficability. During the construction of groomed sea ice roads or airstrips, the ice surface is leveled, often with a bulldozer, and then flooded to create a smooth path. Here, concerns may exist on the centimeter scale as even small cracks and inhomogeneities may negatively impact traffic or the integrity of the surface. Those traversing an unaltered ice cover are impacted by features ranging from light rubble (dimensions $< 1$ m), which presents obstacles as well as promotes snowdrifts capable of concealing cracks, to ridges of sail heights from one to tens of meters (Barker et al. 2006). Industrial operations on offshore structures in arctic landfast ice are interested in how roughness and rubble conditions relate to the efficiency of emergency evacuation plans where people leaving the primary structure must traverse a range of conditions to reach on-ice evacuation shelters (Barker et al. 2006), or to escape to otherwise safe conditions.

As discussed in Section 3.2.8 satellite data has clear limitations for obtaining ice thickness information. However, when using a satellite image of an area where

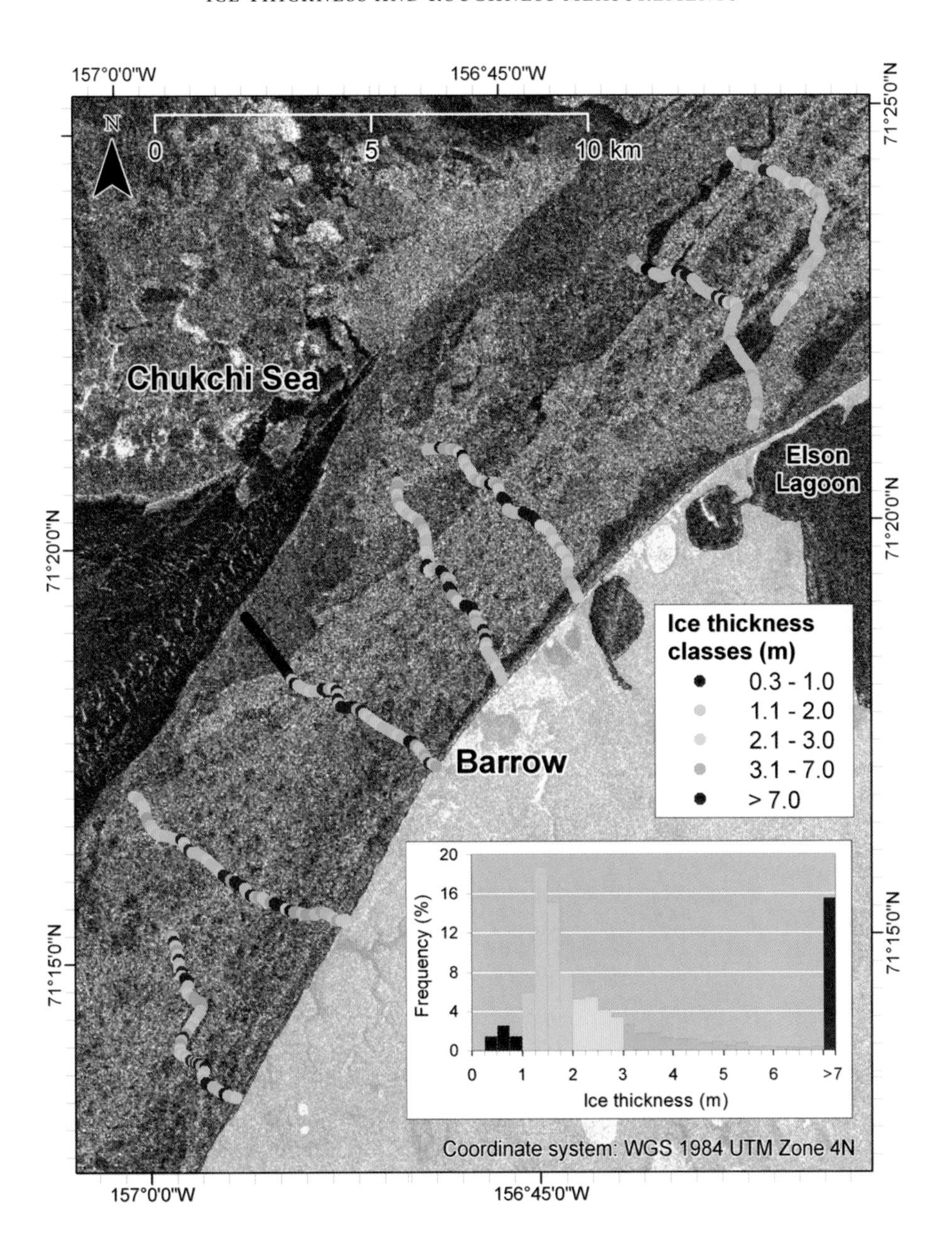


**Figure 3.2.30.** EM-derived ice thickness measurements along trails constructed by Iñupiat whale hunters on the landfast ice off Barrow, Alaska, in spring 2008. Trails originate near land and traverse a range of ice types before terminating at the lead edge where hunting camps are established. (The background of this map is an ERS-2 SAR satellite image from 22 March 2008. The semi-transparent yellow overlay represents land.)

extensive ice thickness measurements have been made using other methods, one can extrapolate thickness information from one area to another based on the information in the satellite image. Figure 3.2.30 presents a spatial plot of EM-derived ice thickness measurements along trails constructed on the landfast ice off Barrow,

Alaska, in spring 2008 during the bowhead whale hunt by Iñupiat whalers. These measurements were made by hauling an EM device, similar to that shown in Figure 3.2.13, on a small sled along the trails. The thickness data presented here shows the thickness distribution of the sampled ice, but also lends insight into how to interpret the SAR satellite image that the data overlies. In general, smooth and relatively thin ice is represented by areas of low radar backscatter (darker areas), such as the region overlaid by measurements of ice thickness less than 1 m (indicated by red dots) in the center left side of the figure. The thicker, rough, and highly deformed ice is represented by the areas of higher backscatter (brighter areas). A stretch of thick ice (indicated by the light and dark blue dots), which is largely composed of ridges thicker than 7 m, parallels the coast and intersects the five southernmost trails. Satellite information coupled with thickness data along trails may inform decisions regarding the trafficability of the ice in areas where trails do not exist. For example, thinner ice of low backscatter may provide for efficient and nonobstructed timely travel as opposed to areas of higher backscatter.

Figure 3.2.31 shows a similar example of an EM profile obtained by hauling an EM instrument with a Ski-Doo between individual observational sites on an ice floe, overlaid on an aerial photograph (Haas et al. 2008c). The thickness measurements were used to delineate individual regions of the ice floe with the same properties

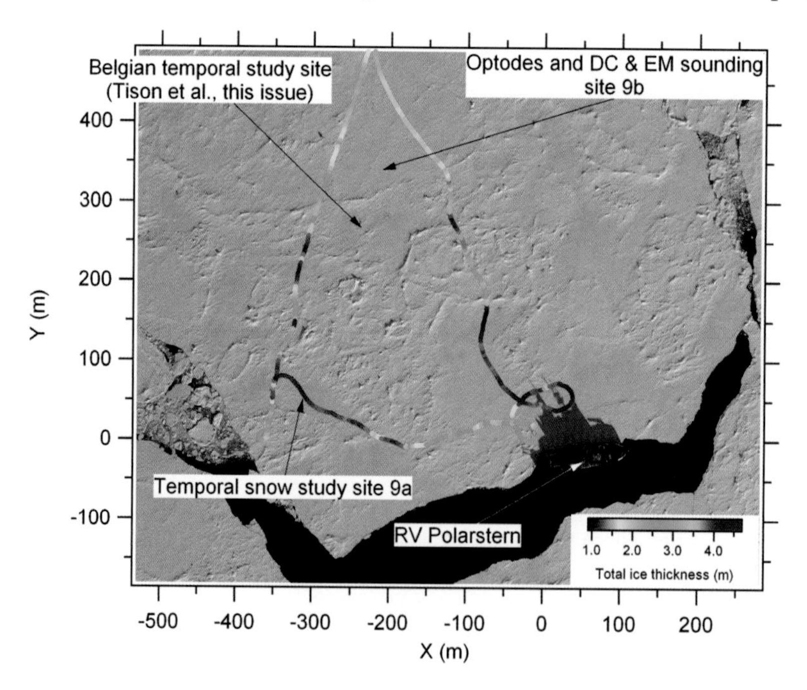


**Figure 3.2.31.** Nadir-looking aerial photograph of an ice floe in the Weddell Sea (Haas et al. 2008c; see also Chapter 3.17). Color-coded line shows ice thickness profile along a Ski-Doo track connecting various sampling sites. Note composition of ice floe of thick rough SYI floe fragments and grayish, smooth thin FYI in between.

and developmental history, to support the interpretation of ice cores and other samples taken at the various sites for biological and glaciological studies (see also Chapter 3.17). Repeat surveys were also performed over a period of five weeks to study temporal changes due to progressing melt. The figure shows the presence of three ice types: smooth, thin first-year ice with total thicknesses of 1.0 to 1.2 m; medium smooth, thick first-year ice with total thicknesses around 2.0 m; and deformed second-year ice thicker than 2.5 m.

Obviously, these measurements are also efficient for presite surveys of ice floes for any kind of activity. They have been used to find thin ice regions for the deployment of buoys or divers, or for biological under-ice studies, where large holes have to be drilled into the ice, and where finding the thinnest ice causing the least trouble for producing those holes is of utmost interest. On the contrary, continuous EM surveys can be used to quickly and accurately profile regions of anticipated ice runways for aircraft operations, for which a homogeneous, thick enough ice cover has to be confirmed.

### 3.2.9.3 Ice Thickness Measurements in Support of Ship Performance Studies

In the design of icebreakers and definition of shipping regulations, knowledge of the performance of ships in ice is a prime requirement. Mostly, icebreakers are evaluated and classified according to the speed they can maintain in level ice of a certain thickness. Full-scale tests are performed in fast ice zones, where ice thickness is relatively uniform and determined by a few drill-hole measurements. However, the results of those tests can be misleading when transferred to real sea ice conditions, which may include a mixture of ice floes of different sizes and age, and pressure ridges. With the same ice thickness, ship performance in a neutral or divergent ice field can be much better than in a convergent ice field under pressure. The ship performance trials of the U.S. Coast Guard Cutter *Healy* in April and May 2000 in individual ice floes in Davis Strait and Baffin Bay were probably among the most comprehensive trials ever held and integrated a number of the methods presented above to measure ice thickness and roughness. These included the rotating laser level, mechanical ice drilling, ground-based EM surveys, hot-water drilling, and snow depth surveys as well as over-the-side video to document ice thickness (Sodhi et al. 2001). However, on scales of an ice field, ice thickness measurements by traditional means are very challenging, as it is difficult to traverse on foot or by Ski-Doo across a mixture of ice floes, brash ice, and open-water leads.

Therefore, ice thickness measurements have been performed by means of shipborne EM sounding in front of icebreakers sailing through ice (e.g., Haas 1998). As with helicopter-borne EM thickness sounding, the EM instrument has been accompanied by a laser altimeter to measure its elevation above the ice surface. Note that for relatively small distances of 4 to 8 m above the ice, sonic range finders can also be used. Here, a Geonics EM31 (Section 3.2.3) has been used, and both

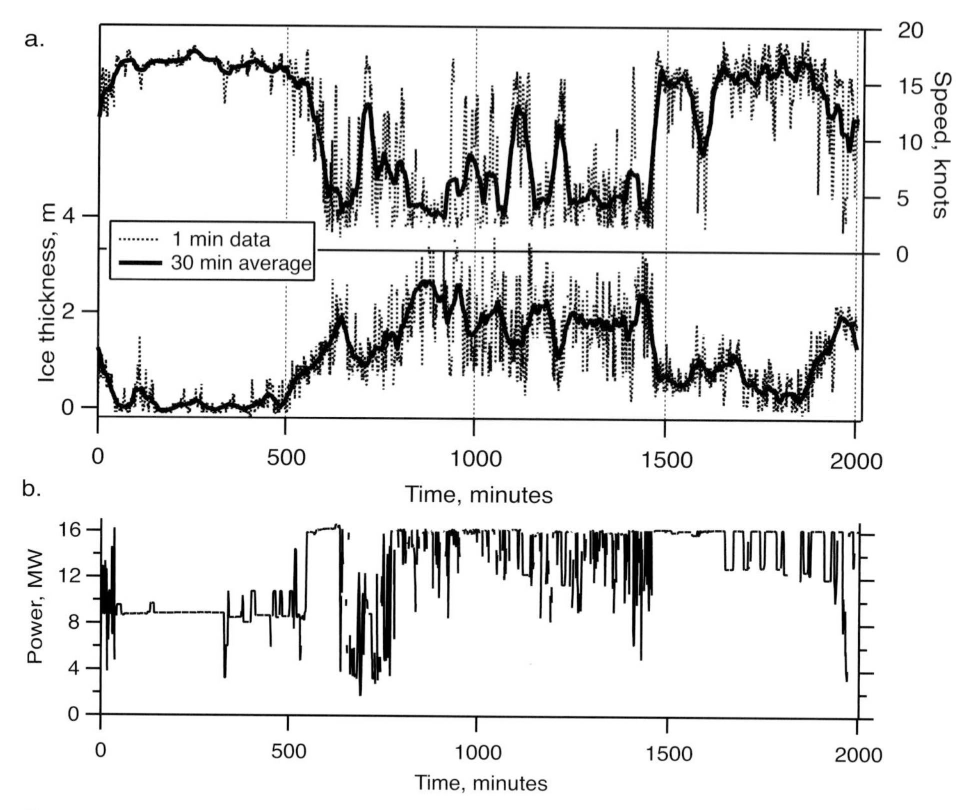


**Figure 3.2.32.** Ice thickness, ships speed (a) and delivered power (b) along a thirty-three hours long profile in the Kara Sea (Haas et al., 1999). Shown are 1-minute data as well as thirty-minute running averages.

instruments were suspended from the ship's bow crane at an elevation of 4 m above the ice surface. Figure 3.2.32 compares ice thicknesses thus derived over a thirty-three-hour period with ship's speed and delivered power (Haas et al. 1999). The data was obtained during the Arctic Demonstration and Exploratory Voyage (ARCDEV) of the Russian icebreaker *Kapitan Dranitsyn* in April 1998 in the first-year ice region of the Kara Sea. The icebreaker had accompanied an ice-capable oil tanker to load crude oil at a port on the Yamal Peninsula, and to deliver it to the European market. It can be seen that the ship speed decreased drastically as the ship entered more severe ice conditions, and was very low for the thickest ice. The relationship is visible in more detail in Figure 3.2.33, showing scatter plots of velocity and thickness for one and thirty minute averaging intervals. The correlation between the data is good ($r = 0.81$ and $r = 0.92$, respectively). However, the scatter for one minute data is very large. Note that with a speed of 10 knots, one minute of data corresponds to a profile length of approximately 300 m (i.e., two to three times the ship length). This scatter is significantly reduced for thirty minute averaging intervals.

There are various reasons for the strong scatter. Even despite the footprint of the EM method, the across-track ice area contributing to the measured ice thickness

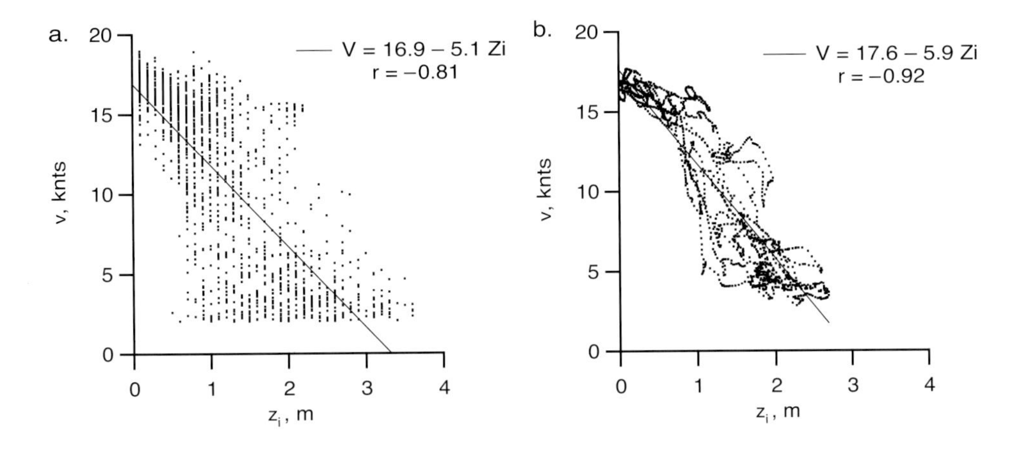


**Figure 3.2.33.** Mean speed V versus mean ice thickness Zi from Figure 3.2.32 for one minute intervals (a) and thirty minute running averages (b) Haas et al. 1999). The lines show a linear regression, and the resulting equations are given as well as the correlation coefficient r.

(~12 m) was less than the width of the ship (~25 m). In addition, ice conditions in front of the ship were extremely variable, and the momentum of the ship made it less sensitive to small-scale variations. However, over longer periods, these variations contribute to a more uniform relationship.

Apart from evaluating the performance and design of an icebreaker based on the results shown in Figure 3.2.33, the relationship can also be used for the development of ship performance models to predict navigation times and speeds under certain ice conditions.

Thickness and properties of the snow can also be determining factors for the performance of ships. Sometimes, one-third of the snow thickness is added as effective, snow-related ice thickness to the true ice thickness. Obviously, snow thicknesses cannot be obtained from the EM measurements, and have to be observed independently.

### 3.2.9.4 Ice Thickness Sounding in Support of Ice Load Measurements

Like icebreakers (Section 3.2.9.3), any offshore structure located in drifting pack ice is subject to variable ice forces and needs to be designed to withstand them. Ice forces on structures are determined by the width of the structure, by the properties and speed of the ice, and by ice thickness. Within the European Union projects Low-Level Ice Forces (LOLEIF) and Structures in Ice (STRICE), attempts were made to relate ice forces to environmental conditions including ice thickness. Therefore, a lighthouse in the drift ice zone of the northern Baltic Sea was equipped with load panels at water level, a ULS, and an EM system (Figure 3.2.34) (Haas and Jochmann 2003). The ULS was located at a depth of 6 m below the water level, and was repeatedly damaged by ridge keel scouring. The EM system comprised an

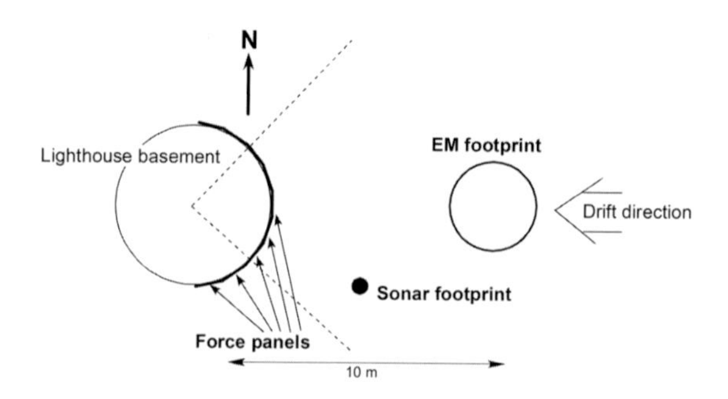


**Figure 3.2.34.** Sketch of an instrument setup to measure ice forces and their relation to ice thickness, showing the cross section of a lighthouse equipped with force panels as well as the locations of EM and ULS thickness measurements (Haas and Jochmann 2003).

EM31 conductivity meter (Section 3.2.3) and a sonic range finder. It was suspended 2 m above the ice by a scaffolding boom. Unfortunately, it was located some 5 m farther away from the lighthouse than the ULS. All data were recorded in real time and were visible on the lighthouse as the ice was passing by, shaking the lighthouse heavily as the ice crushed.

Figure 3.2.35 shows a time series of ice forces and associated ULS and EM ice thicknesses for a typical drift and deformation event. The forces show typical events of low-frequency, long-lasting static loads, with superimposed, up to five times stronger spontaneous loads due to crushing and/or bucking and bending failure. There is strong correlation between static ice forces and thickness in the example. However, peak forces due to crushing and bending failure blur a clear relationship, and can be very high in level ice and even higher than in ridges, as for example, around 07:00 h in the example. Figure 3.2.35 indicates that these peak forces resulted from virtually level zones of about twice the level ice thickness, which have been formed by rafting of the original ice cover and have not developed keels of very thick ice.

The measurement configuration also provided the unique opportunity to compare coincident ULS and EM measurements. As can be seen in Figure 3.2.35, ULS and EM data agree well with each other at level ice thicknesses of typically 0.2 to 0.3 m, and can very well distinguish between level and ridged ice. However, there is large disagreement of up to 60 percent in ridge zones. This is further illustrated in Figure 3.2.36, showing a scatter plot of EM versus ULS thickness from Figure 3.2.35. The correlation between both data sets is very good, however, the slope of the linear regression shows that the EM-derived ice thickness is only 0.4 times the ULS ice thickness. Again, this deviation is due to the footprint of the EM method, and due to the unconsolidated nature of the Baltic ridge keels, which were composed of broken, thin ice blocks. In addition, the sensitivity of EM measurements to thickness variations is limited as seawater in the Baltic is only brackish, leading to small conductivity contrasts between the ice and the water. A similar comparison should be performed under arctic conditions to draw final conclusions about the performance of EM measurements over ridges. In principle, the good correlation shown in Figure 3.2.36 and the linear equation provide a means to correct EM measurements over ridges and to make them comparable with ULS measurements.

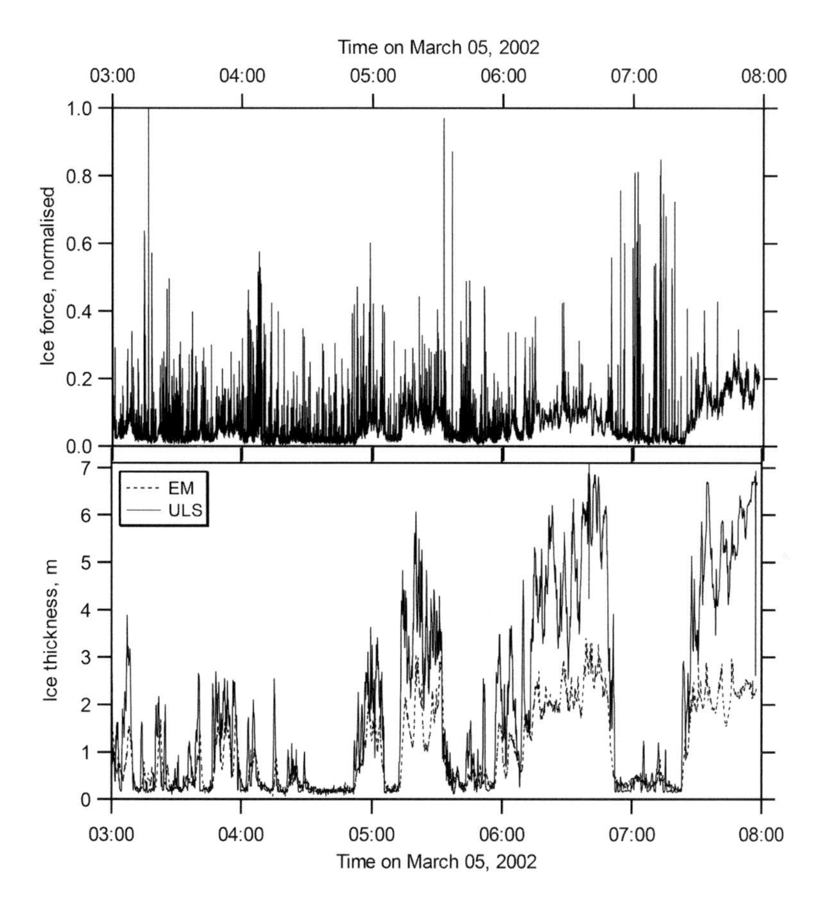


**Figure 3.2.35.** Typical five-hour time series of forces (top) and ULS and EM ice thicknesses (bottom) for a strong drift event on March 05, 2002 (Haas and Jochmann 2003).

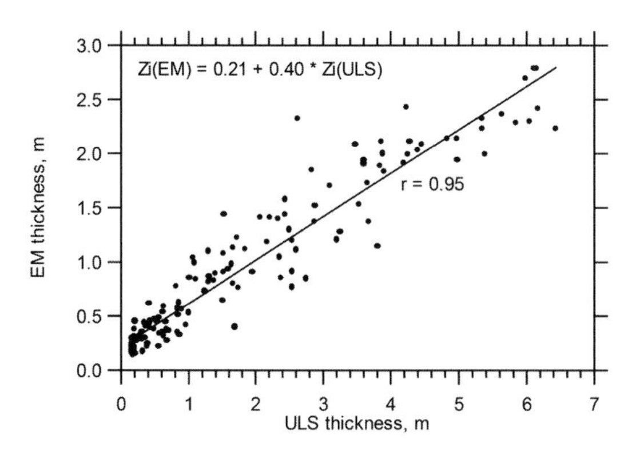


**Figure 3.2.36.** Comparison of coincident, three-minute average EM and ULS ice thickness measurements from Figure 3.2.35 (Haas and Jochmann 2003).

### 3.2.9.5 Monitoring of Ice Thinning and Detection of Weak Points

Understanding of the development of ice thinning during the ablation season is of particular importance for studies of the sea ice mass balance, and for the detection of weak points in the ice cover which could provide a hazard for on-ice travel and the deployment of equipment. Ice mass balance buoys (Section 3.2.6) provide this information with a high accuracy at one location, but many would be required to monitor the thinning along, for example, a Ski-Doo route across the ice. Similarly, ULS profiling could provide information on the thickness change of fast ice, or on the general decrease of mean ice thickness, with limited information on the particular characteristics of the ice, for example, if the melting was stronger over melt ponds or unponded ice. Section 3.2.9.2 has already discussed how extended EM profiles could be obtained to support human activities and various research on the ice. These profiles can of course be repeated throughout the summer season to follow the thinning at different locations along the transects. In addition, EM measurements are nondestructive, leaving hydraulic conditions unchanged once meltponds have developed. In addition, EM measurements are robust against seasonal changes of ice properties and can obtain reliable data even over melt ponds.

Figure 3.2.37 shows an example of such a measurement performed repeatedly along the same profile throughout the summer season (Eicken et al. 2001). The data were obtained during the Surface Heat Budget of the Arctic Ocean (SHEBA) drift station on mixed first- and multiyear ice in the Beaufort Sea. The figure shows the

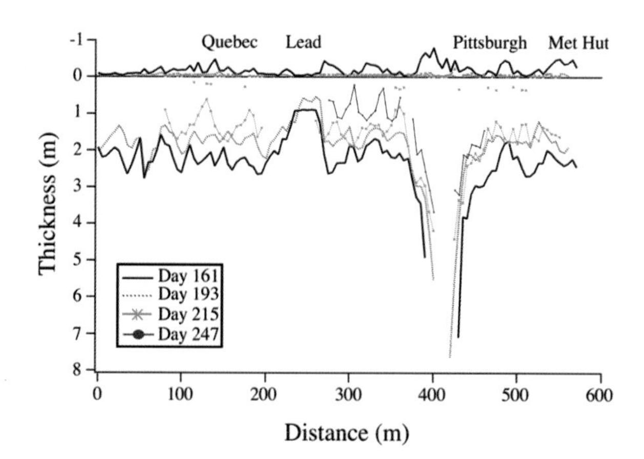


**Figure 3.2.37.** Repeat EM thickness measurements (5 m spacing) along the same profile during the course of one ablation season between Day 161 (June 10) and Day 247 (Sept. 04) (Eicken et al. 2001). Note that thickness axis points downwards so as to roughly match floe bottom topography. Snow (completely gone by day 193) or melt pond depths (individual points visible below zeroline) are plotted at the top: These are drawn to scale but with reversed sign compared to thickness data. Data missing along the profile later in the season are a result of floe disintegration. Reprinted from the Annals of Glaciology with permission of the International Glaciological Society.

results of four surveys between June and September, and shows the variable thinning at different locations with different ice characteristics. The strongest thinning was observed at the ridge flanks. In addition, the roughness of undeformed ice increases during the summer due to deepening of melt ponds and enhanced bottom melt.

Clearly, these data are invaluable for understanding the mechanisms of summer ice decay. However, measurements can also indicate where regions of enhanced melting or generally thinner ice occur along travel routes over the ice, which can for example also be caused by enhanced under-ice currents. Therefore, these measurements can also help to design better routes and to repeatedly assess their safety.

### 3.2.10 OUTLOOK

It seems plausible that the retreat of arctic sea ice observed over the last thirty years will continue into the future, thus permitting a dramatic increase in both arctic transportation systems and arctic resource development (Brigham 2007). Such a transformation, regardless of when it may take place, will require more information on ice thickness and morphology. Data gathering may be heavily driven by an immediate need for information on wide ranges of spatial and temporal scales, unprecedented in the historical data sets discussed in this chapter.

Considering, for example, the likely prospect of increased oil and gas activity in the Arctic (AMAP 2007), a dependence on spatially and temporally resolved ice thickness information for planning and monitoring will emerge. This may be needed for standard operations that may use ice as a platform or view it as an encroaching hazard to structures. From the perspective of oil-spill planning and response, ice surface roughness, as determined from laser altimetry for example, may be used as a proxy for under-ice topography, which is important in determining the fate of migrating or pooled oil under ice (Danielson and Weingartner 2007). Another potential growing need of up-to-date ice thickness information will be for arctic search and rescue efforts (e.g., establishing a safe on-ice runway for a C-130 Hercules). Arctic environmental protection efforts may also emerge as users of ice thickness information as environmental assessments and monitoring become critical in mitigating development impacts. Assessments of ice as a habitat for threatened species, such as the polar bear, may also benefit from available data sets.

While current techniques may become operational and more broadly used, such as shipborne EM sounding from both icebreakers and ice-strengthened vessels, novel approaches to more efficiently employ the geophysical principles for remotely determining ice thickness will be undoubtedly explored.

The small- and large-scale spatial and temporal variability of the ice, its relative thinness compared to water depth or the thickness of glacial ice, as well as its mobility require sophisticated methods for its observation. However, these characteristics also prevent the easy application of many classical methods and reduction of their uncertainties. While the accuracies of most methods may be sufficient for most applications, deployment of sensors is just another challenge as well. Future developments will mainly focus on improved operability, for example by increasing the observational ranges using unmanned aerial and underwater vehicles, or fixed-wing airplanes. Proven methods will be refined, as for example with radar altimetry, which will be operated in high-resolution synthetic aperture radar (SAR) mode of Cryosat (Wingham et al. 2004), to be launched in late 2009. This higher resolution will not only increase the accuracy of sea-surface height retrievals but also result in a higher fraction of ice-only waveforms for the observation of sea ice freeboard (see Section 3.2.8.4). In addition, it can be hoped that space agencies will improve the longevity of observing programs and missions. With this respect, ESA's Sentinel program and NASA's considerations of renewing the ICESat mission are promising efforts. Nevertheless, issues of spatial and temporal resolution have yet to be addressed. In addition, due to both their close physical interaction and their role in determining ice thicknesses from surface elevation or draft, respectively, improved snow thickness estimates are a major task for future developments. The application of any satellite algorithm should be accompanied by extensive ground-truth programs throughout mission lifetimes.

In considering continued needs for ice thickness information and acknowledging that many of the uses of this data have yet to be realized, it is important to accompany performed measurements with as much auxiliary and metadata as possible such that the potential for data use is maximized, especially when considering data that is to be released to the public. Common data protocols and formats should be established, and data storage should be secured in central data archiving centers. Regardless of future arctic climate scenarios, most data collection in the polar regions will continue to prove costly and challenging relative to other areas of the world.

## REFERENCES

AMAP (2007), *Arctic Oil and Gas 2007*, Arctic Monitoring and Assessment Programme, Oslo, Norway.

Barker, A., G. Timco, and B. Wright (2006), Traversing grounded rubble fields by foot—Implications for evacuation, *Cold Reg. Sci. Tech., 46,* 79–99.

Björk, G., C. Nohr, B. G. Gustafsson, and A. E. B. Lindberg (2008), Ice dynamics in the Bothnian Bay inferred from ADCP measurements, *Tellus A, 60*(1), 178–188, doi:10.1111/j.1600-0870.2007.00282.x.

Bourke, R. H., and R. P. Garrett (1987), Sea ice thickness distribution in the Arctic Ocean, *Cold Reg. Sci. Technol., 13,* 259–280.

Brigham, L. W. (2007), Thinking about the Arctic's future: Scenarios for 2040, *The Futurist*, September–October, 27–34.

Danielson, S. E., and T. J. Weingartner (2007), Estimates of Oil Spill Dispersion Extent in the Nearshore Alaska Beaufort Sea Based on In-situ

Oceanographic Measurements, *ADEC Division of Spill Prevention and Response*, Anchorage, Alaska.

Dierking, W. (1995), Laser profiling of the ice surface topography during the Winter Weddell Gyre Study 1992, *J. Geophys. Res., 100*(C3), 4807–4820.

Dowdeswell, J. A., J. Evans, R. Mugford, G. Griffiths, S. McPhail, N. Millard, P. Stevenson, M. A. Brandon, C. Banks, K. J. Heywood, M. R. Price, P. A. Dodd, A. Jenkins, K. W. Nicholls, D. Hayes, E. P. Abrahamsen, P. Tyler, B. Bett, D. Jones, P. Wadhams, J. P. Wilkinson, K. Stansfield, and S. Ackley (2008), Autonomous underwater vehicles (AUVs) and investigations of the ice-ocean interface: Deploying the Autosub AUV in Antarctic and Arctic waters. *J. Glac.*, 54(187), pp. 661–672.Eicken, H., A. L. Lovecraft, and M. L. Druckenmiller (2009), Sea Ice System Services: A Framework to Help Identify and Meet Information Needs Relevant for Arctic Observing Networks, 62(2), in press.

Eicken, H., W. B. Tucker, and D. K. Perovich (2001), Indirect measurements of the mass balance of summer Arctic sea ice with an electromagnetic induction technique, *Ann. Glaciol., 33,* 194–200.

Fetterer, F .M., M. R. Drinkwater, K.C. Jezek, S. W. C. Laxon, R. G. Onstott, and L. M. H. Ulander (1992), Sea ice altimetry, in *Microwave Remote Sensing of Sea Ice, Geophysical Monograph, 68*, pp. 111–135, American Geophysical Union, Washington, DC.

Forsberg, R., and H. Skourup (2005), Arctic Ocean gravity, geoid and sea ice freeboard heights from ICESat and GRACE, *Geophys. Res. Lett., 32,* L21502, doi:10.1029/2005GL023711.

Granberg, H. B., and M. Leppäranta (1999), Observations of sea ice ridging in the Weddell Sea, *J. Geophys. Res., 104*(C11), 25,735–25,745.

George, J. C., H. P. Huntington, K. Brewster, H. Eicken, D. W. Norton, and R. Glenn (2004), Observations on shorefast ice dynamics in Arctic Alaska and the responses of the Iñupiat hunting community, *Arctic, 57,* 363–374.

Haas, C. (1998), Evaluation of ship-based electromagnetic-inductive thickness measurements of summer sea ice in the Bellingshausen and Amundsen Seas, Antarctica, *Cold Regions Science and Technology, 27,* 1–16.

Haas, C. (2004), Airborne EM sea ice thickness profiling over brackish Baltic Sea water,, June 21–25, 2004, vol. 2, pp. 12–17, All-Russian Research Institute of Hydraulic Engineering (VNIIG), Saint Petersburg, Russia.

Haas, C., S. Gerland, H. Eicken, and H. Miller (1997), Comparison of sea ice thickness measurements under summer and winter conditions in the Arctic using a small electromagnetic induction device, *Geophysics, 62*(3), 749–757.

Haas, C., and P. Jochmann (2003), Continuous EM and ULS Thickness Profiling in Support of Ice Force Measurements, in *Proceedings of the 17th International Conference on Port and Ocean Engineering under Arctic Conditions, POAC '03*, Trondheim, Norway, vol. 2, edited by S. Loeset, B. Bonnemaire, and M. Bjerkas, pp. 849–856, Department of Civil and Transport Engineering, Norwegian University of Science and Technology NTNU, Trondheim, Norway.

Haas, C., A. Pfaffling, S. Hendricks, L. Rabenstein, J.-L. Etienne, and I. Rigor (2008a), Reduced ice thickness in Arctic Transpolar Drift favors rapid ice retreat, *Geophys. Res. Lett., 35,* L17501, doi:10.1029/2008GL034457.

Haas, C., J. Lobach, S. Hendricks, L. Rabenstein, and A. Pfaffling (2008b), Helicopter-borne measurements of sea ice thickness, using a small and lightweight, digital EM system, *J. Appl. Geophys.*, http://dx.doi.org/10.1016/j.jappgeo.2008.05.005.

Haas, C., M. Nicolaus, S. Willmes, A. Worby, and D. Flinspach (2008c), Sea ice and snow thickness and physical properties of an ice floe in the western Weddell Sea and their changes during spring warming, *Deep Sea Research II, 55*(8–9), 963–974, doi:10.1016/j.dsr2.2007.12.020.

Haas, C., K.-H. Rupp, and A. Uuskallio (1999), Comparison of along track EM ice thickness profiles with ship performance data, *POAC '99: Proceedings of the 15th International Conference on Port and Ocean Engineering under Arctic Conditions*, edited by J. Tuhkuri and K. Riska, vol. 1, pp. 343–353, Helsinki Univ. Techn., Ship Lab., Espoo, Finland.

Hibler, W. D. (1972), Removal of aircraft altitude variation from laser profiles of the Arctic ice pack, *J. Geophys. Res., 77*(36), 7190–7195.

Hibler, W. D. (1975), Characterization of cold-regions terrain using airborne laser profilometry, *J. Glaciol., 15*(73), 329–347.

Holt, B., P. Kanagaratnam, S. P. Gogineni, V. C. Ramasami, A. Mahoney, and V. Lytle (2008), Sea ice thickness measurements by ultrawideband penetrating radar: First results, *Cold Reg. Sci. Technol.,* doi:10.1016/ j.coldregions.2008.04.007.

Hvidegaard, S. M., and R. Forsberg (2002), Sea ice thickness from airborne laser altimetry over the Arctic Ocean north of Greenland, *Geophys. Res. Lett., 29*(20), 1952, doi:10.1029/2001GL014474.

Hvidegaard, S. M., R. Forsberg, and H. Skourup (2006), Sea ice thickness estimates from airborne laser scanning, in Arctic sea ice thickness: Past, present and future, in *Climate Change and Natural Hazard Series*, vol. 10, edited by P. Wadhams and G. Amanatidis, pp. 193–206, European Commission, Brussels.

Kanagaratnam, P., T. Markus, V. Lytle, B. Heavey, P. Jansen, G. Prescott, and S. P. Gogineni (2007), Ultrawideband radar measurements of thickness of snow over sea ice, *Trans. Geosci. Remote Sens., 45*(9), 2715–2724.

Kovacs, A., J. S. Holladay, and C. J. J. Bergeron (1995), The footprint/altitude ratio for helicopter electromagnetic sounding of sea ice thickness: comparison of theoretical and field estimates, *Geophysics, 60*, 374–380.

Kovacs, A., and R. M. Morey (1986), Electromagnetic measurements of multi-year sea ice using impulse radar, *Cold Reg. Sci. Tech., 12,* 67–93.

Kovacs, A., and R. M. Morey (1991), Sounding sea ice thickness using a portable electromagnetic induction instrument, *Geophysics, 56,* 1992–1998.

Kovacs, A., N. C. Valleau, and J. S. Holladay (1987a), Airborne electromagnetic sounding of sea ice thickness and subice bathymetry, *Cold Regions Sci. and Tech., 14,* 289–311.

Kovacs, A., R. M. Morey, G. F. N. Cox, and N. C. Valleau (1987b), Electromagnetic property trends in sea ice, Part 1, *CRREL Report*, vol. 87–6, U.S. Army Cold Regions and Engineering Laboratory, Hanover, NH.

Kwok, R., G. F. Cunningham, H. J. Zwally, and D. Yi (2006), ICESat over Arctic sea ice: Interpretation of altimetric and reflectivity profiles, *J. Geophys. Res., 111,* C06006, doi:10.1029/2005JC003175.

Kwok, R., and G. F. Cunningham (2008), ICESat over Arctic sea ice: Estimation of snow depth and ice thickness, *J. Geophys. Res., 113,* C08010, doi:10.1029/2008JC004753.

Kwok R., G. F. Cunningham, H. J. Zwally, and D. Yi (2007), Ice, Cloud, and land Elevation Satellite (ICESat) over Arctic sea ice: Retrieval of freeboard, *J. Geophys. Res., 112,* C12013, doi:10.1029/2006JC003978.

Lange, M. A., and H. Eicken (1991), The sea ice thickness distribution in the northwestern Weddell Sea, *J. Geophys. Res., 96*(C3), 4821–4837.

Laxon, S., N. Peacock, and D. Smith (2003), High interannual variability of sea ice thickness in the Arctic region, *Nature, 425,* 947–950, doi:10.1038/nature02050.

Liu, G., and A. Becker (1990), Two-dimensional mapping of sea ice keels with airborne electromagnetics, *Geophysics, 55,* 239–248.

Mahoney, A., H. Eicken, and L. Shapiro (2007), How fast is landfast ice? A study of the attachment and detachment of nearshore ice at Barrow, Alaska, *Cold Reg. Sci. Tech., 47,* 233–255.

Mahoney, A., and S. Gearheard (2008), Handbook for community-based sea ice monitoring, *NSIDC Special Report*, vol. 14, Boulder, CO, U.S. National Snow and Ice Data Center, accessed electronically at http://nsidc.org/pubs/special/nsidc_special_report_14.pdf.

McNeill, J. D. (1980), Electromagnetic terrain conductivity measurements at low inductions numbers, *Technical Note TN-6*, Geonics Limited, Mississauga, Ontario, accessed electronically, http://www.geonics.com/pdfs/technicalnotes/ tn6.pdf.

Melling, H., P. H. Johnston, and D. A. Riedel (1995), Measurement of the topography of sea ice by moored subsea sonar, *J. Atmosph. Oceanic Techn., 12*(3), 589–602.

Melling, H. (1998), Sound scattering from sea ice: Aspects relevant to ice-draft profiling by sonar, *J. Atmosph. Oceanic Technol., 15,* 1023–1034.

Melling, H. (2002), Sea ice of the northern Canadian Arctic Archipelago, *J. Geophys. Res., 107*(C11), 3181, doi:10.1029/2001JC001102, 2002.

Melling, H., D. A. Riedel, and Z. Gedalof (2005), Trends in the draft and extent of seasonal pack ice, Canadian Beaufort Sea, *Geophys. Res. Lett., 32*, L24501, doi:10.1029/2005GL024483.

Multala, J., H. Hautaniemi, M. Oksama, M. Leppäranta, J. Haapala, A. Herlevi, K. Riska, and M. Lensu (1996), An airborne electromagnetic system on a fixed wing aircraft for sea ice thickness mapping, *Cold Reg. Sci. Techn., 16*(24), 355–373.

Mock, S. J., A. D. Hartwell, and W. D. Hibler (1972), Spatial aspects of pressure ridge statistics, *J. Geoph. Res., 77*(30), 5945–5953.

Otto, D. (2004), Validation of ground-penetrating radar measurements of first- and multiyear sea ice and snow thickness in the Arctic and Antarctic, Diploma thesis (in German), Alfred Wegener Institute of Polar and Marine Research, Bremerhaven, Germany, and Geophysical Institute, Technical University Clausthal, Germany.

Peacock, N. R., and S. W. Laxon (2004), Sea surface height determination in the Arctic Ocean from ERS altimetry, *J. Geophys. Res., 109*, C07001, doi:10.1029/2001JC001026.

Perovich, D. K., T. C. Grenfell, J. A. Richter-Menge, B. Light, W. B. Tucker III, and H. Eicken (2003), Thin and thinner: Sea ice mass balance measurements during SHEBA, *J. Geophys. Res., 108*(C3), 8050, doi:10.1029/2001JC001079, 2003.

Pfaffling, A., C. Haas, and J. E. Reid (2007), A direct helicopter EM sea ice thickness inversion, assessed with synthetic and field data, *Geophysics, 72*, F127–F137.

Potter, R. E., and J. T. Walden (1981), Design and Construction of Sea Ice Roads in the Alaskan Beaufort Sea. *Proceedings of the 13th Annual Offshore Technology Conference*, Houston, Texas, pp. 135–140.

Prinsenberg, S. J., and J. S. Holladay (1993), Using air-borne electromagnetic ice thickness sensor to validate remotely sensed marginal ice zone properties, *Port and Ocean Engineering under Arctic Conditions (POAC 93)*, vol. 2, pp. 936–948, edited by Hamburger Schiffbau Versuchsanstalt (HSVA), Hamburg, Germany.

Prinsenberg, S. J., J. S. Holladay, and J. Lee (2002), Measuring Ice Thickness with EISFlowTM, a Fixed-Mounted Helicopter Electromagnetic Laser System, in *Proceedings of the Twelfth (2002) International Offshore and Polar Engineering Conference May 26–May 31, 2002*, vol. 1, pp. 737–740, Kitakyushu, Japan.

Reid, J. E., A. Pfaffling, and J. Vrbancich (2006), Airborne electromagnetic footprints in 1D earths, *Geophysics, 71*, G63–G72.

Richter-Menge, J. A., D. K. Perovich, B. C. Elder, K. Claffey, I. Rigor, and M. Ortmeyer, (2006), Ice mass balance buoys: A tool for measuring and attributing changes in the thickness of the Arctic sea ice cover, *Ann. Glaciol., 44*, 205–210.

Rossiter, J. R., and J. S. Holladay (1994), Ice-thickness measurement, in *Remote Sensing of Sea Ice and Icebergs,* edited by S. Haykin et al., pp. 141–176, John Wiley, Hoboken, NJ.

Rothrock, D. A., D. B. Percival, and M. Wensnahan (2008), The decline in arctic sea ice thickness: Separating the spatial, annual, and interannual variability in a quarter century of submarine data, *J. Geophys. Res., 113*, C05003, doi:10.1029/2007JC004252.

Rothrock, D. A. and M. Wensnahan (2007), The accuracy of sea ice drafts measured from U.S. Navy submarines, *J. Atmos. Oceanic Technol.,* doi:10.1175/JTECH2097.1.

Rothrock, D. A., Y. Yu, and G. A. Maykut (1999), Thinning of the Arctic sea ice cover, *Geophys. Res. Lett., 26*(23), 3469–3472.

Shcherbina, A. Y., D. L. Rudnick, and L. D. Talley (2005), Ice-draft profiling from bottom-mounted ADCP data, *J. Atmosph. Oceanic Techn., 22,* 1249–1266.

Skourup, H., and R. Forsberg (2006), Sea ice freeboard from ICESat: A Comparison with Airborne Lidar Measurements, in *Arctic Sea Ice Thickness: Past, Present, and Future,* Climate Change and Natural Hazard Series, vol. 10, edited by P. Wadhams and G. Amanatidis, pp. 82–92, European Commission, Brussels.

Sodhi, D. S., D. B. Griggs, and W. B. Tucker (2001), Ice Performance Tests of the USCGC Healy, *POAC '01: Proceedings of the 16th International Conference on Port and Ocean Engineering under Arctic Conditions,* vol. 2, edited by R. Frederking, I. Kubat, and G. Timco, pp. 893–907, Canadian Hydraulics Center, National Research Council, Ottawa, Ontario.

Stogryn, A., and G. J. Desargant (1985), The Dielectric Properties of Brine in Sea Ice at Microwave Frequencies, *IEEE Trans. on Antennas and Propagation, AP-33,* 523–532.

Strass, V. H., and E. Fahrbach (1998), Temporal and Regional Variation of Sea Ice Draft and Coverage in the Weddell Sea Obtained from Upward Looking Sonars, in *Antarctic Sea Ice: Physical Processes, Interactions, and Variability,* Antarctic Research Series, vol. 74, edited by M. O. Jeffries, pp. 123–139, American Geophysical Union, Washington, DC.

Sun, B., J. Wen, M. He, J. Kang, Y. Luo, and Y. Li (2003), Sea ice thickness measurement and its underside morphology analysis using radar penetration in the Arctic Ocean, *Sci. China, 46*(11), 1151–1160.

Thorndike, A. S., D. A. Rothrock, G. A. Maykut, and R. Colony (1975), The thickness distribution of sea ice, *J. Geophys. Res., 80*(33), 4501–4513.

Vinje, T., N. Nordlund, and Å. Kvambekk (1998), Monitoring ice thickness in Fram Strait, *J. Geophys. Res., 103*(C5), 10,437–10,449.

Wadhams, P., and N. R. Davis (2000), Further evidence of ice thinning in the Arctic Ocean, *Geophys. Res. Lett., 27*(24), 3973–3975.

Wadhams, P., M. A. Lange, and S. F. Ackley (1987), The ice thickness distribution across the Atlantic sector of the Antarctic Ocean in midwinter, *J. Geophys. Res., 92*(C13), 14,535–14,552.

Wadhams, P., W. B. Tucker III, W. B. Krabill, R. N. Swift, J. C. Comiso, and N. R. Davis (1992), Relationship between sea ice freeboard and draft in the Arctic

Basin, and implications for ice thickness monitoring, *J. Geophys. Res., 97*(C12), 20,325–20,334.

Wadhams, P., J. P. Wilkinson, and S. D. McPhail (2006), A new view of the underside of Arctic sea ice, *Geophys. Res. Lett., 33*, L04501, doi:10.1029/2005GL025131.

Warren, S. G., I. G. Rigor, N. Untersteiner, V. F. Radionov, N. N. Bryazgin, Y. I. Aleksandrov, and R. Colony (1999), Snow depth on Arctic sea ice, *J. Clim., 12,* 1814–1829, doi:10.1175/1520-0442(1999)012<1814:SDOASI > 2.0.CO;2.

Wingham, D. J., L. Phalippou, C. Mavrocordatos, and D. Wallis (2004), The mean echo and echo cross product from a beamforming interferometric altimeter and their application to elevation measurement, *IEEE Trans. Geosci. Rem. Sens., 42*(10), 2305–2323, doi:10.1109/TGRS.2004.834352.

Worby, A. P., C. A. Geiger, M. J. Paget, M. L. Van Woert, S. F. Ackley, and T. L. DeLiberty (2008), Thickness distribution of Antarctic sea ice, *J. Geophys. Res., 113*, C05S92, doi:10.1029/2007JC004254.

Worby, A. P., M. O. Jeffries, W. F. Weeks, K. Morris, and R. Jaña (1996), The thickness distribution of sea ice and snow cover during late winter in the Bellingshausen and Amundsen Seas, Antarctica, *J. Geophys. Res., 101*(C12), 28,441–28,455.

Zwally, H. J., B. Schutz, W. Abdalati, J. Abshire, C. Bentley, A. Brenner, J. Bufton, J. Dezio, D. Hancock, D. Harding, T. Herring, B. Minster, K. Quinn, S. Palm, J. Spinhirne, and R. Thomas (2002), ICESat's laser measurements of polar ice, atmosphere, ocean, and land, *J. Geodyn., 34*, 405–445, doi:10.1016/S0264-3707(02)00042-X.

# Chapter 3.3

# Ice Sampling and Basic Sea Ice Core Analysis

*Hajo Eicken*

## 3.3.1 INTRODUCTION

A deeper understanding of the role of sea ice in the environment and an assessment of the types of services it provides often requires extraction and analysis of ice samples. For example, the importance of in situ testing of mechanical strength, outlined in Chapter 3.5, is enhanced through integration of laboratory measurements of ice-sample compressive strength, ice porosity, and ice microstructure. The same holds true for measurements of ice optical properties, which are governed not so much by the characteristics of the ice itself but rather by the distribution of air, brine, and other impurities throughout the ice matrix. Along the same lines, many studies of sea ice as an important habitat to a broad range of microorganisms require thorough knowledge of the inhabitable pore space, the amount of salt present in the ice, and the temperature at which such life can thrive (Chapters 3.8 and 3.9).

Probably the oldest type of sea ice sampling and property measurement is the use of an ice tester, for example by Iñupiat hunters in coastal Alaska who refer to it as an *Unaaq* (Nelson 1969). It is a long wooden staff, traditionally with an ivory tip, though now mostly equipped with a piece of iron rebar protruding from its base, that is used to tap or jab at the surface of thin, potentially treacherous ice. Apart from determining how many jabs are necessary to penetrate the ice, which in turn is a measure of its strength, experienced hunters can tell a lot about the state of the ice just from the sound of the metal or ivory tapping on the ice. Such ice testers are still in use by indigenous arctic hunters and illustrate the intertwining of sea ice use and sea ice measurements.

One of the first systematic ice sampling and sea ice property measurement campaigns was carried out by Finn Malmgren between 1922 and 1925 during Amundsen's Norwegian North Polar Expedition (Malmgren 1927). Malmgren obtained fundamental data on the density, chemical composition, and thermophysical properties of sea ice that laid the foundation for much of the sea ice research to follow. Russian field studies of sea ice in Siberian waters were driven by shipping and national security interests in the Soviet Arctic and provided a wealth of data (Zubov 1945; Doronin and Kheisin 1977). In the second half of the twentieth century, icebreaker expeditions with mobile cold labs or means to bring back frozen samples greatly increased the sophistication and reach of field studies of sea ice (Weeks and Ackley 1986). Furthermore, increasing interdisciplinarity of sea ice research, exemplified by the book compiled by Thomas and Dieckmann (2003), has resulted in greater demands on sampling and measurement approaches.

This chapter is a brief introduction to basic sea ice sampling and field analysis techniques. It may serve as a starting point to aid with the design of a field sampling program, assembly of field equipment, and planning of basic ice-core analysis techniques that are prerequisite to other studies, such as thermophysical and mechanical ice properties or studies of sea ice ecology described in more detail in subsequent chapters. Emphasis is placed on (1) field studies of ice stratigraphy, which can tell us a lot about the growth history of the ice, potential heterogeneities affecting scatter between parallel samples, and the pore and grain microstructure which controls a range of other ice properties (Eicken 2003); (2) measurements of ice temperature, density, and bulk salinity as important variables describing the state of the sea ice cover; (3) derivation of brine and air volume fractions from these measurements; and (4) discussion of two case studies to illustrate the practical application of such measurements. Subsequent sections of the book, in particular Chapters 3.4, 3.8, 3.9, and 3.17, will expand on these topics and demonstrate how such basic measurements can serve as a foundation to more comprehensive analyses of ice properties and processes. The multimedia DVD that accompanies this book contains footage that provides practical illustrations of (1) sampling site selection and basic ice-core sampling; (2) temperature, density, and salinity measurements; and (3) ice stratigraphic analysis on the entire core, rough thick sections cut on site in the field, and research-grade thin sections produced in a cold laboratory.

## 3.3.2 OVERVIEW OF SAMPLING APPROACHES

### 3.3.2.1 Sampling Site Selection

The criteria that guide selection of a sampling site depend on whether one wishes to sample ice that is representative of the predominant or some specific ice type, or is looking for particular features that are being studied, such as sediment inclusions, surface flooding, or particular deformation features. In selecting a floe or subregion of ice that meets these requirements remote-sensing data, and in particular ship-based or airborne observations, can provide important guidance (Chapters 3.12, 3.15; Section 3.3.4 below). Most process studies or regional assessments focus on the sampling of level sea ice, since it provides the building block of ridges and controls much of the large-scale ice-pack behavior. Model studies and field observations indicate that level ice comprises between two-thirds and three-quarters of the total ice volume and typically more than nine-tenths of the ice surface area in the Arctic (Rothrock and Zhang 2005) and a somewhat smaller proportion in the Antarctic (Worby et al. 2008). While level ice may contain significant amounts of deformed ice in the Antarctic due to the prevailing dynamic ice growth processes, in the Arctic it is mostly composed of undisturbed congelation ice that accretes at the bottom of a floe (Weeks and Ackley 1986; Eicken 2003). However, only ice stratigraphic studies can establish this for certain, as illustrated by the example shown in Chapter 3.12, Figure 3.12.2.

In addition to visual or remote-sensing data, it may be prudent to use thickness drilling or electromagnetic (EM) induction techniques to identify stretches of level ice that may be difficult to detect because of variable snow cover or concealed bottom roughness (Chapter 3.2; Section 3.3.4 below). Once a target area has been identified, removing any snow cover over a larger area can yield additional insight into the integrity and character of the underlying ice (see Figure 3.1.5 in Chapter 3.1, and the video sequence on ice coring on the accompanying DVD). For example, snow cover can conceal cracks, sediments entrained into the ice, roughness elements indicative of deformation processes in the ice growth history, flooding or frozen melt ponds—all of which may be undesirable or the target of a sampling campaign, depending on the research goals. Once a suitable sampling area has been identified and snow has been removed after measuring its depth, the equipment should be set up so as to protect samples from the sun and blowing snow and keep the sampling site uncluttered. Often, a larger number of samples will be taken at one site and it is important to leave enough room and plan the sampling sequence so as not to contaminate the sampling site with seawater or brine splashing onto the surface and percolating into the ice that may be sampled later in the day.

In most cases, samples will be obtained through coring, typically such that a series of cores is taken for different types of analyses: stratigraphy (the distribution of ice crystals and pores containing brine and air and their vertical sequence throughout the ice cover), temperature (requiring holes to be drilled into the core), salinity and other measurements made on melted core samples, etc. A key consideration is the design of the sampling program, in particular with respect to spatial variability of ice properties and the typical aim to obtain a representative sample. To assess the number of samples necessary to test hypotheses related to, for example, seasonal changes in key ice properties, statistical analyses such as those outlined in Chapter 3.9 based on published data for the mean and variances of the measured variable are required. However, at the same time it is also instructive to relate the sampled volume to the heterogeneities present in the ice. Here, we will consider (1) the brine layers and brine pockets permeating the ice crystal matrix (< 1 mm in width and typically few millimeters to at most few centimeters in vertical extent); 2) the ice crystals themselves (in congelation ice at most few centimeters in horizontal cross-section and several centimeters to a few decimeters in vertical extent);

(3) brine channels (mm- to cm-diameter pore networks extending over several centimeters to decimeters in the vertical; Figure 3.3.2); (4) aggregations of different ice blocks due to deformation processes and accumulation of free-floating frazil ice crystals underneath the ice cover (such as the dm-sized blocks frozen into the ice in the cross-section shown in Figure 3.12.2 in Chapter 3.12); and (5) larger-scale variations due to variable snow cover and different ice growth processes.

It is typically impractical to analyze an ice floe in its entirety. The characteristic sample dimension is mostly much smaller and on the order of 10 cm or less. Most ice corers take samples between 7 and 10 cm in diameter and roughly 1 m in length, requiring multiple passes using core-barrel extensions to retrieve cores from thicker ice (Figure 3.3.1). The vertical extent of subsamples taken from such cores also mostly ranges between 5 and 20 cm. Hence, typical sample sizes exceed the size of the smallest pores in sea ice by several orders of magnitude. Typically, ice cores also manage to sample a sufficient number of ice crystal cross-sections for a representative subsample. However, brine channels, spaced on the order of a decimeter or more apart (Figure 3.3.2) and associated with higher salinities and potentially drastically different biogeochemical properties, are a significant source of scatter between individual samples (Tucker et al. 1984; Eicken et al. 1991). Depending on the size of the core sample and the density of the channels, it may

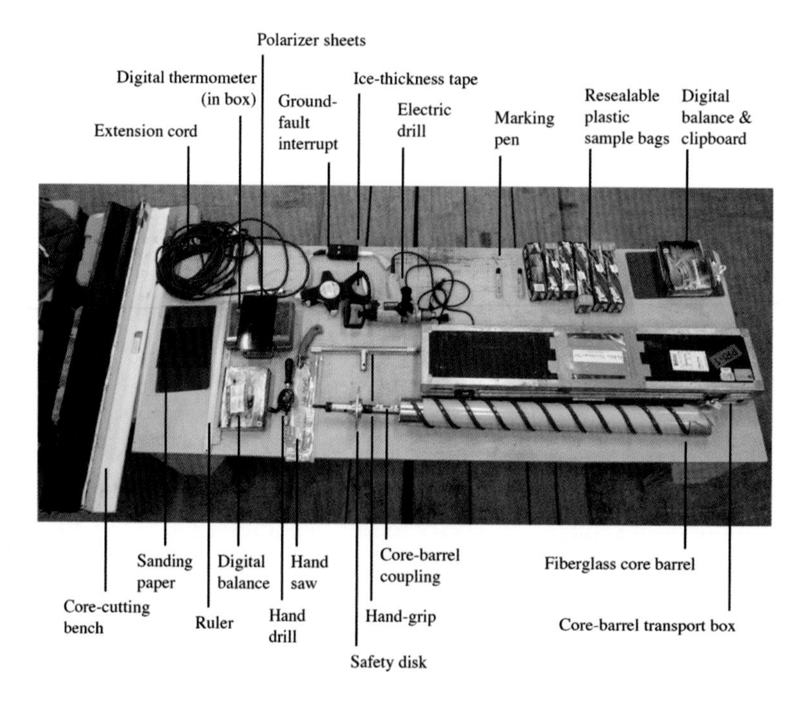


**Figure 3.3.1.** Overview of equipment needed for ice coring and on-site measurements and processing. Shovel to remove snow and generator to power electric drill are not shown.

**Figure 3.3.2.** Vertical sea ice slab cut from landfast ice in the Chukchi Sea near Barrow, Alaska, in April of 1999 with a track-mounted chainsaw (2.1 m long cutting bar). Grid marks on slab are spaced 10 cm apart. Note the brine channel extending from 25 to 80 cm depth, the frazil layer at the top and the horizontal banding (Cole et al. 2004).

require a half-dozen or even a dozen cores to accumulate a large enough sample for investigations of larger-scale or seasonal differences in salinity at the level of a few tenths of a permille (‰) or practical salinity unit (psu). However, fluid dynamics modeling by Petrich et al. (Chapter 3.16) suggests that, averaged over larger footprints than typically achieved with a standard corer, this spatial variability may be substantially smaller.

In particular in the Antarctic, with level ice composed of aggregations of individual ice pancakes (Chapter 3.12), spatial heterogeneity can be considerable at the scale of a meter or more (Eicken et al. 1991), in great contrast with more homogeneous congelation ice found, for example, in some locations in arctic landfast ice (Eicken et al. 2005). The photograph of a composite ice floe in Chapter 3.12 (Figure 3.12.2) illustrates how such heterogeneities are due to the juxtaposition of ice of different origin and growth history. Finally, (pseudo) random variations due to differences in snow depth (Figures 3.1.18; Section 3.1.20 in Chapter 3.1) or local growth-rate variations can drive differences in ice properties at the meter to hundred-meter scale.

Depending on the scientific question or hypothesis, a simple approach of taking a small number of cores (≤3) may be sufficient, or a more sophisticated sampling plan may be required. For the latter, results from the Antarctic (Eicken et al. 1991) suggest that a stratified random sampling approach, which results in acquisition of random sets of subsamples over a range of scales and floe segments, may be most appropriate. Textbooks such as Haining (2003) can provide further guidance on sampling strategies. In arctic multiyear ice, which has been subjected to substantial surface and bottom melt, variability in ice properties can be larger and may require sampling strategies that target a specific type of level ice (such as ice grown from a pond). The case study discussed in Section 3.3.4 below sheds more light on this problem.

Sampling of thick multiyear ice and ridged ice, due to the complicated nature of ridges, is much more challenging, and to date few studies have examined ridges in detail. In addition to the classic work discussed by Weeks and Ackley (1986), Høyland (2002) has recently made progress in examining the structure and properties of ridges. However, much remains to be done to make ridges more accessible to biogeochemical and ecological studies.

### 3.3.2.2 Ice Coring

By far the most common approach to obtaining ice samples across the entire depth of an ice floe is to employ an ice corer (ice-coring auger). Most ice corers used in the field today have their design origins in the ice-coring augers developed at the U.S. Army Corps of Engineers Cold Regions Research and Engineering Laboratory (CRREL) and its precursor, the Snow, Ice, and Permafrost Research Establishment (SIPRE). Detailed design plans and very useful ancillary information can be found in a report by Rand and Mellor (1985). Several glaciological research centers have built their own version of the classic CRREL (or Rand) design. Further refinements to the design have been implemented in a version of the coring system that is shown in Figure 3.3.1 with a plastic core barrel retrieving 9 cm diameter cores and stainless-steel fittings (manufactured and distributed by Kovacs Enterprises in Lebanon, New Hampshire, USA). Use of the corer both by hand using the T-shaped handle or with an electric drill that is connected to the drive head (core-barrel coupling, Figure 3.3.1) is demonstrated on the accompanying DVD.

The advantage of using only the T-handle is that it requires no additional gear beyond the corer itself. Typically, an ice core of 1 m length can be drilled in less than fifteen minutes by hand by experienced personnel. Using either a two-stroke motor (such as those used to drive ice augers to drill holes for ice fishing) or an electric drill powered by a generator can cut down coring time to a few minutes for ice less than 1 m thick. An electric drill is less noisy and reduces potential contamination of the samples by two-stroke motor exhaust. Moreover, in cold conditions electric generators tend to do better in starting up and can be left running during the entire sampling campaign away from the sampling site. In order to avoid the corer slipping through the hole if the connector inadvertently slips out of the drill chuck, use of a so-called safety disk, which free spins during drilling so as not to present a hazard, is advised (Figure 3.3.1).

Running equipment powered by 110 or 220 V alternating current and operating coring and other ice sampling equipment requires both safety precautions common to any laboratory or machine-shop workspace with spinning, hazardous gear (protective gear, proper training, etc.) and furthermore addressing the potential risk of electric shock when working close to highly conducting seawater. For the latter, operation of a generator should always include use of a ground-fault interrupt (GFI; Figure 3.3.1) to minimize electrocution hazards.

Drilling ice cores is straightforward but success improves greatly with practice. Experienced personnel can drill down to around 10 m or so with the type of coring equipment shown in Figure 3.3.1. and in the videos on the DVD. While the DVD provides a broader overview, a few important points to keep in mind when drilling cores include the following:

Starting the corer at perfect right angles to the (level) surface is important both for drilling success and obtaining a good sample; having one or two persons ascertain that the barrel is straight during the initial 10 to 20 cm of drilling is helpful.

Forcing the corer down by leaning on it with excessive force should not be necessary if the core cutters are sharp and well maintained and there is no blockage of the hole.

In order to avoid jam-packing the hole with cuttings, make sure that the cuttings are transported up along the outside of the rotating barrel unhindered and remove the buildup at the top of the hole at regular intervals; in particular in deeper holes, when using extension rods to drill at depths of several meters, it is important to not drill until cuttings build up on the top of the core barrel as this can get the corer stuck in the hole; to avoid buildup, it may be necessary to lift the entire barrel until it comes up out of the hole and sheds excessive cuttings.

When handling the corer while still in the hole (e.g., attaching additional extensions, retrieving the core, etc.), make sure that somebody is holding onto the corer or the connected extensions; corers can vanish down a hole awfully fast.

Once the ice cover has been penetrated and the coring hole fills with water, it is important to very quickly remove the corer from the hole to avoid equipment freezing in, which is an issue both in very cold conditions and in summer when surface freshwater may come into contact with underlying seawater and result in rapid ice formation.

As with all equipment used on the sea ice that routinely comes into contact with highly saline brine and possibly other substances such as entrained sediment particles, thorough rinsing after each use, protection of parts from corrosion, and maintenance of sharp cutter blades is paramount to successful core retrieval.

Operating a corer in cold conditions does prove challenging at times and may require a substantial amount of gear. However, there is a remarkably elegant and—with some practice—highly effective way to extract ice cores from sea ice up to 2 m or so thick, using Russian-designed annular ice augers as described by Cherepanov (1970). These consist of a simple metal ring that has a single cutter inserted into an opening. An extension rod with a hand crank (turning brace) is mounted on the side of the horizontal ring. By using the hand crank in a way that has the cutting ring rotate on the spot, one can cut a circular slot around a cylinder of ice protruding through the cutting ring, removing the cuttings by simply lifting the ring out of the hole at regular intervals. A schematic of this device is also shown in Rand and Mellor (1985). It is easily fashioned and lightweight.

### 3.3.2.3 Cutting Ice Slabs

For some applications, it may be necessary to obtain ice samples over a wider cross-section of ice than that provided by a corer. For example, studies of ice strength and crack propagation have necessitated the extraction of larger slabs of sea ice from the ice cover (Figure 3.3.2; Cole and Shapiro 1998). For these purposes, use of a handheld chain saw with a long bar (up to 1 m length can be handled by a properly trained operator) can be used to extract larger blocks or slabs from the ice cover (Figure 3.3.3). In ice thicker than a meter, the block can be

**Figure 3.3.3.** Extraction of an ice block cut with a handheld chainsaw (at left) from landfast ice near Barrow, Alaska, in early June 2004. Removing blocks of this size from the ice requires ice tongs and a comealong mounted on a hoisting frame.

broken free from its base through the use of wedges driven down into the slot cut by the chain saw. Shapiro has pioneered the use of track-driven chain saws that are equipped with a bar of over 2 m length. These saws can cut through the entire thickness of most level first-year ice and yield slabs (such as the one shown in Figure 3.3.2) that are only a few centimeters thick, revealing the distribution of brine channels and ice stratigraphy in great detail.

## 3.3.3 ICE-CORE ANALYSIS

### 3.3.3.1 Stratigraphic Analysis, Thick and Thin Sections

Many sea ice properties depend on the growth history and pore or grain microstructure of the sea ice (Weeks and Ackley 1986; Eicken 2003). Typically, subsampling of ice cores is also guided at least to some extent by these factors (see Chapters 3.8 and 3.9). Hence, analysis and documentation of a core's stratigraphy immediately after its extraction can be of significant value. Typically, this is best done by placing the core on a dark background (such as the ice-core cutting bench shown in Figures 3.3.1 and 3.3.4a) that accentuates brine and air inclusions and their vertical sequence (layering, brine channels, etc.). Studies of sediment inclusions or ice algae may require photographs of the core against a white background as well (such as shown in Figure 3.8.2 in Chapter 3.8). Typical stratigraphic sequences as related to growth history for sea ice are shown in Figure 3.3.4b, with more detail found in summaries by Weeks and Ackley (1986) and Eicken (2003). The core photograph shown in Figure 3.3.4a shows the transition between granular ice of frazil origin at the top and columnar congelation ice at the bottom. Smaller grain sizes and higher number densities of brine and air inclusions in granular ice typically impart such ice a milky, opaque appearance whereas columnar ice is mostly translucent, revealing brine channels where present (see also Figure 3.3.2). The layering apparent in the core shown in Figure 3.3.4a is indicative of sequential frazil-ice accumulation and can be related to storms or weather patterns resulting in the formation of frazil ice in agitated open water. Interpretation of any other variable measured on these cores would need to take the very different growth history and origin of the upper 0.5 m of the ice floe into account.

Additional information about the growth history of the ice and potential heterogeneities can be obtained from ship-based observations (if the sampling is carried out from an icebreaker; see Chapter 3.12) or from slabs cut from the ice sheet. On occasion, for example, in studies of highly anisotropic ice properties such as its electrical conductivity (Chapter 3.4), it may be necessary to obtain more detailed information about the crystal size and orientation in the field. Using a handsaw, sanding paper, and a pair of polarizer sheets (Figure 3.3.1), one can produce low-quality thick sections from a core fairly quickly, as demonstrated in the video on the accompanying DVD. For this purpose, pruning saws or saws with teeth that

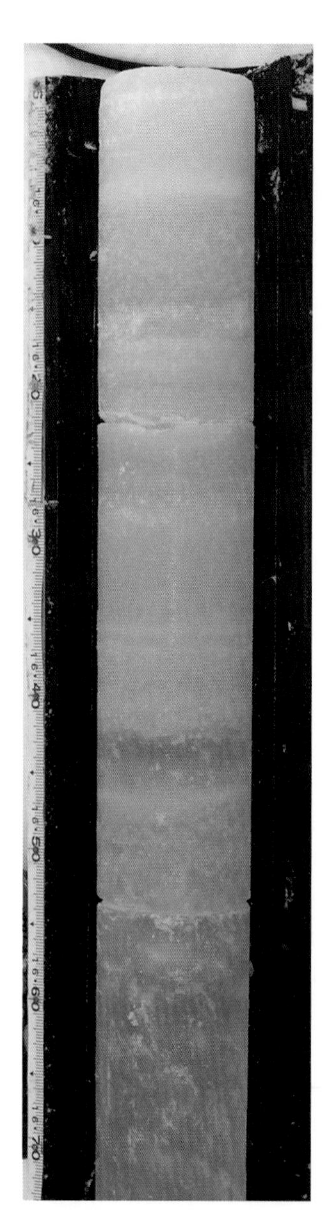

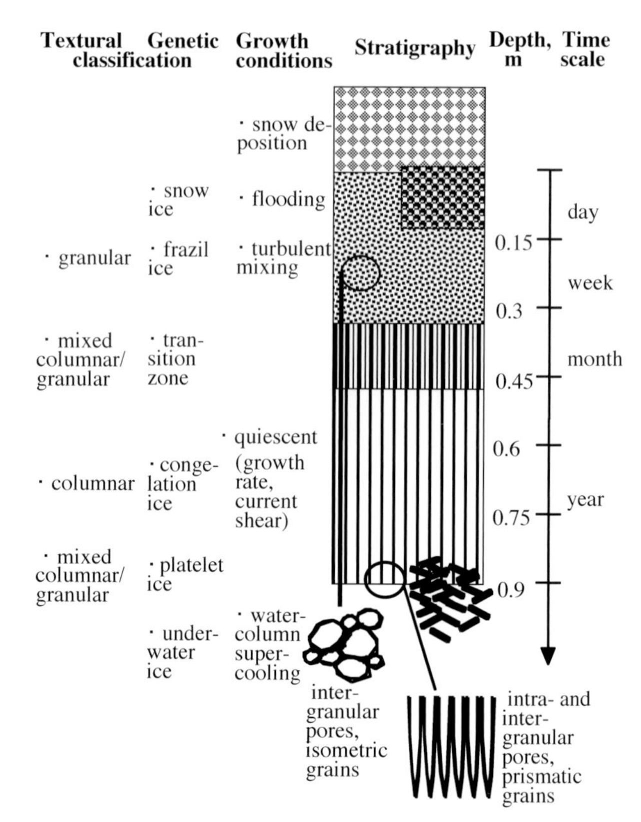


**Figure 3.3.4.** Photograph of the sea ice–core stratigraphy for a core extracted from Chukchi Sea landfast ice in late March 2004 (left; total length shown: 73 cm; cm scale shown at left). Note the layering due to alternating bands of granular ice with different grain sizes and inclusion densities, followed by a transition zone between about 50 and 60 cm depth, underlain by columnar sea ice. Schematic depiction of different growth mechanisms and associated ice textural classes commonly found in sea ice (right).

do not jut out beyond the profile of the blade and cut on both the forward and backward stroke (such as the one shown in Figure 3.3.1) are particularly useful for cutting thin slices of ice, < 1 cm thick. Thinning such slices with sanding paper to about 5 mm thickness and polishing by rubbing the surface with a warm hand can be quite effective in obtaining more information about the size, orientation, and shape of ice crystals (ice texture).

Laboratory-grade thin sections may on occasion also be cut in the field if a properly equipped mobile lab (e.g., on an icebreaker) is available. A separate video on the accompanying DVD provides an introduction to producing research-grade thin sections. Following the approach described by Sinha (1977), a sample slice, ca.

2 cm thick with parallel surfaces, is cut with a chop or band saw and frozen onto a thick glass plate with water. The top of this sample is shaved off with a microtome until a clean, planar surface results. This surface is then frozen onto a second glass plate (holding the sample in place with vises) and again attached with a ring of water frozen around the entire perimeter of the sample. Using a band or chop saw, the sample is now cut in the middle, leaving the microtomed portion with an ice sample of less than 1 cm thickness stuck to the glass slide. The new, freshly cut surface is now microtomed down to the final thickness, which can be as little as 200 to 300 μm for some studies. Between crossed polarizing sheets, the different ice crystals appear in different interference colors and more detailed microstructural analysis can follow (Eicken 2007).

### 3.3.3.2 Sea Ice State Variables: Temperature, Salinity, and Density

The most fundamental ice properties measured on a core are the ice temperature, its density, and its salinity. These can be thought of as "state variables" for sea ice, such that knowledge of the three allows the physical state of the ice cover to be described to the extent that other key ice properties can be directly derived from these variables with a model or estimated from empirical relationships. For example, the models of ice strength discussed in Chapter 3.5 depend on the pore (brine and air) volume fraction in sea ice, which in turn requires accurate measurement of the three aforementioned properties. The thermodynamic phase relations that govern the coexistence of different phases such as liquid brine or solid ice dictate that for an isothermal volume of sea ice in thermodynamic equilibrium at atmospheric pressure, the relative volume fraction of brine is fixed and depends solely on the ice temperature T and its bulk composition or, rather, in the case of "standard" sea ice, its bulk salinity $S_{si}$ (Weeks and Ackley 1986; Eicken 2003). The salinity $S_b$ of the brine contained within such an ice volume in turn also depends only on the temperature of the bulk-ice volume. The coupling between these different variables is a key aspect of sea ice as a geophysical material as well as a biological habitat for microorganisms, since any temperature change directly affects the porosity and pore microstructure of the ice as well as the salinity of the brine (Petrich and Eicken 2009). As direct measurement of these properties in the field is difficult, if not impossible, commonly the in situ brine volume fraction and other properties are derived from the bulk salinity of an ice sample, its density, and its temperature. However, due to difficulties in obtaining accurate density data, many studies have to resort to assuming zero or some constant value for the air volume fraction.

Measurements of ice temperature on the core should be performed immediately after extraction of the core (within five minutes at the very most; see Chapter 3.4.2.5 for a more detailed discussion). The core should be shaded and ideally placed into or under an insulated cover, allowing holes to be drilled into the core with an electric or hand drill, into which a thermistor with a digital thermometer readout is placed (see Figure 3.3.1 for equipment). With typical accuracies of around 0.1 to 0.2 K, such measurements cannot be compared to dedicated in situ temperature recordings discussed in Chapter 3.4, but they provide sufficient detail and accuracy for most practical applications.

Accurate density measurements are difficult to make on small samples in the field, and ideally require careful sample preparation (machining) and dimensioning to obtain accurate data on total sample volume in relation to its mass. Timco and Frederking (1996) provide an excellent overview of the errors and challenges associated with sea ice density measurements. If care is taken to produce high-quality core samples with uncompromised ice surfaces, and parallel saw cuts are used to generate smooth ends for the core, then for sample lengths of 20 cm or more reasonably accurate density data can be obtained. Mass is easily determined in the field using a digital balance, with samples placed in resealable plastic bags to prevent loss of brine during the measurements. To obtain accurate dimensions on the core diameter and length a caliper is indispensable. For multiyear sea ice, which typically exhibits low-density layers near the surface (< 700 kg $m^{-3}$), such density measurements can be highly informative. First-year sea ice, on the other hand, typically does not contain large amounts of air and hence determination of any significant differences between samples is often hampered by lack of precision in the measurement approach. These problems are discussed in more depth in the supplementary document on results from a sea ice field course, edited by S. Farrell, provided on the accompanying DVD.

For salinity measurements, it is important to section the core as soon as possible after drilling and transfer to sealed plastic containers or leak-proof plastic bags in order to minimize the amount of brine drainage from the ice. Containers should be prelabeled, and since plastic bags tend to leak in particular when handled in cold weather, using double bags is prudent. Brine drainage is a particular problem in the summer, when ice permeability is high (Chapter 3.4.4). While some brine loss is unavoidable, quick, careful work can help minimize artifacts due to brine loss. Once samples are safely stowed away, they can be transferred to the laboratory for further analysis.

Salinity is most commonly determined by measuring the electrical conductivity of a melted ice-core sample. For seawater of standard composition, the salinity is defined in terms of the conductivity of a sample relative to a reference standard, expressed as a dimensionless value on the practical salinity scale (pss). Commonly, salinities are reported in practical salinity units (psu) or permille (‰) since salinities on the pss are approximately the same as the concentration of salt expressed in grams of solute per liter of solution. The pss's range of validity extends from 2 to 42 and does not cover temperatures below –2°C. Furthermore, sea ice can deviate significantly in composition from standard seawater, which may affect direct comparisons with seawater salinities in some cases. Typically, commercially available conductivity probes for water-sample analysis are capable of converting conductivity data directly to salinity. For low-salinity, multiyear ice samples, it may be appropriate to also report the (temperature-corrected) conductivity for greater accuracy.

#### 3.3.3.3 Deriving Brine and Air Volume Fractions from Sea Ice State Variables

The thermodynamic phase relations determine the liquid volume fraction of brine in equilibrium with sea ice of a given salinity. From the compilation of experimental data by Assur (1960) for the phase relations and based on the continuity equations for a multiphase sea ice mixture, Cox and Weeks (1983) derived an extremely useful set of equations describing the brine volume fraction as a function of ice temperature and salinity:

$$\frac{V_b}{V} = \left(1 - \frac{V_a}{V}\right) \frac{\rho_i S_{si}}{F_1(T) - \rho_i S_{si} F_2(T)} \quad \text{(Equation 3.3.1).}$$

Here, $V_b/V$ is the volume fraction of brine, while $V_a/V$ is the volume fraction of air in a sample. The density of pure ice is given as

$$\rho_i = 0.917 - 1.403 \times 10^{-4} T \quad \text{(Equation 3.3.2),}$$

with $\rho_i$ in g cm$^{-3}$ and $T$ in °C. $F_1(T)$ and $F_2(T)$ are empirical polynomial functions $F_i(T) = a_i + b_i T + c_i T^2 + d_i T^3$, based on the phase relations. The coefficients for different temperature intervals are listed in Table 3.3.1, including a set of coefficients derived by Leppäranta and Manninen (1988) for temperatures between −2 and 0°C. However, as the temperature increases towards 0°C, the derived data become increasingly inaccurate, due to both difficulties in accurate measurements of temperature and salinity and the impact of compositional changes in the brine and other factors.

The brine salinity and density can be approximated for temperatures above −23°C as

$$S_b = \left(1 - \frac{54.11}{T}\right)^{-1} \times 1000‰ \quad \text{(Equation 3.3.3),}$$

$$\rho_b = 1 \quad 8 \times 10^{-4} S_b \quad \text{(Equation 3.3.4),}$$

with $T$ in °C. The volume fraction of air $V_a/V$ can be derived from a measurement of the density of a sea ice sample ρ:

$$\frac{V_a}{V} = 1 - \frac{\rho}{\rho_i} \quad \rho S_{si} \frac{F_2(T)}{F_1(T)} \quad \text{(Equation 3.3.5).}$$

Equations 3.3.1–3.3.5 thus provide us with a tool to derive key quantities of importance for a wide range of physical, biological, and chemical studies of sea ice, given that properties such as ice strength or its electric conductivity depend to first order on the volume fraction of liquid and ice (and gas).

**Table 3.3.1.** Coefficients for functions $F_1(T)$ and $F_2(T)$ for different temperature intervals, from Leppäranta and Manninen (1988), Cox and Weeks (1983)

| T, °C | $a_1$ | $b_1$ | $c_1$ | $d_1$ |
|---|---|---|---|---|
| $0 \geq T > -2$ | -0.041221 | -18.407 | 0.58402 | 0.21454 |
| $-2 \geq T \geq -22.9$ | -4.732 | -22.45 | -0.6397 | -0.0174 |
| $-22.9 > T \geq -30$ | 9899 | 1309 | 55.27 | 0.7160 |
| | $a_2$ | $b_2$ | $c_2$ | $d_2$ |
| $0 \geq T > -2$ | 0.090312 | -0.016111 | $1.2291 \times 10^{-4}$ | $1.3603 \times 10^{-4}$ |
| $-2 \geq T \geq -22.9$ | 0.08903 | -0.01763 | $-5.330 \times 10^{-4}$ | $-8.801 \times 10^{-6}$ |
| $-22.9 > T \geq -30$ | 8.547 | 1.089 | 0.04518 | $5.819 \times 10^{-4}$ |

## 3.3.4 CASE STUDIES: SAMPLING-SITE SELECTION, STRATIGRAPHY, AND DATING OF SEA ICE CORES

### 3.3.4.1 Sampling Multiyear Ice off the Coast of Northern Alaska

The following case study illustrates how remote-sensing imagery and other ancillary data can help identify suitable sampling sites and provide context to basic ice-core data. As part of an International Polar Year (IPY) network aiming to assess changes in the (sub-) arctic seasonal ice zone, thickness transects with an EM instrument were flown out of Barrow, Alaska, in spring of 2008 (Figure 3.3.5; the methodology of airborne thickness measurements is detailed in Chapter 3.2). In order to be able to attribute interannual and regional changes in ice thickness to key processes, such as reduced ice growth, increased deformation, changes in melt patterns, etc., the prevailing ice types were sampled along the thickness profiles. QuikSCAT radar scatterometer data, available on a daily basis through several data providers, helped in developing a flight and sampling plan. As shown in Figure 3.3.5 (and explained in more detail in Chapter 3.15.3), multiyear ice exhibits higher backscatter coefficients than first-year ice due to the important contribution of volume scattering by air and brine inclusions in low-salinity, "transparent" (at radar wavelengths) multiyear ice (e.g., Nghiem et al. 2007). High-resolution synthetic aperture radar (SAR) imagery was also used to further constrain the flight pattern and identify suitable sampling sites.

After the thickness overflights and brief examination of the resulting data set, and taking into consideration visual observations from these flights, a target sampling

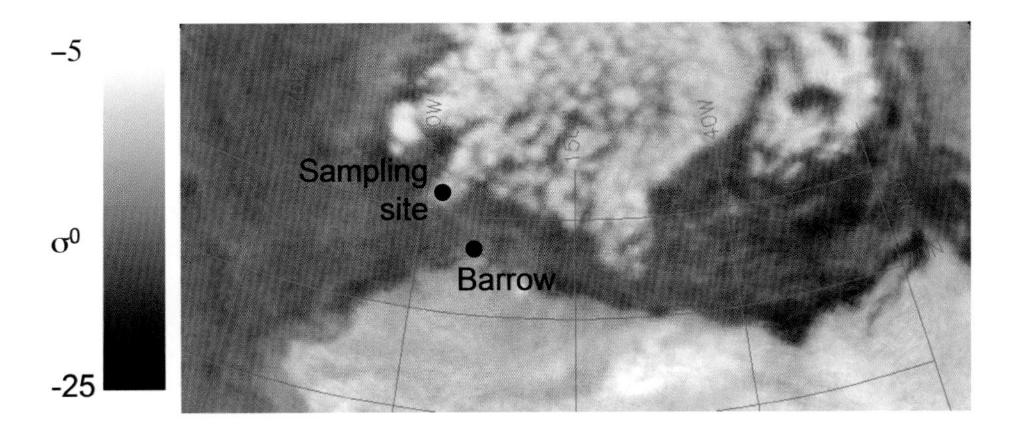


**Figure 3.3.5.** QuikSCAT radar scatterometer scene for 15 April 2008 for the Chukchi and Beaufort Seas north of Alaska (image provided by the National Oceanic and Atmospheric Administration's National Environmental Satellite Data and Information Service). Shown at left is the backscatter coefficient scale for the scene. Note the higher backscatter of multiyear ice advected from the High Canadian Arctic, compared to first-year ice with low backscatter values. Linear, ground-projected pixel dimensions for QuikSCAT data are 50 km, hence the coarse appearance of the image shown here.

region for multiyear ice was identified. The rolling topography—somewhat concealed by deep snow cover, absence of newly formed ridges in the interior of floes, and high freeboard—helped guide the coring site selection (see aerial and surface-based photographs in Figure 3.3.6). At the landing site, a ground-based EM was deployed to obtain the thickness profile shown at the bottom in Figure 3.3.6. While the data were not analyzed on-site, they provide an important reference, and examination of the raw apparent conductivity values measured on site also gives a rough indication of ice thickness and its variability, even if in direct measurements with a thickness auger would additionally be required for more accurate assessments. In combination with an examination of the ice surface excavated at several sites through deep snow cover, it became apparent that the rolling topography reflected alternating patches of frozen melt ponds (smooth, completely flat surface of clear ice with few inclusions) and weathered white ice (rough, low-density surface layer; see Perovich et al. 2002 for more details on surface evolution of multiyear sea ice). Since surface ponds often correspond to locations of thin ice with reduced winter growth and enhanced bottom summer melt (Eicken et al. 2001), sampling the weathered white ice was more appropriate to obtain information about ice properties and its growth history.

The core drilled on site was 3.42 m in length, which corresponded roughly to the thickness mode identified in the frequency distribution of ice thickness data collected from the air and on the ground. A much shorter or longer core would not have been representative of the prevailing ice type and may have required obtaining further cores at this site. Careful on-site examination of the core stratigraphy provided no evidence of deformation features, and indicated (to be verified by further

**Figure 3.3.6.** Aerial photograph (top) (several hundred meters across at front), and surface photograph (middle) taken of a multiyear sea ice floe sampled in the eastern Chukchi Sea (see Figure 3.3.5 for location) on 15 April 2008. Note the undulating surface, concealed by deep snow cover. A surface EM-derived (see Chapter 3.2) thickness profile (bottom) obtained over the sampling site (the core shown in Figure 3.3.7 was obtained at roughly 200 m along the profile), extending along the view of the ground-based photograph indicates significant variability in ice thickness.

analysis in the laboratory) that the ice had mostly formed through bottom accretion. The core exhibited horizontal, straight, or partly undulating milky layers in three locations (Figure 3.3.7). As described in more detail below, such layers can be indications of transitions between spring growth and fall ice accretion, representing the location of the ice bottom during the summer months. Moreover, near the bottom, a sequence of clear ice, a bubbly layer, and ordinary, somewhat translucent sea ice was observed (Figure 3.3.7). Such ice is indicative of freshwater buildup underneath the ice cover in summer, resulting in infiltration of the overlying ice layers by low-salinity water, and occasionally (as apparently in this case) formation of a low-salinity ice cover at the interface between the freshwater lens and underlying seawater ("false bottom") that later in the subsequent fall congeals into ordinary sea ice (Schwarzacher 1959; Eicken et al. 2002). At the sampling site, such interpretation was formulated as a vague hypothesis, to be tested later in the lab through salinity measurements (which in fact support this conclusion, as shown by the profile in Figure 3.3.7) and to be confirmed through measurements of stable-isotopic composition of the ice samples.

While on-site interpretation of core stratigraphy can only go so far, it can provide important guidance to evaluate whether the goals of the sampling activity have been met—in this case to obtain congelation ice that would allow potential dating of the ice floe and reconstruction of its origin based on ancillary data—or whether additional samples in different locations may have to be taken. It is important to note that some of these later measurements, such as the determination of

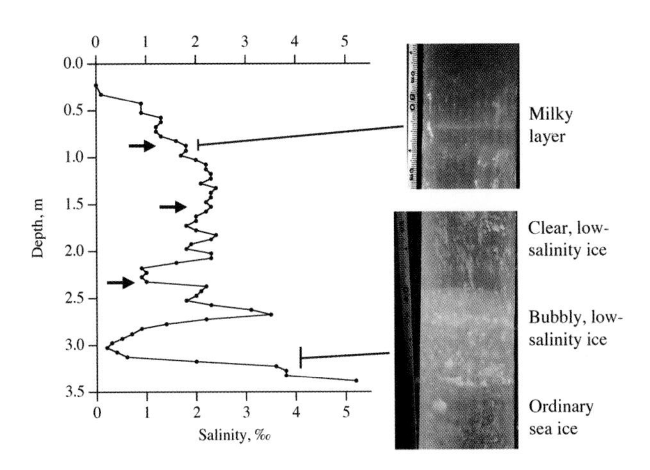


**Figure 3.3.7.** Salinity profile (left; salinities measured on slices of 5 cm vertical extent) and core stratigraphy photographs for an ice core obtained across the entire thickness (3.42 m) of the multiyear ice floe shown in Figure 3.3.5 on 15 April 2008. The arrows show the positions of so-called milky layers, indicative of the location of the ice bottom during summer (potential annual markers, see photograph at top right, cm-scale shown to the left of the core; photograph at top right has been contrast-enhanced to highlight milky layer). The photograph at the bottom right shows a stratigraphic sequence indicative of meltwater buildup at the bottom of the ice, as confirmed by the salinity profile.

concentrations of the heavy isotopes of oxygen and hydrogen in the ice, are not as susceptible to pore-scale variability in the ice cover, hence allowing for a relaxation of the requirement of several parallel samples. Since dating of sea ice cores can be of considerable value even during the course of a field campaign, we will briefly discuss this topic in the subsequent section.

The core shown in Figure 3.3.7 does in fact allow us to draw some tentative conclusions (to be confirmed through detailed stable-isotope analysis and examination of ice drift patterns) about the age of the ice floe. Hence, with four annual layers identified, the floe is at least five years old. Taking into account that growth rates at the base of such thick ice are very small due to reductions in conductive heat flux (Chapter 3.4), with accretion of only a few decimeters of ice at most in one growth season, and considering substantial surface melt rates, it appears likely that the ice is probably older, possibly seven or even eight years old. However, it needs to be kept in mind that dating of ice cores and stratigraphic analysis is still very much an inexact science, in the sense that spatial variability and the absence of annual layers under certain conditions or the presence of layers that may be mistaken for seasonal markers can render such dating approaches challenging. Nevertheless, as demonstrated below, the combined use of different indicators and tracers can help place ice sampled at a specific site into a broader context.

### 3.3.4.2 Annual Layers and Dating of Sea Ice Cores

Increasing sophistication in the study of sea ice samples, including detailed biogeochemical and biological analyses described in Chapters 3.8 and 3.9; geochemical tracer studies of sea ice (Masque et al. 2003; Lannuzel et al. 2007); or advanced studies of ice thermal and dielectric properties described in Chapter 3.4, call for greater scrutiny of the ice stratigraphy in conjunction with other ice properties. Of particular interest is the detection of annual layering and dating of ice cores, especially if it can be combined with ice growth and ice drift models to provide a more comprehensive view of the environmental history of sea ice (Macdonald et al. 1999; Pfirman et al. 2004). A more detailed discussion of the broader approach can be found in the studies of Pfirman et al. (2004) and Eicken et al. (2005). Here, we want to briefly focus on the identification of annual layers in the field, as this may provide useful guidance for other ice sampling efforts.

Identification of second- or multiyear ice, in particular in areas where snow cover and lack of surface melt (such as in the Southern Ocean) blur the distinction between ice that has survived one melt season and first-year ice, can be challenging. High-resolution salinity profiles of the uppermost ice layers can provide some indication whether ice has undergone even small amounts of surface melt, since the melt cycle does reduce initially elevated salinities near the top of the ice cover to values well below the mean salinity of the ice interior (shown in the extreme in the old ice pictured in Figure 3.3.7).

Two other types of annual markers that typically conform to the position of the ice bottom during the summer melt period are of potential use in dating ice cores in the field. The pioneering work by Cherepanov (1957) and Schwarzacher (1959) already pointed out the presence of annual horizons that appear as changes in ice crystal size (requiring thin-section analysis in the laboratory) or milky layers in the core stratigraphy (Figure 3.3.7). Schwarzacher attributed these mostly to infiltration by fresher summer water from below or formation of a layer of smaller-grained crystals at the top of the new ice horizon grown in fall. The former corresponds to the sequence of clear, low-salinity ice—bubbly, low-salinity ice—ordinary sea ice shown in Figure 3.3.7. While salinity measurements typically provide some insight into this process, it requires samples of high vertical resolution (≤ 5 cm vertical extent) to fully detect the corresponding features. In the cores, the presence of isometric or vertically oriented tubular gas inclusions arranged in a layer below or above clear ice is also indicative of freshwater lenses, since dissolved gases are not retained within the liquid brine in such fresher ice. While an in-depth discussion is beyond the scope of this chapter (more details can be found in Eicken et al. 2002 and Pfirman et al. 2004), stable-isotope data may also help identify such annual layers. Freshwater infiltrating from below often contains significant fractions of snowmelt, depleted in the heavy oxygen and hydrogen isotopes. Alternatively, one can find large offsets in isotopic composition due to signals recorded from two different water masses, one prior to and the other after summer melt. This effect is illustrated in an ice core obtained between Svalbard and Franz Josef Land in the northern Barents Sea shown in Figure 3.3.8, where the ice below 2 m depth has an isotopic composition commensurate with the Barents Sea sector of the Arctic.

The core shown in Figure 3.3.8 also helps illustrate results of an analysis of a larger number of ice cores from different sectors of the Arctic (Pfirman et al. 2004), indicating that most milky layers are in fact of biogenic origin and provide a direct link between ice accretion and melt processes, ice physical properties, and sea ice ecology. The core shown in Figure 3.3.8 exhibited two milky layers at 0.795 and 1.94 m depth. These layers are typically a few millimeters thick, with quite sharp lower boundaries (Figure 3.3.7). Often they have a slightly undulating appearance (with wavelengths of several centimeters). They are most easily detected by examining the core against a dark background, tilting the sample in different directions to view directly along the plane of the milky layer. Measurements of chlorophyll *a* as a measure of ice-algal biomass demonstrate that these layers very often correspond to internal submaxima in the chlorophyll profile, as shown in Figure 3.3.8. Moreover, for the example shown here, counts of diatom cells and empty frustules by R. Gradinger on samples cut at 1 cm vertical resolution demonstrate that the milky layer corresponds to a horizon of very high ice-algal densities of more than ten thousand cells or frustules per milliliter of melted ice. Within 1 or 2 cm below this horizon, concentrations drop by one to two orders of magnitude. This suggests

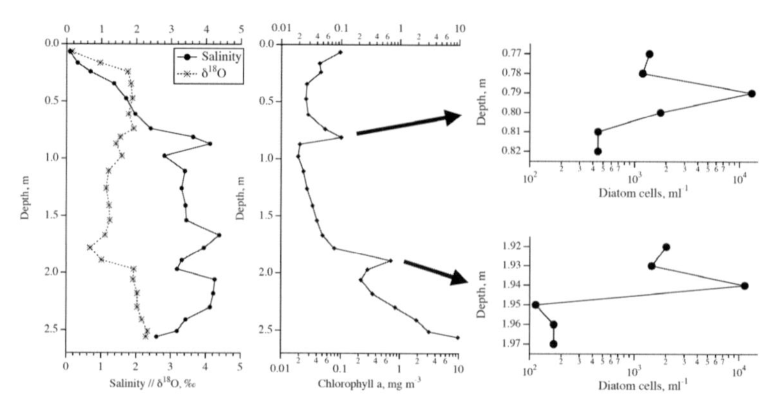


**Figure 3.3.8.** Salinity and stable-isotope ($\delta^{18}O$) profile (left) for a sea ice core taken in the Barents Sea sector of the Arctic at 82°10'N 42°03'E on 20 August 1993. The chlorophyll *a* profile as a measure of algal biomass is shown at center. The submaxima in chlorophyll concentrations correspond to milky layers identified in the ice core stratigraphy at 0.795 and 1.94 m. High-resolution diatom cell and frustule (empty silica housing) counts across these layers are shown at right. Diatom and chlorophyll data kindly provided by Rolf Gradinger.

that milky layers are in fact the remnants of the bottom ice-algal layers growing at the base of the ice cover in spring and summer, discussed in more depth in Chapters 3.8 and 3.9. While they may not always be present or their presence may be obliterated by substantial bottom melt (in which case freshwater infiltration horizons are likely to be found), their identification in the field can help with preliminary dating of ice and guide further sampling.

## 3.3.5 CONCLUSION AND OUTLOOK

As illustrated by the two case studies introduced above, comprehensive analysis of ice cores requires more than the basic measurements made in the field. However, with increasing degrees of sophistication of field and laboratory studies it can be argued that the importance of these basic ice property investigations will increase rather than diminish. Many studies of sea ice require an ever more detailed picture of the ice growth history and its stratigraphy to put into context a range of variables determined on melted ice samples in the laboratory. Since much of this work is carried out long after the end of a field campaign, completing a comprehensive on-site characterization program can be of great value in later data interpretation and hypothesis testing, in particular if no archived cores are available for follow-up investigations.

At the same time, it needs to be recognized that field measurements are evolving rapidly due to the availability of automated measurement platforms and increasing use of remote-sensing data. Hence, as outlined in Chapter 3.13, drifting buoys may well replace many of the standard ice temperature and ice growth/ice melt measurements made at drifting research stations in the past. When deploying such buoys in drifting ice, ice cores taken on location may provide valuable information on the past ice growth and melt history, possibly in more detail than even a drifting buoy can provide.

Recently, the potential for automated measurements of ice salinity and brine volume fraction has been advanced substantially in a study by Notz et al. (2005) that employed a capacitance probe, frozen into new ice, to determine the ice and liquid volume fractions in conjunction with measurements of ice temperature. Work by Pringle et al. (2009) has demonstrated the promises and limitations of deriving brine volume fractions and pore-microstructural parameters from in situ impedance measurements of the complex permittivity. Further developments, in particular to make such probes more robust and easily deployed into solid sea ice, may help reduce the need for ice-core salinity measurements and minimize errors due to, for example, brine loss from samples.

Another area that is poised for significant advances is an improved understanding of the geochemistry of sea ice, in particular in relation to the sea ice phase relations. Laboratory studies have advanced ice sampling methodology to examine gases and precipitates in sea ice in more detail (Tison et al. 2002). More important, geochemical modeling of phase equilibria in aqueous solutions has now reached a point where it may enable important revisions of the "standard" sea ice and seawater composition model (Marion et al. 1999) and provide the basis for more accurate derivations of liquid-phase fractions in sea ice of different origin and age (Marion and Kargel 2008).

The full range of what is possible with methodological advances is perhaps best demonstrated in a series of studies by Callaghan et al. (2007) of brine volume fractions and brine movement in sea ice using Earth's field nuclear magnetic resonance (NMR). Looking out towards the next International Polar Year, it may well be such a combination of technological advances and increasing collaboration over a range of different disciplines that will make studies of fundamental ice properties more useful, more powerful, and more efficient in the future.

## ACKNOWLEDGMENTS

Obtaining high-quality samples of sea ice usually requires substantial logistic support by a wide variety of people, ranging from an Iñupiat guide and ice expert to a Coast Guard helmswoman or from a helicopter mechanic to a mess cook. Advances in our understanding of sea ice are often directly tied to the level of backing extended by this support network. This is to express my deep appreciation of this support through the years, which has made fieldwork safe, successful, and fun! This chapter reflects knowledge that has been freely shared with the author by a range of international experts, both with and without "formal" training, including teachers, students, mentors, reviewers, and colleagues too numerous to mention individually. The international community of those with an interest in and knowledge of sea ice is blessed with a high level of collegiality and mutual support; I am most grateful to all those who have shared so freely and generously, and encourage those who read this chapter and feel that something important is missing or misrepresented to hold me accountable in this same communal spirit.

## REFERENCES

Assur, A. (1960), Composition of sea ice and its tensile strength, *SIPRE Res. Rep.*, *44*, 1–54.

Cherepanov, N. V. (1957), Using the methods of crystal optics for determining the age of drift ice (in Russian), *Probl. Arktik.*, *2*, 179–184.

Cherepanov, N. V. (1970), A new annular ice auger, *Prob. Arctic Antarct.*, *32*, 506–510.

Cole, D. M., H. Eicken, K. Frey, and L. H. Shapiro (2004), Observations of banding in first-year Arctic sea ice, *J. Geophys. Res.*, *109*, doi:10.1029/2003JC001993.

Cole, D. M., and L. H. Shapiro (1998), Observations of brine drainage networks and microstructure of first-year sea ice, *J. Geophys. Res.*, *103*, 21,739–21,750.

Cox, G. F. N., and W. F. Weeks (1983), Equations for determining the gas and brine volumes in sea ice samples, *J. Glaciol.*, *29*, 306–316.

Doronin, Y. P., and D. E. Kheisin (1977), *Sea Ice*, Amerind Publ. Co., New Delhi.

Eicken, H. (2003), From the Microscopic to the Macroscopic to the Regional Scale: Growth, Microstructure and Properties of Sea Ice, in *Sea Ice: An Introduction to Its Physics, Biology, Chemistry, and Geology*, edited by D. N. Thomas and G. S. Dieckmann, pp. 22–81, Blackwell Science, London.

Eicken, H. (2007), Sea ice: Crystal Texture and Microstructure, in *Encyclopedia of the Antarctic*, edited by B. Riffenburgh, pp. 840–845, Routledge, New York.

Eicken, H., I. Dmitrenko, K. Tyshko, A. Darovskikh, W. Dierking, U. Blahak, J. Groves, and H. Kassens (2005), Zonation of the Laptev Sea landfast ice cover and its importance in a frozen estuary, *Global Planet. Change*, *48*, 55–83.

Eicken, H., H. R. Krouse, D. Kadko, and D. K. Perovich (2002), Tracer studies of pathways and rates of meltwater transport through Arctic summer sea ice, *J. Geophys. Res.*, *107*, 10.1029/2000JC000583.

Eicken, H., M. A. Lange, and G. S. Dieckmann (1991), Spatial variability of sea ice properties in the northwestern Weddell Sea, *J. Geophys. Res.*, *96*, 10,603–10,615.

Eicken, H., W. B. I. Tucker, and D. K. Perovich (2001), Indirect measurements of the mass balance of summer Arctic sea ice with an electromagnetic induction technique, *Ann. Glaciol.*, *33*, 194–200.

Haining, R. (2003), *Spatial Data Analysis: Theory and Practice*, Cambridge University Press, Cambridge.

Høyland, K. V. (2002), Consolidation of first-year sea ice ridges, *J. Geophys. Res.*, *107*, 3062, doi:10.1029/2000JC000526, 000515,000521–000515.

Lannuzel, D., V. Schoemann, J. de Jong, J.-L. Tison, and L. Chou (2007), Distribution and biogeochemical behaviour of iron in the East Antarctic sea ice, *Marine Chem.*, *106*, 18–32.

Leppäranta, M., and T. Manninen (1988), *The Brine and Gas Content of Sea Ice with Attention to Low Salinities and High Temperatures*, Finnish Inst. Marine Res. Internal Rep. 88–2, Helsinki.

Macdonald, R. W., E. C. Carmack, and D. W. Paton (1999), Using the $\partial^{18}O$ composition in landfast ice as a record of arctic estuarine processes, *Mar. Chem.*, *65*, 3–24.

Malmgren, F. (1927), On the Properties of Sea Ice, *Norweg. North Pol. Exped. "Maud" 1918–1925, vol. 1, no. 5*, 1–67.

Marion, G. M., and J. S. Kargel (2008), *Cold Aqueous Planetary Geochemistry with FREZCHEM*, Springer, Berlin.

Marion, G. M., A. J. Komrowski, and R. E. Farren (1999), Alternative pathways for seawater freezing, *Cold Reg. Sci. Technol.*, *29*, 259–266.

Masque, P., J. K. Cochran, D. Hebbeln, D. J. Hirschberg, D. Dethleff, and A. Winkler (2003), The role of sea ice in the fate of contaminants in the Arctic Ocean: Plutonium atom ratios in the Fram Strait, *Environ. Sci. Technol.*, *37*, 4848–4854.

Nelson, R. K. (1969), *Hunters of the Northern Ice*, University of Chicago Press, Chicago.

Nghiem, S., and G. Neumann (2007), Arctic Sea Ice Monitoring, *McGraw-Hill Yearbook of Science and Technology*, *1*, 12–15.

Notz, D., J. S. Wettlaufer, and M. G. Worster (2005), A non-destructive method for measuring the salinity and solid fraction of growing sea ice in situ, *J. Glaciol.*, *51*, 159–166.

Perovich, D. K., T. C. Grenfell, B. Light, and P. V. Hobbs (2002), Seasonal evolution of the albedo of multiyear Arctic sea ice, *J. Geophys. Res.*, *107*, 10.1029/2000JC000438.

Petrich, C., and H. Eicken (2009, in press), Growth, Structure, and Properties of Sea Ice, in *Sea Ice: An Introduction to Its Physics, Biology, Chemistry, and Geology, 2nd ed.*, edited by D. N. Thomas and G. S. Dieckmann, Blackwell Science, London.

Pfirman, S., W. Haxby, H. Eicken, M. Jeffries, and D. Bauch (2004), Drifting Arctic sea ice archives changes in ocean surface conditions, *Geophys. Res. Lett.*, *31*, doi:10.1029/2004GL020666.

Pringle, D., G. Dubuis, and H. Eicken (2009), Impedance measurements of the complex dielectric permittivity of sea ice at 50 MHz: Pore microstructure and potential for salinity monitoring, *J. Glaciol.*, *55*, 81–94.

Rand, J., and M. Mellor (1985), Ice-coring augers for shallow depth sampling, *CRREL Rep.*, *85-21*, 1–22.

Rothrock, D. A., and J. Zhang (2005), Arctic Ocean sea ice volume: What explains its recent depletion? *J. Geophys. Res.*, *110*, C01002, doi:01010.01029/02004JC002282.

Schwarzacher, W. (1959), Pack-ice studies in the Arctic Ocean, *J. Geophys. Res.*, *64*, 2357–2367.

Sinha, N. K. (1977), Technique for studying structure of sea ice, *J. Glaciol.*, *18*, 315–323.

Thomas, D. N., and G. S. Dieckmann (2003), *Sea Ice: An Introduction to Its Physics, Biology, Chemistry, and Geology*, Blackwell Science, London.

Timco, G. W., and R. M. W. Frederking (1996), A review of sea ice density, *Cold Reg. Sci. Technol.*, *24*, 1–6.

Tison, J.-L. et al. (2002), Tank study of physico-chemical controls on gas content and composition during growth of young sea ice, *J. Glaciol.*, *48*, 177–191.

Tucker, W. B., A. J. Gow, and J. A. Richter (1984), On small-scale variations of salinity in first-year sea ice, *J. Geophys. Res.*, *89*, 6505–6514.

Weeks, W. F., and S. F. Ackley (1986), The Growth, Structure and Properties of Sea Ice, in *The Geophysics of Sea Ice*, edited by N. Untersteiner, pp. 9–164, Plenum Press, New York (NATO ASI B146).

Worby, A. P., C. A. Geiger, M. J. Paget, M. L. Van Woert, S. F. Ackley, and T. L. DeLiberty (2008), Thickness distribution of Antarctic sea ice, *J. Geophys. Res.*, *113*, C05S92, doi:10.1029/2007JV004254.

Zubov, N. N. (1945), *Arctic Ice*, Izd. Glavsevmorputi, Moscow.

# Chapter 3.4

# Thermal, Electrical, and Hydraulic Properties of Sea Ice

*Daniel Pringle and Malcolm Ingham*

## 3.4.1 INTRODUCTION

The flow of electric charge, heat, and fluids are described by the dc electrical conductivity $\sigma$ [S/m], thermal conductivity $k$ [W/m] and fluid permeability $\Pi$ [$m^2$], respectively. They are defined in Table 3.4.1. In this chapter we consider the measurement of these and related properties and processes. The dependence of these properties on temperature, salinity, and microstructure is needed to understand and predict a range of related processes such as heat fluxes, the fate of surface meltwater, and nutrient supply to sea ice biota.

**Table 3.4.1.** Transport property definitions

| Property | Definition | Flux | Driving gradient |
|---|---|---|---|
| Thermal conductivity, $k$ ($W\ m^{-1}K^{-1}$) | Fourier's law, $J_Q = -k\nabla T$ | Heat flux, $J_Q$ ($W\ m^{-2}$) | Temperature, $\nabla T$ ($K\ m^{-1}$) |
| Electrical conductivity, $\sigma$ ($S\ m^{-1}$) [(a)] | Ohm's law, $J = -\sigma\nabla V$, *i.e.*, $J = \sigma E$ | Current density, $J$ ($A\ m^{-2}$) | Electric potential, $\nabla V$ ($V\ m^{-1}$), Electric field, $E$ ($N\ C^{-1}$ or $V\ m^{-1}$) |
| Fluid permeability, $\Pi$ ($m^2$) | Darcy's law, [(b)] $q = -\frac{\Pi}{\mu}\nabla P$ | Specific discharge, [(c)] $q$ ($m\ s^{-1}$) | Pressure, $\nabla P$ ($Pa\ m^{-1}$) |

*(a) The units of $\sigma$ are Siemens per meter, the reciprocal of a resistivity of 1 $\Omega$-m. (b) $\mu$ is dynamic viscosity ($kg\ m^{-1}s^{-1}$). (c) Specific discharge units derive from the volume discharge flux, $m^3\ s^{-1}m^{-2} = ms^{-1}$.*

These properties are related to the composition and anisotropic microstructure of sea ice, with an increasing complexity (Yen 1981): heat capacity depends essentially on only the ice and brine volume fractions; density depends also on the gas volume faction; thermal conductivity depends on the geometric arrangement of these phases; the electrical conductivity depends critically on the connectedness of the brine phase; and the fluid permeability on the size of these connections (Eicken

2003; Golden et al. 2007). The dielectric permittivity is also sensitive to brine layer orientation with respect to electric field polarization (e.g., Golden and Ackley 1980; 1981; Morey et al. 1984; Hallikainen and Winebrenner 1992) and potentially to the ice-brine interface surface area (de Loor 1968; Santamarina et al. 2001; Pringle et al. 2009).

Important questions to achieve accurate measurements are: (1) Is the microstructure well characterized and undisturbed by the measurement? and (2) What is the length scale to which the measurement is sensitive, and how does this compare with length scales of typical structural variations?

The overview we present in this chapter relates primarily to level ice. Beyond our scope are potential effects of deformation and the presence of impurities such as entrained sediment and biological matter and environmental pollutants.

References are given for more technical or comprehensive reviews of these properties and their measurement. An expanded version of this chapter is available in the accompanying DVD.

## 3.4.2 THERMAL PROPERTIES

### 3.4.2.1 Introduction

The thermal properties of sea ice govern its growth and decay, and its mediating effect between ocean and atmosphere. The use of thermal properties is largely confined to engineering and modeling applications, in which they are calculated as functions of ice and snow temperature and bulk salinity. Thermal properties are not routinely measured because previous measurements are in good agreement with theoretical predictions—with exceptions noted below. For each thermal property, we give definitions and address theoretical predictions and measurement methods and results. Detailed reviews and summaries of sea ice thermal properties can be found in Yen (1981), Weeks and Ackley (1986), and Fukusako (1990). We also discuss the measurement of temperature itself, one of the most common and useful sea ice field measurements.

### 3.4.2.2 Specific Heat

The presence of brine inclusions complicates the definitions of specific and latent heats of sea ice (e.g., Eicken 2003). Raising the temperature of sea ice requires the differential melting of ice around brine inclusions, and lowering its temperature requires differential freezing. The specific heat of sea ice $c(S,T)$ is defined as the amount of heat required to raise the temperature of unit mass of sea ice by one degree Kelvin, with the differential melting factored in. The latent heat $L(S,T)$ is defined as the amount of heat required to completely melt a piece of ice initially with salinity $S$ and temperature $T$ (Yen 1981). "Completely melt" means raised to a melt temperature $T_m$ equal to the temperature at which brine with the same bulk salinity $S$ would have frozen (Yen 1981). We now focus on $c(S,T)$ from which $L(S,T)$ can be found by integrating from $T$ to $T_m(S)$.

The specific heat depends only on the mass fractions of different phases present and not on their spatial arrangement. This gives:

$$\begin{aligned} c_{si} &= m_i c_i + m_b c_b + m_{ss} c_{ss} + m_a c_a \\ &= \rho_i v_i c_i + \rho_b v_b c_b + \rho_{ss} v_{ss} c_{ss} + \rho_a v_a c_a \end{aligned} \quad \text{(Equation 3.4.1)},$$

where ρ is density, $v$ volume faction, and $i$, $b$, $ss$, $a$ refer respectively to ice, brine, solid salts, and air (here used synonymously with gas). In practice, the air and solid salt terms can be neglected (Yen 1981; Pringle 2005). Analytical and numerical expressions for $c(S,T)$ can be derived from Equation 3.4.1. Schwerdtfeger (1963) and Ono (1967) gives $c$ in units of J $kg^{-1}$ $K^{-1}$ as

$$c = 2113 + 7.5 - 3.4\ S + 0.08\ S\ T + 18040\ \frac{S}{T^2} \quad \text{(Equation 3.4.2)},$$

for $S$ [ppt] and $T$ [°C]. The differential melt/freezing contribution appears mainly in the last term. The apparent discontinuity at $T = 0$°C never eventuates for a salinity-dependent melting point $T_m(S) < 0$ and this term is absent when derived for $S = 0$ ppt.

*Measuring Specific Heat*

Specific heat is traditionally measured by applying a known amount of heat $Q$ to a mass $m$, measuring the change in equilibrium temperature $\Delta T$, and calculating the specific heat from

$$Q = mc\Delta T \quad \text{(Equation 3.4.3)}.$$

Samples are contained in calorimeters with the aims of ensuring thermal equilibrium before and after heating, precise control of the applied heat, and no heat loss. The specific heat of the sample can be determined from measurements with and without a sample loaded. Specific heat measurements have been made on natural sea ice and laboratory-grown saline ice in the laboratory (Schwerdtfeger 1963; Johnson 1989) and in the field (Malmgren 1927; Nazintsev[1] 1959; 1964; Johnson 1989; Trodahl et al. 2000).

Laboratory measurements were made by Johnson (1989) with equipment designed for later use in the field (Trodahl et al. 2000). Their robust, field-worthy

1 See also Doronin and Kheisin (1977).

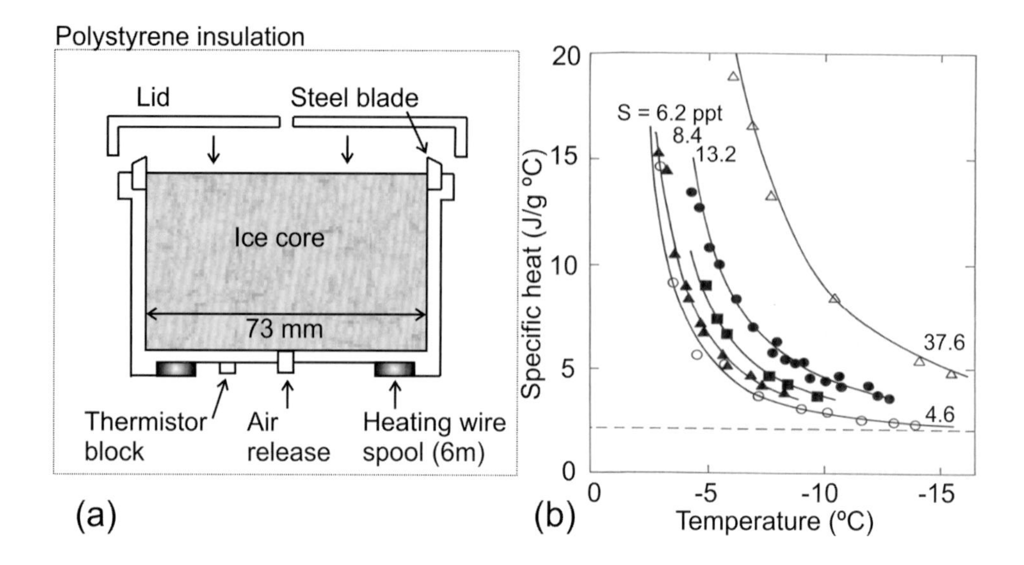


**Figure 3.4.1.** Calorimeter sketch and specific heat results from Johnson (1989) and Trodahl et al. (2000) in excellent agreement with predicted values. Open symbols are artificial NaCl ice, $S = 4.6$ and 37.6 ppt; solid symbols are FY sea ice with salinity $S = 6.2$, 8.4, and 13.2 ppt from different depths within 1.7 m thick FY fast ice in McMurdo Sound, Antarctica. Lines are Equation (3.4.2).

calorimeter is sketched in Figure 3.4.1. Heat was applied with a resistive coil of heating wire attached to the base of a calorimeter made of aluminum alloy and insulated with polystyrene. Core samples with a diameter of 7.6 cm were fit tightly in the 7.3 mm diameter calorimeter by tamping past a circular steel blade on the calorimeter rim. A quasi-static analysis was developed to account for slowly varying background temperatures (Johnson 1989). The sample size and applied heat were selected to ensure the time to reach thermal equilibrium was much shorter than times associated with thermal leaks. Figure 3.4.1 shows an excellent agreement between predictions from Equation 3.4.2 and the measurements of Johnson (1989) and Trodahl et al. (2000).

### 3.4.2.3 Thermal Conductivity

The bulk thermal conductivity of sea ice depends on the volume fraction and spatial arrangement of brine and gas inclusions. This is because the thermal conductivity of ice (~ 2 W $m^{-1}$ $K^{-1}$) is several times that of brine (~ 0.6 W $m^{-1}$ $K^{-1}$). The conductivity of air is much lower (~ 0.025 W $m^{-1}$ $K^{-1}$) so the bulk conductivity is almost inversely proportional to density. Theoretical models of sea ice thermal conductivity developed for idealized geometries of air and brine inclusions can be found in Yen (1981) and Pringle (2005).

A weakness of these models is that they do not provide simple analytical expressions for bulk sea ice thermal conductivity $k$ in terms of bulk salinity and temperature. However, fitting simple appropriate analytical expressions to the predictions of these models gives, in units of W m$^{-1}$ K$^{-1}$ (Pringle et al. 2007):

$$k = \frac{\rho}{\rho_i}\ 2.11 - 0.011T + 0.09\frac{S}{T} - 0.01(\rho - \rho_i) \quad \text{(Equation 3.4.4).}$$

Here $\rho$ and $\rho_i$ are the density of sea ice and fresh ice, respectively (kg m$^{-3}$). When $\rho > 890$ kg m$^{-3}$ this can be simplified, with a loss in accuracy of less than 2 percent, to:

$$k = \frac{\rho}{\rho_i}\left(2.11 - 0.011T + 0.09\frac{S}{T}\right) \quad \text{(Equation 3.4.5).}$$

Collated results are in reasonable agreement with these predictions; see Figure 3.4.2 discussed below.

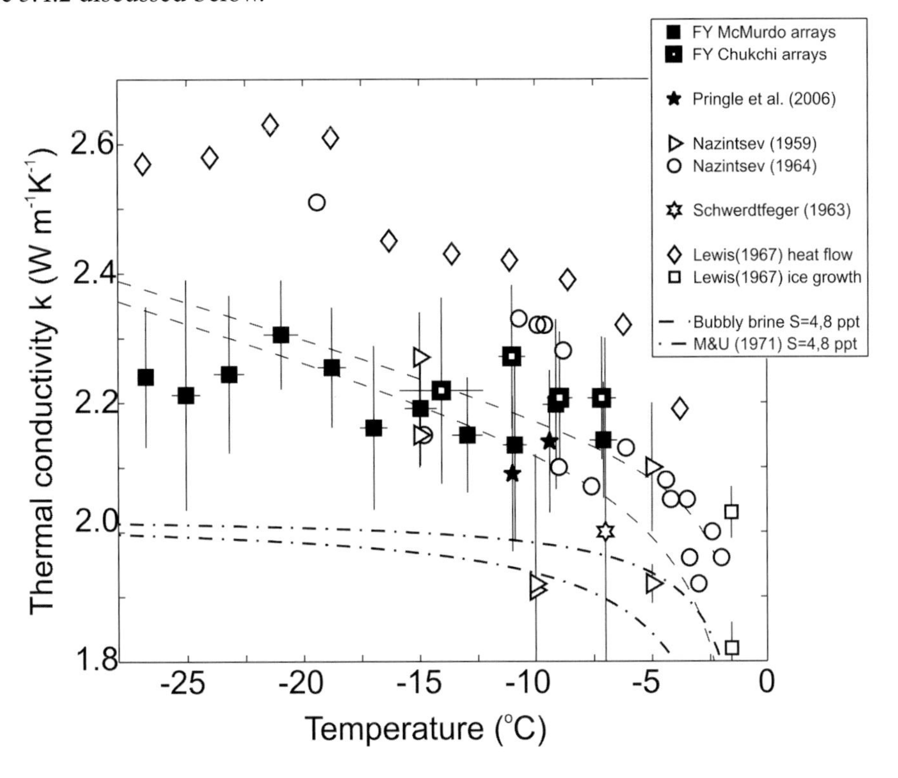


**Figure 3.4.2.** Results for sea ice thermal conductivity in FY and artificial ice. All reliable measurements are shown, including thermistor array analysis and laboratory measurements on natural and artificial samples. The high scatter reflects measurement difficulty. A much better fit is found for the "bubbly brine" effective medium model from which Equation (3.4.4) was derived, than the expression from Maykut and Untersteiner (1971) proposed largely with MY ice in mind but still commonly used in sea ice models for MY and FY ice, $k = 2.03 + 0.117\ S/T$. (Pringle et al. 2007)

### *Measuring Thermal Conductivity*

Field and laboratory measurements of sea ice thermal conductivity have been made since the growth-rate measurements of Stefan (1891). The two paramount measurement considerations are constraining the heat flow and minimizing disturbance to the microstructure.

As temperatures are easier to measure than heat flow itself, a common approach has been to make thermal diffusivity measurements and calculate the conductivity using either measured or predicted specific heats. This approach has been used in laboratory measurements of the thermal diffusivity of arctic and artificial saline ice samples (Nazintsev 1959), the analysis of antarctic sea ice temperatures (Weller 1967), and the temperature response of ice in Mombetsu Harbor due to an artificial heat sink on the surface (Ono 1968). An in situ hot-wire method was used by Malmgren (1927).

Schwerdtfeger extracted a 1.5 × 1.5 × 0.3 m block of first-year (FY) sea ice, refroze the hole with freshwater, and measured the temperature gradients in this fresh ice ($\nabla T_i$) and the adjacent sea ice ($\nabla T$). Sea ice thermal conductivity was calculated as $k = (\nabla T_i/\nabla T_{si})\, k_i$ where $k_i$ is the thermal conductivity of fresh ice at the measured temperature. The conductivity has also been calculated from in situ temperature measurements of $T(z,t)$ with various heat flow and ice-growth analyses (e.g., Stefan 1891; Malmgren 1927; Lewis 1967).

The most recent example of the heat flow approach was a conservation of energy analysis of high-precision thermistor string temperature measurements. Initial results suggested two intriguing results: a conductivity reduction in the upper 20–50 cm, tentatively attributed to enhanced phonon scattering for small crystals in the granular surface ice, and a possible enhancement by brine convection in warm ice near the ice-ocean interface (McGuinness et al. 1998; Trodahl et al. 2000; 2001). These effects were later found to be artifacts of the analysis, and under conditions in which this analysis is accurate, results were in good agreement with predictions of Equation 3.4.4 (Pringle et al. 2006; 2007). Direct field-laboratory measurements on extracted cores (10 cm × 2.5 cm diameter) of FY and multiyear (MY) ice were also consistent with these predictions and showed no surface reduction (Pringle et al. 2006). Figure 3.4.2 shows collated results from the above measurements.

### *Details of Thermistor Array Conductivity Measurements*

In the thermistor array measurements made by the Victoria University of Wellington program (McGuinness et al. 1998; Trodahl et al. 2000; 2001; Pringle et al. 2007), a conservation of energy analysis was applied, using Equation 3.4.2 for the specific heat to convert a "diffusivity measurement' into a conductivity measurement. For one-dimensional heat flow, the temperature response is given by:

$$\rho \frac{\partial U}{\partial t} = k \frac{\partial^2 T}{\partial z^2} + \frac{\partial k}{\partial z} \frac{\partial T}{\partial z} \quad \text{(Equation 3.4.6)},$$

where ρ is density, and $U(S,T)$ is the sea ice internal energy per unit mass found by integrating the specific heat in Equation 3.4.2 with respect to temperature. After low-pass filtering, the derivatives in Equation 3.4.6 were determined by finite differences. The conductivity was then determined as the gradient of scatter plots of $\rho\partial U/\partial t - (\partial k/\partial z)(\partial T/\partial z)$ vs. $\partial^2 T/\partial z^2$, as in Figure 3.4.3.

Due to the small temperature differences involved in these derivatives, the precision of temperature measurements is paramount here. To avoid numerical artifacts attention must be paid to the effect of finite sampling intervals in space and time. This analysis returns profiles of $k(z)$. Temperature-binned results from landfast FY ice in McMurdo Sound and the Chukchi Sea at Barrow, Alaska, are shown in Figure 3.4.2.

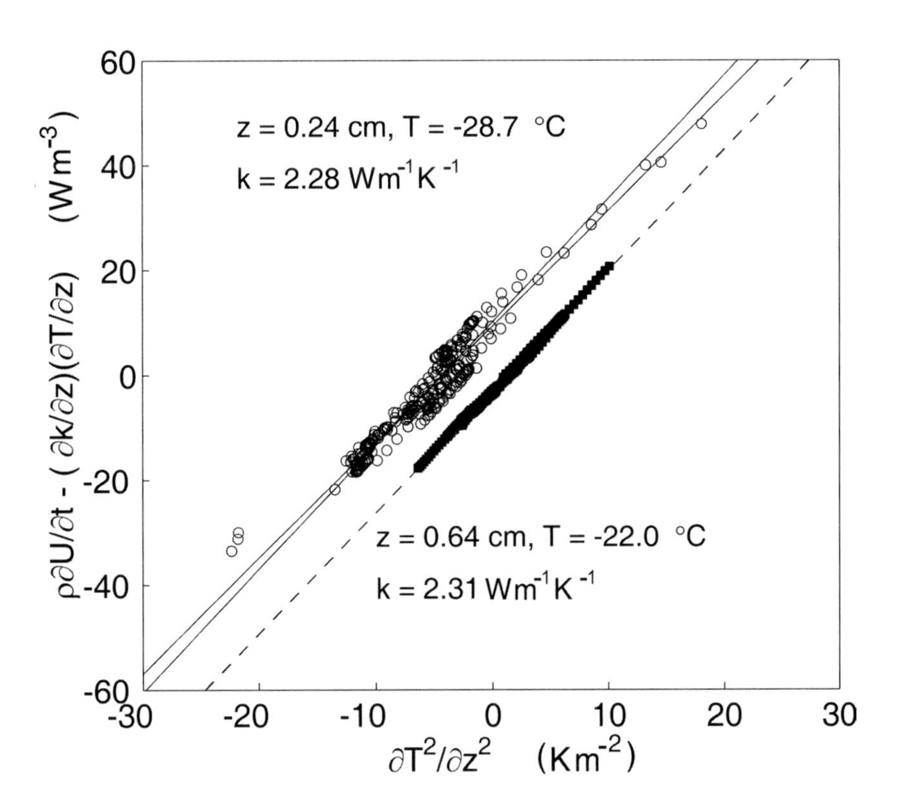


**Figure 3.4.3.** Determination of thermal conductivity $k$ from finite difference analysis of natural sea ice temperature variations. Shown are data from two different depths. These are offset from each other due to small thermistor calibration offsets, but this does not affect $k$ given by the gradient. (Pringle et al. 2007)

*Convective Heat Flow*

Convection and advection of brine can enhance conductive (i.e., molecular-diffusive) heat flow. A convective event in which cold, saline brine travels downwards and is replaced by warmer, less saline brine from below results in a net transfer of sensible and latent heat from ocean to atmosphere. The relative size of this contribution depends on location and ice type. From ten seasons of in situ temperature records in landfast FY ice, Pringle et al. (2007) estimated the convective contribution to the spring/winter heat flux to be a few percent of the total winter heat flow. For cold, low-salinity MY arctic pack ice, the contribution will be less. For FY antarctic ice with a sufficiently high snow loading to suppress the ice-snow interface below sea level, widespread flooding can occur with a significant contribution to the heat transfer (Lytle and Ackley 1996).

### 3.4.2.4 Thermal Diffusivity

The thermal diffusivity describes the propagation of temperature variations in a material. It is a compound quantity that factors in the rate at which heat is conducted ($k$) and the heat required for a temperature increase per unit volume ($\rho c$),

$$\alpha = \frac{k}{\rho c} \qquad \text{(Equation 3.4.7).}$$

As the brine volume fraction increases in sea ice, $c$ increases and $k$ decreases. Therefore $\alpha$ is increasingly dependent on salinity as temperature increases. This reduction of two parameters to one is not particularly useful in the case of sea ice. However, because temperature variations are easier to measure than heat flows, $k$ has generally been calculated either from measurements of diffusivity or by "diffusivity-like" measurements of temperature with various heat flow analyses.

### 3.4.2.5 Temperature Measurements

Temperature measurements are frequently made for ice characterization and monitoring. Coupled with salinity profiles, they enable calculation of brine volume profiles and other ice properties. The two common approaches to measure sea ice temperature profiles are: (1) data loggers connected to in situ thermistor strings, typically used in drifting buoys and landfast monitoring sites (e.g., Richter-Menge et al. 2006; Druckenmiller et al. 2009; Chapter 3.13); and (2) handheld thermometers, used on extracted cores.

*In Situ Ice Temperature Measurements*

The first in situ sea ice temperature profiles were made by Malmgren (1927) using galvanometers and nickel thermometers attached to wooden boards frozen into drifting ice. The present approach of automated measurements of thermistor strings was first employed by Lewis (1967). Thermistors are devices whose electrical resistance varies strongly with temperature and whose small size (~ 2 mm diameter) gives them a small thermal mass and rapid response to temperature changes. Thermistors dissipate very little power (~ μW) so self-heating is typically not a concern. Measurement precision is limited by analog to digital conversion (ADC) in the data logger used to measure electrical resistance, *R*. This can be optimized by attention to detail in the bridge circuits typically used. For the current industry-standard 13-bit ADC Campbell CR10X data loggers, this is approximately ± 0.02°C. The accuracy in measured temperature is limited by the inversion from resistance to temperature. Each type of thermistor has a characteristic $R(T)$ relationship and a manufacturer-specified inversion to calculate $T(R)$ over the temperature range of choice. Variability between thermistors is quoted as an "interchangeability." typically about ± 0.1°C. This uncertainty can be reduced by one-point calibrations of individual thermistors, for example, in an ice-water bath or in well-mixed seawater at the time of deployment. In practice, an absolute accuracy to better than 0.01°C is difficult. We recommend calibrating individual sensors to ± 0.05°C (or better) if possible.

To optimize accuracy, any embedded array should cause minimal disturbance to the temperature field and heat flow in the ice. "Thermal short circuiting" in which heat is preferentially conducted up the array can be avoided by matching the total thermal conductance of the array to the ice it displaces (Trodahl et al. 2000). Figure 3.4.4 (a,b) shows two approaches to optimize thermal coupling of

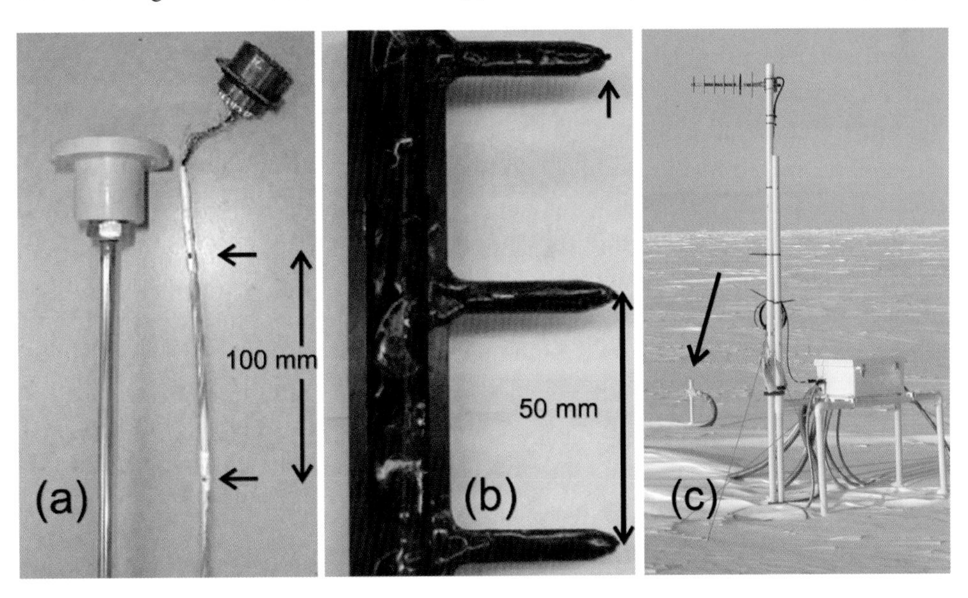


**Figure 3.4.4.** Thermistor string designs and installation. (a) VUW 6.35 mm wide tube with thermistors mounted on removable Teflon "spine." (b) UAF "finger" design with protruding thermistors; (c) CRREL-designed, 2.5 cm diameter strings in UAF wireless mass balance site, at Barrow, Alaska. Note conduit to protect cables from arctic foxes, and instrument box elevated to reduce snow buildup.

thermistors to the ice. The Victoria University of Wellington (VUW) design has thermistors mounted on a removable Teflon spine inserted into a thin-walled (0.125 mm), 6.35 mm diameter stainless-steel tube. The void space is filled with sunflower oil, which congeals at ice temperatures. This design minimizes size and closely matches total thermal conductance of the array to the ice it displaces. The University of Alaska Fairbanks (UAF) design-mounts thermistors on protruding fingers for intimate contact between thermistors and surrounding ice. Thermistors and wiring are sealed in epoxy, complicating repairs. The modular CRREL-developed thermistor strings used in drifting ice mass buoys are well suited when robustness or reusability are paramount (e.g., Richter-Menge et al. 2006). These are 1 m long, 2.5 cm diameter, epoxy-filled PVC pipes with thermistors embedded close to the side surface. The lengths can be coupled together for use in thick ice. The top extension of one such array is seen in Figure 3.4.4c.

*Temperature Measurements on Extracted Cores*

Temperature profiles are commonly made in the field on extracted cores with handheld electronic thermometers. The typical procedure is to secure the extracted core on a clean, flat surface; drill small holes from the perimeter to the center of the core; and move down the core, placing the sensor in the holes. The core should be shaded from direct sun and the temperature profile measured quickly before it changes. As sea ice thermal diffusivity is highest at low salinities and low temperatures, cold ice responds fastest in this regard. Fortunately, such cold ice is often close to the ambient temperature, except under thick snow. A common rule of thumb is to make measurements in less than five minutes from extraction (H. Eicken, personal communication, 2008).

This rule of thumb is supported by conduction calculations using $T$, $S$ profiles for cores extracted in Barrow. We conclude that if solar heating can be avoided, then temperature profiles on extracted cores can easily be accurate to better than 0.5°C. Eicken et al. (2004) quote an accuracy of 0.2 K. Our calculations suggest the recommendation to start measurements at the coldest end of the core and move towards the warmer end. This may seem counterintuitive. However, the rate of temperature change is determined not just by the temperature difference between the ice and the air but also by the thermal diffusivity. The diffusivity of warm, saline ice is much lower than cold, fresh ice, and therefore the rate of temperature change (for a given temperature difference) is faster for cold, fresh ice than it is for warm, saline ice.

The characteristic thermal response time scales as $t_T \sim d^2 / \alpha$ where $d$ is a characteristic length scale and $\alpha$ the thermal diffusivity. Thermal response times for $d = 7.5$ cm diameter cores will be nearly half of those for 10 cm cores: the smaller the core, the faster the measurement should be made. As to the optimum vertical separation between measurement holes, it is common to use 2, 5, or 10 cm.

The overarching considerations are to minimize measurement time and to capture important variations in the temperature profile. In general, vertical spacing of 5 cm is a good starting point.

#### 3.4.2.6 Thermal Properties Outlook and Summary

The specific heat and latent heats and conductive heat flow in sea ice are well understood, with good comparison between predictions and measurements. An important remaining question is to assess convective and advective heat transport, and their parameterization for inclusion in large-scale models. A related area of current research is the evolution of melt ponds, whose impact on the sea ice heat budget through formation, persistence, drainage, and superimposed ice formation is neither fully understood nor incorporated in large-scale models (Eicken et al. 2004).

Equations 3.4.2 and 3.4.4 provide an accurate all-purpose means to calculate the specific heat and thermal conductivity. However, Equation 3.4.4 differs from that typically used in both small-scale ice growth and large-scale models, many of which use either constant values (i.e., salinity and temperature independent) or a conductivity expression derived for MY ice (Untersteiner 1961; Maykut and Untersteiner 1971). Reasons likely include the need for parameterization changes to be easily implemented, and their effect on output, stability, and parameter tuning to be assessed.

### 3.4.3 ELECTRICAL PROPERTIES

#### 3.4.3.1 Introduction

The electrical properties of sea ice describe its interaction with electromagnetic (EM) fields. The complex dielectric permittivity controls reflection, scattering, absorption, and transmission of incident EM radiation, and the dc conductivity describes the current induced by a static electric field. These properties are used principally for remote sensing and ice characterization, and they underlie the optical properties impacting heat and mass balance (see Chapter 3.6).

The permittivity of sea ice at GHz frequencies has long been a focus of attention because of its importance for sea ice remote sensing (see Chapter 3.15). At MHz frequencies, for which the wavelength is significantly greater than the scale of features within the ice, the permittivity is related to both the geometry of brine inclusions and the brine volume fraction. Measurement of the permittivity at these frequencies has been examined for its possible use for automated sea ice salinity measurements. The geometry and connectivity of brine inclusions have a significant effect on the dc electrical conductivity, and measurements of this parameter aim to unravel how the ice microstructure evolves with temporal changes in temperature and salinity.

The single biggest challenge in measuring the dielectric permittivity and the electrical conductivity of sea ice is the need to make measurements with the ice in a natural, undisturbed state.

### 3.4.3.2 Definitions

*Dielectric Permittivity*

The dielectric permittivity of a material describes the manner in which electromagnetic waves are reflected and/or transmitted when they are incident on its surface. It is related to the ability of the material to polarize in response to an applied electric field. The dielectric permittivity is generally expressed in the complex form

$$\varepsilon = \varepsilon' - j\varepsilon'' \quad \text{(Equation 3.4.8)},$$

in which both $\varepsilon'$ and $\varepsilon''$ are functions of frequency and $j = \sqrt{-1}$. The real part, $\varepsilon'$, gives the contrast relative to free space and is often referred to as the dielectric constant. The imaginary part, $\varepsilon''$, represents a loss factor related to the loss of electromagnetic energy in the material.

The bulk permittivity of sea ice depends on the volume fractions, arrangement, and permittivity of its components. The permittivity of solid ice and brine, the principal constituents of columnar first-year sea ice, both obey the Debye equation (Debye 1929):

$$\varepsilon = \varepsilon_{\infty} + \frac{\varepsilon_l - \varepsilon_{\infty}}{1 + j\omega\tau} - \frac{j\sigma}{\varepsilon_0\omega} \quad \text{(Equation 3.4.9)}.$$

Here $\varepsilon_l$ is the static (i.e., dc) permittivity, $\varepsilon_{\infty}$ is the permittivity at high frequency, $\tau$ is the relaxation time of the material (the characteristic time over which the material loses the polarization induced by an applied electric field), $\omega$ is the angular frequency, and $\varepsilon_0$ is the permittivity of free space. In a good electrical conductor the third term on the right-hand side of Equation 3.4.9 dominates the imaginary part of the complex permittivity, and $\varepsilon''$ may be expressed as

$$\varepsilon'' = \frac{\sigma}{\omega\varepsilon_0} \quad \text{(Equation 3.4.10)}.$$

The contrast between the permittivity of solid ice and brine makes the bulk permittivity of sea ice dependent on their relative volume fractions. Figure 3.4.5 shows $\varepsilon'$ and $\varepsilon''$ for 10–16 GHz. The GHz response of $\varepsilon'$ in particular can be related to the brine volume fraction (Morey et al. 1984; Arcone et al. 1986), while $\varepsilon''$ is also dependent on the geometry of brine inclusions. Ice with a high alignment of brine layers is highly anisotropic in the microwave with $\varepsilon$ dependent on the orientation of the applied electric field with respect to the brine layers (Kovacs and Morey 1979; Morey et al. 1984).

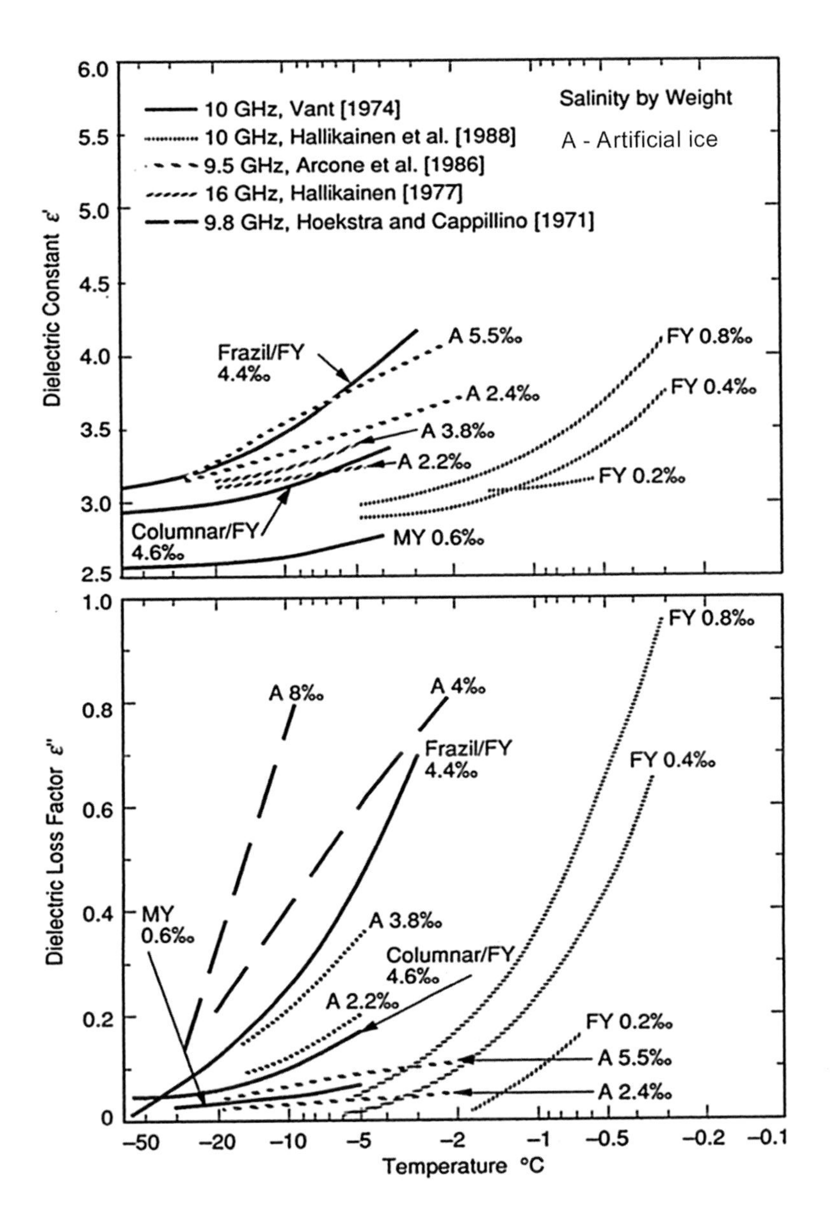


**Figure 3.4.5.** Experimental values of sea ice permittivity at 10 to 16 GHz, as a function of temperature and salinity by weight (Hallikainen and Winebrenner 1992)

Methods for sea ice permittivity measurements described below derive ε primarily from its effect on the transmission/reflection amplitudes of EM waves and on capacitance/impendence. Useful references for further theoretical details and previous measurements are Tinga et al. (1973), Morey et al. (1984), Arcone et al. (1986), Stogryn (1987), Hallikainnen and Winebrenner (1992), and Pringle et al. (2009).

*dc Conductivity and Resistivity*

The dc electrical conductivity $\sigma$ [S/m] measures the electric current that passes through the material when a potential difference (voltage) is applied across it. It is defined by Ohm's law as in Table 3.4.1. Geophysics applications typically consider the inverse of the conductivity, the resistivity $\rho$ [$\Omega$m]. In general, sea ice has a much higher resistivity than seawater, which due to the presence of salt ions has one of the lowest resistivities of any substance occurring naturally in bulk ($\rho \approx 0.2$–$0.5\ \Omega$m). The resistivity is even lower for concentrated brine inclusions. Due to the high resistivity contrast between brine and ice, the bulk resistivity of sea ice is highly dependent on both the volume fraction of brine inclusions and their connectivity. The preferential vertical alignment of brine inclusions make the resistivity anisotropic, with a horizontal resistivity greater than the vertical resistivity, $\rho_H > \rho_V$.

The resistivity is an intrinsic property, as opposed to the electrical resistance $R$ [$\Omega$], which depends on sample geometry. For the simple case of a wire of length $L$ and cross-sectional area $A$ the resistance is $R = \rho L/A$. For the measurement geometries addressed below, numerical inversions are required to determine the resistivity from the measured ratios of potential difference to current, $\Delta V/I$.

The dc resistivity of sea ice has been measured primarily with surface-based "resistivity soundings" (also made in vertical pits) and more recently with cross-borehole resistivity tomography between in situ vertical strings of electrodes.

### 3.4.3.3 Dielectric Permittivity Measurements

*Transmission and Reflection Permittivity Measurements*

Several methods have used the permittivity-dependence of the transmission and reflection of EM waves. The partial reflection and transmission of an EM wave incident on a boundary between two different materials is governed by the discontinuity in their EM properties. The situation can be analyzed using wave theory, by considering an alternating signal to be part reflected and part transmitted by circuit elements depending upon their impedance mismatch. The reflection coefficient from a load $Z_L$ when a signal is sent through a circuit of impedance $Z$ is (e.g., Bowick 1997):

$$Re = \frac{Z_L - Z}{Z_L + Z} \quad \text{(Equation 3.4.11).}$$

The permittivity can therefore be calculated from reflection measurements provided that its relationship to the load impedance is known.

*50 MHz Hydraprobes*

Such a relationship between permittivity and load impedance is known for the dielectric probes originally developed by Campbell (1988; 1990) to measure water content and salinity of soils. A commercial version of these probes is the Stevens Water Monitoring System Hydra Probe (hydraprobe) and measures the complex permittivity at 50 MHz.

Campbell's original design of has a central tine of about 5–10 cm length surrounded at a radius of 1–2 cm by six other parallel tines. The six outer tines are electrically connected through a circular guard ring, while the central tine is insulated from this but used, via a BNC connector, for input of an ac signal. This arrangement approximates a concentric coaxial transmission line terminated by an open circuit in parallel with a stray capacitance. As such, the probes can be used to measure the permittivity of the material between the central and outer tines.

The ratio of reflected to incident voltage can be used to calculate the total impedance of the probe, $Z_T$. This is given by

$$Z_T = \frac{1}{\frac{1}{Z_P} + \frac{1}{Z_S}} \quad \text{(Equation 3.4.12)},$$

$Z_P$ being the impedance of the coaxial transmission line and $Z_S$ the constant stray impedance (determined by calibration in, e.g., air and water), allowing $Z_P$ to be found from any measurement of $Z_T$. The impedance $Z_P$ takes the standard form for a coaxial transmission line of length ($L$), inner ($a$) and outer ($b$) diameters, and filled with a dielectric material of relative permittivity ε (Campbell 1988):

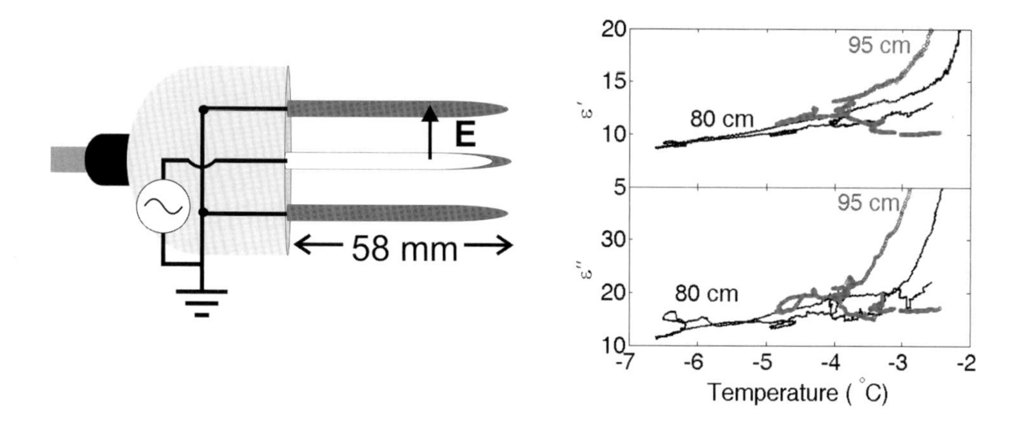


**Figure 3.4.6.** Schematic of 50 MHz hydraprobe and permittivity results from FY ice in Barrow. Dielectric constant ε′ and loss factor ε″ measured at 80 and 95 cm below upper ice surface. (Pringle et al. 2008)

$$Z_P = \frac{2}{c\sqrt{\varepsilon}} \log(b/a) \operatorname{cotanh}\left\{ j\frac{\omega\sqrt{\varepsilon}}{c} L \right\} \quad \text{(Equation 3.4.13)},$$

where $c$ is the speed of light. For small arguments, the cotan term can be expanded, giving

$$Z_P = \frac{2\log(b/a)}{j\omega L\varepsilon} \left\{ 1 - \frac{\varepsilon\omega^2 L^2}{3c^2} + \frac{\varepsilon^2\omega^4 L^4}{45c^4} + \ldots \right\} \quad \text{(Equation 3.4.14)}.$$

Thus measurement of $Z_p$ allows the permittivity to be calculated, although the solution for $\varepsilon$ is only analytical if terms beyond the second in this expansion can be ignored. Practically, this renders the probes unreliable when $\varepsilon' >> \varepsilon''$ and, of more relevance to sea ice, when $\varepsilon'' >> \varepsilon'$ (Seyfried et al. 2004; 2005; Pringle et al. 2009).

Vertical arrays of hydraprobes have been deployed below growing landfast FY sea ice with tines protruding from the convex side of a half-pipe of 10 cm diameter PVC (Backstrom and Eicken 2006; Pringle et al. 2009). Probe output was converted into values of $\varepsilon'$ and $\varepsilon''$ with proprietary software. These measurements did show brine motion in warming ice, but are ultimately unreliable for automated salinity measurements because of: subrepresentative sample volumes, not always large compared with the scale of brine features; the intrinsic dependence of $\varepsilon'$ and $\varepsilon''$ on sample microstructure when the microstructure is unknown; and occasional malfunction at the low temperature, high-salinity conditions outside manufacturer specifications (Pringle et al. 2009). Nevertheless, purpose-built probes based on the same principle but with closer measurement control may hold promise.

*Other Transmission and Reflection Methods*

The earliest widely referenced measurements on natural sea ice were reported by Vant et al. (1978) at frequencies up to 40 GHz. Measurements were made of the transmission coefficient of ice cores taken from the Beaufort Sea. To reduce brine drainage, cores were cut, handled, and stored at temperatures below –20°C. Results were interpreted both in terms of an empirical correlation of $\varepsilon$ with brine volume fraction, and in terms of the ellipsoidal inclusion model of Tinga et al. (1973).

The first attempt at in situ measurements of the permittivity of artificial sea ice (grown in a large, 12.2 × 5.2 m pool) was that of Arcone et al. (1986). A slot cut in the ice allowed a microwave transmitter to be placed beneath the ice with a receiver antenna at an appropriate height above the ice.

Campbell and Orange (1974) and Kovacs and Morey (1978) used on-ice radar and measured the reflection of short 100–300 MHz pulses from the base of the ice. These measurements showed sea ice to behave as a microwave polarizer, with very low basal reflection when highly aligned ice lamellae were parallel with the polarization of the pulses. These observations in fact identified c-axis alignment induced by prevailing ocean currents (Kovacs and Morey 1978).

Related ground penetrating radar (GPR) measurements in the hundreds of MHz range have been explored as a means to detect oil in ice based on their permittivity contrast (Dickens et al. 2005; 2006; Steinbronn et al. 2007).

*Impedance-Based Permittivity Measurements*

In situ permittivity can be measured via the impedance of a capacitor frozen into the ice. This is essentially the method reported by Addison (1969) for laboratory-grown NaCl ice between 10 Hz and 30 MHz, and later by other authors up to 30 GHz (e.g., Addison 1970; Hoekstra and Cappillino 1971).

The impedance of a parallel plate capacitor with plate area $A$ and separation $L$ depends upon the dielectric permittivity of the material between the plates, according to

$$C = \varepsilon\varepsilon_0 \frac{A}{L} \qquad \text{(Equation 3.4.15).}$$

When $\varepsilon$ is complex, the impedance has resistive and reactive parts $Z = R + jX$ from which both $\varepsilon'$ and $\varepsilon''$ can be measured as

$$\varepsilon' = -\frac{X}{R^2 + X^2} \frac{L}{A\omega\varepsilon_0} \qquad \text{(Equation 3.4.16a),}$$

$$\varepsilon'' = \frac{R}{R^2 + X^2} \frac{L}{A\omega\varepsilon_0} \qquad \text{(Equation 3.4.16b).}$$

Addison (1969) used this approach for laboratory measurements on ice grown around arrays of parallel plates. The plates were made of nylon mesh, coated by evaporation with gold and fixed across a 7.6 cm diameter rigid Lucite tube that was open at the bottom to allow circulation of seawater. In situ permittivity measurements of natural sea ice have not been made with this method. It holds promise but would require equipment modifications to minimize the difference between the ice grown between the plates and typical ice.

*Parallel-Wire Impedance Measurements*

A capacitance-based approach to in situ salinity measurements has been pursued recently by Notz and coworkers (Notz 2005; Notz et al. 2005; Notz and Worster 2008). They made 2 kHz impedance measurements in artificial and natural sea ice grown around parallel pairs of horizontal platinum wires with a lateral separation of 5 mm. The impedance is dominated by the capacitive contribution from charged double layers, making the measurement sensitive to the brine volume fraction surrounding the wires.

In their notation, they determined the liquid fraction 1-$\phi$ (where $\phi$ is the solid fraction) from the ratios of brine conductivity $\gamma$ and impedance $Z$ at and after freeze in. Notz et al. (2005) derived an empirical relationship for $\gamma(S_b, T)$ for their equipment. The liquid fraction is calculated from this as:

$$1-\phi) = \frac{\gamma_0 Z_0}{\gamma(S_b, T)Z} \quad \text{(Equation 3.4.17)},$$

where $\gamma_0$ and $Z_0$ are measured at initial ice growth, and $\gamma(S_b,T)$ and $Z'$ in subsequent measurements. The bulk salinity can then be calculated from the liquid fraction and the brine liquidus relationship.

While the low-profile of the small-diameter wires does minimize disturbance to the growing ice, the method assumes that the brine fraction surrounding the wire is equal to that away from the wires, and deployment of the "harp" of wires is problematic for ice covers that have already formed. It is, nevertheless, the field method with the most success towards automated salinity measurements.

### 3.4.3.4 dc Resistivity Measurements

*Resistivity Soundings*

In many geophysical applications the subsurface resistivity is measured by passing a current $I$ between two surface electrodes and measuring the resulting potential difference $V$ between two other surface electrodes. A wider separation of the current electrodes forces more of the current to a greater depth and therefore samples the deeper resistivity.

A resistivity sounding is a series of such measurements made by systematically varying the electrode spacings. Several standard electrode geometries exist for which the resistivity structure can be found via numerical modeling and inversion techniques. Surface-based resistivity soundings are easily made, although current injection may be difficult or impossible through superimposed ice in the melt season or in desalinated ice. They return the geometric mean resistivity and underestimate ice thickness.

We sketch the theory underlying resistivity soundings for isotropic media. Further details can be found in the supplementary material on the accompanying DVD. Consider an electric current $I$ injected into a half-space of uniform resistivity $\rho$ at a single surface point. The potential due to this current varies as

$$V = \frac{\rho I}{2\pi r} \quad \text{(Equation 3.4.18)},$$

(assuming that $V \to 0$ as $r \to \infty$). For the geometry in Figure 3.4.7, in which the current is injected into the ground at electrode $C_1$ and taken out at electrode $C_2$, the potential difference $\Delta V$ measured between electrodes $P_1$ and $P_2$ is

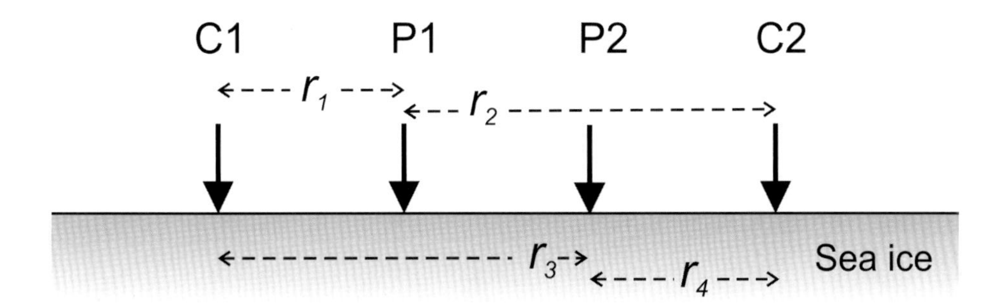


**Figure 3.4.7.** Generalized electrode geometry for resistivity sounding.

$$\Delta V = V_1 - V_2 = \frac{\rho I}{2\pi}\left(\frac{1}{r_1} - \frac{1}{r_2} - \frac{1}{r_3} + \frac{1}{r_4}\right) \text{(Equation 3.4.19).}$$

The uniform ground resistivity ρ can therefore be calculated by measuring $\Delta V$ for a known current $I$, and electrode positions. A half-space of uniform resistivity does not generally occur in nature. However, it is still standard practice to measure $\Delta V$, $I$, $r_1$, $r_2$, $r_3$, and $r_4$ and to express the result in terms of an *apparent resistivity*

$$\rho_a = \frac{2\pi R}{G} \quad \text{(Equation 3.4.20),}$$

where $R$ is the ratio of the measured potential difference and the current( $\Delta V/I$ ) and $G$ the geometric factor $\left(\frac{1}{r_1} - \frac{1}{r_2} - \frac{1}{r_3} + \frac{1}{r_4}\right)$ depending on electrode positions.

A dc resistivity sounding involves a series of measurements of $\rho_a$ at different electrode positions, typically by gradually moving the current electrodes $C_1$ and $C_2$ further apart to sample the deeper resistivity. Results are usually plotted as $\rho_a$ vs. electrode separation $a$, with a log-log scale as in Figure 3.4.8, and forward modeling (see below) is used to fit a resistivity structure to the data.

The above analysis applies for resistivity structures which vary in space but are isotropic, that is, the same in all directions. The detailed discussion of resistivity measurements in anisotropic media, like sea ice, can be found in Bhattacharya and Patra (1968). In this case Equation 3.4.18 is modified to

$$V = \frac{I\rho_m}{2\pi r} \quad \text{(Equation 3.4.21),}$$

where $\rho_m = (\rho_H \rho_V)^{1/2}$ is the geometric mean resistivity. Surface soundings made over a uniformly anisotropic half-space then return the apparent value of $\rho_m$. When the ice has a uniform value of $\rho_m$ with depth a sounding will underestimate the thickness by the anisotropy factor $\lambda = (\rho_H / \rho_V)^{1/2}$ (Ingham et al. 2008). This effect

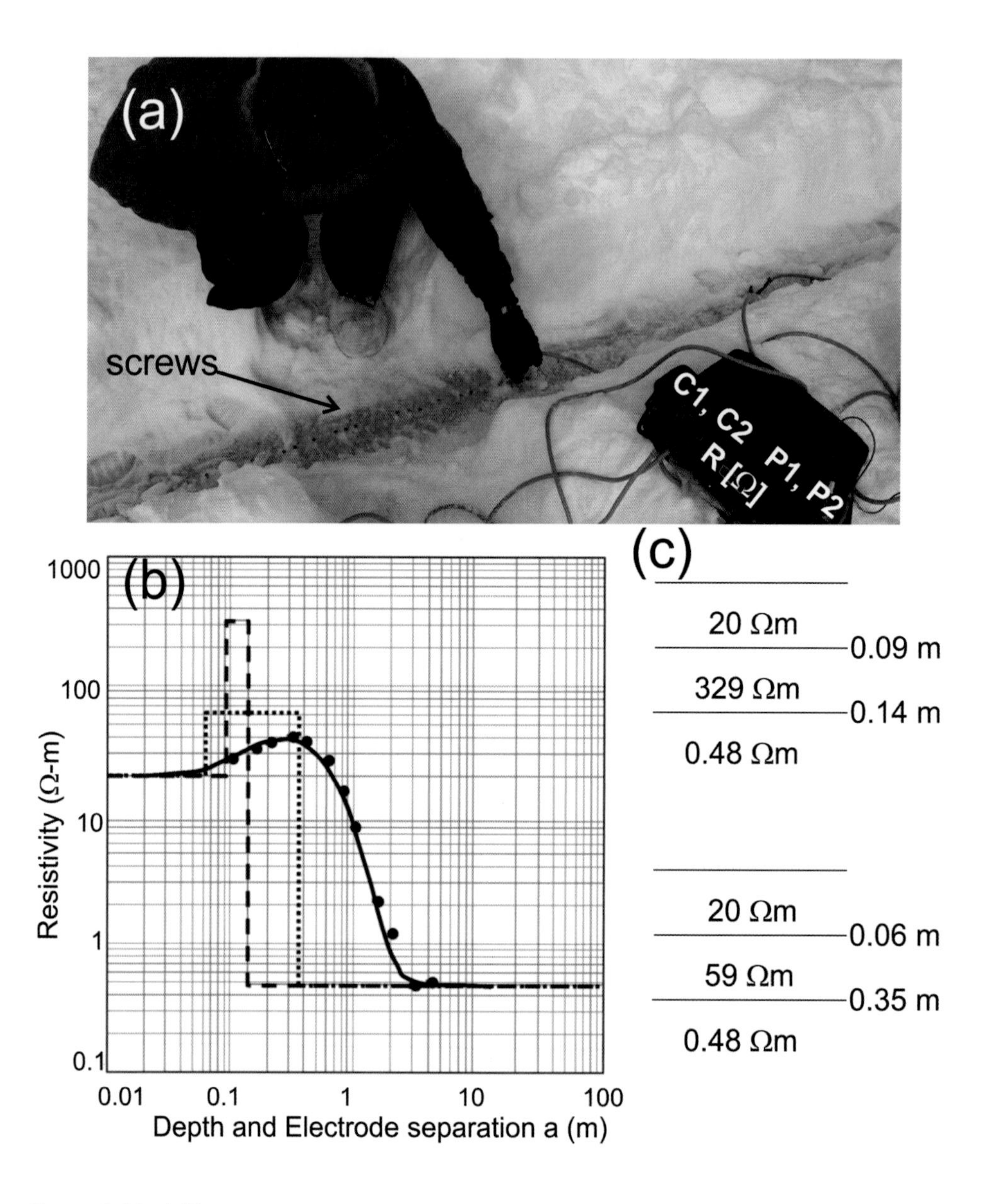


**Figure 3.4.8.** A Wenner array resistivity sounding in Barrow, Alaska. (a) Measurements on sea ice with an ABEM resistivity meter and screws for electrodes. Two current and two potential electrodes are connected to the meter, which displays the measured resistance. (b) Results and model fits. Dots are measured $\rho_a$ as a function of electrode spacing. Dashed and dotted lines are 1D layered resistivity models (shown on right) plotted against depth. Solid line is resistivity curve from models fit to the data. (c) Resistivity models. The base layer of 0.48 Ωm is the underlying seawater. Actual ice thickness was 1.45 m. (Ingham et al. 2008.)

was observed in the early soundings discussed by Fujino and Suzuki (1963), Thyssen et al. (1974), Timco (1979), and Buckley et al. (1986) who found ice thickness underestimated by between 35 and 75 percent.

The results in Figure 3.4.8 are from measurements by Ingham et al. (2008) on 1.45 m thick first-year sea ice near Barrow, Alaska. A Wenner array configuration was used in which electrodes are collinear, in order C1, P1, P2, C2 with constant spacing $a$ between adjacent electrodes. For FY ice, $a = 0.1$ m is suitable to sample the very near-surface, and separations out to $a = 4$ m are sufficient to ensure current penetrates to the underlying seawater for ice 1.5 m thick. To avoid confusion in the field, it is helpful to draft a table of electrode positions for a range of $a$ values from 0.1 to 4 m using logarithmic steps. For electrodes with good electrical contact with the ice, we recommend stainless-steel screws drilled into the ice to a depth of about 2 cm. Such screws can be seen in Figure 3.4.8, which also shows a linear trench cleared of snow and an ABEM resistivity meter. This meter injects a selectable current (typically 10 mA), and the measured potential difference is displayed on the screen. Newer resistivity meters write data to files for later retrieval.

It is possible to calculate theoretically how $\rho_a$ will depend on electrode positions for any model resistivity structure, for example, layers of different resistivity. Forward modeling is then used to derive a resistivity structure by fitting the model response to the measured variation of $\rho_a(a)$. The data in Figure 3.4.8 can be fit by a number of simple three-layered models of which two are shown. The deepest layer of 0.48 Ωm represents seawater, the depth to which is severely underestimated due to the anisotropic resistivity. Differences in the two models illustrate the general non-uniqueness associated with the modeling of data in any potential-field geophysical method. Models that fit the data with an increased number of layers may also be found, but may also underestimate ice thickness.

### *Cross-Borehole Resistivity Tomography*

An alternative approach to measure the resistivity structure and the accurate thickness of sea ice is cross-borehole resistivity tomography (CBRT), in which several vertical strings of electrodes are frozen into the ice. Current is passed between two electrodes in different strings and the resulting potential difference measured between two further electrodes.

The detailed theory behind the use of CBRT measurements in an anisotropic medium is presented in Ingham et al. (2008) and in the supplementary material on the accompanying DVD. The important result is that appropriate combinations of electrodes allows either $\rho_m$ or $\rho_H$ to be measured. Inversion of these measurements returns, respectively, 2-D or 3-D images of $\rho_m$ or $\rho_H$ between the electrode strings. Only two vertical electrode strings are needed to measure $\rho_H$ but a minimum of four are needed to measure $\rho_m$. The bulk values of $\rho_H$ (and $\rho_V$) obtained from $\rho_H$ and $\rho_m$ depend critically on the horizontal (and vertical) connectivity of brine

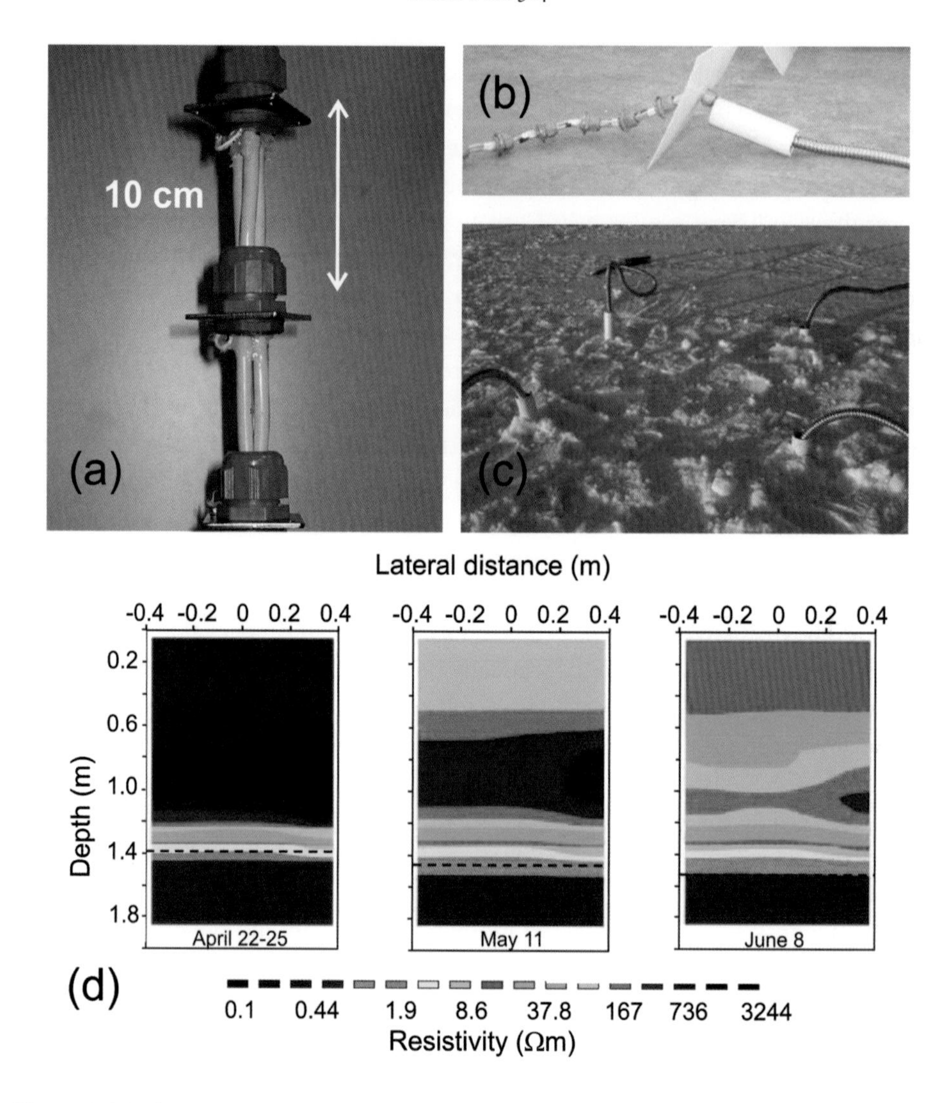


**Figure 3.4.9.** Cross-borehole resistivity tomography. (a) Marine-grade stainless-steel washers as electrodes; note holes drilled out of upper washer to improve contact with ice. (b) Polyethylene top plate, and PVC pipe and conduit for fox protection. (c) Four strings installed at corners of a 1 x 1 m square, Barrow 2008. (d) Vertical slices through 3D resistivity models derived from cross-borehole resistivity measurements made on three different dates in 2006. The horizontal dashed line marks the measured ice-water interface (after Ingham et al. 2008).

inclusions. Therefore CBRT measurements have the potential to measure microstructural evolution during the growth and melt season.

An electrode string used by Ingham et al. (2008) in landfast FY ice in Barrow, Alaska, is shown in Figure 3.4.9. Electrodes are 5 × 5 cm square marine-grade stainless-steel washers spaced at 10 cm intervals along a multicore cable and held in place by cable clamps. Takeouts from the cable are bolted onto a corner of each washer, and holes are drilled in the other corners to aid electrical conduct with the ice as it freezes on. Electrode strings are deployed though small-diameter auger holes in thin FY ice. A solid plate (Figure 3.4.9) holds the string in place relative to the surface of the ice and weights suspended from the end keep it vertical as electrodes progressively freeze in as the ice thickens. With four strings deployed at corners of a 1 × 1 m square, the measured ice is largely undisturbed. This reduces the sensitivity to any atypical ice formed around the electrodes. Metal conduit and PVC piping in Figure 3.4.9c are for protection against arctic foxes, a lesson learned from experience!

Results of these measurements are shown in Figure 3.4.9d. Inversion was performed with Res2D/3DInv proprietary software. Each panel is a section through the 3-D inversion for $\rho_H$ measured between two boreholes 1 m apart. Measurement at different times as the ice warmed through the spring season show marked changes in the resistivity structure. The general decrease in resistivity is associated both with penetration of surface meltwater and with microstructural changes associated with increased connectivity of brine pockets (Ingham et al. 2008).

The above methods are often referred to as dc (direct current) resistivity soundings, suggesting a direct current is applied between the current electrodes. However, the resistivity meters used typically inject a current that is actually a somewhat randomized square wave, reversing in direction at a low frequency (~10 Hz). This is done because of the "self-potential" of the subsurface—some materials show a voltage between the electrodes $P_1$ and $P_2$ even if there is no current induced through $C_1$ and $C_2$.

### 3.4.3.5 Electrical Properties Outlook and Summary

The permittivity of sea ice at microwave (GHz) frequencies, important for remote sensing, is satisfactorily understood. Much present work centers on the relationship between the electrical properties and sea ice microstructure at radio frequencies (kHz to MHz). This area holds promise for advancing our understanding of the connection between sea ice properties and microstructure, including surface effects and anisotropy.

A key aim is to be able to make automated measurements of salinity and brine volume fraction using electrical measurements. Reaching this goal will require improved measurement of the complex dielectric permittivity in the frequency range where $\varepsilon'' >> \varepsilon'$, either through the development of next-generation sensors or significant improvements to traditional capacitive and transmission line designs. Significant progress has been made with the parallel-wire 2 kHz impedance measurements of salinity profiles in growing ice by Notz et al. (2005) and Notz and Worster (2008). It remains to be seen whether this method is broadly suitable for fully automated, remote measurements. One issue is the deployment in already-formed ice covers. This problem might in principle be avoided with capacitance measurements in which parallel plates are deployed around an undisturbed ice volume.

The relatively new technique of cross-borehole resistivity tomography has generated new interest in dc resistivity measurements previously regarded, due to the resistivity anisotropy, to be too complex to yield useful structural information. This work may benefit from the recent development of inversion techniques able to account for anisotropy. These techniques have so far been applied only on length scales much larger than relevant for sea ice (Herwanger et al. 2004). Our own work in progress suggests that cross-borehole resistivity tomography can resolve vertical and horizontal anisotropy in the resistivity. Extension to ac tomography should provide additional information on the relationship between physical properties and microstructure, and is a viable method to measure the complex permittivity. Such extensions will require both new experimental and inversion techniques.

## 3.4.4. HYDRAULIC PROPERTIES AND FLUID PERMEABILITY

### 3.4.4.1. Introduction

Fluid flow within sea ice is important in many geophysical and biological contexts at the top, interior, and bottom of the ice. Each location has a fluid reservoir (surface water at the top, brine inclusions in the interior, and seawater at the bottom of the ice), a pressure gradient to drive flow, and a connected fluid pathway. We here focus on the pathways as measured by the fluid permeability $\Pi$.

The near-surface permeability is an important control on the fate of surface meltwater, underpinning the formation, evolution, and drainage of surface melt ponds. These ponds strongly influence the surface heat balance through the contrast in albedo between highly absorbing ponds and highly reflective bare ice beneath drained ponds (see Chapter 3.6). Extensive surface ponding also impacts over-ice travel in the spring; local residents are said to prefer the drier early-morning conditions caused by the diurnal cycle of melt and drainage (Parry 1828; R. Glenn personal communication; see Eicken et al. 2004). Within the ice, brine motion is relevant to the stable salinity profile, and therefore salinity-dependent ice properties, the ice-ocean salt flux, heat transport, and heat budget (Notz 2005; Petrich et al. 2006; and references within). The high-porosity, high-permeability basal ice enables fluid exchange with the ocean, providing nutrient delivery to resident microorganisms (Dieckmann et al. 1991; Eicken 1991; Chapter 3.8).

Brine is not the only pore fluid of interest. Heightened interest in offshore arctic oil and gas exploration and development in an era of increasing environmental awareness and accountability have rekindled questions of oil-ice interaction and remediation. Crude oil is less dense ($\rho_{oil} \approx 800$ kg/m$^3$) than seawater and sea ice, so oil spilled under ice is buoyant and will migrate into the pore space as allowed by the ice permeability (and surface tension considerations). Areas of recent research include field testing for oil spread and radar-detection methods (Dickins et al. 2005; 2006; Steinbronn 2007).

Early laboratory experiments showed sea ice to be essentially impermeable to fluid flow for brine volume fractions $v_b < 5\%$ (Cox and Weeks 1975). The permeability increases dramatically above this porosity, with a measured range in natural sea ice covering six orders of magnitude (Eicken et al. 2003). The permeability-porosity relationship is complicated by changes to the pore space through melting and refreezing associated with the flow itself, the development of secondary porosity features, deformation, and temperature cycling.

Important to the fate of surface meltwater is the development of thin layers of impermeable superimposed ice as snowmelt refreezes in the upper layers of the cold ice. A thorough understanding of sea ice hydrology therefore requires field-based measurements of permeability as well as observations of local ice characterization, snow depth, surface roughness, and other variables related to melt-season processes.

### 3.4.4.2 Background and Definitions

Fluid flow through porous media is described by Darcy's law. It relates the specific discharge $q$ (m $s^{-1}$) to the pressure gradient $\nabla P$ (Pa $m^{-1}$) through the fluid's dynamic viscosity $\mu$ (kg $m^{-1}$ $s^{-1}$ = Pa s) and the medium's fluid permeability $\Pi$ ($m^2$):

$$q = -\frac{\Pi}{\mu}\nabla P \quad \text{(Equation 3.4.22).}$$

Assuming that the area fraction available for flow is equal to the porosity, $\phi$, the average discharge velocity is given by $v = q/\phi$.

Darcy's law can be recast in terms of the hydraulic head, $h$ (m) and the "hydraulic conductivity," $K$ (m $s^{-1}$) (e.g., Freeze and Cherry 1979):

$$K = \frac{\rho g}{\mu}\Pi \quad \text{(Equation 3.4.23),}$$

$$q = -K\nabla h \quad \text{(Equation 3.4.24),}$$

where $\rho$ (kg $m^{-3}$) is fluid density and $g$ (m$s^{-2}$) is acceleration due to gravity.

Permeability is a property only of the porous medium, whereas the hydraulic conductivity depends also on the fluid through density and viscosity. The permeability is a second-order tensor, relating 3-D flows to 3-D pressure gradients, although a distinction is made for sea ice only between the vertical permeability $\Pi_v$ and the effective isotropic permeability $\Pi$. A distinction is also made between the total porosity $f_t$ and the effective porosity contributing to flow $f_e$ (Petrich et al. 2006).

Darcy's law applies to laminar flow. At sufficiently fast flow turbulence occurs, resulting in energy dissipation and a discharge less than predicted by Darcy's law. The transition between laminar and turbulent flow is characterized by a critical value of the dimensionless Reynolds number, the ratio of inertial to viscous forces,

$$Re = -\frac{\rho U l}{\mu} \quad \text{(Equation 3.4.25)}.$$

Here ρ, μ, and $U$ are fluid density, dynamic viscosity, and mean velocity, respectively, and $l$ is a characteristic length scale of the medium. In a tube, $l = 2r$ and turbulence occurs above the critical Reynolds number, $Re_C = 2300$. For complex porous media like sea ice, $l$ is an average pore diameter and $Re_C$ is in the range 1 to 10. (Bear 1972). Laminar and turbulent flow have both been analyzed in sea ice permeability measurements.

There is a length-scale dependence in the permeability that does not appear in the electrical and thermal conductivities. For a cylindrical tube, the permeability for viscous, incompressible flow (Hagen-Poiseuille flow) is $\Pi = r^2 / 8$, whereas the electrical and thermal conductivities are intrinsic material constants. This $r^2$ dependence means that permeability is very sensitive to the largest features allowing flow. A practical consequence of this is that rapid drainage of a large melt pond may occur through a large crack, seal breathing hole, or inopportunely placed core hole!

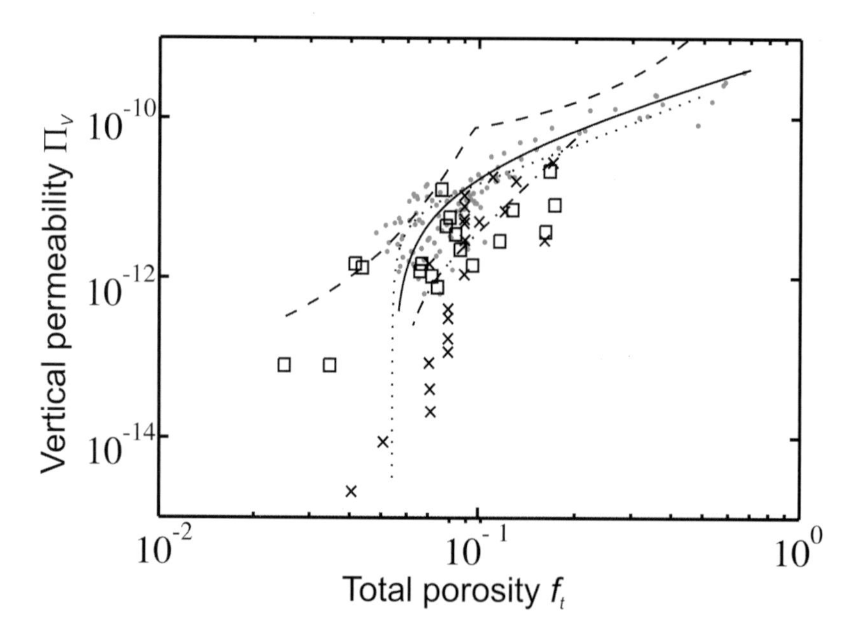


**Figure 3.4.10.** Vertical permeability as a function of total porosity (Petrich et al. 2006). Symbols are experimental results: Squares, Saeki et al. (1986); crosses, data of Ono and Kasai (1985), and Saito and Ono (1978) scaled by Maksym and Jeffries (2000); and gray dots, scaled data of Cox and Weeks (1975). Lines are analytical expressions: Solid, Petrich et al. (2006); dashed, Eicken et al. (2004), dash-dot scaled Freitag (1999) young ice; and dotted, scaled Freitag (1999) old ice.

Measurement, numerical, and analytical results for the permeability are shown in Figure 3.4.10, from Petrich et al. (2006). Note the three orders of magnitude range in permeability. A table of analytical expressions proposed for the permeability-porosity relationships can be found in the supplemental material on the accompanying DVD.

To address the threshold-like behavior in the permeability, Golden et al. (1998; 2007) have advanced a percolation theory approach. Golden et al. (1998) applied an excluded-volume model to explain the apparent threshold for fluid permeability at $\phi \approx 5\%$. For a typical FY ice salinity of 5 ppt, this corresponds to a critical temperature of approximately –5°C, giving rise to the so-called "rules of fives." Analytical and numerical excluded-volume models, random resistor networks, and Monte Carlo simulations have been applied to characterize the transition (Golden et al. 2007).

### 3.4.4.3 Field Measurements of Permeability and Fluid Flow

Field measurements of fluid flow in sea ice include tracer methods and "bail tests." In tracer methods, the concentration of some nonreactive tracer is measured at different positions and times. These measurements are passive and process oriented in that the measurements both rely on and reveal naturally occurring fluid flow. In bail tests, the permeability is measured by imposing above-normal pressure gradients.

*Bail Tests*

In situ sea ice permeability measurements have been performed by Freitag (1999), Freitag and Eicken (2003), and Eicken et al. (2004) by adapting hydrological bail tests (e.g., Freeze and Cherry 1979). Figure 3.4.11 shows a sketch of the experimental

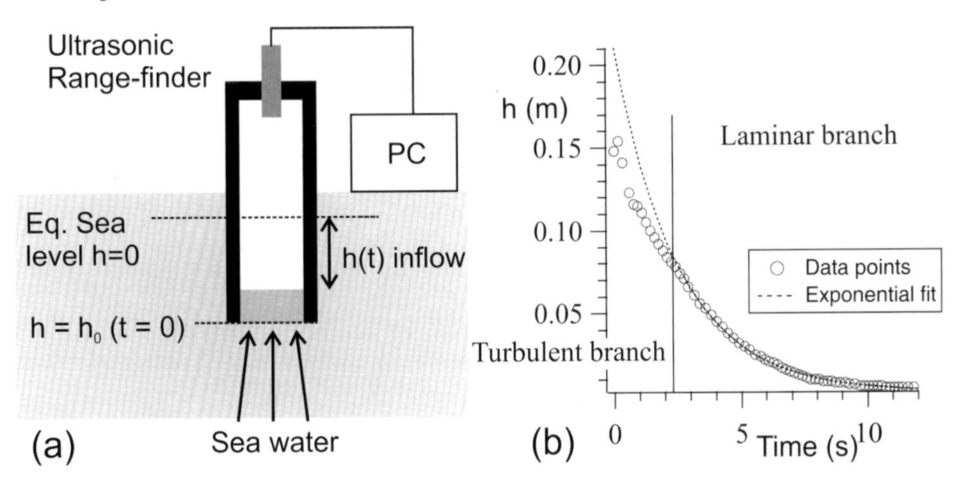


**Figure 3.4.11.** Bail tests in sea ice, after Freitag and Eicken (2003). (a) A pressure head is created in a "blind hole" by evacuating it after sealing the sides. (b) Results and fitted curve from Freitag and Eicken (2003); turbulent and laminar flow regions are discussed in the text.

approach. A sonic ranger measures the refill rate in an evacuated core hole. The hole is sealed around the sides to allow inflow only through its base. Inflow is driven by the hydraulic head $h$, being the difference between the instantaneous water level and the equilibrium seawater level.

Figure 3.4.11b shows that after some initial turbulence (discussed below) the recovery curve $h(t)$ follows an exponential decrease. This is expected from Darcy's law because the specific discharge is proportional to the hydraulic head driving force; that is, $dh(t)/dt \propto h(t)$. For inflow from directly below the hole, the recovery curve is given by (Freitag and Eicken 2003):

$$h(t) = h_0 \exp\left(-\Pi_a \frac{g}{\mu L} t\right) \qquad \text{(Equation 3.4.26).}$$

As in Figure 3.4.11, $L$ is here the ice thickness below the hole and $h_0$ is the hydraulic head at $t = 0$ (Freitag and Eicken 2003). An apparent permeability $\Pi_a$ can then be determined from the gradient of a plot of $\ln(h(t)/h_0)$ against time. However, Equation 3.4.26 holds only if the lateral permeability $\Pi_l = 0$. A correction for nonzero $\Pi_l$ was determined numerically by Freitag (1999). For a ratio of lateral to vertical permeability, $\Pi_l / \Pi_v = 0.1$, the vertical permeability is given by:

$$\Pi_v = \frac{\Pi_a}{0.17 + 10.7L} \qquad \text{(Equation 3.4.27).}$$

The induced flows widen pore space, but can be corrected for. Repeat measurements with the same hole showed the effective permeability to increase by about 7 percent per measurement cycle.

Figure 3.4.12 shows the distribution of permeability values between $10^{-13}$ and $10^{-7}$ m$^2$ found by applying Equations 3.4.26 and 3.4.27 to data from the laminar flow regime. Back extrapolation was applied to calculate the permeability for undisturbed ice. The measurements return an integration of $\Pi_v$ under the hole, and are therefore sensitive to natural variability. See Eicken et al. (2002; 2004) for a full discussion of measurements uncertainties, resulting in an absolute measurement uncertainty of less than 50 percent (Eicken et al. 2004).

In twenty-nine of their forty-six bore holes, Freitag and Eicken (2003) identified turbulent flow, as in Figure 3.4.11. Assuming transitions to have taken place in large secondary pores, they estimated effective tube radii of 0.5–2.5 mm, consistent with upper limits of previous work (Eide and Martin 1975; Martin 1979; Wakasutchi and Saito 1985).

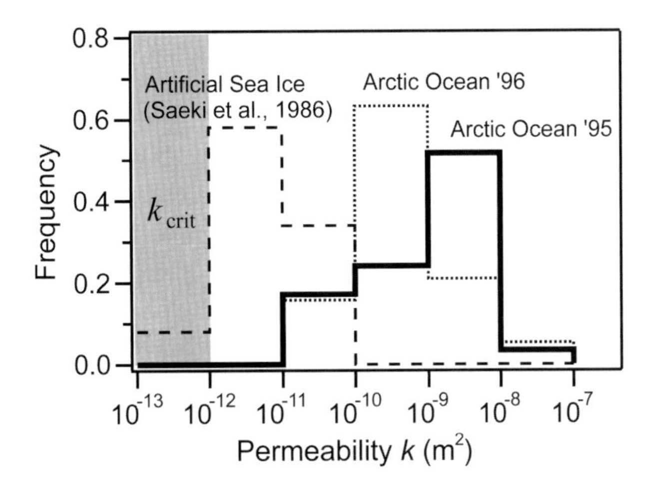


**Figure 3.4.12.** Frequency distributions of permeability from in situ bail tests and laboratory measurements. Meltwater percolates downwards for $k > k_{crit}$. After Freitag and Eicken (2003).

*Tracer Methods*

Tracer dispersion provides another means to measure sea ice permeability and to assess fluid flow pathways and processes (Freitag 1999; Freitag and Eicken 2003; Eicken et al. 2002; 2004). Tracer concentration is measured as a function of space and time (see Figure 3.4.13). The tracer must be nonreactive in sea ice and readily measured to low concentrations. Natural isotopic measures can be used to quantify the flux of snow and ice meltwater, and fluorescent dyes used to calculate

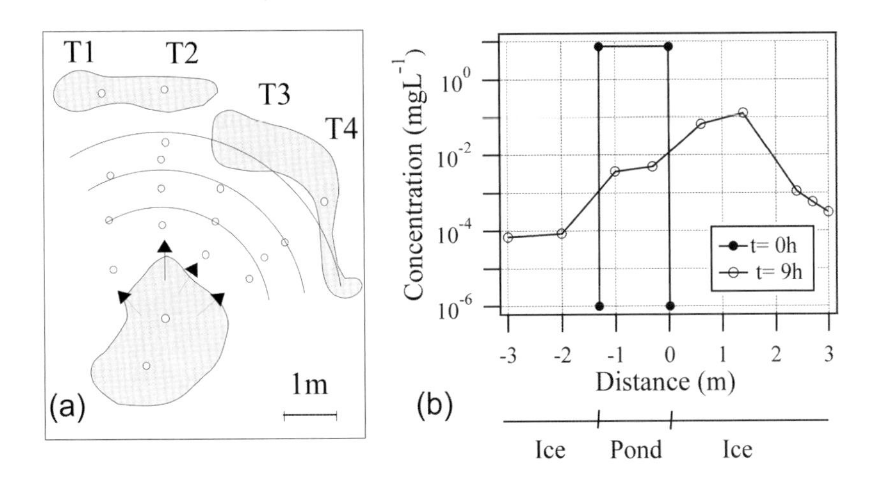


**Figure 3.4.13.** (a) Sketch of tracer study transects: Circles mark sampling locations spreading radially from the central release location, gray areas indicate melt ponds. (b) An example of wind-driven flow in level ice: Flouresceine concentration profiles at 1 m depth along the prevailing wind direction. From Freitag and Eicken (2003).

permeability and reveal pathways. These approaches require careful fieldwork and contamination-free sample collection for later laboratory analysis, details of which can be found in the references below.

Natural tracers suitable for tracking surface meltwater into and under the ice require a strong contrast between snowmelt-derived water (water of meteoric origin) and sea ice formed from seawater. These include the stable isotope $^{18}O$ and cosmogenic radionuclide $^{7}Be$ (Eicken et al. 2002). Due to fractionation effects, snow falling on sea ice is significantly depleted in heavier isotopes $^{18}O$ and $^{2}H$ compared with seawater and sea ice. Mass spectrometry can be performed on samples from different depths in the ice to determine the fraction of meteoric water, and hence the degree of surface infiltration. Concentrations of $^{18}O$ are reported in terms of $\delta^{18}O$:

$$\delta^{18}O = \left( \frac{^{18}O/^{16}O_{sample}}{^{18}O/^{16}O_{VSMOW}} - 1 \right) 1000\,ppt \qquad \text{(Equation 3.4.28).}$$

VSMOW refers to Vienna Standard Mean Ocean Water, an isotopic water standard. The total fraction of meteoric water (rain or snow) $f_m$ in a sample is then:

$$f_m = \frac{\delta^{18}O_{sample} - \delta^{18}O_{seaice}}{\delta^{18}O_{meteoric} - \delta^{18}O_{seaice}} \qquad \text{(Equation 3.4.29).}$$

These measures can identify meltwater infiltration and also under-ice entrapment of surface melt, which can occur due to the freezing of freshwater lenses under the ice (Eicken 1996; Eicken et al. 2002).

A proof-of-concept study using $^{7}Be$ is presented in Eicken et al. (2002). Large sample volumes are required with chemical treatment to precipitate out $^{7}Be$ and gamma spectroscopy to measure concentration. The fifty-three day half-life of $^{7}Be$ decay is not large compared with time scales of sea ice growth. Therefore, the analysis of $^{7}Be$ concentration profiles must account for the ice growth history to determine the time since exposure to the atmosphere.

Fluorescence-based studies using artificial tracers have been used to study fluid pathways (Eicken et al. 2002; 2004) and photosynthetic activity in sea ice (Arrigo 2003). When fluorescent materials are exposed to light at a specific excitation frequency, they emit light at a lower emission frequency, with the energy difference absorbed as heat. Following the release of a tracer in a melt pond or at depth in a core hole, the dispersion may be tracked by measuring tracer concentrations of ice core samples taken at various positions and times (see Figure 3.4.17). Samples are analyzed with a fluorometer using the excitation and emission frequencies of the specific tracer. Flourescein (FLC, $C_{20}H_{10}O_5Na_2$) and sulforhodamine (SRB, $C_{27}H_{29}O_7N_2S_2Na$) have been used in sea ice. They both show a linear relationship between strength of fluorescence and concentration over certain concentration ranges. For FLC this range is 1 mg/L to $5 \times 10^{-6}$ mg/L, and for SRB it is 1 mg/L to $5 \times 10^{-4}$ mg/L. A third tracer, Rhodamine (RB, $C_{28}H_{31}ClN_2O$) is adsorptive, reducing loss from samples and making it well suited to core sampling in longer-term studies of flow fields (Krembs 2000; Eicken et al. 2002). All of these dyes allow visual tracking: FLC is bright green, SRB is bright red, and RB is violet.

A low-tech tracer, effective at least for demonstration purposes, is bright-red beet juice, whose spread through thick sections can clearly illustrate connected brine features.

### 3.4.4.4 Laboratory Permeability Measurements on Natural Samples

Recent laboratory measurements have been made on samples from field campaigns to examine sea ice hydrology (Freitag 1999) and ice algae biology (Merkel 2008). Earlier laboratory measurements were made by Saito and Ono (1978) and Kasai and Ono (1984). Equipment and methods can be found in the supplementary material.

Steps used by Merkel (2008) for sample extraction and preparation were: (1) extraction using an 8.5 cm internal diameter corer; (2) cutting into 5 cm thick slices; (3) centrifuging to remove brine (Beckmann CPR Rotor G3.7 at 1200–1400 rpm for 4–5 minutes); (4) wrapping in airtight aluminum foil and sealing in airtight plastic bags; (5) storage at −40°C; (6) machining at −20°C by turning down to 8.3 mm diameter with a lathe and cutting flat top and bottom surfaces.

### 3.4.4.5 Connection with Traditional and Local Knowledge: Salt Harvesting

Fluid flow in sea ice provides an interesting connection between scientific inquiry and local and traditional knowledge. Kasai and Ono (1984) reported measurements on thin ice grown in refreezing ponds within the fast-ice cover in Saroma-ko Lagoon, Hokkaido. When the thin ice layer was loaded with a weight, brine pooled on the surface with a salinity of 72.5 ppt, more than twice as saline as seawater. This was surely brine driven from the surface ice by the depression below equilibrium sea level. The salinity corresponds to a liquidus equilibrium temperature of −4.3°C. With decreasing ice temperature the brine formed by such a process would increase, and below −20°C the equilibrium brine salinity would exceed 250 ppt. A related observation made in Barrow, Alaska, by Richard Glenn is recounted in Wohlforth (2004, p. 127). A wintertime compression event on the fast ice at Barrow in 1990 deformed an area of thin ice in a recently refrozen lead into an undulating surface of hummocks with brine pools in the depressions. Recovered brine was "one quarter salt" and a cup of it evaporated overnight, leaving only solid salts. This phenomenon did not surprise Iñupiaq elders in Barrow—it was "how they gathered salt in the old days" (Wohlforth 2004, p. 128).

#### 3.4.4.6 Outlook for Fluid Flow Studies

The fluid permeability, and sea ice hydrology in general, warrant increased future attention. Of the three transport properties reviewed here, the permeability is the most sensitive to microstructural variations, and displays the highest natural variability in its values and driving gradients. Through snow cover–permeability linkages, it also exerts great influence on important large-scale ice behavior and feedbacks. As outlined in Eicken et al. (2004), it nevertheless remains a difficult challenge to measure and understand sea ice hydrology, let alone incorporate it into large-scale predictive models. More field data for permeability in different ice types might enable a statistical representation for porosity-permeability relationships for different ice types. Relative contributions to flow in features of different length scales, and a means to apply this at field scales, deserves attention. Numerical and laboratory methods also hold promise in this regard; both also suffer from a scarcity of previous results.

As part of Fourth International Polar Year project SIZONet, topographical studies of surface hydrology are being undertaken on scales up to kilometers. Measurements conducted in the melt season in Barrow, Alaska, have included visual and infrared aerial photography, and repeat profiling of albedo and pond and snow depths. An integration of these results will surely benefit future progress in understanding sea ice hydrology.

### 3.4.5. SUMMARY AND FUTURE OUTLOOK

Of the properties reviewed in this chapter, thermal properties are the most well understood, but some questions remain. The specific heat is well parameterized by Equation 3.4.2 and the thermal conductivity by Equation 3.4.4, at least when conductive heat flow dominates. There is, however, a need to assess heat flow in warm ice in which convective contributions will become increasingly significant. This will require new experimental approaches. With predictions of increasing fractions of FY ice, there should also be an effort to consider more closely differences between modeling parameterizations derived from MY arctic ice and those most relevant to particularly antarctic FY ice. Looking forward to the next International Polar Year, envisaged for 2032, it is reasonable to expect the widespread use in models of new, updated parameterizations for heat flow in different types of sea ice which include convective contributions.

Electrical properties are an area of active research. Fundamental questions remain with regard to microstructural dependence of the permittivity, especially of interfacial effects at radio frequencies. Progress in this area may enable automated salinity measurements using dc conductivity, capacitance, and/or frequency-dependent dielectric permittivity measurement. It may also allow in situ quantitative characterization of brine inclusion connectivity and geometry at various scales—an important step towards a more realistic model of sea ice microstructure. Impetus for this work may derive from interest in ice-oil interaction and detection, and to improve remote sensing including sea ice thickness surveying. Present initiatives include the development of multifrequency EM induction methods, dc and ac (i.e., permittivity) resistivity tomography, and the development of accurate permittivity measurements on extracted cores. Another area of interest is modeling to reduce uncertainties in aerial and on-ice thickness surveys, including the need for ice-type calibrations and loss of information due to comparatively large instrument footprints.

Sea ice hydrology and fluid permeability deserves the increased attention during IPY-4. The permeability is highly sensitive to microstructural variations, with order-of-magnitude differences between the permeability of cold, low-salinity MY ice and warm, saline FY ice. Melt-season hydrology affects the permeability of the upper ice surface and hydraulic heads and is important to understand. The aim of including sea ice hydrology in large-scale models will require progress on statistical treatment of the surface hydrology and addressing the scarcity of permeability measurements from both the field and laboratory. Integrated field campaigns are already addressing some of these questions. Sea ice represents an extreme example in vadose zone hydrology, and it may be that real benefit can be derived from interdisciplinary collaboration with those in more traditional hydrology fields already concerned with statistical and upscaling methods.

## ACKNOWLEDGMENTS

DP acknowledges support during IPY from the Arctic Region Supercomputing Center, Geophysical Institute, and the University of Alaska International Polar Year Office. Special thanks to the field support team at the Barrow Arctic Science Consortium for help, stories, and sharing their ice. MI acknowledges the support of the Geophysical Institute while on Research and Study leave from Victoria University.

## REFERENCES

Addison, J. R. (1969), Electrical properties of saline ice, *J. Appl. Phys.*, *40*, 3105–3114.

Addison, J. R. (1970), Electrical relaxation in saline ice, *J. Appl. Phys.*, *41*(1), 54–63.

Arrigo, K. (2003), Primary production in sea ice, in *Sea Ice: An Introduction to Its Physics, Biology, Chemistry, and Geology*, edited by D. N. Thomas and G. S. Dieckmann, pp. 22–81, Blackwell Science, London.

Arcone, S. A., A. J. Gow, and S. McGrew (1986), Structure and dielectric properties at 4.8 and 9.5 GHz of saline ice, *J. Geophys. Res.*, *91*(C12), 14,281–14,303.

Backstrom, L. G. E., and H. Eicken (2006), Capacitance probe measurements of brine volume and bulk salinity in first-year sea ice, *Cold. Reg. Sci. Technol.*, *46*, 167–180.

Bear, J. (1972), *Dynamics of Fluids in Porous Media*, American Elsevier, New York..
Bhattacharya, P. K., and H. P. Patra (1968), *Direct Current Geoelectric Sounding; Principles and Interpretation*, Elsevier, Amsterdam.
Bowick, C. (1997), *RF Circuit Design*, Newnes, Boston, MA.
Buckley, R. G., M. P. Staines, and W. H. Robinson (1986), In situ measurements of the resistivity of Antarctic sea ice, *Cold Reg. Sci. Tech.*, *12*, 285–290.
Campbell, J. E. (1988), Dielectric properties of moist soils at RF and microwave frequencies, Ph.D. thesis, Dartmouth College, Hanover, NH.
Campbell, J. E. (1990), Dielectric properties and influence of conductivity in soils at one to fifty megahertz, *Soil Sci. Am. J.*, *54*(2), 332–341.
Campbell, K. J., and A. S. Orange (1974), The electrical anisotropy of sea ice in the horizontal plane, *J. Geophys. Res.*, *79*(33), 5059–5063.
Cox, G. F. N., and W. F. Weeks (1975), Brine drainage and initial salt entrapment in sodium chloride ice, *CRREL Research Report, 345*, Cold Regions Research and Engineering Laboratory, Hanover, NH.
Debye, P. (1929), *Polar Molecules*, Dover Publications, New York.
de Loor, G. P. (1968), Dielectric properties of heterogeneous mixtures containing water, *J. Micr. Power*, *3*(2) 67–73.
Dickins, D., B. Hirst, G. Gibson, L. Liberty, J. Bradford, V. Jones, E. Owens, L. Zabilinsky, and D. De Vitas (2005), New and innovative equipment and technologies for the remote sensing and surveillance of oil in and under ice, United States Department of Interior, Minerals Management Service.
Dickins, D., P. J. Brandvik, L. Faksness, J. Bradford, and L. Liberty (2006), Svalbard experimental spill to study spill detection and oil behavior in ice, U.S. Department of Interior, Minerals Management Service, CGISS Technical Report #06-02.
Dieckmann, G. S., M. A. Lange, S. F. Ackley, and J. C. Jennings (1991), The nutrient status in sea ice of the Weddell Sea during winter—effects of sea ice texture and algae, *Polar Biol.*, *11*, 449.
Doronin, Y. P., and D. E. Kheisin (1977), *Sea Ice* (translated from Russian), Amerind, New Delhi.
Druckenmiller, M. L., H. Eicken, D. J. Pringle, C. C. Williams, and M. A. Johnson (2009), Towards an integrated coastal sea ice observatory: system components and a case study at Barrow, Alaska, *Cold Reg. Sci. Technol., 56,* 61–72.
Eicken, H. (2003), From the Microscopic to the Macroscopic to the Regional Scale: Growth, Microstructure and Properties of Sea Ice, in *Sea Ice: An Introduction to Its Physics, Biology, Chemistry, and Geology*, edited by D. N. Thomas and G. S. Dieckmann, pp. 22–81, Blackwell Science, London.
Eicken, H., S. F. Ackley, J. A. Richter-Menge, and M. A. Lange (1991), Is the strength of sea ice related to its chlorophyll content?, *Polar Biol.*, *11,* 347–350.
Eicken, H., R. Gradinger, B. Ivanov, A. Makshtas, and R. Pác (1996), Surface Melt Puddles on Multi-year Sea Ice in the Eurasian Arctic, in *World Climate*

*Research Programme WCRP-94: Proceedings of the ACSYS Conference on the Dynamics of the Arctic Climate System* (Göteborg, Sweden, 7–10 November 1994), WMO/TD 760, pp. 267–271, World Meteorological Organization, Geneva.

Eicken, H., T. C. Grenfell, D. K. Perovich, J. A. Richter-Menge, and K. Frey (2004), Hydraulic controls of summer Arctic pack ice albedo, *J. Geophys. Res., 109,* C08007, doi:10.1029/2003JC001989.

Eicken, H., H. R. Krouse, D. Kadko, and D. K. Perovich (2002), Tracer Studies of pathways and rates of meltwater transport through arctic summer sea ice, *J. Geophys. Res., 107*(10), 10.1029/2000JC000583.

Eide, L., and S. Martin (1975), The Formation of brine drainage features in young sea ice, *J. Glaciol., 14,* 137–154.

Freeze, R. A., and J. A. Cherry (1979), *Groundwater,* Prentice-Hall, Englewood Cliffs, NJ.

Freitag, J. (1999), Untersuchungen zur Hydrologie des arktischen Meereises: Konsequenzen für den kleinskaligen Stofftransport (The hydraulic properties of Arctic Sea ice: Implications for the small scale particle transport) *Ber. Polarforsch.* (Reports on polar research), *325,* Bremen.

Freitag J., and H. Eicken (2003), Melt water circulation and permeability of Arctic summer sea ice derived from hydrological field experiments. *J. Glaciol., 49*(166), 349–358.

Fujino, K., and Y. Suzuki (1963), An attempt to estimate the thickness of sea ice by electrical resistivity method II, *Low Temp. Sci., A21,* 151–157.

Fukusako, S. (1990), Thermophysical properties of ice, snow, and sea ice, *Int. J. Thermophys., 11*(2), 353–373.

Golden, K. M., and S. F. Ackley (1980), Modeling of anisotropic electromagnetic reflection from sea ice, *CRREL Report, 80-23,* U.S. Army Corps of Engineers, Hanover, NH.

Golden, K. M., and S. F. Ackley (1981), Modeling of anisotropic electromagnetic reflection from sea ice, *J. Geophys. Res., 86*(C9), 8107–8116.

Golden, K. M., H. Eicken, A. L. Heaton, J. E. Miner, D. J. Pringle, and J. Zhu (2007), Thermal evolution of permeability and microstructure in sea ice, *Geophys. Res. Lett., 34,* L16501, doi:10.1029/2007GL030447.

Hallikainen, M. T. (1977), Dielectric properties of NaCl ice at 16 GHz. Report S-107, Helsinki University of Technology, Radio Laboratory, Epoo, Finland.

Hallikainen, M. T., M. Toikka, and J. Hyyppä (1988), Microwave dielectric properties of low-salinity sea ice, in *Proceedings of the International Geoscience and Remote Sensing Symposium 1988* (IGARSS'88), pp. 419–420, Edinburgh, UK.

Hallikainen, M., and D. P. Winebrenner (1992), The physical basis for sea ice remote sensing, in *Microwave Remote Sensing of Sea Ice,* edited by F. D. Carsey, *Geophysical Monograph Series, 68,* pp. 29–46, American Geophysical Union, Washington, DC.

Herwanger, J. V., M. H. Worthington, R. Lubbe, A. Binley, and J. Khazaehdari (2004), A Comparison of cross-hole electrical and seismic data in fractured rock, *Geophysical Prospecting, 52*, 109–121.

Hoekstra, P., and P. Cappillino (1971), Dielectric properties of sea and sodium chloride ice at UHF and microwave frequencies, *J. Geophys. Res., 76,* 4922–4931.

Ingham, M., D. Pringle, and H. Eicken (2008), Cross-borehole resistivity tomography of sea ice, *Cold Reg. Sci. Technol., 52*, 263–277.

Johnson, H. L. (1989), The specific heat of sea ice, B.Sc. honour's thesis, Victoria University of Wellington, New Zealand.

Kasai, T., and N. Ono (1984), An experimental study of brine upward migration in thin sea ice, *Low. Temp. Sci., 43A*, 49–155.

Kovacs, A., and R. M. Morey (1978), Radar anisotropy of sea ice due to preferred azimuthal orientation of the horizontal c-axes of ice crystals, *J. Geophys. Res., 83*(12), 171–201.

Kovacs, A., and R. M. Morey (1979), Anisotropic properties of sea ice in the 50- to 150-MHz range, *J. Geophys. Res., 84,* 5749–5759.

Krembs, C., R. Gradinger, and M. Spindler (2000), Implications of brine channel geometry and surface area for the interaction of sympagic organisms in Arctic sea ice, *J. Exp. Mar. Biol. Ecol., 243*, 55–80.

Lewis, E. L. (1967), Heat flow through winter ice, in *Physics of Snow and Ice: International Conference on Low Temperature Science 1966,* edited by H. Oura, *1*(1), pp. 611–631, Institute of Low Temperature Science, Hokkaido University, Sapporo, Japan.

Lytle, V. I., and S. F. Ackley (1996), Heat flux through sea ice in the Western Weddell Sea: Convective and conductive transfer processes, *J. Geophys. Res., 101*(C4), 8853–8868.

Malmgren, F. (1927), On the properties of sea ice, in *The Norwegian North Polar Expedition with the "Maud", 1918–1925, 1,* 5, pp. 1–67, Geophysical Institute, Bergen, Norway.

Martin, S. (1979), A field study of brine drainage and oil entrainment in first-year sea ice, *J. Glaciol., 22*(88), 473–502.

Maksym, T., and M. O. Jefries (2000), A one-dimensional percolation model of flooding and snow ice formation on Antarctic sea ice, *J. Geophys. Res., 105*(C11), 26,313–26,331.

Maykut, G., and N. Untersteiner (1971), Some results from a time-dependent thermodynamic model of sea ice, *J. Geophys. Res., 76*, 1550–1576.

Merkel, H. (2008), Impact of ice permeability and under-ice water currents on nutrient dynamics in Arctic sea ice, M.S. thesis, University of Alaska Fairbanks.

Morey, R. M., A. Kovacs, and G. F. N. Cox (1984), Electromagnetic properties of sea ice, *CRREL Report* 84-2, U.S. Army Corps of Engineers., Hanover, NH.

McGuinness, M. J., K. Collins, H. J. Trodahl, and T. G. Haskell (1998), Nonlinear thermal transport and brine convection in first year sea ice, *Ann. Glaciol., 27*, 471–476.

Nazintsev, Y. L. (1959), Eksperimental'noe opredelenie teploemkosti i tem-pera-turoprovodnosti moskogo l'da (Experimental determination of the specific heat and thermometric conductivity of sea ice), *Probl. Arkt. Antarkt.*, *1*, 65–71 (English translation available from *Am. Meterol. Soc.*, Boston)

Nazintsev, Y. L. (1964), Nekotorye dannye k raschetu teplovykh Svoistv morskogo l'da. (Some data on the calculation of thermal properties of sea ice), *Tr. Arkt. Antarkt.*, Nauchlo Issled Inst., *267*, 31–47.

Notz, D. (2005), Thermodynamic and fluid-dynamical processes in sea ice, Ph.D. thesis, University of Cambridge, UK.

Notz, D., J. S. Wettlaufer, and M. G. Worster (2005), A non-destructive method for measuring the salinity and solid fraction of growing sea ice in situ, *J. Glaciol.*, *51*, 159–166.

Notz, D. and M. G. Worster (2008), In situ measurements of the evolution of young sea ice, *J. Geophys. Res.*, *113*, C03001, doi: 10.1029/2007JC004333.

Ono, N. (1967), Specific heat and fusion of sea ice, in *Physics of Snow and Ice: International Conference on Low Temperature Science 1966*, edited by H. Oura, *1*(1), pp. 599–610, Institute of Low Temperature Science, Hokkaido University, Sapporo, Japan.

Ono, N. (1968), Thermal properties of sea ice IV. Thermal constants of sea ice, *Low-Temp. Sci.*, *A26*, 329–349.

Ono, N., and T. Kasai (1985), Surface layer salinity of young sea ice, *Ann. Glaciol.*, *6*, 298–299.

Parry, W. E. (1828), *Narrative of an Attempt to reach the North Pole,* John Murray, London.

Petrich, C., P. J. Langhorne, and Z. F. Sun (2006), Modelling the interrelationships between permeability, effective porosity, and total porosity in sea ice, *Cold Reg. Sci. Technol.*, *44*(2), 131–44, doi:10.1016/j.coldregions.2005.10.001.

Pringle, D. J. (2005), Thermal conductivity of sea ice and Antarctic permafrost, Ph.D. thesis, Victoria University of Wellington, New Zealand.

Pringle, D., G. Dubuis, and H. Eicken (2009), Impedance measurements of the complex dielectric permittivity of sea ice at 50 MHz: Pore microstructure and potential for salinity monitoring. *J. Glaciol.*, *55*, 81–94.

Pringle, D. J., H. Eicken, H. J. Trodahl, and L. G. E. Backstrom (2007), Thermal conductivity of landfast Antarctic and Arctic Sea Ice, *J. Geophys. Res.*, *112,* C04017, doi:10.1029/2006JC003641.

Pringle, D. J., H. J, Trodahl, and T. G. Haskell (2006), Direct measurement of sea ice thermal conductivity: No surface reduction, *J. Geophys. Res.*, *111*, C05020, doi:10.1029/2005JC002990.

Richter-Menge, J. A., D. K. Perovich, B. C. Elder, K. Claffey, I. Rigor, and M. Ortmeyer (2006), Ice mass balance buoys: A tool for measuring and attributing changes in the thickness of the Arctic sea ice cover, *Ann. Glaciol.*, *44*, 205–210.

Saeki, H., T. Takeuchi, M. Sakai, and E. Suenaga (1986), Experimental study on permeability coefficients of sea ice, *Ice Technology, Proceedings of the 1st International Conference, Cambridge, Mass. USA*, edited by T. K. S. Murthy, J. J. Connor, and C. A. Brebbia, pp. 237–246, Springer-Verlag, Berlin.

Saito, T., and N. Ono (1978), Percolation of sea ice. I – Measurements of kerosene permeability of NaCl ice, *Low Temp. Sci. Ser., A, 37*, 55–62.

Santamarina, J. C., K. A. Klein, and M. A. Fam (2001), *Soils and Waves*, Wiley & Sons, New York.

Schwerdtfeger, P. (1963), The thermal properties of sea ice, *J. Glaciol., 4*, 789–807.

Seyfried, M. S., L. E. Grant, E. Du, and K. Humes (2005), Dielectric loss and calibration of the Hydra Probe Soil Water Sensor, *Vadose Zone Journal, 4*, 1070–1079.

Seyfried, M. S. and M. D. Murdock (2004), Measurement of soil water content with a 50-MHz soil dielectric sensor, *Soil Sci. Soc. Am. J., 68*, 394–403.

Stefan, J. (1891), Über Die Theorie der Eisbildung, Insbesondere über die Eisbildung im Polarmeere, *Ann. Physik, N.F., 3rd Ser., 42*, 269–286, 1981 (Results given in Malmgren 1927).

Steinbronn, L., J. Bradford, L. Liberty, D. Dickins, and P. J. Brandvik (2007), Oil Detection in and under sea ice using ground-penetrating radar, *Eos Trans. AGU, 88*(52), Fall Meet. Suppl., Abstract H31G-0740.

Stogryn, A. (1987), An analysis of the tensor dielectric constant of sea ice at microwave frequencies, *IEEE Trans. Geosci. Rem. Sens., GE-25*(2), 147–158.

Thyssen, F., H. Kohnen, M. V. Cowan, and G. W. Timco (1974), DC resistivity measurements on the sea ice near pond inlet, N.W.T. (Baffin Island), *Polarforschung., 44*, 117–126.

Timco, G. W. (1979), An analysis of the in-situ resistivity of sea ice in terms of its microstructure, *J. Glaciol., 22*, 461–471.

Tinga, W. R., W. A. G. Voss, and D. F. Blossey (1973), Generalized approach to multiphase dielectric mixture theory, *J. Appl. Phys., 44*, 3897–3902.

Trodahl, H. J., M. McGuinness, P. Langhorne, K. Collins, A. Pantoja, I. Smith, and T. Haskell (2000), Heat transport in McMurdo Sound first-year fast ice, *J. Geophys. Res., 105*, 11,347–11,358.

Trodahl, H. J., S. Wilkinson, M. McGuinness, and T. Haskell (2001), Thermal conductivity of sea ice: Dependence on temperature and depth, *Geophys. Res. Lett., 28*, 1279–1282.

Untersteiner, N. (1961), On the mass and heat balance of Arctic sea ice, *Arch. Meteorol. Geophys. Bioklimatol., Ser. A, 12*, 151–182.

Vant, M., R. Gray, R. O. Ramseier, and V. Makios (1974), Dielectric properties of fresh and sea ice at 10 and 35 GHz, *J. Appl. Phys., 45*, 4712–4717.

Vant, M. R., R. O. Ramseier, and V. Makios (1978), The complex-dielectric constant of sea ice at frequencies in the range 0.1–40 GHz, *J. Appl. Phys., 49*, 1264–1280.

Wakatsuchi, M., and T. Saito (1985), On brine drainage channels of young sea ice, *Ann. Glaciol.*, 6, 200–202.

Weeks, W. F., and S. F. Ackley (1986), The growth, structure, and properties of sea ice, in *The Geophysics of Sea Ice,* NATO ASI Series, Series B, Physics Vol., edited by N. Untersteiner, pp. 9–153, Plenum Press, New York.

Weller, G. (1967), The effect of absorbed solar radiation on the thermal diffusion in Antarctic fresh-water ice and sea ice, *J. Glaciol.*, *6,* 859–878.

Wohlforth, C. (2004), *The Whale and the Supercomputer: On the Northern Front of Climate Change*, North Point Press, New York.

Yen, Y.-C. (1981), Review of thermal properties of snow, ice and sea ice, *CRREL Report, 81-10,* U.S. Army Cold Reg. Res. and Eng. Lab., Hanover, NH.

# Chapter 3.5

# Ice Strength: In Situ Measurement

*D. M. Masterson*

## 3.5.1 INTRODUCTION

Knowledge of the physical strength of materials is fundamentally necessary in all areas of engineering. Steel and concrete structures require a knowledge of the material strength in order that they may be designed to resist environmental loads and to carry service loads, including their own weight, safely and without failure or excessive deformations. In fact, deformations beyond a preset acceptable limit usually govern the design of conventional structures. No one likes to be on a bridge that sways when someone walks on it, or to be on a floor that noticeably moves up and down when people walk.

The methods used to determine the strength of materials such as steel, concrete, soils, and rocks have been established for 100 years or more and are relatively well developed (Seely 1935; Timoshenko 1983; Goodman 1989; Terzaghi et al. 1996). Even so there is often controversy over the values to be used for design and the methods of obtaining the strengths. It is often asked what the compressive strength of a cylinder of concrete measuring 20 cm in length by 10 cm in diameter has to do with the strength of a large building, which contains thousands of cubic meters of concrete and steel. Certainly there is controversy regarding the determination of the strength of "geotechnical" materials such as soil and rock. Unlike steel and concrete, which are "controlled" materials and are manufactured to a specification using carefully controlled ingredients, soil and rock must be taken as found with all of their variability and nonhomogeneity. The strength of a foundation or a mine shaft is thus the integration of many materials with different strengths and composition.

Ice is a geotechnical material, found as is in nature, and must be dealt with accordingly. Unlike the methods of soil and rock mechanics, the history of ice strength determination is much shorter and is not so well developed (Dykins 1967; Michel 1978). Many of the procedures used to determine the strength of ice have been borrowed and adapted from those used in soil and rock mechanics. Procedures for testing steel and concrete have also been adapted. Ice has a unique characteristic unlike that of any of the other engineering materials commonly used. Ice is formed of frozen water and has a density roughly 10 percent less than that of water. Thus, once formed on a body of water it floats on the surface and either becomes a menace to objects placed in the water, such as structures or ships, or serves as a load-bearing medium, allowing the support and passage of people and other loads over it. Floating ice has been used to support oil-drilling rigs weighing up to 2,000 tonnes and people all over the world like to skate on it. Whether we are trying to determine the forces ice will exert on an offshore structure or ice-resistant ship or to determine its ability to safely support a desired load, we need to know its strength.

Ice exists in nature at a temperature relatively close to its melting temperature, and thus its behavior is affected by the ambient temperature. At temperatures well below 0°C it is very rocklike, having a brittle nature and fracturing suddenly at small deformation. It could be described as glasslike at temperatures below about −10°C. At temperatures close to the melting point, ice changes and becomes quite ductile or plastic. When stress or load is applied, it deforms like butter. This attribute is called *ductility*, and is in marked contrast to brittle behavior, which exhibits no ductility.

Another attribute affecting strength of many materials, but especially soil, rock, and ice, is scale (Iyer 1983; Jordaan 1989). The strength of a sample of material 10 cm on a side will be different than the strength of a sample 1 m on a side, with the larger sample having a lower unit strength. This phenomenon is referred to as *scale* or *size effect*. The more nonhomogeneous a material is, the more the size/scale effect is noticed. Ice as found in nature is not a homogeneous material, having entrained salt, air, and debris as it grows. In addition it has a random distribution of flaws, which contribute to the scale effect on strength. This is a very important phenomenon when considering the force ice will exert on a large offshore structure and failure to account for it will lead to gross overconservatism in the calculated loads, as will be discussed later in Section 3.5.4 on scale effects.

The material herein is presented from a civil engineering point of view and relates to the needs of engineers for ice strengths required for design of offshore operations and facilities. The coverage, although focusing on sea ice, is relatively broad and presents an overview. It would be impossible to cover in depth this very diverse topic, which has been investigated by many researchers and practitioners over a period of fifty years or more.

The specific topics covered are as follows:

- Principles of strength of materials from a civil engineering perspective;
- Types of small-scale tests, including both laboratory and in situ;
- Medium- and large-scale tests performed in situ;
- Full-scale pressure and load measurements on deployed structures;
- Example pressure/load calculation;
- Load-bearing structures such as ice roads.

## 3.5.2 STRENGTH OF MATERIALS BASICS FROM A CIVIL ENGINEER'S PERSPECTIVE

Ice is a brittle or semi-brittle material and thus is relatively weak in tension (Dykins 1967). Most of the strength testing on ice is done in the compressive mode as this is the type of test most readily performed. Tension strength values are usually obtained indirectly using beams and are generally found to be about one-third of the unconfined compressive strength of ice. This discussion will concentrate on compressive strengths, presented in terms of the unconfined and confined compressive strengths. Details of tensile strength testing are not presented in this chapter, although they form part of the design criteria for icebreaking ships.

### 3.5.2.1 Unconfined Compressive Strength (UCC)

The basic test of compressive strength is unconfined compression (Spencer et al. 1993). The test is illustrated in Figure 3.5.1a where a load or stress ($\sigma$ = stress = load/loaded area) is applied in one direction, usually along the long axis of a prismatic or cylindrical specimen. The strength obtained is a function of the rate at which the load is applied to the specimen, or, in technical terms, of the strain rate or stress rate. Strain is defined as the change in length of the test specimen/the original length. Strain rate is the change in strain per unit of time, usually seconds. Likewise, stress rate is the change in stress per second.

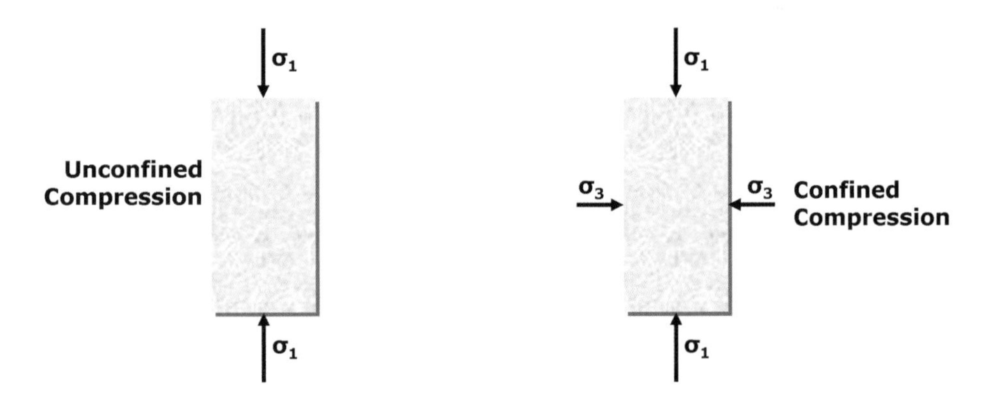


**Figure 3.5.1.** (a) Unconfined compression test; (b) Confined compression test

### 3.5.2.2 Confined Compressive Strength (CC)

Confined compressive strength is obtained by means of triaxial testing in the laboratory (Spencer et al. 1993; Singh et al. 1996). The stresses applied are illustrated in Figure 3.5.1b and are compared to those for unconfined compressive strength.

These tests are more complex to perform since the second principal stress ($\sigma_2 = \sigma_3$) has to be applied hydrostatically (same stress in all directions) to the specimen which is enclosed inside a steel cylinder—in other words, the same stress is applied in two directions. The ratio of primary stress ($\sigma_1$) to secondary stress ($\sigma_2$ or $\sigma_3$) can be varied in different tests. This makes it possible to obtain a strength "envelope" for the ice. The results of these tests are useful for assisting in interpretation of field results from pit tests and borehole jack tests as will be illustrated subsequently.

## 3.5.3 SMALL-SCALE IN SITU TESTING OF ICE

Ice cores can be tested in situ using a simple compression machine to obtain UCC (Masterson et al. 1997). This avoids transporting ice cores to a laboratory. However, this is still time-consuming and often is not practical under the limitations imposed in the field. Also, the stiffness of the test machine can affect the strengths obtained and it may not be feasible to transport a machine of the required stiffness to the test site. A flexible or nonstiff test machine stores energy during the test and releases this energy into the specimen near the end of the test, when the maximum strength of the specimen is being reached. This release of energy into the specimen will result in a lower apparent strength and thus distorts the outcome of the test. It is thus very important to ensure that the test machine is not biasing the strength obtained.

Alternatively, other tests are useful and readily done in situ and do not tax the often-limited field support. These include "pit" tests, flatjack tests, and borehole jack tests.

### *Pit Tests*

"Pit" tests are confined compression (CC) tests performed in a pit cut in the ice with a chain saw as illustrated in Figure 3.5.2. For this test, a hydraulic cylinder or ram is placed between two plates mounted against opposing walls of the pit. Oil is pumped into the ram and the plates are forced against the ice. One plate is larger than the other and is described as the base plate while the other is the test plate. Failure of the ice occurs at the test plate when the maximum strength of the ice is reached. The movement or displacement of the test plate is measured using a mechanical dial gauge mounted independently on the base of the pit. Alternatively, an electronic device such as a sliding resistor or LVDT (linear variable differential transformer) can be used to measure displacement of the test plate. From this test a load or pressure vs. displacement curve is obtained from which the ultimate strength and stiffness (elastic modulus) of the ice are obtained.

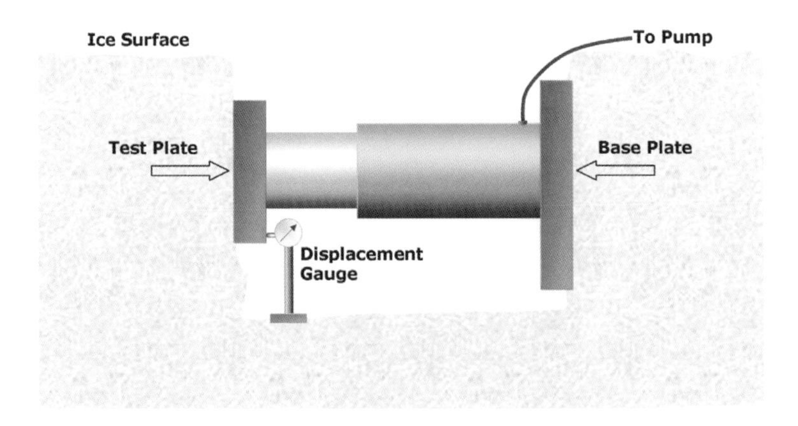


Figure 3.5.2. Ice pit test

*Flatjack Test*

*Elastic modulus* is defined as the ratio of stress/strain (Seely 1935; Timoshenko 1970). The flatjack test (Figures 3.5.3a and 3.5.3b) is similar in principle to the pit test (Iyer and Masterson 1987). For this test, a thin-walled, edge-welded flatjack is placed vertically in a chain-saw cut in the ice. Once placed, the flatjack may or may not be frozen into the slot. Hydraulic fluid is pumped into the flatjack, thus loading the ice and eventually causing it to fail. The pressure and volume of fluid pumped into the jack are measured, and thus a pressure vs. displacement graph is obtained as shown in Figure 3.5.3c.

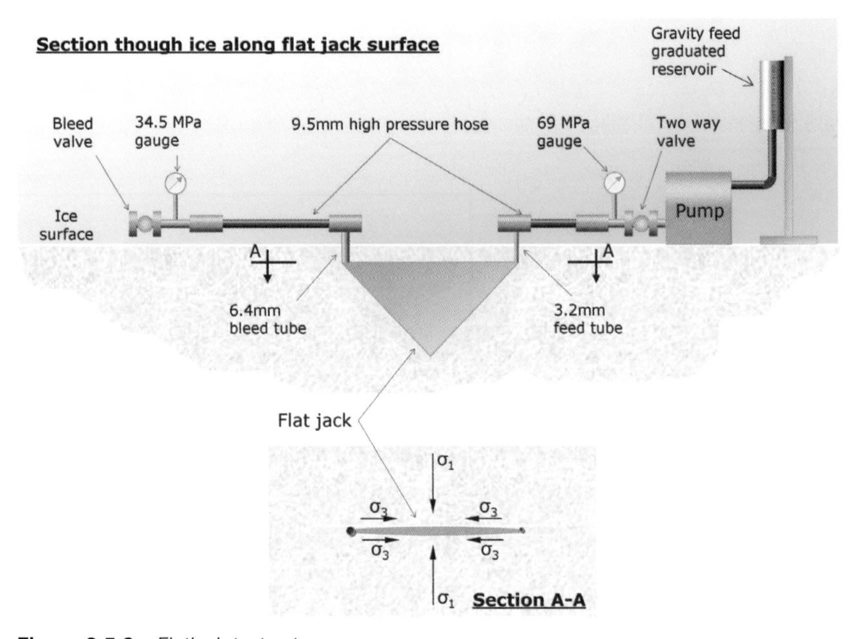


Figure 3.5.3a. Flatjack test setup

Figure 3.5.3b. Inserting a flatjack into saw cut

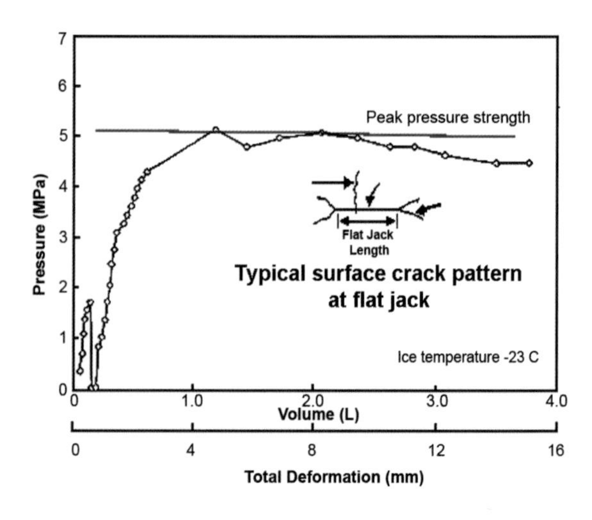


Figure 3.5.3c. Typical pressure vs. deformation plot for flatjack test

*Borehole Jack Test*

Borehole jack (BHJ) tests are a convenient means of obtaining ice strength through the thickness of an ice sheet or feature (Masterson and Graham 1992; Masterson 1996). The tests are performed in a 150 mm diameter hole drilled in the ice. Plates mounted on opposing sides of the downhole ram are forced against the ice by hydraulic pressure from a pump at the ice surface. The device is shown in Figure 3.5.4a. The jack exerts pressure on the opposing plates until failure or ultimate strength of the ice is reached. The stress state at the ice surface is shown in Figure 3.5.4b for a borehole jack and for a pressure meter, a device that applies pressure uniformly over a 360-degree arc. Test devices like the borehole jack are used for ice and rock testing because of their relatively high strength while the pressure meter is used for soils, which are weaker. The borehole jack is easily deployed in the field and tests are quickly completed. The device is now accepted as an ISO standard for obtaining in situ ice strengths. The borehole jack produces a strength profile through the ice depth and identifies weak and strong layers. Testing in this manner results in strengths and stiffnesses obtained on undisturbed material, as opposed to cores, which must be retrieved and may not be totally representative of the native ice.

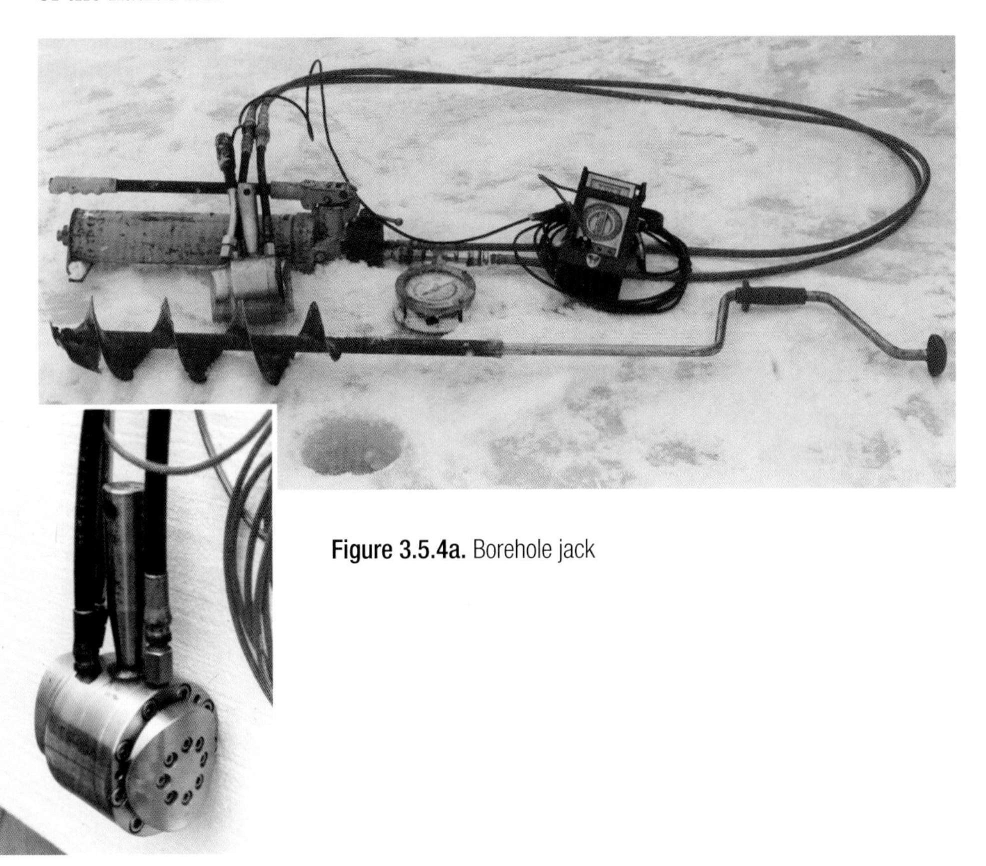

**Figure 3.5.4a.** Borehole jack

An example of the link between triaxial tests performed in the laboratory and borehole jack tests is illustrated by Figure 3.5.5. This figure contains a plot of major principal stress ($\sigma_1$) on the vertical axis and minor principal stress ($\sigma_3$) on the horizontal axis. See Figure 3.5.1b for illustrations of these stresses. The major principal stress ($\sigma_1$) from BHJ tests is known since this is the stress applied to the front plate by the hydraulic pump as shown in Figure 3.5.4b. The minor stress ($\sigma_3$) is not known and it is a stress in the ice perpendicular to the plate. Both stresses are known for the laboratory tests. Comparison on a stress plot such as Figure 3.5.5 with triaxial test results enables us to estimate the minor principal stress at ultimate strength for the borehole jack. In this case an upper and lower bound stress envelope have been drawn using the laboratory tests. Once these bounds are established, the borehole jack strength is placed midway between the upper and lower bound curves. Its y coordinate is the strength obtained from the test, in this case 30 MPa. Placing the test point at 30 MPa on the vertical axis and then moving it right until it is between the upper and lower bound curves tells us that the minor principal stress for this test can be estimated as about 23 MPa.

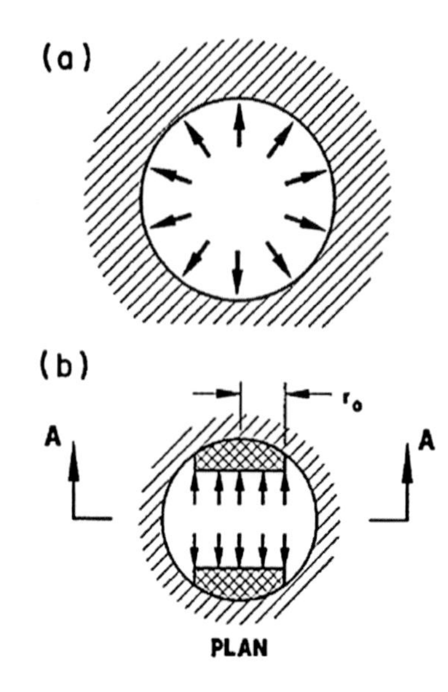


**Figure 3.5.4b.** Borehole Jack Stress States

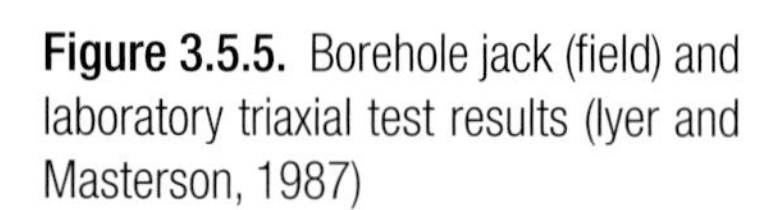

**Figure 3.5.5.** Borehole jack (field) and laboratory triaxial test results (Iyer and Masterson, 1987)

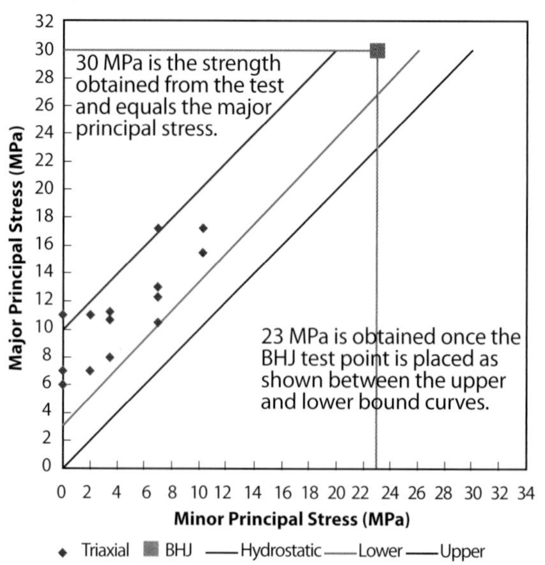

In situ testing apparatuses vary from the simple to the complicated. Depending on the features of the unit, either rate of loading or rate of deformation can be varied. Load and pressure can be applied using a hand hydraulic pump or one powered by an electric motor. Recording of the pressure and displacement can be done by hand or by using data acquisition equipment. Thus tests can be very much hands-on or can be quite automated.

### 3.5.4 SCALE EFFECTS IN ICE STRENGTH

Most materials in their natural state exhibit a scale or size effect on strength (Iyer 1983; Jordaan 1989). As the size of the loaded area on a material like rock or ice increases, the failure stress decreases. This is often attributed to more frequent encounters with flaws and cracks as the size increases. It can also result from a nonuniformity in boundary conditions, which causes the failure mode to change, for example from crushing of the ice to buckling (see Figures 3.5.11a and 3.5.11b). Small-scale tests will tend to miss this size or scale effect, which results in higher strengths than obtained from medium- and full-scale tests. Samples taken at random may encounter the flaws, and therefore the scatter in strengths can be an indication of the effect of flaws, nonhomogeneity, and cracks. Scatter in the strengths obtained from these tests can also be a function of the test conditions, further confusing the issue.

### 3.5.5 MEDIUM-SCALE INDENTOR TESTS

At Pond Inlet, Baffin Island, Canada, four tunnels were cut into the side of a grounded iceberg, the tunnels being 3 m × 3 m × 15 m deep (Figure 3.5.6a) (Masterson et al. 1992). A hydraulically operated test apparatus was inserted into the tunnels. The apparatus, having four cylinders each capable of exerting a 4.5 MN

**Figure 3.5.6a.** Tunnels in iceberg at Pond Inlet

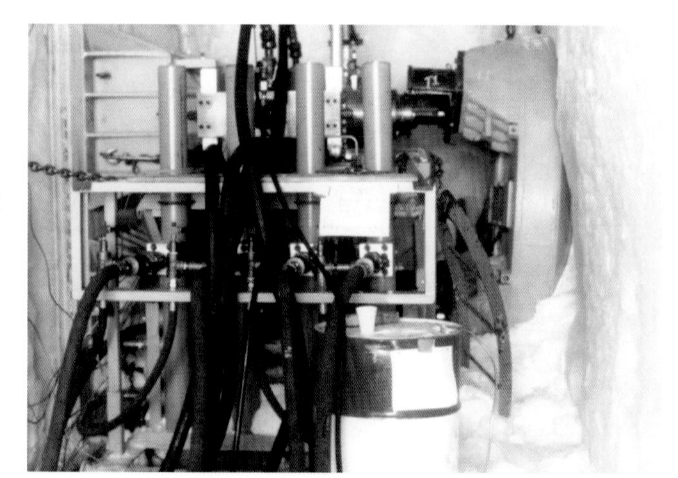

**Figure 3.5.6b.** Hydraulic test apparatus in tunnel

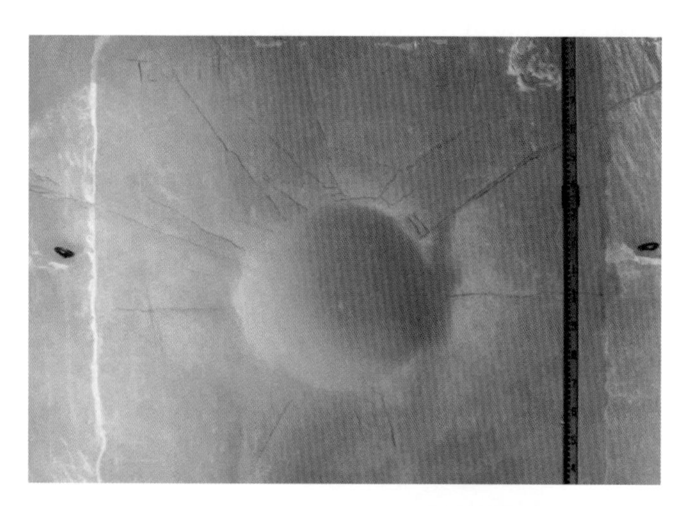

**Figure 3.5.6c.** Ductile failure V=0.3 mm/sec, 1.0 $m^2$ spherical

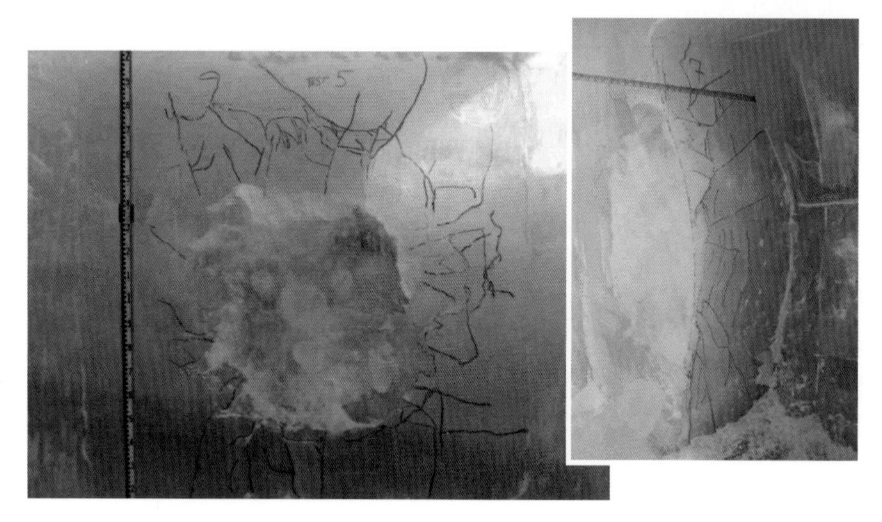

**Figure 3.5.6d.** Brittle failure V=10 mm/sec, 1.0 $m^2$ spherical

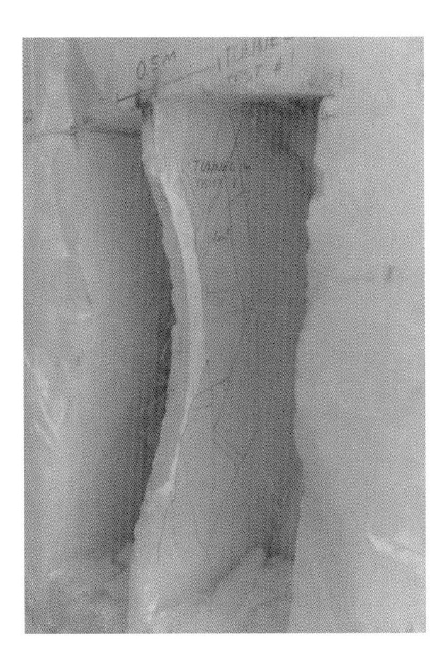

**Figure 3.5.6e.** Double section through test area

**Figure 3.5.6f.** Indentor test in multiyear ice trench

(450 tonne) force, were used to apply a load on the walls of the tunnels, as shown in Figure 3.5.6b. Thus ice strengths and pressures on areas up to 3 m$^2$ were obtained using mainly spherical indentors. Typical failures of the indented ice are shown in Figures 3.5.6c–3.5.6e. An additional set of medium-scale tests was performed in trenches excavated in multiyear ice in the Arctic Islands of Canada as shown in Figure 3.5.6f (Masterson et al. 1999).

Load vs. time traces from the Pond Inlet tests are shown in Figure 3.5.6g. The data obtained made it possible to quantify the effects of size of loaded area on maximum pressures as shown in Figure 3.5.7. In addition, effects of velocity on ice strength were quantified as shown in Figure 3.5.8. The figure shows an increase in

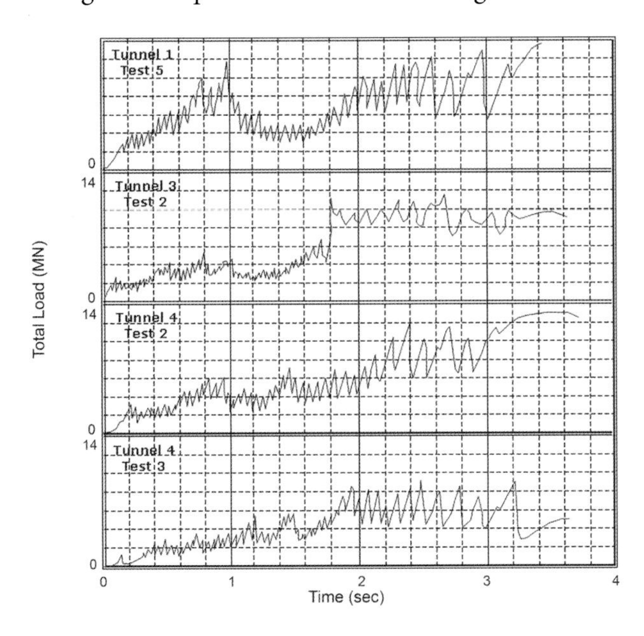


**Figure 3.5.6g.** Load-time traces from Pond Inlet tests

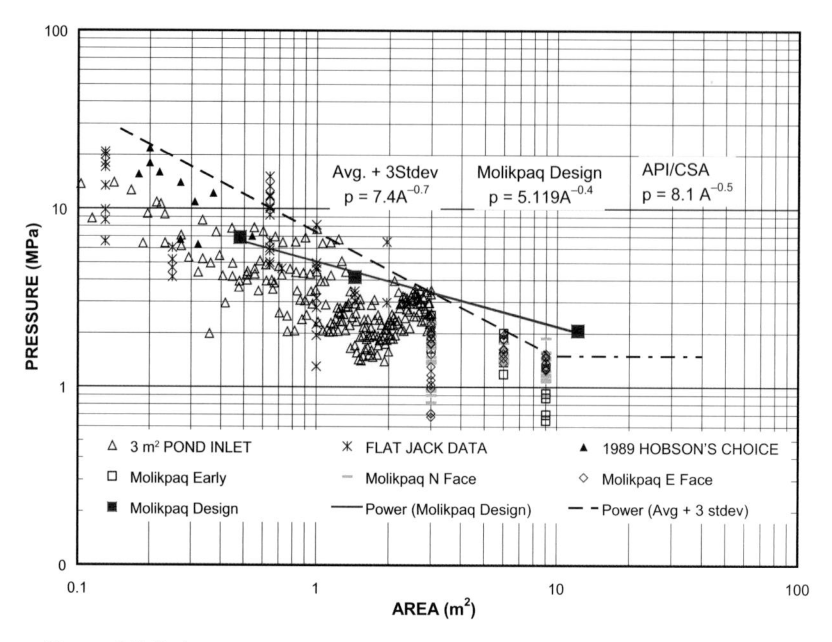


**Figure 3.5.7.** Ice pressure vs. area curve

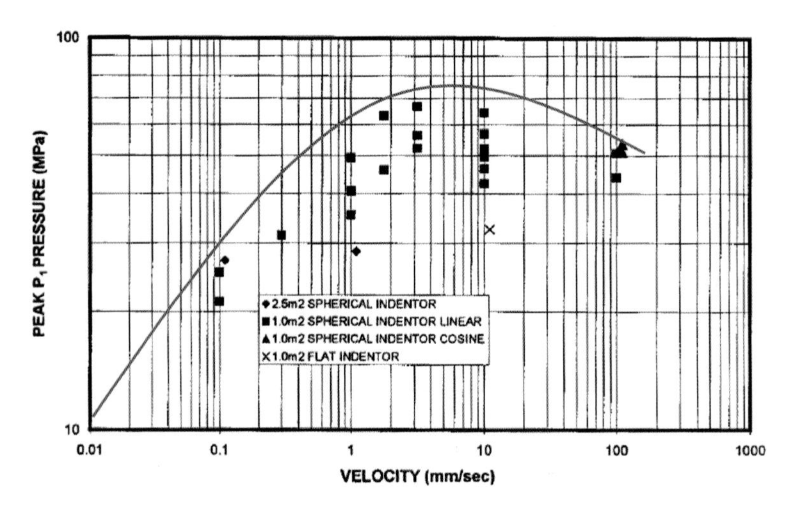


**Figure 3.5.8.** Ice pressure vs. impact velocity

ice strength from the low application rates at the left of the graph to an impact speed of about 3 mm/s, after which the strength decreased. The peak strength occurred at the ductile to brittle transition, with the ice being in ductile or creep mode at slower speeds and in brittle mode at higher speeds.

Figure 3.5.7 is a very important figure that deserves further explanation. This figure is a composite of pressures (pressure, similar to stress, is force/loaded area) obtained from different loaded areas and from different sources. An outline of the issues and data follows.

The design of offshore or near-shore structures to be placed in waters covered by ice during winter requires two types of loadings to be defined: global (total) loads and local loads. Global loads need to be specified so the overall stability of the structure can be assured with an adequate margin of safety. Local loads, which are applied over small areas of the hull, also need to be determined. They are used to design the shell plate, stiffeners, scantlings, etc. for offshore and near-shore structures. Local pressures are also very important in the design of ships required to ply ice-covered waters, particularly since pressures on the ship's hull (especially the side shell or side of the ship) continue to be quite controversial.

Prior to 1980, local ice pressures were obtained solely from the results of small-scale laboratory tests. As a result there tended to be both over- and under-conservatism in the recommended pressures, depending on the interpretation of available data. During the 1980s field tests were conducted at a larger scale whereby measurements were made on the hulls of structures, specifically the steel structure Molikpaq (described later) and on ship hulls during dedicated ramming tests (described later). These data strongly indicated a size or area affect, whereby the pressure was observed to decrease as the loaded area increased. The relationship is usually expressed in the form (Masterson et al. 2007):

$$p = kA^{n} \quad \text{(Equation 3.5.1)},$$

where $p$ = ice pressure (MPa), $A$ = loaded area (m$^2$), and $k$, $n$ = constants, $n$ being less than 1.

The pressure relationship found in the design guidelines API RP 2N and CSA S471 is given as:

$$p = 8.1\, A^{-0.5} \quad \text{(Equation 3.5.2)}.$$

The data employed to arrive at this relationship were derived from a combination of medium-scale impact tests, dedicated ship ramming tests (CANMAR 1982; CANMAR 1985; German and Milne/VTT 1985; ARCTEC 1986) and measurements taken from ice load panels on the Molikpaq (Masterson and Frederking 1993). All of the data were obtained from ice in the Beaufort Sea and Arctic Islands, much of it multiyear ice. Thus the curve is applicable to those (and other) geographical regions with similar ice types. Design pressures in more temperate regions will be lower. For instance, the $k$ value used for the design of some of the Sakhalin structures was 4 (Masterson et al. 2007).

## 3.5.6 ICE STRENGTH AND PRESSURES OBTAINED FROM FULL-SCALE MEASUREMENTS ON DEPLOYED STRUCTURES

It has long been realized that there is a need to measure full-scale ice pressures and loads on deployed structures. This type of information is especially important for calculating the stability of offshore structures and for calculating loads and pressures on ships' hulls. In response to this need, structures placed in the Beaufort Sea, the Baltic, and some bridges were instrumented. Thus the range of applicability and reliability of ice load calculations was greatly extended. This information has also been used for iceberg impact-load determination.

### 3.5.6.1 Global Ice Pressure

The global ice pressure is used to obtain the ice load on an entire structure, where:

$$\text{Load} = p_g \times \text{contact area} \quad \text{(Equation 3.5.3)},$$

*where* $p_g$ = global ice pressure (MPa) and contact area = width of contact × ice thickness.

The global ice pressure is found to be a function of the aspect ratio of the loaded area, as shown schematically in Figure 3.5.9, where:

$$\text{aspect ratio} = \text{width of contact / ice thickness} \quad \text{(Equation 3.5.4)}.$$

A calculation of the aspect ratio effect is given in Figure 3.5.10 (Masterson et al. 2007). Two situations are shown, one where the aspect ratio is less than 1 and another where the aspect ratio is greater than 1. Narrow structures will experience

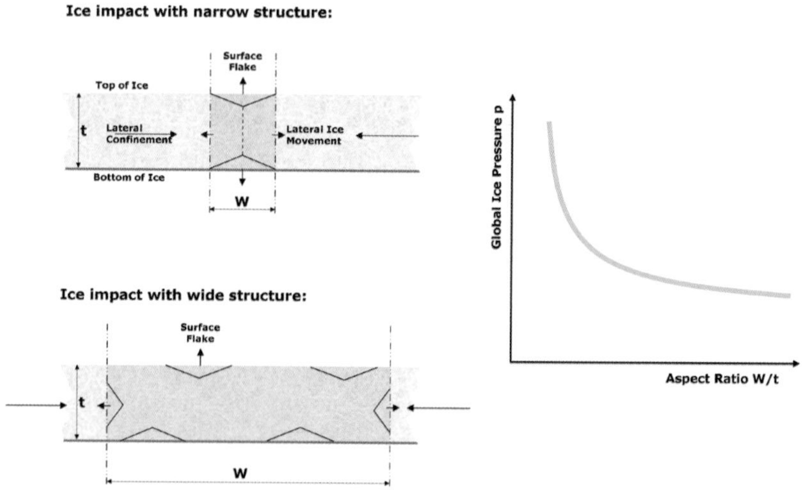

**Figure 3.5.9.** Illustration of Aspect Ratio Effect

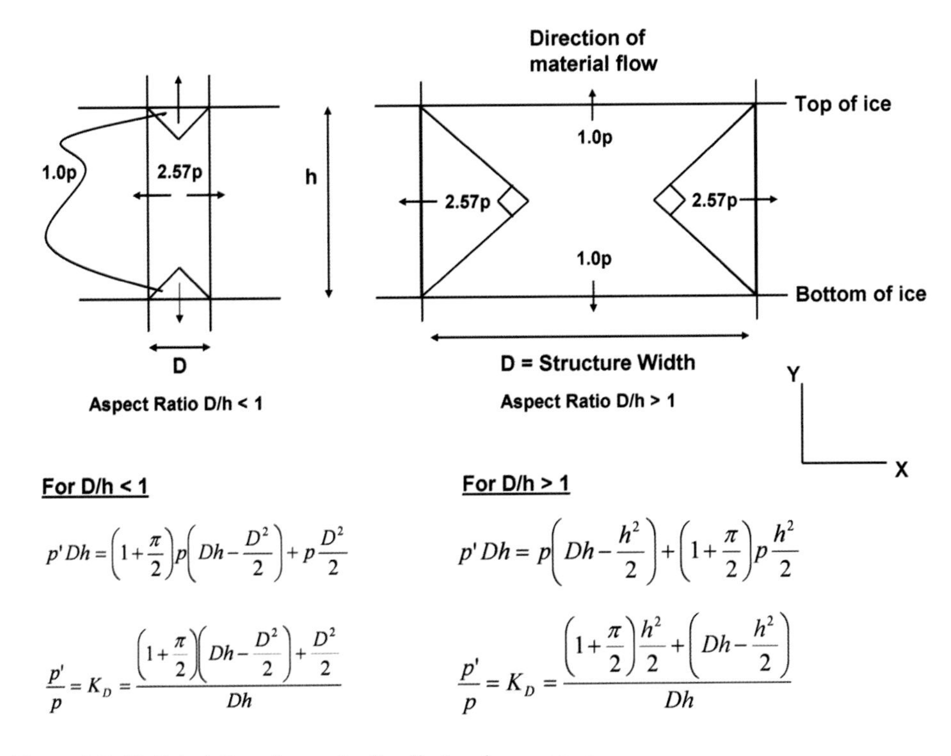


**Figure 3.5.10.** Calculation of aspect ratio effect on ice pressure

higher pressures because the majority of the material is flowing horizontally and is in confined compression, with resulting higher failure stress. Wider structures, on the other hand, will experience predominately upward and downward flaking in unconfined compression at a lower failure stress. The wider the structure, the more material fails at the lower, unconfined compressive stress.

The global ice pressure as calculated above considers that the ice fails only in compression when it contacts the structure (Figure 3.5.11a). For wide structures the ice will seldom fail at the same time at all points along the width of contact and thus there is nonsimultaneous failure, with the ice reaching its peak strength at different times at different points along the width of the structure (Rogers et al. 1986; Jeffries and Wright 1988; Wright and Timco 1994). This results in a reduced time-average pressure. It occurs because of two principal realities:

The ice strength is variable, because as discussed, the ice is nonhomogeneous and is randomly flawed with cracks and asperities.

The contact between the ice and the structures is variable in space and time and thus while one point may be fully contacted and near (or at) failure, other points will be in lesser contact and either not yet near failure or past failure. This is a boundary condition effect. The wider and larger a contact area, the more nonuniform the contact is likely to be, leading to a variance in contact pressure.

Thus the contact pressure exerted by the ice along the width or over the area of contact, especially where these are large (for example over a 100 m structure

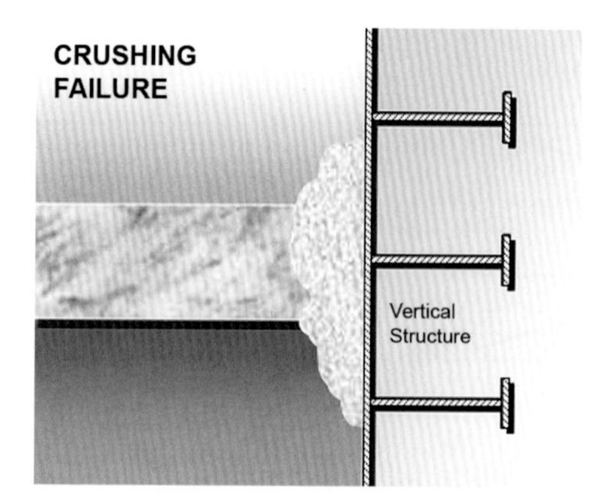


**Figure 3.5.11a.** Crushing failure. Section of hull structure is shown at left with stiffeners. Advancing, crushing (compressive failure) ice is shown on the right. Below the ice is water and the ice is floating on the water.

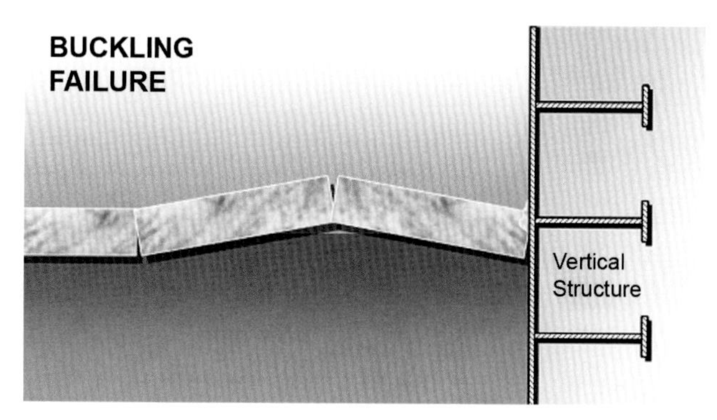


**Figure 3.5.11b.** Buckling failure. Section through hull structure is shown on the right with stiffeners. Ice impacting the hull is shown on the left with a typical buckling failure. Below the ice is water and the ice floats on the water.

width or on a contact area of 100 $m^2$), will be variable and will have a statistical distribution. In statistical terms, the average contact pressure may be 1 MPa and the standard deviation may be 0.2 MPa. Assuming a normal distribution (bell curve or Gaussian), this means that 50 percent of the pressures will be less than 1 MPa and 50 percent are greater. Also 84 percent of the pressures will be less than 1.2 MPa and 97.7 percent will be less than 1.4 MPa. Sixty-eight percent of the pressures will lie between 0.8 and 1.2 MPa.

A further phenomenon results in nonuniformity of pressure and contact. This is known as *mixed modal failure* whereby the ice fails in crushing at some points and at other points of contact it fails in bending or buckling (Figure 3.5.11b). The forces associated with bending or flexural failure, and likewise buckling or stability failure (a closely related mode of failure to bending), are lower and thus mixed modal failure usually results in lower average ice pressures. Under mixed-mode conditions, the average pressure might be 0.8 MPa and the standard deviation about equal to that of pure crushing.

The variability of the ice, including its strength and physical makeup, lends itself to statistical analysis. For this reason, probabilistic ice loads are calculated using Monte Carlo methods, whereby various physical parameters, such as strength, thickness, and floe size, are each assigned statistical distributions derived from existing databases of ice physical properties. Pressures can be assigned statistical distributions using the results of full-scale measurements. Random picks of each parameter are effected over and over again, load calculations being done in each case. The maximum loads then provide the design load required for structural analysis.

To quantify the proportion of the different types of events, and to quantify the pressures involved, full-scale measurements have been required.

### 3.5.6.2 Full-Scale Measurements

Full-scale measurements of loads and pressures have included the following projects. Loads due to very large ice floes hitting Hans Island, located between Greenland and Ellesmere Island, have been back calculated (Danielewicz and Blanchet 1987). An ice floe interacting with the island is shown in Figure 3.5.12. During the interactions, deceleration of the large impacting ice floes was measured and observations were made of failure process. In some interactions, splitting of the floes was observed, as seen in Figure 3.5.12. The failures were mixed mode, involving crushing, buckling, flexure, or bending and splitting. Generally the loads were relatively low.

**Figure 3.5.12.** Multiyear ice floe impacting Hans Island

A steel structure called the Molikpaq, a MODU or mobile offshore drilling unit, located in the Canadian Beaufort Sea at different sites, including one called Amauligak I-65, was instrumented at deployment for ice load and pressure measurements (Jefferies and Wright 1988; Wright and Timco 1994). This structure produced the first measured loads during encounters with moving multiyear ice. There were several encounters with multiyear ice during the 1985–1986 winter season, resulting in the first recording of major ice loads and pressures. Figure 3.5.13 shows the Molikpaq in ice and typical interactions. Ice pressure panels measuring 1 m by 2 m, strain gauges placed on the steel hull, extensometers to measure movement, and video records resulted in the collection of time series of loads. The data

Figure 3.5.13. Ice interacting with the *Molikpaq*

obtained resulted in a unique full-scale database of loads and pressures which is the underpinning of arctic codes today. Analysis of the data continues, since it was voluminous and at times also controversial.

Global ice pressures of 1 MPa were interpreted as acting on the Molikpaq hull during simultaneous crushing events while lower pressures were interpreted during mixed-mode failure. These measurements have enabled development of specific design pressures for Canadian Standards Association, American Petroleum Institute, and ISO codes. Subsequent data from the Baltic have been added to provide pressures and loads for thinner first-year ice. These Baltic projects are known as LOLEIF (low-level ice forces) and STRICE (measurements on structures in ice) (Schwarz and Jochmann 2001).

Molikpaq is now at Piltun-A offshore Sakhalin, north of Japan, having been reinstrumented for the deployment (Weiss et al. 2001). Load and pressure measurements of first-year ice interactions only have been made since 1999 and continue to be made. Typical load vs. time traces, obtained from both pressure panels and strain gauges, are shown in Figure 3.5.14. These traces show the variability of load on a structure with time. Realistic fatigue loads and cycle numbers were derived from the time series. Analyzed data has been used for designs at Sakhalin.

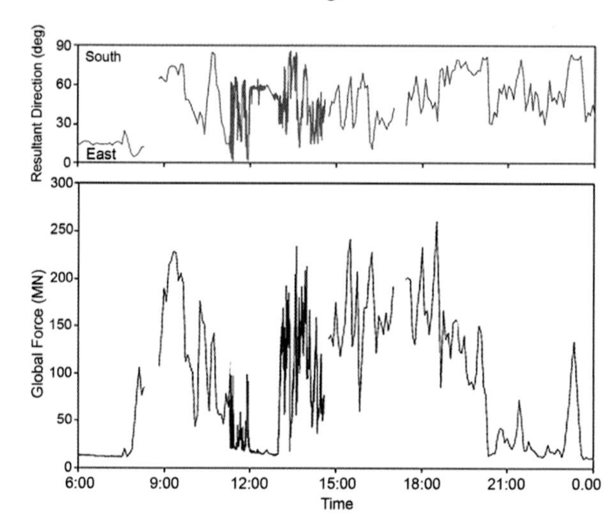


Figure 3.5.14a. Load-time trace from Sakhalin

Figure 3.5.14b. Load-time trace from Beaufort Sea

### 3.5.7 HOW ICE LOADS ARE CALCULATED

"Global" ice pressures are based on data from medium- to large-scale tests and from full-scale measurements. The data obtained reflect the effects of scale and aspect ratio as well as regional differences in ice composition and strength (Masterson et al. 2000). The process of determining the applicable procedures has been complex, especially where the development of codes and recommended practices is concerned. No codes existed for the calculation of ice loads on offshore structures before the early 1980s. Codes did exist for the design of bridges in rivers of Canada, the United States, Russia, and other northern countries, but they were limited in scope and could not be readily applied to offshore structures and to ship design. Again, classification societies such as Lloyd's Register had procedures for designing ice-strengthened hulls, but these "rules" were often arbitrary and based on limited ice-structure interaction information.

Experience has shown that there are regional effects on the maximum ice pressure. This results from length of time the ice is exposed to temperatures below freezing and from the magnitude of the temperature. In areas such as the Beaufort Sea and Arctic Archipelago, the winters are long and the temperatures are cold, causing the ice to have a high strength over a considerable thickness. In more temperate, subarctic zones, such as the Sakhalin region of eastern Siberia or Labrador off the east coast of Canada, the winters are less severe and the total amount of freezing, expressed as freezing degree-days, is less. Thus, ice which is formed, and especially deformed ice which has been rafted (one layer on top of another), will not have an opportunity to reach the same strength as in the Arctic. Ice of more southerly or temperate climates will develop even less strength.

One way of looking at this is to consider that ice whose temperature is well below the freezing point will lose its latent heat and will have a solid, crystalline matrix. Ice in temperate zones which is at sea has constant exposure to seawater and limited opportunity to lose heat to a cold atmosphere. Thus the bulk of the ice will remain at a temperature near the freezing point, an appreciable amount being on the verge of phase change, and thus will have a lower physical strength.

Generally, three regions are defined as follows:

**Arctic**, having about four thousand freezing degree-days per winter season (e.g., Beaufort Sea);
**Subarctic**, having about two thousand freezing degree-days per winter season (e.g., northeast Sakhalin and Labrador); and
**Temperate**, having about one thousand freezing degree-days per winter season (e.g., Aniva Bay, Northumberland Strait, Cook Inlet, Baltic Sea, and Bohai Bay).

Freezing degree-days (FDD) are a measure of how cold it has been and how long it has been cold; the cumulative FDD is usually calculated as a sum of average daily degrees below freezing for a specified time period (ten days, one month, a season, etc.).

The horizontal load applied to an offshore structure due to ice failure against the structure is calculated as follows (Draft ISO standard 19906):

$$F_G = p_G\, h\, w \quad \text{(Equation 3.5.5)},$$

where $F_G$ = global ice load in MN, $p_G$ = global ice pressure in MPa, $h$ = average ice thickness, and $w$ = contact width of the structure.

The global ice pressure is defined in the draft ISO code as:

$$p_G = C_R\, h^n \left(\frac{w}{h}\right)^m \quad \text{(Equation 3.5.6)},$$

where $p_G$ = global average ice pressure, in MPa; $w$ = width of the structure, in m; $h$ = thickness of the ice sheet, in m; $m$, $n$ = empirical exponents to take account of the size effect ($m = -0.16$; $n = -0.50 + h/5$ for $h < 1.0$ m and $n = -0.30$ for $h \geq 1.0$ m); and $C_R$ = ice strength coefficient, in units that depend on the values of $m$ and $n$.

For first-year and multiyear ice, e.g., Beaufort Sea, where $h \geq 1.5$ m:

$$C_R = 2.8 \quad \text{(Equation 3.5.7)}.$$

For thin first-year ice, e.g., Baltic Sea where $h < 1.5$ m:

$$C_R = 1.8 \quad \text{(Equation 3.5.8)}.$$

Considering a 100 m wide structure placed in the Beaufort Sea impacted by multiyear ice with an average thickness of 10 m, then the global ice pressure is calculated to be:

$$p_G = 1.0 \times 2.8 \times 10^{-0.3} \times (100/10)^{-0.16} = 0.97 \text{ MPa} \quad \text{(Equation 3.5.9)}.$$

Thus the force to crush the ice in contact with the structure is $0.97 \times 100 \times 10 = 970$ MN (100,000 tonnes).

This force may be reduced if it can be shown that there is not sufficient “driving force” to cause crushing of the thick multiyear ice at the structure interface (Croasdale et al. 1992). The calculation of driving force uses the same principles to determine ice pressures but in addition information is needed on the size of the

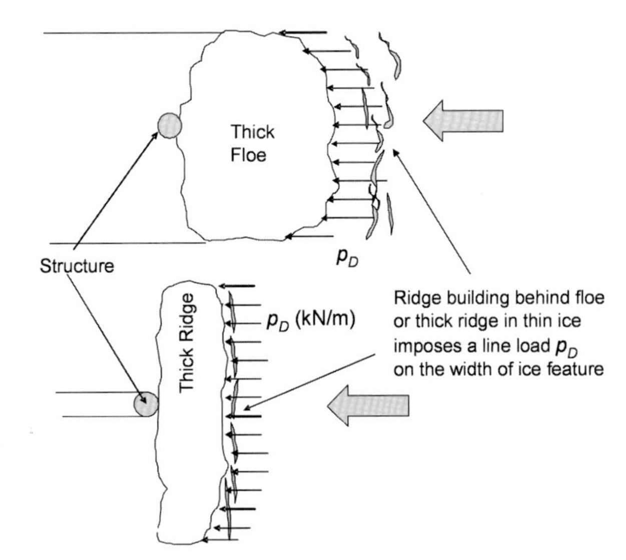


Figure 3.5.15. Ridge building behind a thick floe or ridge

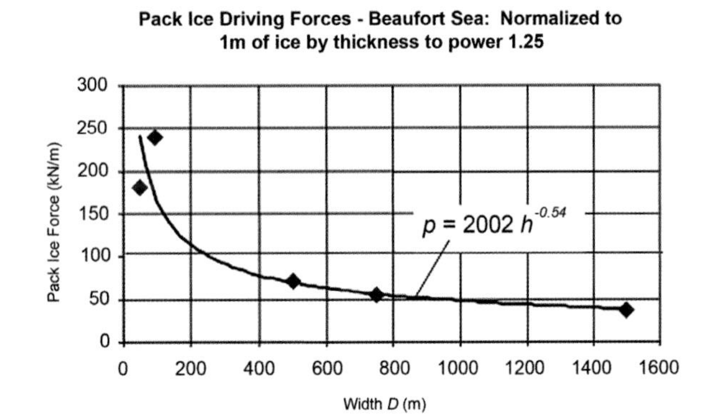


multiyear ice floe and on the thickness of the ice pack surrounding the multiyear floe. Driving force is illustrated in Figure 3.5.15. If the multiyear floe is not surrounded by pack ice of very large (infinite) extent, then the force can be limited by the available kinetic energy of the multiyear floe. It is not the intent to perform these further calculations here.

## 3.5.8 SHIP IMPACT TESTS

Data obtained from dedicated ramming tests with ice breakers and ice-capable ships comprises a very important part of the information on ice pressure and area effects on those pressures. Also, ship rams occur at relatively high speed, so it is possible to determine from these tests whether there are any significant strain rate effects on ice pressures exerted on hulls at areas up to several or tens of square meters.

Ships such as the *MV Arctic* (German and Milne/VTT 1985), the *Kigoriak* (CANMAR 1982; CANMAR 1985), the *Polar Star*, *Polar Sea* (ARCTEC 1986), *Polarstern*, *Oden*, and Canadian Coast Guard icebreakers have been instrumented with

**Figure 3.5.16.** *Kigoriak* ramming ice

strain gauges and pressure sensors for dedicated ice-ramming trials. Figure 3.5.16 shows the *Kigoriak* icebreaker ramming both an individual ice floe and continuous multiyear pack ice. During these trials, horsepower requirements plus pressures were measured. Much work was done in the Baltic on propulsion through ice and the associated shaft horsepower required. As a result of the Baltic measurements, the Finns published a signal paper on friction measurements at Icetech in Calgary in 1994 (Mäkinen et al. 1994).

## 3.5.9 LOCAL ICE PRESSURES

The design of structure or ship hull requires that pressures and loads on different parts and sections of hull be determined. In this manner the shell plate, stiffeners, and bulkheads can be designed. A pressure vs. area curve (previously shown in Figure 3.5.7) is used for this purpose. Figure 3.5.17a shows a typical framing arrangement for a steel hull of a structure or ship, and Figure 3.5.17b shows the local loads applied to critical areas.

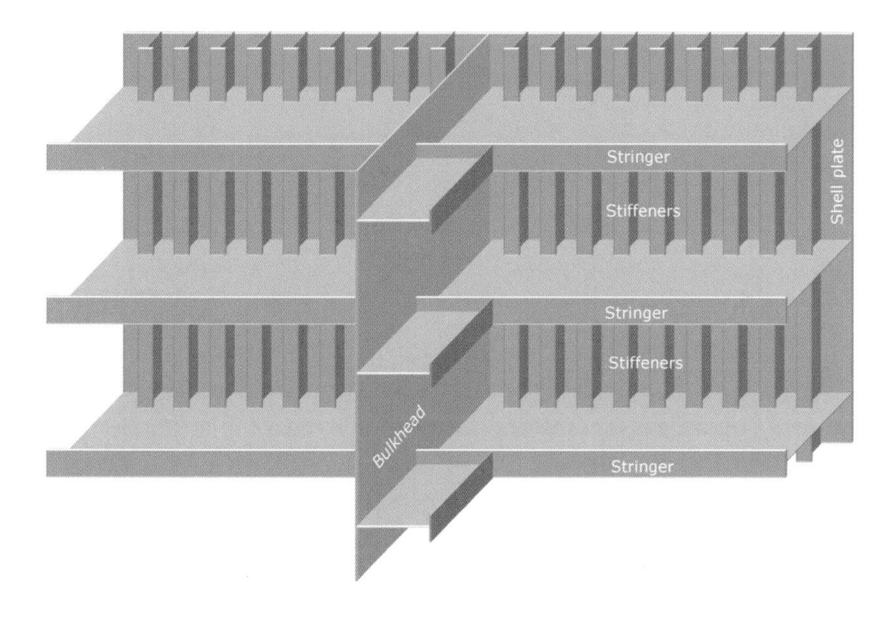


**Figure 3.5.17a.** Typical hull framing arrangement

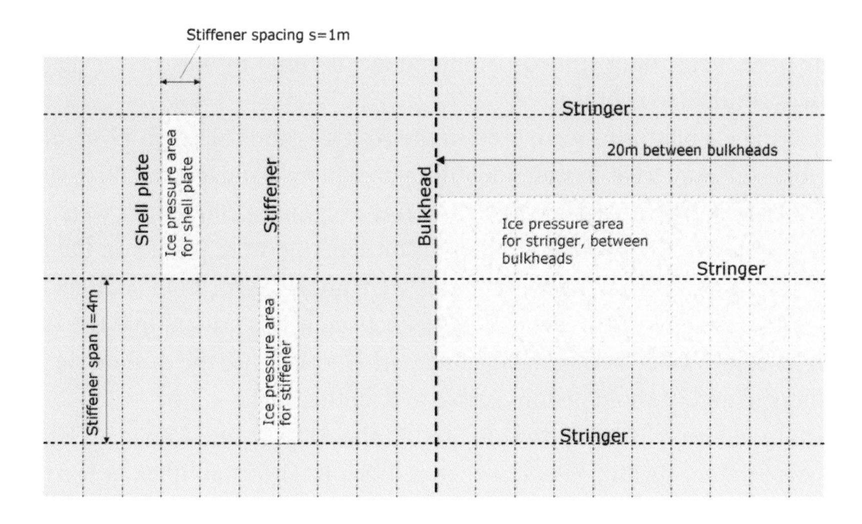


**Figure 3.5.17b.** Hull loaded areas for design

Examples of local ice pressure determination, using an updated but equivalent pressure-area relationship from the one of equation 3.5.2, are as follows (Masterson et al. 2007):

1. Pressure and load required for shell plate and stiffener:
   Area = 1 m × 4 m = 4 $m^2$
   pressure = 7.4 × $4^{-0.7}$ = 2.8 MPa (406 psi)
   load = 1142 tonnes

2. Pressure and load required for stringer:
   Area = 4 m × 20 m = 80 $m^2$
   pressure = 7.4 × $80^{-0.7}$ = 0.34 MPa (49 psi)
   load = 2800 tonnes

## 3.5.10 LOAD-BEARING ICE STRUCTURES

Crossings of lakes and rivers have been achieved using the naturally formed ice cover in winter for centuries (Gold 1960; 1971). The main requirement to cross safely is to have continuous ice of sufficient thickness to support the intended loads. There also are numerous examples of the use of floating ice to form roadways over rivers and lakes for the transportation of heavy loads (Kivisild et al. 1975; Haspel and Masterson 1979). Ice roads have been used for the transportation of heavy loads in northern Quebec, arctic Canada, and the North Slope of Alaska. There is a mine in the Northwest Territories of Canada which is resupplied each year using a 400 km ice road. Recently an offshore production island and pipeline to shore were installed using floating ice roads on the North Slope of Alaska (Masterson and Spencer 2001). During this project twenty thousand loads were transported over a single road over a one-month period, each load being approximately 35 tonnes. The resupply of St. Petersburg (called Leningrad at the time) by Russia during World War II was conducted over ice.

Floating ice platforms were used in the Arctic Islands between 1974 and 1986 to drill forty offshore wells in water depths up to 600 m (Baudais et al. 1976; Hood et al. 1979; Masterson et al. 1987). Oil drilling rigs weighing 1,200 tonnes were placed on ice thickened to 6 m by free flooding or by spraying were placed on the ice for periods of 90 to 120 days. These structures were analyzed for both sudden, brittle failure and for creep or time-dependent deflection. It was critical that the ice surface not go below water during the drilling period, otherwise the drilling rig would be submerged and it would be impossible to drill the well.

Grounded spray ice islands were used in the Alaskan and Canadian Beaufort Sea for exploratory drilling (Funegard et al. 1987; Bugno et al. 1990; Weaver et al. 1997; Masterson et al. 2004). They have been used in water depths of up to 12 m. Two wells were drilled from them in Canada and five wells were drilled in Alaskan waters. The last wells were drilled in 2003. These structures are constructed using very large pumps with a nozzle pressure of 1,400 kPa which, by spraying water into the air, cause it to be atomized and to thus freeze much more rapidly than by conventional flooding methods. Both floating ice platforms and grounded ice islands allow the seasonal drilling of exploratory wells, with the well being plugged with cement at the end of the program. The rig and all other material is removed and the ice melts during summer, leaving no residue.

The design of load-bearing structures such as ice roads and ice islands for drilling require knowledge of the consistency and continuity of the ice comprising them. In situ strength testing with the borehole jack provides a means of determining average strength and of identifying weak areas or layers. This type of testing is consistent with in situ testing used in soils to determine their average properties. Scale and absolute strength are not so much a concern as is the identification of weak areas. If the ice structure has more than about 15 percent weak areas, then it becomes unacceptably probable that either it will collapse under its own weight or that of the applied loads, or else it will not be able to resist the horizontal loads due to the surrounding ice pack movement.

### 3.5.10.1 Construction of Ice Structures

Preparation of the ice surface for flooding, if it is used to thicken the ice, varies according to local practice and experience. If flooding is conducted then the thicker snow is removed using mechanical means such as a blade or snowblower before flooding commences. Typically 5 cm to 10 cm of snow may be left on the surface before flooding begins. Some projects do not have the means available to remove the snow and the deeper snow cover on the ice is soaked with water and allowed to freeze. In this case it is important to ensure that the water-soaked snow has completely frozen before continuing with the application of water. This can delay the construction schedule and thus it is always preferable to remove thick snow.

To prepare an ice cover for vehicular traffic, especially for wheeled traffic, the snow can either be removed by blade or blower, or it can be packed down to a relatively high density to ensure the wheeled vehicles do not become stuck. The snow, where possible, should be well removed from the road with the snowbanks along the side being sloped or "feathered" back. This avoids stress concentrations at the road edge and excessive cracking of the ice, plus it discourages the accumulation of further drifting snow.

Thickening of the ice is usually accomplished using low-head, high-volume pumps, which pump water from beneath the ice onto the surface. The water is applied in layers about 3 cm thick and is allowed to freeze before another layer is applied. In this way the accumulation of ice thickness averages about 3 cm per day for most roads as a job average. During very cold periods when the winds are most likely to be low, the rate can be higher but the average considering the ability to cover existing ice and delays due to weather and equipment is as quoted. It is best practice not to dyke or confine the water but to allow it to flow freely and achieve a tapered cross section of the road or pad. This avoids sharp transitions and the formation of cracks at the edge of the flooded area. It is sometimes necessary to use snow dykes to prevent water from escaping and to achieve the required ice accumulation over a reasonable length of time. Judgment is required at all times.

The equipment used for thickening the ice varies and locally available equipment often is the most convenient and effective. The pumps should be capable of operating in very cold temperatures. They should be submersible or Archimedes screw type (augers), which have no hoses to freeze (Figure 3.5.18a). These types of pumps are self-contained and drain readily when shut down. On road construction, the pumps will be moved from hole to hole, which have been previously drilled, as the road is flooded. The crews operating the pumps require means of transportation and of moving the pumps, and they also require a drill or auger to make a hole of the required diameter.

Spraying techniques can also be used to construct roads as this will accelerate the accumulation of ice thickness. Spraying water into the air using high-pressure pumps (1.5 MPa pressure at the nozzle) causes it to be lofted into the air tens of meters and also causes it to be atomized (Figure 3.5.18b). The air travel results in rapid cooling of the water droplets and, if the spray nozzle is properly sized and aimed, the water droplets will be about 80 percent frozen when they fall to the existing ice surface. It is desirable not to have the water droplets 100 percent frozen at contact with the surface as this leads to the formation of corn ice or snow that has very little cohesion; this is not a good bearing surface as it lacks shear strength. Also, it is not desirable to spray too wet or the resulting slush that lands on the ice will require a long time to freeze. If the spraying is planned and managed properly, optimum buildup or ice accumulation rates are achieved with maximum density and shear strength. The density of spray ice is usually about 0.6 tonnes/m$^3$. This density and shear strength are very suitable for resisting lateral ice loads exerted on grounded ice islands from the surrounding moving pack ice and for supporting heavy loads such as a drilling rig on grounded islands or floating ice structures. A primary and reliable means of measuring the shear strength of the ice is to use a static cone penetrometer (Bugno et al. 1990). Density is measured by means of core samples that are cut into appropriate lengths after which their volume and weight are measured. Density is then weight divided by volume. Borehole jacks and pressure meters have limited used since it is very difficult to deploy them in the mass of the spray ice.

**Figure 3.5.18a.** Flooding an ice road

**Figure 3.5.18b.** Spraying an ice structure

### 3.5.10.2 Ice-Bearing Capacity

Throughout industry and government agencies the floating ice sheet strength equation is employed to predict the load that an ice sheet is capable of carrying. The basic ice sheet strength equation was evaluated through the compilation of observed failures in the 1950s and 1960s (Gold 1960; 1971). This equation is given below:

$$P = A \times h^2 \quad \text{(Equation 3.5.10)},$$

where $P$ is the ice sheet capacity, in tonnes; $A$ is the coefficient for ice road operation, in tonnes/m², in the range 343 tonnes/m² $\leq A \leq$ 686 tonnes/m²; and $h$ is the ice thickness, in m.

For complex loading conditions, more rigorous analysis of the load and stress levels is required. Equations and procedures for this type of analysis are contained in the ISO 19906 offshore arctic code and in various publications.

### 3.5.10.3 Ice Strength Profiles

Figure 3.5.19 shows a profile of borehole strengths plus temperature, salinity, and brine volume. In this example a 150 mm core was taken and the temperature and salinity were measured at different depths of the core. Brine volume is then calculated from these measurements. Strength tests were conducted in the hole at 300 mm (1 ft.) intervals. In this manner strength can be compared with other physical properties of the ice.

There were no voids or weak layers in this ice, but there is a variation in strength. The average strength of the ice in any borehole must be greater than a specified minimum value. The condition was more than met for this hole since the specified minimum was set at 8 MPa.

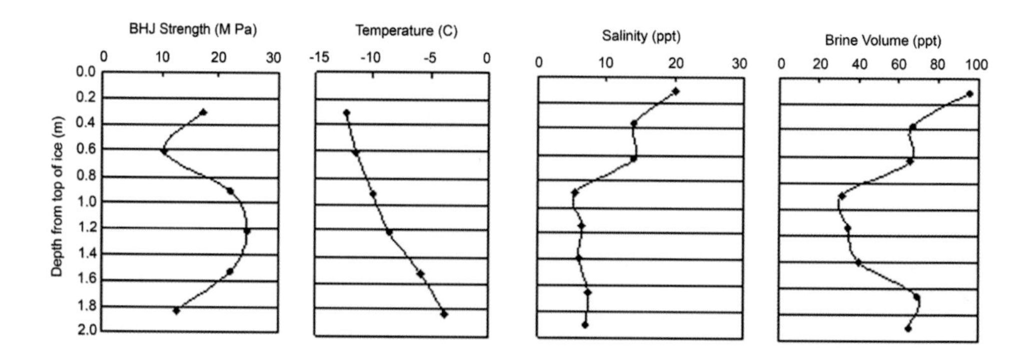


**Figure 3.5.19.** Ice strength profile with related parameters

### 3.5.11 SUMMARY AND CONCLUSION

The material presented takes a civil engineering point of view and relates to the needs of engineers for ice strengths required for design of offshore operations and facilities. The coverage, although focused on sea ice, is relatively broad and presents an overview. It would be impossible to cover in depth this very diverse topic which has been investigated by many researchers and practitioners over a period of fifty years or more. Tensile strength-testing methods were not covered as they are a separate topic and are not germane to the discussion as a whole.

In summary, the following key aspects have been covered, with some conclusions on the use of field methods in strength testing:

- Principles of strength of materials from a civil engineering perspective
  - This includes unconfined compressive strength and confined compressive strength and the basic relation between the two plus their use in engineering.
- Types of small-scale tests, including both laboratory and in situ
  - Laboratory tests include uniaxial and triaxial tests in cylindrical or prismatic specimens. Small-scale field or in situ tests include field unconfined compressive tests, pit tests, borehole jack tests, and flat-jack tests. Generally the latter three types of tests are very reliable and most suited to field conditions.
- Medium- and large-scale tests performed in situ
  - Medium-scale tests have been performed on iceberg ice, multiyear ice, and first-year ice in diverse locations. These tests were conducted using steel indentors pushed against the ice with hydraulic cylinders. Relationships between pressure and contact area were determined accurately from such tests and the results have been used extensively in the derivation of structure design codes. It was from these tests that scale or size effects on the strength of ice were first reliably quantified.
- Full-scale pressure and load measurements on deployed structures
  - Full-scale measurements of ice loads and pressures on the Molikpaq structure, on ships in ice with hull instrumentation, and on Hans Island have provided the most valuable insights into the global loads that structures placed in ice-covered waters will experience. Such measurements are expensive to conduct and often difficult to interpret, but are the backbone of reliable ice load determination. No design code will proceed without including this type of information.
- Example pressure/load calculation
  - Pressure and load calculations have been included to illustrate the principles and techniques involved. A typical steel-framed structure has been used for the example.
- Load-bearing structures such as ice roads
  - Load-bearing structures such as ice roads and ice islands have been

described, along with basic design and construction methods. In the Arctic Islands and the Beaufort Sea these structures have played an important role in the exploration of oil and gas. Ice roads are extensively used by the mining and oil and gas industry to transport heavy loads.

Future advances in sea ice engineering will come from the measurement of ice loads and pressures at full scale on structures deployed in water covered by ice. The importance of such instrumentation and measurement cannot be overemphasized. Ice thickness and type of cover measurements are equally important, and these measurements, along with the pressure and load data, enable the reliable calculation of deterministic and probabilistic ice loads and pressures. In situ testing is a useful tool to be used in combination with the above efforts.

Ice-bearing capacity will benefit from continued reporting of performance and related physical properties of the ice structure, coupled with accurate knowledge of the loads placed on the facility.

## REFERENCES

ARCTEC (1986), *Consolidation of Local Ice Impact Pressures Aboard the USCGC Polar Sea*, A final report to Ship Structure Committee, U.S. Coast Guard and Transport Canada, Transportation Development Centre by TP8533E.

Baudais, D. J., J. S. Watts, and D. M. Masterson (1976), *A System for Offshore Drilling in the Arctic Islands*, Offshore Technology Conference, Houston, Paper No. OTC 2622, May.

Bugno, W., D. M. Masterson, J. Kenny, and R. Gamble (1990), *Karluk Ice Island, Proceedings of the Ninth International Conference of OMAE*, Book No. 10296F.

CANMAR (1982), *Final Report on Full Scale Measurements of Ice Impact Loads and Response of the Kigoriak*, A report by Canadian Marine Drilling Ltd. to Coast Guard Northern TP5871E .

CANMAR (1985), *Kigoriak and Robert Lemeur 1983 Ice Impact Tests, Refinement of Model, Ship/Ice Interaction Energies*, A final report for Coast Guard Northern by Canadian Marine Drilling Ltd. TP6813E.

Croasdale, K., R. Frederking, B. Wright, and G. Comfort (1992), Size effect on pack ice driving forces, in *Proceedings of the 11th International Symposium on Ice*, vol. 3, pp. 1481–1496, Banff, Alberta, June 15–19.

Danielewicz, B., and D. Blanchet (1987), Measurements of multi-year ice loads on Hans Island during 1980 and 1981, in *Proceedings of the Ninth Conference on Port and Ocean Engineering under Arctic Conditions (POAC)*, vol. 1, p. 465, University of Alaska Fairbanks.

Dykins, J. E. (1967), Tensile properties of sea ice grown in a confined system, in *Physics of Snow and Ice*, vol. 1, pp. 523–537, Bunyeido Printing Co., Japan.

Funegard, E. G., R. H. Nagel, and G. G. Olson (1987), *Design and Construction of the Mars Spray Ice Island*, Offshore Mechanics and Arctic Engineering (OMAE), Houston, TX.

German and Milne/VTT, M. V. (1985), Arctic, test results and analysis, final report, *A report by German and Milne and Technical Research Centre of Finland*, Ship Laboratory to Transport Canada, Coast Guard Northern TP6270E.

Gold, L. W. (1960), Field study on the load bearing capacity of ice covers, Woodlands Review, *Pulp and Paper Magazine Canada, 61*, 153–154, 156–158.

Gold, L. W. (1971), Use of ice covers for transportation, *Canadian Geotechnical Journal, 8*, 170–181.

Goodman, R. E. (1989), *Introduction to Rock Mechanics,* 2nd ed., John Wiley and Sons, New York.

Haspel, R. A. and D. M. Masterson (1979), Reindeer Island Floating Ice Road Project, in *Proceedings of Workshop on Winter Roads*, National Research Council of Canada, 18–19 October 1979, Ottawa, Ontario.

Hood, G. L., D. M. Masterson, and J. S. Watts (1979), Installation of a subsea completion in the Canadian Arctic Islands, Petroleum Society of CIM, Paper No. 79–30–20, Presented at the 30th Annual Meeting of the Petroleum Society of CIM, Banff, Alberta, May 8–11.

Iyer, S. H. (1983), Size effects and their effect on the structural design of offshore structures, in *Proceedings of POAC '83*. Seventh International Conference on Port and Ocean Engineering Under Arctic Conditions, Technical Research Center of Finland, Helsinki, Finland, April 5–9, 1983.

Iyer, S. H., and D. M. Masterson (1987), Field strength of multi-year ice using thin-walled flat jacks, in *9th International Conference on Port and Ocean Engineering under Arctic Conditions*, University of Alaska Fairbanks.

Jefferies, M. G., and W. H. Wright (1988), Dynamic response of "Molikpaq" to ice-structure interaction, in *Proceedings of OMAE-88,* vol. 4, pp. 101–220, Houston, TX.

Jordaan, I. J. (1989), Scale effects in ice-structure interaction: Analytical aspects, in *Proceedings of Workshop on Ice Properties, St. John's, National Research Council Canada Technical Memorandum No. 144, NRCC 30358*, pp. 161–181, June.

Jordaan, I. J., M. Fuglem, and D. G. Matskevitch (1996), Pressure-area relationships and the calculation of global ice forces, in *Proceedings, IAHR Symposium on Ice,* vol. 1, pp. 166–175, Beijing, China.

Kivisild, H. R., G. D. Rose, and D. M. Masterson (1975), Salvage of heavy construction equipment by a floating ice bridge, *Canadian Geotechnical Journal, 12*(1), 58–69.

Mäkinen, E., S. Liukkonen, A. Nortala-Hoikkanen, and A. Harjula (1994), Friction and hull coatings in Ice Operations, in *Proceedings ICETECH '94, Society of Naval Architects and Marine Engineers, March 1984,* Calgary, Alberta, Paper E.

Masterson, D. M. (1996), Interpretation of in situ borehole ice strength measurement tests, *Canadian Journal of Civil Engineering, 23*(1), 165–179.

Masterson, D. M., D. J. Baudais, A. Pare, and M. Bourns (1987), Drilling of a well from a sprayed floating ice platform Cape Allison C 47, in *Proceedings, OMAE '87,* Houston, TX, Paper No. 162.

Masterson, D. M., R. Cooper, P. A. Spencer, and W. P. Graham (2004), Thetis spray/chip ice islands for Harrison Bay, Alaska, ice in the environment, in *Proceedings of the 17th IAHR International Symposium on Ice,* St. Petersburg, Russia, June 21–25.

Masterson, D. M., R. M. W. Frederking, B. Wright, T. Karna, and W. P. Maddock (2007), A revised ice pressure-area curve, in *Recent Development of Offshore Engineering in Cold Regions* (POAC-07, Dalian, China, June 27–30), edited by Yue, Dalian University of Technology Press, Dalian, pp. 305–314. The paper is contained in the proceedings of the conference, the 19th Conference on Port and Ocean Engineering under Arctic Conditions. Masterson, D. M., and R. M. W. Frederking (1993), Local contact pressures in ship/ice and structure/ice interactions, *Cold Regions Science and Technology, 21,* 169–185.

Masterson, D. M., R. M. W. Frederking, and P. A. Truskov (2000), Ice force and pressure determination by zone, in Proceedings of *6th International Conference on Ships and Marine Structures in Cold Regions, ICETECH*. St. Petersburg Russia.

Masterson, D. M., and W. Graham (1992), Development of the original ice borehole jack, in *Proceedings, IAHR Ice Symposium 1992,* Banff, Alberta.

Masterson, D. M., W. P. Graham, S. J. Jones, G. R. Childs (1997), A comparison of uniaxial and borehole jack tests at Fort Providence ice crossing, 1995, *Canadian Geotechnical Journal, 34,* 471–475.

Masterson, D. M., D. E. Nevel, R. C. Johnson, J. J. Kenny, and P. A. Spencer (1992), The medium scale iceberg impact test program, in *Proceedings, IAHR Ice Symposium 1992,* Banff, Alberta, pp. 930–953.

Masterson, D. M., and P. A. Spencer (2001), The Northstar on-ice operation, *POAC 01,* Ottawa, Ontario, August, Paper No. 105. The sixteenth conference was hosted by the Canadian Hydraulics Centre, NRC in Ottawa, Canada from August 12–17, 2001.

Masterson, D. M., P. A. Spencer, D. E. Nevel, and R. P. Nordgren (1999), Velocity effects from multi-year ice tests, in *18th International Conference on Offshore Mechanics and Arctic Engineering (OMAE 99),* St. John's, Newfoundland, July 11–16.

Michel, B. (1978), *Ice Mechanics,* Pressses de l'Universite Laval, Quebec.

Rogers, B. T. M. D. Hardy, V. Neth, and M. Metge (1986), Performance monitoring of Molikpaq while deployed at Tarsuit P-45, in *Proceedings of 3rd Canadian Conference on Marine Geotechnical Engineering.*

Schwarz, J., and P. Jochmann (2001), Ice force measurements within the LOLEIF Project, in *Proceedings of POAC*, Ottawa, Ontario. The sixteenth conference was hosted by the Canadian Hydraulics Centre, NRC in Ottawa, Canada from August 12–17, 2001.

Seely, F. B. (1935), *Resistance of Materials,* John Wiley & Sons, Inc., New York.

Singh, S. K., and I. J. Jordaan (1996), Triaxial tests on crushed ice, *Journal of Cold Regions Science and Technology, 24,* 153–165.

Spencer, P. A., D. M. Masterson, and M. Metge (1993), The Flow properties of crushed ice: Crushing Plus extrusion tests, in *OMAE '93 Conference,* Glasgow, Scotland, Vol. 4. Proceedings of the Offshore Mechanics and Arctic Engineering Conference (OMAE).

Terzaghi, K., R. B. Peck, and G. Mesri (1996), *Soil Mechanics in Engineering Practice,* 3rd ed., John Wiley & Sons, Inc., New York.

Timoshenko, S. P. (1970), *Theory of Elasticity,* McGraw-Hill, New York.

Timoshenko, S. P. (1983), *History of Strength of Materials,* Dover Publications.

Weaver, J. S., and J. P. Poplin (1997), Case history of the Nipterk P-32 spray ice island, *Can. Geotech Journal, 34,* 1–16.

Weiss, R. T., B. Wright, and B. Rogers (2001), In-ice performance of the Molikpaq off Sakhalin Island, in *16th International Conference on Port and Ocean Engineering under Arctic Conditions (POAC '01),* Ottawa, Ontario, August 12–17.

Wright, B. D., and G. W. Timco (1994), Review of ice forces and failure modes on the Molikpaq, in *12th International Association for Hydraulic Research (IAHR) Symposium on Ice, Trondheim, Norway,* vol. 2, pp. 816–825.

# Chapter 3.6

# Sea Ice Optics Measurements

*Don Perovich*

## 3.6.1 INTRODUCTION

It all begins with the sun. It is the source of heat for the global climate system and the primary source of energy for most terrestrial and marine food webs. Understanding how sunlight interacts with sea ice is critical for such large-scale issues of global significance as climate change and the behavior of the arctic marine ecosystem.

Sea ice optics is also important on local and regional scales in terms of the services provided by sea ice. Snow-covered sea ice serves as a platform for over-ice transportation. Sunlight absorbed in the surface layer is a key term in the surface energy budget and has a direct impact on snow and ice melt. Surface melt, particularly once ponds form, substantially degrades transportation on the ice surface. Sunlight is also absorbed into the interior of the ice, causing warming and increasing the brine volume. This weakens the ice and enhances its permeability. Finally, light is transmitted to the bottom of the ice and the upper ocean, where it contributes to the biological activity that is the foundation of the polar marine food web.

A detailed background on the optical properties of sea ice is available in several articles, including Perovich (1996), Grenfell (1991), and Light et al. (2003). The focus of this chapter will be on optical measurements. It is useful, however, to make a few quick theoretical points that bear directly on optical observations. *Optical* refers to the solar spectrum, the wavelength region from 250 to 2500 nm. This includes ultraviolet (250–400 nm), visible (400–750 nm, i.e., blue to red), and the near infrared (750–2500 nm). Figure 3.6.1 shows the incident solar spectrum measured in the Arctic under sunny skies (Grenfell and Perovich 2008). Snow has a major impact on the optical properties of a sea ice cover and must be included in optical measurements.

The interaction of sunlight with ice and snow is governed by two processes: absorption and scattering. Scattering is when the light is redirected into a different

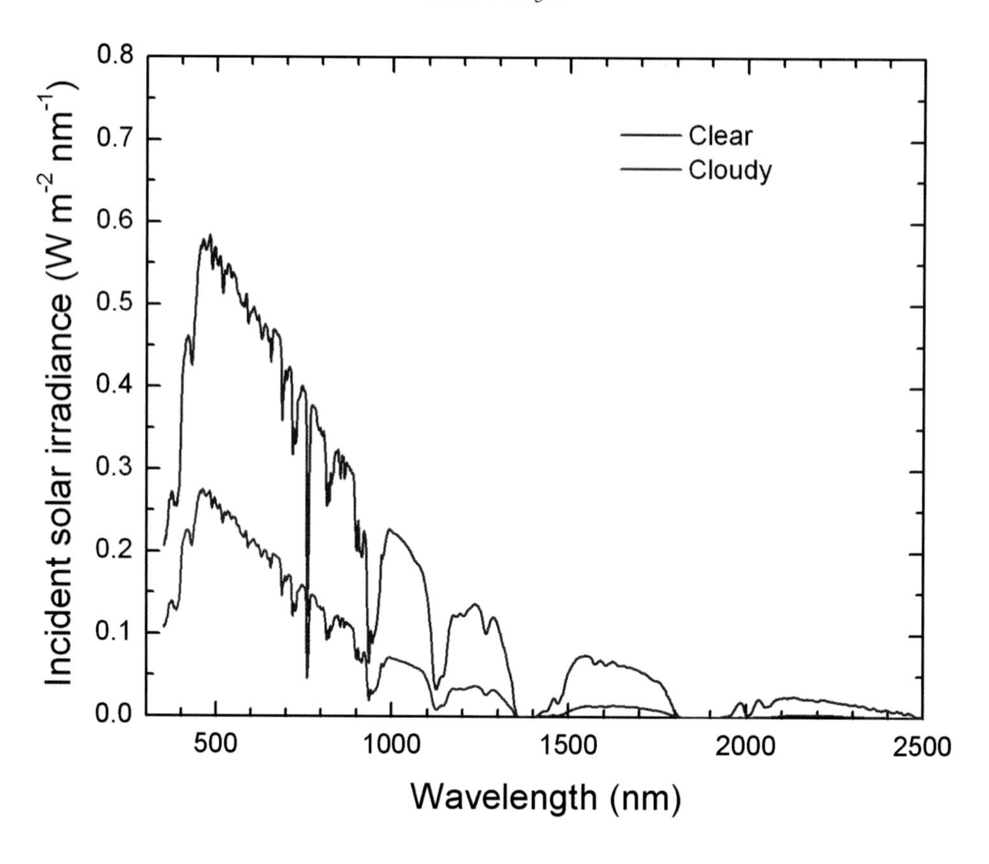


**Figure 3.6.1.** Incident solar irradiance on a sunny day in the Arctic. The measurements were in August 2005 aboard the USCG *Healy* in the Arctic Ocean. The solar zenith angle was 65°.

direction and absorption is when the light is absorbed by the medium. Sea ice is composed of ice with inclusions of brine pockets and air bubbles. These inclusions cause scattering. Scattering is large in snow and sea ice and exhibits little wavelength dependence. It affects the magnitude of the optical properties. For example, snow albedos are large because there is a tremendous amount of scattering (Grenfell et al. 1994). Water albedos are small, because there is little scattering (Pegau and Paulson 2001). Absorption exhibits strong wavelength dependence and is largely responsible for spectral differences in observed optical properties. The absorption coefficient is a measure of the energy lost to absorption. For a purely absorbing medium, with no scattering, the absorption of light depends on $e^{-kz}$, where k is the absorption coefficient and z is the thickness of the medium. There is a quantity called the e-folding length, which is the distance it takes to reduce the radiance to 1/e (i.e., to reduce to 37 percent). Absorption coefficients for sea ice range over 5 orders of magnitude (Figure 3.6.2a) (Perovich and Govoni 1991; Warren et al. 2006). At 400 nm the e-folding length for pure ice is 1350 m, while it is only 6 mm at 1400 nm.

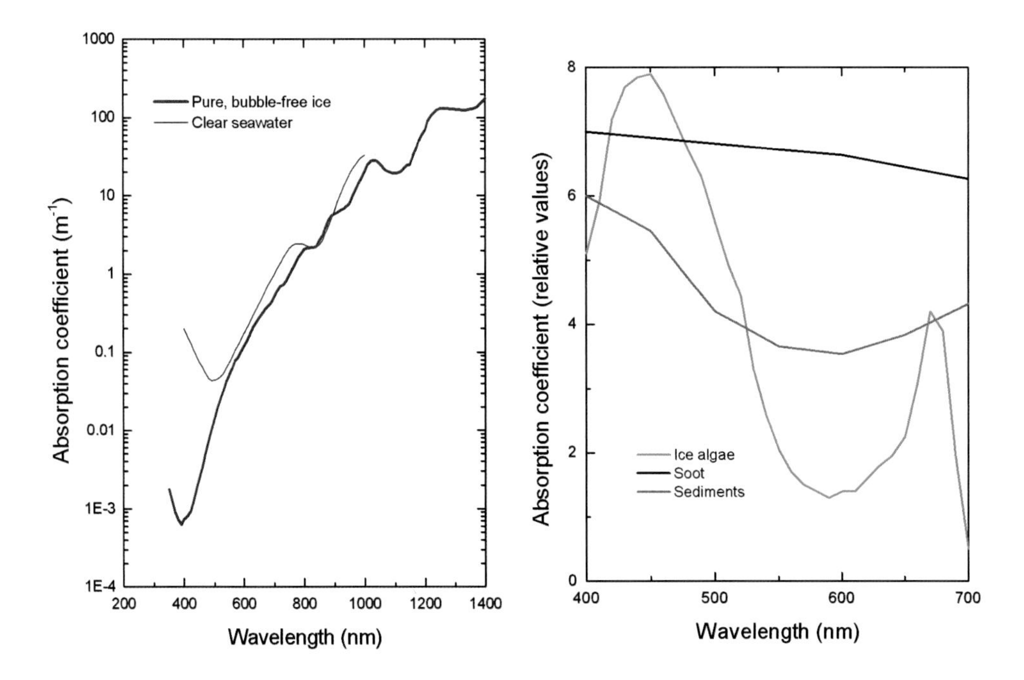


**Figure 3.6.2.** Absorption coefficients of (a) water and pure bubble-free ice and (b) particulates in ice including ice algae, soot, and sediments

Sea ice can also contain particulates, such as soot, sediments, and algae, which can influence both the spectral albedo and transmittance. Relative spectral absorption coefficients for these particulates are presented in Figure 3.6.2.b. Algae have a strong spectral absorption peak between 450 and 500 nm, where ice and water are most transparent. Because of this difference it is possible to use spectral transmission measurements to infer the amount of biological activity in the ice from transmission measurements (Perovich et al. 1993).

## 3.6.2. APPROACHES AND TECHNIQUES

There are many different optical properties and in some cases, many different names for the same property depending on the research area (Mobley 1994). Table 3.6.1 summarizes a few sea ice optical parameters and properties. These properties can be measured as bulk values integrated over wavelength or spectrally at individual wavelengths. Ultimately the property of interest and the type of measurement made depends on the scientific question being asked. For climate change studies the central question is, how is the solar radiation partitioned by the ice cover? The parameter of prime interest is the albedo, the fraction of the incident solar radiation that is reflected by the surface. A perfect reflector has an albedo of 1 and a perfect absorber has an albedo of 0. Arctic sea ice covers most of this range with cold, snow-covered ice having an albedo of 0.85, bare ice 0.65, melt ponds ranging from

0.15 to 0.40, and leads 0.07 (Grenfell and Maykut 1977; Perovich et al. 2002). Measurements of transmittance are also of interest to determine the relative amounts of sunlight that are absorbed in the ice and in the ocean (Perovich 2005; Light et al. 2008). Information on the amount and spectral composition of transmitted light is also important for biological studies. Of particular interest is the amount of photosynthetically active radiation (PAR), which is the number of photons between 400 nm and 700 nm. PAR has interesting units of $\mu E\ m^{-2}\ s^{-1}$, where an Einstein (E) is equal to one mole ($6.02 \times 10^{23}$) of photons. Measurements of the angular dependence of light reflected from the ice are important for the interpretation of visible and near infrared remote sensing data.

**Table 3.6.1.** Selected optical terms and properties

| Optical terms | Properties |
|---|---|
| Solar radiation | Sunlight |
| Wavelength integrated | Property averaged over a selected wavelength range |
| Spectral | Property as a function of wavelength |
| Radiance | Flux in a narrow beam ($W\ m^{-2}\ steradian^{-1}$) |
| Irradiance | Flux integrated over a hemisphere ($W\ m^{-2}$) |
| Albedo | Fraction of incident irradiance that is reflected (dimensionless) |
| Transmittance | Fraction of incident irradiance that is transmitted (dimensionless) |
| Absorption coefficient | Amount of absorption ($m^{-1}$) |
| Scattering coefficient | Amount of scattering ($m^{-1}$) |
| Phase function | Angular distribution of scattered light |
| Extinction coefficient | Attenuation due to scattering and absorption ($m^{-1}$) |

An optical instrument typically consists of a few fundamental components that define the field of view (foreoptics), determine the wavelength range (filters, prisms, diffraction gratings), and measure the solar energy (detector). The earliest optical devices integrated measurements over the entire solar spectrum. This was due to the importance of total solar energy for heat budget considerations and also to the relative simplicity of building an all-wave instrument. During the 1957 International Polar Year, all-wave radiometers were used to measure the incident irradiance and the albedo (Figure 3.6.3). Similar instruments are currently being used for the same purpose in the 2007–2009 International Polar Year, and due to their strengths and the importance of this information, they may well be used fifty years from now in the 2057 International Polar Year. They are simple in design, rugged in use, inexpensive in cost, and accurate in performance. Figure 3.6.4 shows a close-up of an all-wave radiometer made by Eppley Laboratories. The instrument has a black thermopile covered by a quartz hemisphere and produces a voltage that depends on the irradiance. The photograph also illustrates a pervasive problem with the instrument. The domes ice up and fog up and therefore must be regularly checked and kept clean for accurate measurements.

**Figure 3.6.3.** The albedometer has two all-wave radiometers measuring incident and reflected irradiance. The albedo is simply the reflected irradiance divided by the incident irradiance. The photograph was taken in April 1998 at Ice Station SHEBA in the Beaufort Sea.

**Figure 3.6.4.** Eppley shortwave radiometer on right showing buildup of hoar frost. Check out the interesting young ice in the background. The ice was approximately 10 cm thick and was rafting. The photograph was taken in September 2005 during the *Healy/Oden* Trans-Arctic Expedition.

For many applications spectral information on the optical properties is needed. For these measurements a different type of instrument is needed—a spectroradiometer. Spectroradiometers divide the solar spectrum into individual wavelength bands and then measure the energy in each of those bands. There are several different ways to split the spectrum. Early devices that were in use during the 1957 IPY used individual filters that could examine the optical properties at a few discrete wavelength bands. Prisms and diffraction gratings are other methods for splitting the spectrum. Today instruments primarily use filters if only a few channels are needed and diffraction gratings if more detailed spectral information is desired. The diffraction grating spreads the spectrum onto a multichannel detector. Usually the detector is a photodiode array, but in some instruments a charged coupled device array (CCD) may be used. Many field instruments use fiber optics to transmit light from the observation site to the instrument. A field-portable spectroradiometer is shown in Figure 3.6.5.

**Figure 3.6.5.** Spectral measurements of albedo on a cold (–25°C) clear day and during melt season. The instrument uses a fiber optics probe. A turret-type cosine collector was used on the left and reflector type on the right.

Fiber-optic probes usually have foreoptics attached to define the field of view. Figure 3.6.5 shows several examples of foreoptics used in radiance and irradiance measurements. For radiance measurements a lens is used as foreoptics to provide a narrow field of view. Different lenses afford of a wide range of fields of view. Irradiance is the light integrated over a hemisphere and requires a special foreoptics called a *cosine collector*. This means that the incident light will be weighted by the cosine ($\theta$), where $\theta$ is the angle to the foreoptics measured from normal. It is difficult to create a foreoptics that has a good cosine response for large values of $\theta$ (sun close to the horizon). There are many variations of cosine collector, but two basic types—turret and reflector, which are shown in Figure 3.6.5. The turret foreoptics reduces the light transmitted through the foreoptics by different amounts at different angles to generate a cosine response. Turret-type foreoptics are typically used in under-ice and ocean instruments. The reflector collector uses a white diffuser to reproduce a cosine response and has greater sensitivity than the turret collector.

### 3.6.3 METHODS AND PROTOCOLS

While there are many different measurements and numerous instruments, there is a standard protocol that can be widely applied to optical observations. Following a consistent protocol is important to ensure that measurements are made in an accurate and reproducible fashion. Seven simple rules for excellent optical measurements are presented here.

***Seven Simple Rules for Excellent Optical Observations***

1. Know your instrument.
2. Know your footprint.
3. Measure towards the sun.
4. Keep the measurement site pristine.
5. Keep irradiance detectors level.
6. Characterize the sky.
7. Characterize the medium.

It is an excellent time to be doing optical work, since there are many superb laboratory and field instruments available. This easy availability makes it critical to obey the first rule—"Know your instrument." Know how the hardware and the software work. Before you ever go into the field take time to practice making measurements at home. Measure the albedo of your yard, your driveway, and your cat. Get to know how to operate the instrument, with your eyes closed, because there will be days in the cold or in the sun when the computer screen is unreadable. Know the limitations of your instrument. Don't take the specifications at face value. Figure 3.6.7 shows spectral albedos for bare and ponded sea ice measured using an instrument rated from 350 to 2500 nm. There are intriguing features in certain wavelength regions of rapid oscillation in albedo. These variations engender complex explanations involving absorption bands and molecular rotations and vibrations of ice, brine, or solid salts. However, there is a simpler, and more accurate, explanation of the observations—noise. Measured light levels are too low to be used. Know the instrument well enough to be able to discern between good and bad data.

It is important to know what you are measuring. How big is the footprint of your instrument and where is the instrument pointed? Are you measuring the melt pond or the bare ice next to it? Does the cosine collector have a clear field of view or a good look at your boots? For radiance measurements, with a lens as foreoptics,

**Figure 3.6.6.** Examples of foreoptics. From left to right: A one-degree field of view lens, an aiming device for radiance measurements, a turret-type cosine collector, and a reflector-type cosine collector. Note that the foreoptics are black to minimize spurious reflections.

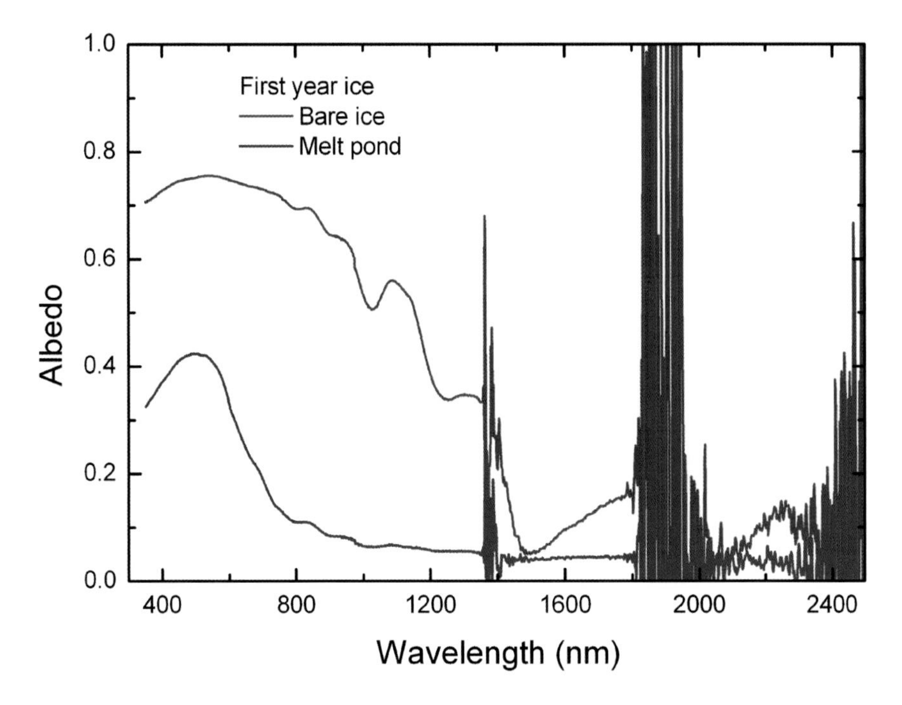


**Figure 3.6.7.** Albedos for bare and ponded first year ice. Be aware of the signal to noise ratio. These measurements were made on shorefast sea ice near Barrow, Alaska, in early June.

some simple geometry can determine the footprint size. If the foreoptics is pointed straight down at the surface, the footprint is a circle of radius R, where R = H tan (φ/2). H is the height of the foreoptics and φ is the field of view (FOV) of the fore-optics. Table 3.6.2 shows the footprint radius as a function of the field of view for an instrument 1 m above the surface.

**Table 3.6.2.** Footprint size as a function of foreoptics field of view at instrument heights of 1 m above surface

| Field of view (°) | Footprint radius (cm) |
|---|---|
| 1 | 0.9 |
| 2 | 1.7 |
| 5 | 4.4 |
| 10 | 8.7 |
| 15 | 13.2 |
| 20 | 17.6 |
| 25 | 22.2 |

For a cosine collector the situation is a bit more complex. In principle, light from the entire hemisphere is contributing to the measurement giving a footprint of infinite horizontal extent. However, the contributions are weighted by the cosine of the incidence angle. Consider the case of measuring reflected irradiance, with a cosine collector looking down at a uniform surface from a height of 1 m. The percent contribution to the reflected irradiance as a function of footprint radius is plotted in Figure 3.6.8. Fifty percent of the contribution comes from a 58 cm radius footprint, while approximately 90 percent of the signal is from a 2 m radius circle. Selecting a site with a footprint of about 2 m radius provides a reasonable compromise between accuracy and practicality. The height of the instrument can also be lower if a smaller footprint is required. When measuring incident and reflected irradiance, keep the detector level, otherwise the irradiance will be a mix of sky and surface, resulting in an erroneous measurement.

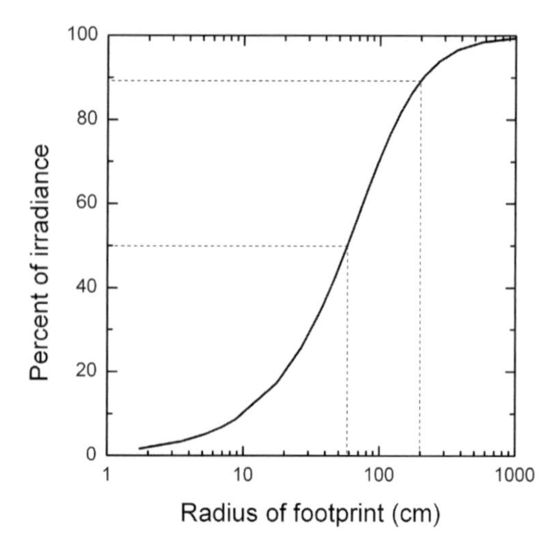


**Figure 3.6.8.** Percent of total irradiance from a footprint of radius R for an radiometer with cosine collector that is 1 m high. The lines denote the footprint radius corresponding to 50% and 90% of the radiance.

The measurement site must be in good shape, unencumbered by shadows or artifacts. Face towards the sun to keep shadows off the site. Keep the site pristine. Don't walk on it or put equipment on it. Scientifically, there is little interest in the albedo of footprint-covered snow. Finally, the light conditions and the site must be characterized. This characterization is invaluable when interpreting the observations. The degree of characterization needed depends on the nature of the scientific questions. A description of the sky conditions could be as simple as selecting one the following four choices: (1) clear skies; (2) partly cloudy; (3) complete cloud cover, solar disk visible; and (4) complete cloud cover, solar disk not visible. If the focus is on incident irradiance, a detailed description of the sky including cloud type may be needed.

The level of detail needed to characterize the snow and ice conditions also depends on the scientific question being addressed. Photographs of the surface are always a useful and simple component of the characterization. These photographs should include the footprint of the instrument and, to provide context, some of the adjacent area. For albedo surveys covering hundreds of meters, a simple description of the surface state (dry snow, melting snow, bare ice, melt pond) and ice type (young ice, first-year ice, multiyear ice, ridged ice) is sufficient. Measurements of snow depth, pond depth, and surface granular layer thickness may also be recorded. In some cases more detail about the ice may be required. This can include vertical profiles of ice temperature, salinity, brine volume, density, and crystallography.

Including size distributions of air bubbles and brine pockets may also be of interest. These distributions can be determined by analyzing thin sections (see Chapter 3.3).

### 3.6.4 AN APPLICATION

There is a wealth of possible optical observations and applications. The observations can be as simple as a single wavelength-integrated albedo or as complex as a regional description of the partitioning of solar radiation by a spatially varying and temporally evolving ice cover. Let us consider a particular application to illustrate the principles and methods discussed in this chapter.

Improved sea ice models can greatly enhance our understanding of the ice cover and our ability to take full advantage of sea ice as a platform. Consider an integrated snow-ice-ocean-atmosphere model that can determine such parameters as the strength of the ice cover, the trafficability of the snow cover, and the primary productivity under the ice. This model would be a powerful tool. Much information would be needed to create such a model, including detailed observations about the dependence of the optical properties on the state and structure of the ice cover.

A comprehensive picture of ice optical properties begins with measurements at a single site. The selected site will be uniform in thickness and surface conditions over a distance of several meters. First the spectral albedo of the site will be measured. Then an ice core will be carefully removed from the center of the site, causing as little disruption to the ice surface as possible. Then a detector will be lowered into the hole. Measurements will be made in the hole at 1 cm spacing for depths of 0–10 cm, every 2 cm for the next 10–20 cm, every 5 cm for the next 20–50 cm, then every 10 cm for the remainder of the ice. Depending on the foreoptics used, this will provide vertical profiles of upwelling or downwelling irradiance or radiance. Since sea ice is a scattering medium, the impact of the hole on the light field in the ice is limited. The hole does disrupt the light field in the ice, observations of downwelling irradiance in the top 10 to 20 cm, but beyond that effects are small (Light et al. 2008). Transmission measurements can avoid the hole by using an under-ice arm that extends a meter from the hole. Finally, using an irradiance foreoptics attachment, the downwelling irradiance through the entire ice will be measured. Another instrument will be monitoring the incident conditions during the transmission measurements to correct for changes in the incident irradiance.

Those are just the optical measurements. As always sky conditions will be described. There will also be a detailed description of the ice. Immediately upon removal the ice core will be photographed and the temperature profile measured. The core will be returned to the lab and cut into 5 cm thick sections. Densities of these sections will be measured. Vertical and horizontal thin sections will be prepared. The remainder of the sections will be melted to determine the salinity profile of the ice. A biological, geochemical, and oxygen isotopic analysis will be performed on the meltwater. Chapter 3.4 provides the details of characterizing the

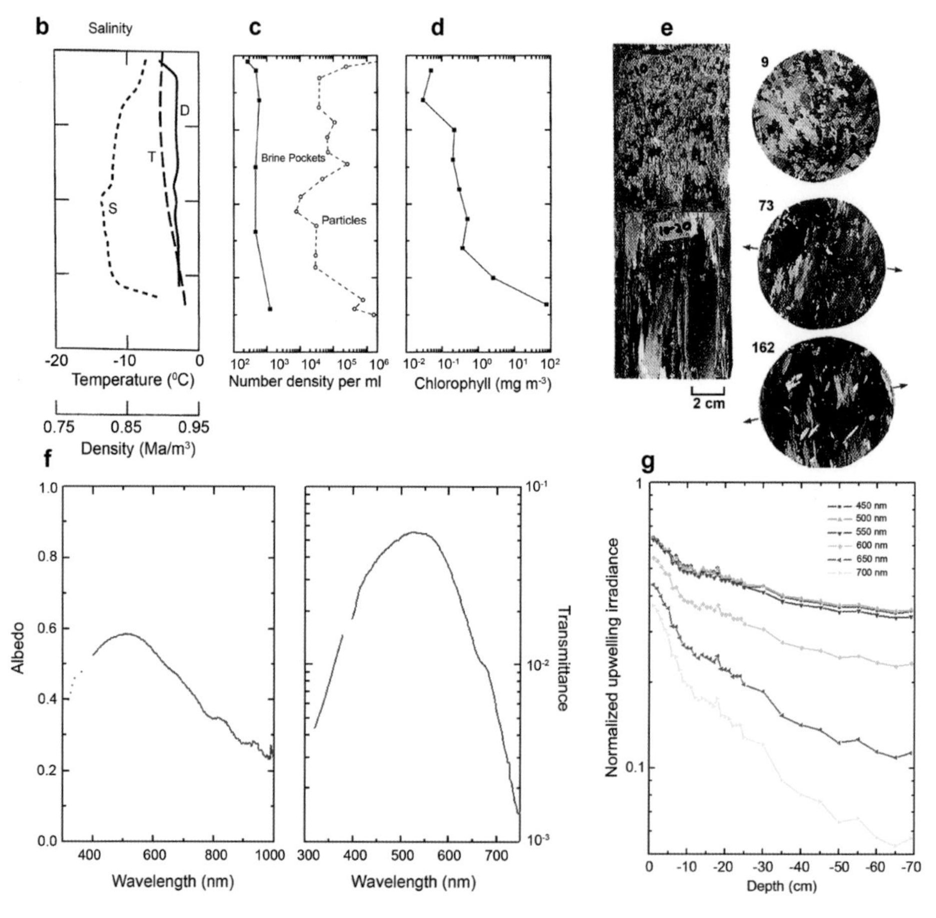


**Figure 3.6.9.** An ensemble of physical and optical measurements of first-year ice at Barrow, Alaska, during May 1994. This includes (a) a photograph of the site; (b) vertical profiles of ice temperature, salinity, and density; (c) vertical profiles of number densities of brine pockets and air bubbles; (d) vertical profile of chlorophyll a as a measure of algal biomass; (e) vertical and horizontal ice thin sections photographed using polarized light; (f) spectral albedo and transmittance; and (g) vertical profiles of spectral downwelling irradiance.

ice physical properties and structure. Figure 3.6.9 shows the results from such a measurement suite (Grenfell et al. 1998; Perovich et al. 1998).

All of these observations will be combined to determine the fundamental optical properties defining scattering and absorption in the ice. The measurement program will be repeated for different ice types, building a library of optical properties that can be incorporated into the radiative transfer component of an integrated sea ice model.

### 3.6.5 FUTURE DIRECTIONS

What sea ice optics measurements will be made during the 2057 International Polar Year? It is easy to conjure up images of autonomous underwater vehicles cruising under the ice measuring the spatial distribution of spectral transmittance, working in conjunction with unmanned aerial vehicles mapping large-scale albedo and surface conditions. And to imagine the results transmitted to data centers where it is integrated with results from satellites, field camps, and other sources and made instantly available to researchers and other interested parties. But perhaps there is a fundamental question to consider: Will there be any arctic sea ice in the summer of 2057? Some current predictions indicate that by 2057 arctic summers will either be sea ice–free or have only a small amount of perennial sea ice near the Canadian Archipelago.

While it is difficult to predict the key sea ice optical issues of 2057, we can consider the key questions of today. There are critical interdisciplinary problems regarding sea ice and climate, and sea ice and the marine environment. The arctic sea ice cover shows a clear trend towards a greater proportion of first-year ice. Much of the past sea ice research has focused on perennial ice, and now first-year ice needs greater attention. What are the optical properties of first-year ice? Most importantly, how does the albedo evolve as the ice thins? More information is needed about the evolution of the physical and optical properties of sea ice as it thins from 1 m thick to 0.5 m to water. The optical measurements would be straightforward, but the logistics difficult.

Melt ponds are another key issue that affects both the albedo and the transmittance. A lack of information on pond fraction is a major stumbling block in estimating large-scale ice albedo. We need to understand, and to model, the evolution of melt pond fraction, depth, and albedo over the summer melt season for both first-year and perennial ice.

Finally the arctic sea ice cover is undergoing significant changes; summer ice extents are declining; ice thickness is decreasing; and the melt season is lengthening. How will these changes affect the partitioning of solar radiation? How will they impact the marine ecosystem? What will be the impact of changing seasonality? Consider this scenario. An earlier start to the melt season means a lower albedo in late May and June when the incident sunlight is large. Consequently there will be an increase in the solar heat input to the ice ocean system and enhanced melting, thereby accelerating the decline of the ice cover. From a biological perspective an earlier melt onset coupled with thinner ice will result in more solar energy deposited at an earlier time into the ecosystem living at the bottom of the ice and the upper ocean. But will the snow cover change? How will such changes impact the intricate timing between the many components of the ecosystem?

To address these issues a systems approach is needed that integrates the optical, physical, and biological properties of the ice, snow, ocean, atmosphere, and ecosystem. There are tremendous opportunities for exciting, interdisciplinary science.

## REFERENCES

Grenfell, T. C. (1991), A radiative transfer model for sea ice with vertical structure variations, *J. Geophys. Res., 96*(C9), 16,991–17,001, doi:10.1029/91JC01595.

Grenfell, T. C., et al. (1998), Evolution of electromagnetic signatures of sea ice from initial formation to the establishment of thick first-year ice, *IEEE Trans. Geosci. Remote Sens., 36,* 1642–1654.

Grenfell, T. C., and G. A. Maykut (1977), The optical properties of ice and snow in the Arctic Basin, *J. Glaciol., 18,* 445–463.

Grenfell, T. C., and D. K. Perovich (2008), Incident spectral irradiance in the Arctic Basin during the summer and fall, *J. Geophys. Res., 113,* D12117, doi:10.1029/2007JD009418.

Grenfell, T. C., S. G. Warren, and P. C. Mullen (1994), Reflection of solar radiation by the Antarctic snow surface at ultraviolet, visible, and nearinfrared wavelengths, *J. Geophys. Res., 99,* 18,669–18,684.

Light, B., T. C. Grenfell, and D. K. Perovich (2008), Transmission and absorption of solar radiation by Arctic sea ice during the melt season, *J. Geophys. Res., 113,* C03023, doi:10.1029/2006JC003977.

Light, B., G. A. Maykut, and T. C. Grenfell (2003), Effects of temperature on the microstructure of first-year Arctic sea ice, *J. Geophys. Res., 108*(C2), 3051, doi:10.1029/2001JC000887.

Mobley, C. D. (1994), *Light and Water: Radiative Transfer in Natural Waters,* Academic Press, San Diego.

Pegau, W. S., and C. A. Paulson (2001), The albedo of Arctic leads in summer, *Ann. Glaciol., 33,* 221–224.

Perovich, D. K. (1996), *The Optical Properties of Sea Ice,* CRREL Monograph, 96-1, May.

Perovich, D. K. (2005), On the aggregate-scale partitioning of solar radiation in arctic sea ice during the SHEBA field experiment, *J. Geophys. Res., 110,* C03002, 10.1029/2004JC002512.

Perovich, D. K., et al. (1998), Field observations of the electromagnetic properties of first-year sea ice, *IEEE Trans. Geosci. Remote Sens.*, *36,* 1705–1715.

Perovich, D. K., G.F. Cota, G.A. Maykut, and T.C. Grenfell, (1993), Bio-optical observations of first-year Arctic sea ice, *Geophys. Res. Lett.*, 20, 1059–1062.

Perovich, D. K., and J. W. Govoni (1991), Absorption coefficients of ice from 250 to 400 nm, *Geophys. Res. Lett., 18,* 1233–1235, doi:10.1029/91GL01642.

Perovich, D. K., T. C. Grenfell, B. Light, and P. V. Hobbs (2002), Seasonal evolution of multiyear Arctic Sea ice, *J. Geophys. Res.*, *107*(C10), 8044, doi:10.1029/2000JC000438.

Warren, S. G., R. E. Brandt, and T. C. Grenfell (2006), Visible and near ultraviolet absorption spectrum of ice from transmission of solar radiation into snow, *Appl. Opt.*, *45*(21), 5320–5334, doi:10.1364/AO.45.005320.

# Chapter 3.7

# Measurements and Modeling of Ice-Ocean Interaction

*Kunio Shirasawa and Matti Leppäranta*

## 3.7.1 INTRODUCTION

Sea ice strongly interacts with the underlying ocean, and it directly impacts on the order of 50 m of water column under the ice. The presence of the phase boundary constrains the temperature of the ice bottom to the freezing point of the surface water. With sea ice a mixture of variable fractions of ice and brine with no fixed melting point, the interfacial ice-water boundary is of finite thickness, often referred to as the skeletal layer. Ice growth and accretion is controlled by the release and transfer of latent heat through the ice into the atmosphere. The ice-ocean heat flux from the underlying water body into the bottom of the ice sheet (hereafter referred to as the oceanic heat flux) reduces the growth rate or may induce bottom melt. The salinity of the surface layer is influenced by salt rejection from growing ice and by freshwater input from melting ice. Momentum exchange between ice and ocean is controlled by the drag force due to the velocity difference between the fluid and the solid. This force is usually parameterized as a quadratic drag law. The exchanges of heat and salt/freshwater are approximated by similar bulk exchange formulas. The magnitude of the exchange depends on the stratification of the surface layer, and hence the exchanges of momentum, heat, and salt/freshwater are interdependent.

The thermal evolution of the ice cover and its growth or melt are forced by the heat fluxes at its upper and lower boundaries and penetration of solar radiation into the ice sheet. Heat exchange at the upper surface is well understood. In contrast, the energy balance at the lower boundary, in particular the oceanic heat flux, is less well studied, because measurements and process studies of ice-ocean interaction present a greater challenge. One such study, the Surface Heat Budget of the Arctic Ocean (SHEBA) program, consisted of a field experiment conducted from October 1997 through October 1998 in the Chukchi Sea (e.g., Perovich et al. 2003). The oceanic heat flux during the SHEBA summer ranged from 10 to 40 W $m^{-2}$, with an average summer bottom melt rate of 0.50 cm $d^{-1}$, equivalent to an average oceanic heat flux of 17.5 W $m^{-2}$ (Perovich et al. 2003). Some undeformed landfast ice covers such as in lakes, lagoons, fjords, bays, or along sheltered coastlines can serve as stable platforms for measurements of the oceanic heat flux, in the absence of deformation processes with an ice cover thickening through bottom accretion. Such measurements under landfast ice have yielded oceanic heat flux values in the range of 1–100 W $m^{-2}$, indicating that heat transfer to the bottom of the ice can substantially retard ice growth (e.g., Shirasawa et al. 2006). Uusikivi et al. (2006) observed very small heat fluxes of 1 W $m^{-2}$ or less in laminar and transitional laminar/turbulent flow regimes under sheltered coastal landfast ice in the Baltic Sea.

Sea ice provides a unique habitat for polar microbial assemblages in ice-covered waters (e.g., Arrigo 2003). The ice-ocean interface and its underlying oceanic boundary layer (OBL) can serve as habitats for algal communities, with light levels sufficient for net growth and nutrients supplied through turbulent and convective exchange in the skeletal layer and the pore space of the ice cover. The turbulent fluxes of momentum, heat, and salt along with nutrients and other compounds of biogeochemical importance can be hydrodynamically controlled by the current regime and stratification in the OBL beneath sea ice, with strong seasonality due to changing ice conditions.

In this section, ice-ocean interaction is discussed in the context of not only physical but also biogeochemical processes in the OBL beneath sea ice. First, we present the general theoretical framework applicable to the OBL. Methods to estimate turbulent fluxes are also discussed. Then field measurement technologies are introduced along with examples of resulting data sets. Specifically, the influence of the oceanic heat flux on the evolution of ice thickness is considered in more detail. Various approaches, including eddy correlation technique, as well as bulk, temperature-profile, and residual methods to estimate the oceanic heat flux are explained in the context of measurements made at Saroma-ko Lagoon, on the Okhotsk Sea coast of Hokkaido (Shirasawa et al. 2006; Leppäranta and Shirasawa 2007). Since the information from Saroma-ko Lagoon provides a good example for the seasonal evolution of undeformed level ice growth and decay, we have also applied mathematical models to examine the influence of the oceanic heat flux on the ice thickness evolution.

## 3.7.2 OCEANIC BOUNDARY LAYER (OBL) BENEATH SEA ICE

### 3.7.2.1 Neutral Conditions

Planetary boundary layers (PBL) consist of three main parts (e.g., Tennekes and Lumley 1972): a thin viscous layer, a surface layer (or logarithmic layer) where the stress is approximately constant, and an Ekman layer where the velocity rotates due to the Coriolis effect and the stress decreases to zero. The full solution across the PBL is derived by matching these sublayers, normally assuming steady-state conditions. The thickness of the planetary boundary layer scales with the square root of the fluid density is ~1 km for the atmosphere over the ice-atmosphere interface and ~50 m for the ocean below the ice-ocean interface.

Using complex variable notation the horizontal velocity is $U = U_1 + iU_2$ where $i = \sqrt{-1}$. The planetary boundary layer dynamics equations are (e.g., Cushman-Roisin 1994):

$$ifU = ifU_g + \frac{1}{\rho}\frac{\partial \tau}{\partial z} \quad \text{(Equation 3.7.1a)},$$

$$\frac{\partial p}{\partial z} = -g\rho \quad \text{(Equation 3.7.1b)},$$

where $U_g$ is the geostrophic velocity, $f$ is the Coriolis parameter, $z$ is depth, and $\tau$ is the vertical shear stress (the Coriolis parameter is defined as $f = 2\Omega\sin\phi$, where $\Omega$ is the angular velocity of the Earth and $\phi$ is latitude [please consult the Appendix at the end of this chapter for a complete list of symbols and constants]). The vertical shear stress is expressed as:

$$\tau = \rho K\frac{\partial U}{\partial z} \quad \text{(Equation 3.7.1c)},$$

where $\rho$ is water density and $K$ is the kinematic eddy viscosity. For the vertical direction the hydrostatic equation applies, and in barotropic (nonstratified) flow the geostrophic velocity is independent of depth. At the bottom of the ocean surface boundary layer the stress vanishes and, therefore, $U = U_g$, while at the top of the boundary layer, the velocity must be equal to the velocity of the surface boundary, $U = U_0$.

The thickness of the OBL is on the order of 50 m. An excellent presentation of the OBL physics is provided by McPhee (2008). Other publications on the OBL beneath sea ice are by, for example, McPhee (1986), McPhee (1990), Shirasawa and Ingram (1991a), and Leppäranta (2005). The oceanic heat flux beneath sea ice has been measured in various regions (e.g., McPhee 1986; Shirasawa 1986; McPhee 1990; Wettlaufer 1991; Shirasawa and Ingram 1991a; 1991b; Omstedt and Wettlaufer 1992; McPhee et al. 1996; Lytle and Ackley 1996; Shirasawa et al. 1997; Shirasawa and Ingram 1997). As for the atmospheric boundary layer over sea ice, field investigations include, for example, Rossby and Montgomery (1935), Brown (1980), Joffre (1984), Overland (1985), Shirasawa and Aota (1991), Guest and Davidson (1991), Andreas (1998), and Andreas et al. (2005).

### 3.7.2.2 Surface Layer

In the surface layer, the vertical shear stress is constant, as is the velocity direction. Hence the velocity vector can then be aligned with the real axis of the coordinate system. A velocity scale $u_*$, referred to as the *friction velocity*, can be defined by

$\tau = \rho u_*^2$ for the surface layer. The length scale derives from Prandtl's *mixing length hypothesis* stating that the size of turbulent eddies scales with the distance from the boundary, that is, the length scale is $z$ (e.g., Tennekes and Lumley 1972). Therefore $K \sim u_* z$, and for the stress to be constant, the vertical velocity gradient is $\sim u_*/z$, in exact terms:

$$\frac{\partial U}{\partial z} = \frac{u_*}{\kappa z} \quad \text{(Equation 3.7.2)},$$

where $\kappa \cong 0.4$ is the von Karman constant. In neutral stratification, a logarithmic velocity profile results:

$$U(z) - U_0 = \frac{u_*}{\kappa} \log \frac{z}{z_0} \quad \text{(Equation 3.7.3)},$$

where $z_0$ is *the roughness length.* For undeformed sea ice the roughness length is of the order of 0.1 to 1 cm, and a characteristic friction velocity is 2 cm $s^{-1}$ for the ocean. For deformed sea ice the effective roughness length is of the order of 1 to 10 cm due to the impact of hummocks and ridges on skin drag. Data for the roughness of the ice bottom are mostly lacking (see McPhee 2008). Work during the Arctic Ice Dynamics Joint Experiment (AIDJEX) in the Beaufort Sea gave $z_0 \approx 20$ cm, which includes hummocks and ridges (McPhee 1986). The local value for an undeformed, level ice bottom is much less, down to 0.1 cm. Shirasawa (1986) and Shirasawa and Ingram (1997) obtained $z_0 \approx 0.5$–2.0 cm from turbulence measurements near the bottom of landfast ice in the Canadian Arctic Archipelago.

Undeformed ice surfaces are fairly smooth; that is, the roughness length is much smaller (in the range of 0.1–1.0 cm, e.g., Shirasawa and Ingram 1991a) than the boundary layer thickness, while in ridged sea ice fields the ridge keels penetrate to significant depths (some 5–20 m) into the OBL. In the presence of sea ice, its velocity ($u$) serves as the boundary condition for the OBL.

### 3.7.2.3 Ekman Layer

In the Ekman layer the eddy viscosity coefficient $K$ is taken as constant. The Coriolis acceleration is important, and consequently turning of the velocity takes place. We have:

$$ifU = ifU_g + K\frac{d^2U}{dz^2} \quad \text{(Equation 3.7.4)}.$$

The general solution is $U = U_g + C_1\exp(\lambda_1 z) + C_2\exp(\lambda_2 z)$, where $C_1$ and $C_2$ are constants to be determined from the boundary conditions and $\lambda_{1,2} = \pm\sqrt{if/2K}$. The boundary conditions are the no-slip condition ($U = U_0$) at the surface and the stress-free geostrophic velocity at the bottom. A more simple condition may be used for the deep boundary, $U \to U_g$ as $z \to \infty$, resulting in:

$$U = U_g + \left(U_o - U_g\right)\exp\left\{-\left[1+\operatorname{sgn}(f)\,i\right]\frac{\pi z}{D}\right\} \quad \text{(Equation 3.7.5)},$$

where $D = \pi\sqrt{2K/|f|}$ is the *Ekman depth*, by definition. At $z = D$, the latter term on the right-hand side has decreased to $e^{-\pi} \approx 4\%$ from the boundary level and rotated clockwise the angle of $\pi$.

Figure 3.7.1 illustrates the velocity distribution in the atmospheric and oceanic Ekman layers. If $|U_{wg}| << |u|$, the well-known Ekman spiral results, while for $U_{wg} \sim u$, the boundary layer modification is less distinct but the high-velocity medium is always driving that with a lower velocity.

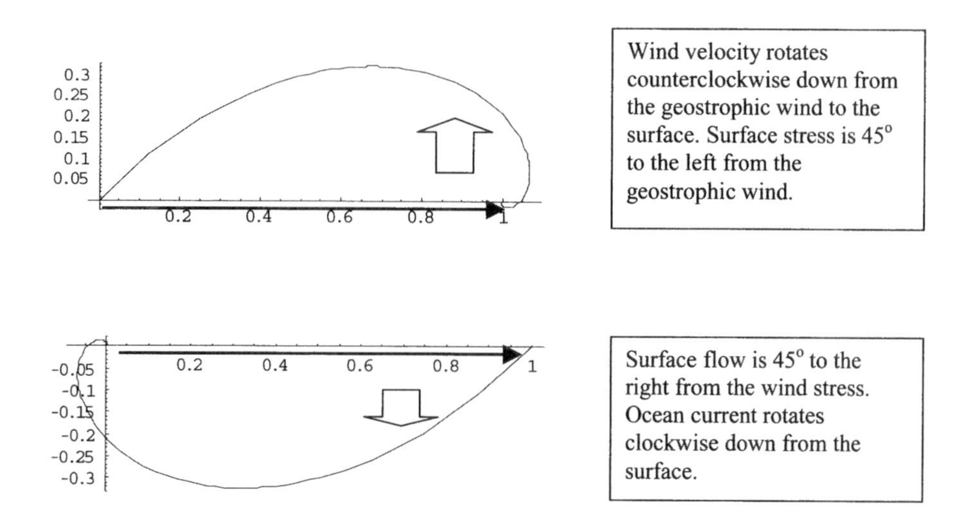


**Figure 3.7.1.** Theoretical form of the atmospheric and oceanic Ekman layers above and beneath sea ice (northern hemisphere). Note that oceanic velocities are two orders of magnitude less than atmospheric velocities. From Leppäranta (2005).

In neutral conditions the OBL is in practice homogeneous, and only ice-ocean momentum transfer takes place. In near-neutral conditions, heat and salt/freshwater transfer also take place, but as these become large stratification results and needs to be taken into account to quantify transfer.

Ice is an excellent platform for OBL investigations. It represents the case with pure shear stress forcing without disturbances from surface waves, and it provides a stable site for measurements. Ekman made observations himself in winter 1901 on the ice of Oslo Fjord and for the first time recorded the Ekman profile in the ocean. In fact, it took a long time before the Ekman theory was validated in open ocean conditions. Figure 3.7.2 shows the Ekman spiral recorded beneath drifting sea ice in April 1975 in the Baltic Sea (Leppäranta 1990; 2005).

**Figure 3.7.2.** Averaged velocities of wind, sea ice, and currents (depths 7, 19, and 40 m), 8–15 April 1975 in the Baltic Sea. For the velocity scale, the ice velocity averaged 3.1 cm s$^{-1}$ (at maximum it was 35 cm s$^{-1}$). From Leppäranta (2005).

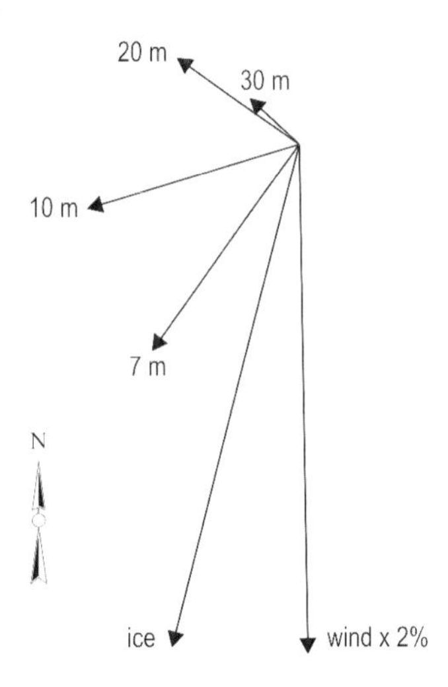


### 3.7.2.4 Stratified Conditions

In general, temperature and salinity of the OBL depend on depth, and the evolution of the boundary layer is an interconnected temperature-salinity-velocity problem. Sverdrup (1928) emphasized the role of stratification in the planetary boundary layers. He claimed that over the Siberian shelves in summer a 25–40 m low-density upper layer exists, which moves almost uniformly with the ice and with very weak momentum transfer to the deeper water. Rossby and Montgomery (1935) examined the air-ice and ice-water stresses using the data of Sverdrup (1928) and Brennecke (1921). They criticized Sverdrup's assumption of a bulk drift of the upper ocean with ice, and instead showed that there is always a shear layer beneath the ice. However, they agreed with Sverdrup in that the presence of a stable shear layer in summer significantly reduces ice-water friction.

For stratified non-neutral conditions, the surface layer is treated with Monin-Obukhov theory (e.g., Andreas 1998):

$$\frac{dU}{dz} = \frac{u_*}{\kappa z}\phi_m\left(\frac{z}{L_{MO}}\right) \quad \text{(Equation 3.7.6a)},$$

$$\frac{dT}{dz} = \frac{T_*}{\kappa z}\phi_T\left(\frac{z}{L_{MO}}\right) \quad \text{(Equation 3.7.6b)},$$

$$\frac{dS}{dz} = \frac{S_*}{\kappa z}\phi_S\left(\frac{z}{L_{MO}}\right) \quad \text{(Equation 3.7.6c)},$$

where $\phi_m$, $\phi_T$ and $\phi_S$ are the so-called universal functions, $T_*$ and $S_*$ are characteristic temperature and salinity scales in the surface layer, and $L_{MO}$ is the *Monin-Obukhov length*, which is equal to the ratio of turbulence production by shear and buoyancy.

For a detailed presentation of the Monin-Obukhov theory to sea ice application, we refer the reader to McPhee (2008).

To extend the Ekman layer for stratified flows, the Rossby number similarity theory is used with the roughness Rossby number $Ro_* = u_*/(fz_0)$ as the scaling parameter (McPhee 1986). Omstedt (1998) has used a second-order turbulence model to examine the boundary layer beneath sea ice. This approach includes the equations of turbulent energy and its dissipation and therefore avoids the need for drag law parameters. However, this model is one-dimensional and its applicability consequently limited to local boundary layer problems only (Leppäranta and Omstedt 1990; Omstedt et al. 1996).

## 3.7.3 TURBULENT FLUXES AT THE ICE-OCEAN INTERFACE

Turbulent flux quantities are obtained using averages and fluctuations; for example, $T' = T - \langle T \rangle$, $T$ is the measured value, and $\langle \cdot \rangle$ is the averaging operator. Using the formulation of Shirasawa and Ingram (1991a), turbulent fluxes of momentum $\tau$, heat $Q_w$, and salt $S_w$ are calculated as:

$$\tau = \rho\sqrt{< u'w' >^2 + < v'w' >^2} = \rho u_*^2 \quad \text{(Equation 3.7.7)},$$

$$Q_w = \rho c_w \langle w'T' \rangle \quad \text{(Equation 3.7.8)},$$

$$S_w = \langle w'S' \rangle \quad \text{(Equation 3.7.9)},$$

where $u'$ is the down-flow fluctuation, $v'$ is the cross-flow fluctuation, $w'$ is the vertical flow fluctuation, $c_w$ is the specific heat of seawater, $T'$ is the seawater temperature fluctuation, and $S'$ is the seawater salinity fluctuation.

These fluxes need to be parameterized for the mean quantities. The usual approach is to employ *bulk formulas*, that is, for the flux $F_P$ of a property $P$, we have:

$$F_P = \rho C_P(z)[P(z) - P(0)] \times [U(z) - U(0)] \quad \text{(Equation 3.7.10)},$$

where $C_P$ is the transfer coefficient and $z$ is depth. For momentum transfer, the exchange coefficient is called the *drag coefficient*.

### 3.7.3.1 Momentum Flux

In general fluid dynamics, by dimensional analysis the drag force of the fluid on a solid plate is given as $\tau = \rho C U^2$, where $\rho$ is the density of the fluid, $U$ is the velocity of the fluid relative to the boundary, and $C$ is the (dimensionless) drag coefficient (e.g., Li and Lam 1964). The drag coefficient depends on the Reynolds number

Re = $UL/\nu$, $\nu$ being the kinematic viscosity of the fluid, and the surface roughness. In laminar flow $C \propto \mathrm{Re}^{-1}$ and therefore $\tau \propto U$, but in turbulent flow the Reynolds number becomes very large and the drag coefficient is then independent of it, therefore $\tau \propto U^2$.

For the PBL the geostrophic flow serves as the natural undisturbed reference velocity. The geostrophic current is usually a good approximation for the current velocity at the bottom of the boundary layer. Beneath drift ice the boundary layer flow is turbulent, and the drag coefficient depends on the surface roughness and the stratification. In planetary boundary layers the Coriolis effect causes turning of the flow, and therefore a second stress law parameter is required: the *turning angle* or *Ekman angle* between the geostrophic flow and the surface stress. This angle also depends on the surface roughness and the stratification.

Thus the oceanic drag law is written as:

$$\tau_w = \rho C_w \left| U_{wg} - u \right| \exp(i\theta_w)(U_{wg} - u) \quad \text{(Equation 3.7.11a)},$$

where $C_w$ is the water drag coefficient and $\theta_w$ is the boundary layer turning angle. The factor $\exp(i\theta_w)$ simply rotates the velocity vector by the angle $\theta_w$; in the northern hemisphere the angle is positive and counterclockwise turning results but in the southern hemisphere the opposite is true. The drag coefficient consists of skin friction and form drag due to ridges, $C = C^S + C^F$. *Note*: In vector form the water stress is written as:

$$\tau_w = \rho C_w \left| U_{wg} - u \right| (\cos\theta_w + \sin\theta_w \, \mathrm{k} \times)(U_{wg} - u) \quad \text{(Equation 3.7.11b)}.$$

The geostrophic drag parameters have been widely used. In coupled ice-ocean modeling the oceanic top layer velocity is used as the reference, and then the drag parameters depend on the thickness of this top layer.

Table 3.7.1 shows a collection of oceanic drag parameters for several regions (Leppäranta 2005). Analyzing ice drift based on the Ekman theory, Shuleikin (1938) assumed $\theta_w \approx 30°$, a constant. There is a natural variation of ±50 percent in the drag coefficients due to the stability of the stratification and the roughness. Nansen (1902) stated that floating ice increases the velocity of the OBL because no energy is needed for wave formation and ice is rougher than open water to absorb more energy from the wind—if the internal ice resistance to motion is not too strong. The reference depth of 1 m has been used in several turbulence measurements beneath sea ice. In Table 3.7.1 the study of Shirasawa and Ingram (1997) gives a local drag coefficient of smooth ice, while the others refer to the geostrophic flow and include the form drag by hummocks and ridges. A specific case is the so-called *dead water* (Ekman 1904), where a large amount of ice momentum is used to generate internal waves. This necessitates the existence of a shallow, stable surface layer, at the bottom of which the internal waves form. The representative drag coefficient may then be five times the normal ice-ocean drag coefficient (Waters and Bruno 1995).

**Table 3.7.1.** The parameters of the quadratic ice-ocean drag laws for neutral oceanic stratification. gsc = geostrophic current. From Leppäranta (2005).

| Region | Drag coefficient | Turning angle (deg) | Level | Reference |
|---|---|---|---|---|
| Barrow Strait | $5.4 \cdot 10^{-3}$ | 0 | 1 m | Shirasawa and Ingram (1997) |
| Beaufort Sea | $5.0 \cdot 10^{-3}$ | 25° | gsc | McPhee (1982) |
| Baltic Sea | $3.5 \cdot 10^{-3}$ | 17° | gsc | Leppäranta and Omstedt (1990) |
| Weddell Sea | $1.6 \cdot 10^{-3}$ | 15° | gsc | Martinson and Wamser (1990) |

Representative arctic—the best-known region—values for the geostrophic drag coefficient and turning angle are $C_w = 5 \cdot 10^{-3}$ and $\theta_w = 25°$, resulting from the AIDJEX field experiments in the Beaufort Sea (McPhee 1982). Due to different roughness of the ice in the Baltic Sea the drag coefficients are 25 to 50 percent smaller than in the Arctic. In the Weddell Sea the water drag coefficient appears to be as low as $1.6 \cdot 10^{-3}$ (Martinson and Wamser 1990). The most likely explanation is that the bottom of antarctic sea ice is very smooth or the frictional oceanic boundary layer is thin in comparison with the Arctic or the Baltic Sea.

In general the drag coefficient and turning angle may be defined for an arbitrary velocity reference level, and their values would then depend on the chosen level. If the velocity profile and stratification are known, then transformation of the drag law parameters between different levels is straightforward. In particular, in measurement campaigns the depth level of the instrumentation may define the reference level. In shallow waters, where the depth is much less than the Ekman depth, the OBL covers the whole water body, and then the drag coefficient depends on the depth of the water and the turning angle can be ignored. Taking the logarithmic profile for the surface layer, in neutral conditions the roughness length and drag coefficient are related by $C_w(z) = [\kappa/\log(z/z_o)]^2$.

### 3.7.3.2 Oceanic Heat Flux

The bulk formulation for the heat exchange is parameterized in terms of a heat transfer coefficient, $C_H$. The oceanic heat flux then reads:

$$Q_w = \rho c_w C_H (T - T_0) U_w \quad \text{(Equation 3.7.12).}$$

The oceanic heat flux beneath sea ice $Q_w$ has been measured in various regions (e.g., McPhee 1986; Shirasawa 1986; McPhee 1990; Wettlaufer 1991; Shirasawa and Ingram 1991a; 1991b; McPhee et al. 1996; Lytle and Ackley 1996; Shirasawa et al. 1997; Shirasawa and Ingram 1997; McPhee 2008). Heat transfer coefficients can be obtained from, for example, Omstedt and Wettlaufer (1992), Shirasawa et al. (1997), Shirasawa et al. (2006), and Leppäranta and Shirasawa (2007). Here, we use $C_H = 3{\cdot}10^{-4}$ based on measurements at Saroma-ko Lagoon (Shirasawa et al. 2006; Leppäranta and Shirasawa 2007).

At the ice-ocean interface (the bottom of the ice sheet) the empirical freezing temperature is taken as:

$$T_0(S_0) = -0.057S_0 + 1.710523{\cdot}10^{-3}S_0^{3/2} - 2.154996{\cdot}10^{-4}S_0^{2} \quad \text{(Equation 3.7.13)},$$

where $S_0$ is water salinity at the ice bottom.
The specific heat of seawater, $c_w$, is defined as:

$$c_w = 4217.4 - 7.644S + 0.17709S^{3/2} \quad \text{(Equation 3.7.14)}.$$

In numerical models of sea ice thermodynamics, the oceanic heat flux has commonly been treated as a fixed tuning factor. For example, Maykut and Untersteiner (1971) employed a constant oceanic heat flux of 2 W m$^{-2}$ to obtain the best-fit equilibrium thickness cycle for multiyear ice in the Arctic Ocean in their classic simulations. Model results for the Southern Ocean indicate that in the Antarctic oceanic heat fluxes can be one order of magnitude larger. In the ice-covered Saroma-ko Lagoon, on the Okhotsk Sea coast of Hokkaido, Japan, Shirasawa et al. (2005) obtained the best-fit 10-year statistics for the ice and snow thickness using a fixed oceanic heat flux of 5 W m$^{-2}$. However, results from analytic and numerical models, as well as field measurements, illustrate the consequences of oceanic heat flux variability on ice thickness and structure (Leppäranta 1993; Shirasawa et al. 2005; Shirasawa et al. 2006; Leppäranta and Shirasawa 2007).

Another approach to estimate the oceanic heat flux is the profile method:

$$Q_w = \kappa_w \left.\frac{\partial T}{\partial z}\right|_{z=0-} \quad \text{(Equation 3.7.15)},$$

where $\kappa_w$ is the thermal conductivity. The lower limit of the thermal conductivity is the molecular thermal conductivity, $\kappa_w = 0.6$ W/(m °C).

*Example.* Taking the vertical temperature gradient as 0.6–1.4°C/m yields an oceanic heat flux of 0.4–0.8 W m$^{-2}$ (Figure 3.7.3). This is too low, about one-tenth of the values obtained from the bulk method, and therefore the heat conduction beneath the ice cover is much stronger than molecular conduction. The Reynolds number was ~$10^4$.

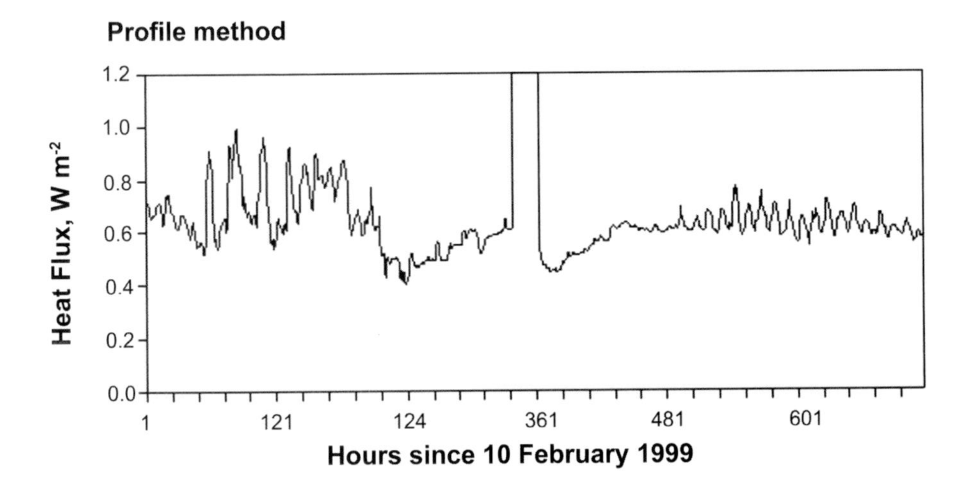


**Figure 3.7.3.** The oceanic heat flux at the ice-ocean interface estimated by the profile method for the period from 10 February to 11 March 1999 at Saroma-ko Lagoon. The gap in the time series shows redeployment of the current meter.

### 3.7.3.3 OBL and Sea Ice Microbial Communities

Sea ice provides a unique habitat for polar microbial assemblages in ice-covered waters (e.g., Arrigo 2003). The ice-ocean interface is an optimal habitat for algal communities, where light is sufficient for net growth and nutrient supply high. Sea ice modulates radiative transfer between the atmosphere, snow, the ice cover, and the ocean: The incident light is scattered back from the snow/ice cover surface, absorbed within the snow/ice column, and penetrates into the ocean depending on the relative contribution of absorption and scattering or reflection along the path of individual photons (Eicken 2003). The amount of light at the ice-ocean interface is sufficient for sea ice algal blooms to develop in early spring at the ice bottom (Arrigo 2003). When sea ice grows at the ice-ocean interface, it is controlled by the boundary conditions of the in situ temperature and salinity, and the phase fractions and characteristics of ice, brine, salts, and gas inclusions change accordingly (Eicken 2003). The transport of nutrients along with other biologically relevant compounds across the ice-ocean interface can be hydrodynamically controlled by the current regime (i.e., laminar or turbulent flow, or transition from laminar to turbulent flow) and stratification in the OBL beneath sea ice, where the oceanic environment is subject to large seasonal variations. Suspended and dissolved matter in the under-ice seawater can also be transported into sea ice through brine channels by hydrodynamic forcing such as upward flushing of seawater through sea ice due to snow accumulation and loading (Hudier et al. 1995).

## 3.7.4 MEASUREMENTS IN THE OBL BENEATH SEA ICE

### 3.7.4.1 Equipment for Measurements

OBL field investigations are conducted to examine the oceanic heat flux and its link with the growth/melt rates of sea ice at the ice-ocean interface, current regimes, and stratification in the OBL beneath sea ice. The growth/melt rates of sea ice can be measured as the difference in the sea ice thickness over some certain time difference. An *ice thickness gauge* is generally used to measure the in situ thickness of sea ice through a drill hole. A *current meter* can be used to measure the current speed and direction at the reference depth from the ice-ocean interface and/or current profiles in the OBL beneath sea ice. A *Conductivity-Temperature-Depth (CTD) profiler* can be used to measure the temperature and salinity regimes in the OBL beneath sea ice. Stratification can be evaluated by considering the current, temperature, and salinity regimes in the OBL. Typical equipment for OBL measurements in the field is described below.

*Ice Thickness Gauge*

An ice thickness gauge is a simple device to measure the thickness of sea ice. Figure 3.7.4 shows an example of thickness measurements with a gauge consisting of a steel bar connected to a measurement tape at the center and a string at the end of

**Figure 3.7.4.** Ice thickness measurements by a thickness gauge. A thickness gauge consists of a steel bar fixed with a tape scale at the centre and a string at the edge of the bar on the left. One person lowers the bar through a drill hole on the right, and then pulls up the scale to measure a total thickness of the ice and a draft thickness (between the water surface and the ice bottom).

the bar. In the photograph one person drops the bar into a drill hole and pulls up on the tape to measure the total thickness and draft of the ice between the water surface and the ice bottom. Ice thickness measurements are discussed in detail in Chapter 3.2.

*Conductivity-Temperature-Depth (CTD) Profiler*

A CTD profiler is used to obtain conductivity-temperature-depth profiles through the water column beneath sea ice, providing information on the under-ice ocean stratification. Figure 3.7.5 shows an example of measurements with a CTD Lite (Model ACTD-DF, Alec Electronics Co., Ltd., Japan). This instrument consists of a temperature sensor with accuracy ±0.02°C, a conductivity cell with accuracy ±0.03 mS/cm and a depth sensor with accuracy 0.3 percent. In the photograph one person lowers the CTD sensor through a drill hole while monitoring the display of the data logger in a blue box.

**Figure 3.7.5.** CTD measurements by a CTD Lite (Model ACTD-DF, Alec Electronics Co., Ltd., Japan). One person lowers the CTD meter through a drill hole along with monitoring the readings of the meter on the display of the data logger (a blue box).

*Current Meter*

A current meter is used to measure the current regime in the under-ice oceanic boundary layer. Figure 3.7.6 shows an example of current measurements with a three-dimensional electromagnetic current meter (Model ACM32M, Alec

Electronics Co., Ltd., Japan). This instrument consists of a current sensor with accuracies of ±1 cm/s in velocity and ±2° in direction, a temperature sensor with ±0.05°C accuracy, a conductivity sensor with ±0.05 mmho/cm accuracy, and a depth sensor with accuracy ±0.15 m. In the photograph one person lowers the current meter through a drill hole and fixes it with a tripod at a depth of 1 m (as the reference level) below the ice bottom in the oceanic boundary layer.

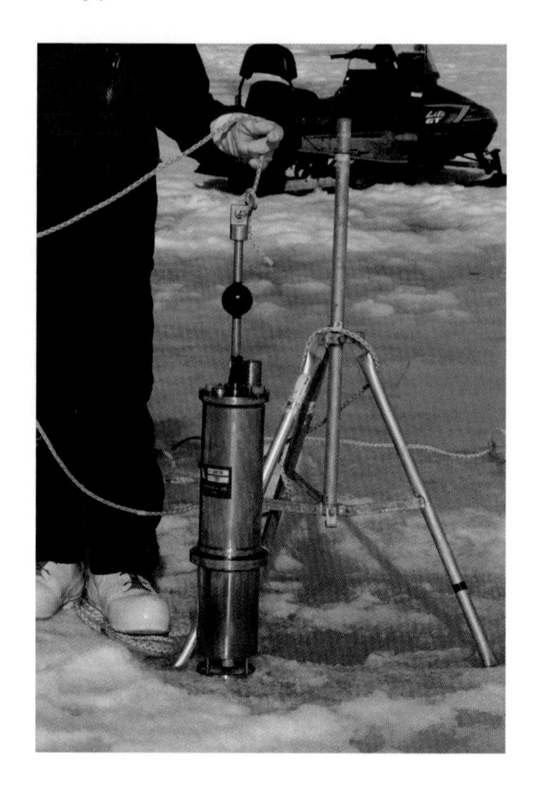

**Figure 3.7.6.** Current measurements by a three-dimensional electromagnetic current meter (Model ACM32M, Alec Electronics Co., Ltd., Japan). One person lowers the current meter through a drill hole and fixes it with a tripod at the depth of 1 m (as the reference level) below the ice bottom in the oceanic boundary layer.

## 3.7.5 FIELD EXPERIMENTS

Below, we will discuss observations of air–sea ice interactions across the ice cover of Saroma-ko Lagoon located on the Okhotsk Sea coast of Hokkaido (Figure 3.7.7). Since the data from Saroma-ko Lagoon provide a good example of the seasonal evolution of conditions under undeformed level ice, mathematical models can be used to examine the influence of the oceanic heat flux on the ice thickness evolution (Shirasawa et al. 2006; Leppäranta and Shirasawa 2007). This approach is in principle applicable to any ice-covered waters.

The surface area of the lagoon is 149.2 km$^2$, maximum depth is 19.5 m, and mean depth is 14.5 m. The lagoon is connected to the Sea of Okhotsk via two inlets, through which the water exchanges mainly by tidal forcing. Freshwater input by two major rivers keeps the salinity to less than 32 psu (practical salinity units). The average freeze-up date is January 8 (Shirasawa and Leppäranta 2003). At the beginning of the freeze-up, atmospheric forcing keeps the lagoon open and frazil ice is formed. Once nearly the entire surface of the lagoon is frozen, ice grows mainly through thermal processes, which provides an ideal case for sea ice thermodynamic modeling. Ice reaches a maximum thickness of 35–62 cm with inclusion of a large portion of snow ice annually in the southeastern part of the lagoon. There are large spatial variations in ice thickness due to the oceanic heat flux from the rivers and inlets.

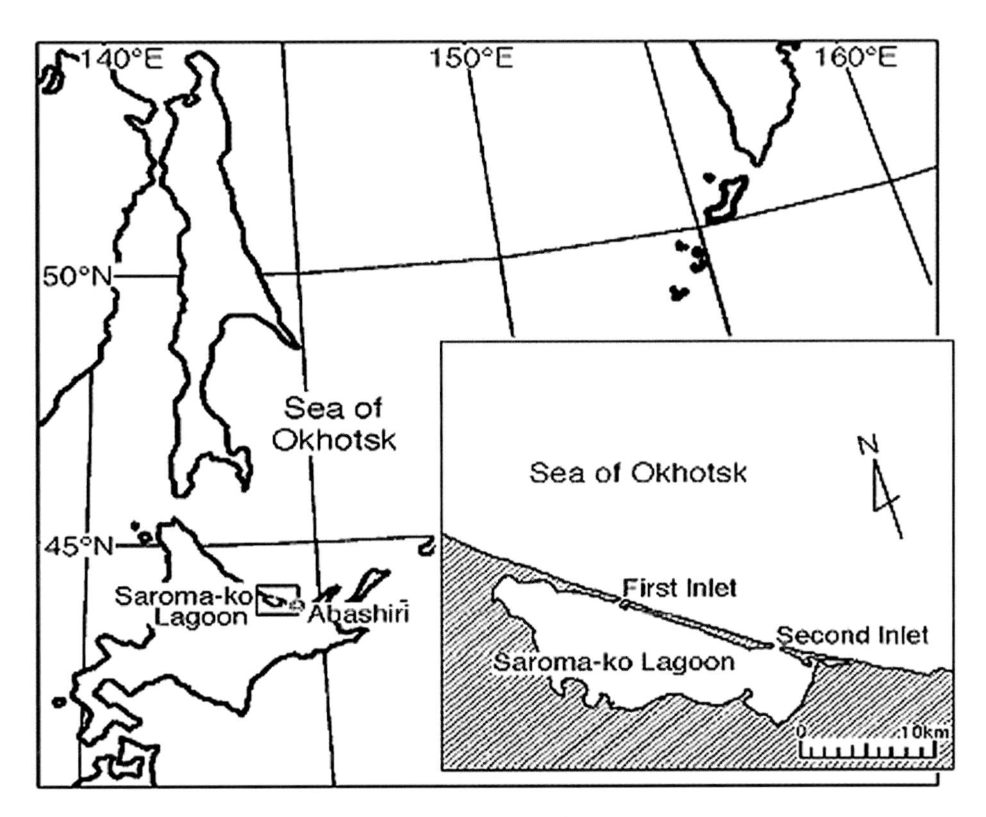


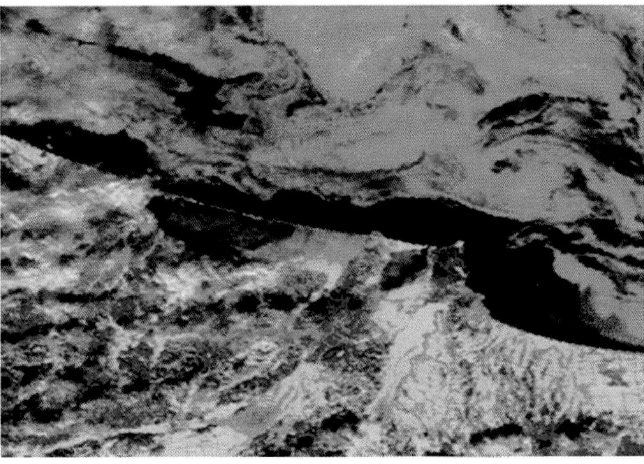

**Figure 3.7.7.** A location map of Saroma-ko Lagoon. The photograph shows sea ice areas in Saroma-ko Lagoon and in the coastal region of the Sea of Okhotsk.

An example of time-series records of the current velocity, water temperature, and salinity measured at a fixed depth in the central area of the lagoon is shown in Figure 3.7.8 (Shirasawa and Leppäranta 2003). The period of observations with freeze-up and breakup dates is included. The deployment of a three-dimensional electromagnetic current meter (ACM 32M, Alec Electronics Co., Ltd., Japan) was made on December 6, 1999, when the lagoon was still open and the water temperature was about 3°C. Ice freeze-up started in late January, and almost the entire surface of the lagoon was covered with ice until the onset of ice breakup in early

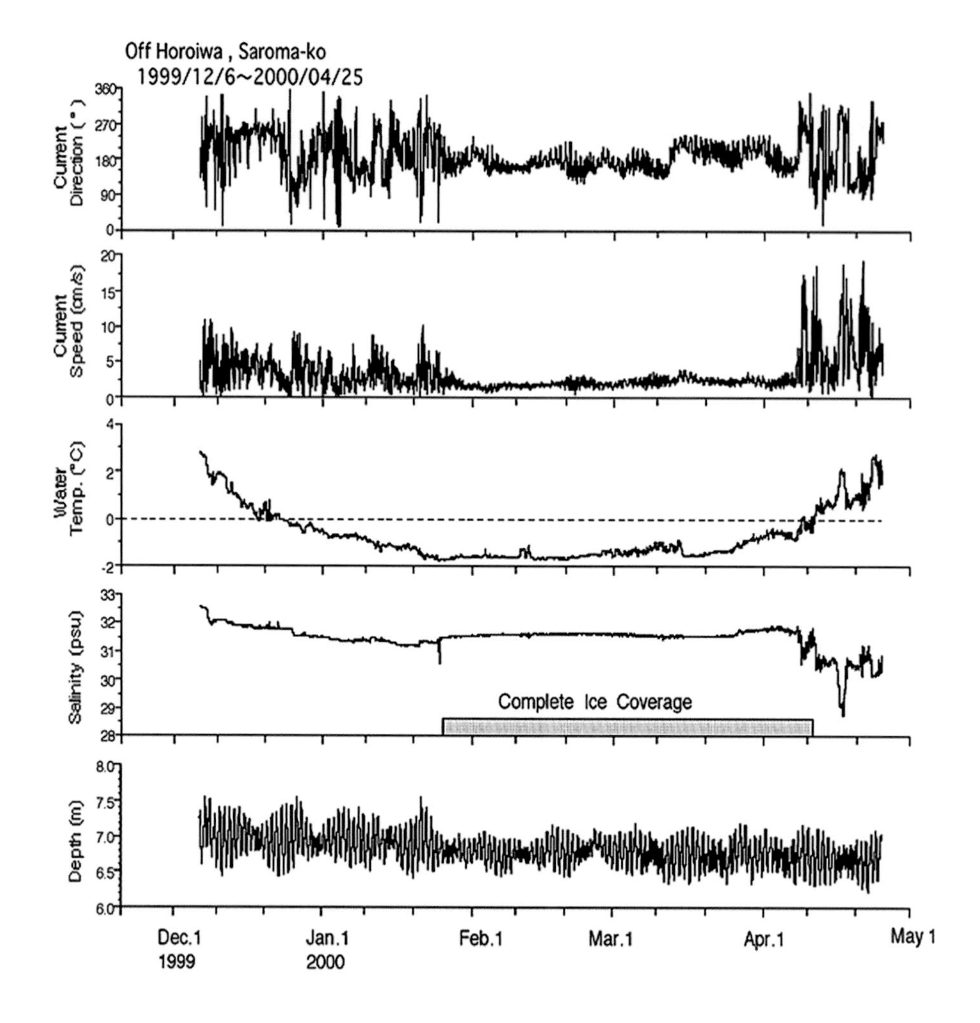


**Figure 3.7.8.** Time series of currents, temperature, and salinity obtained from the mooring station at the central area of Saroma-ko Lagoon during the period from 6 December 1999 through 25 April 2000. The depth of the mooring was 5 m below sea ice. From Shirasawa and Leppäranta (2003).

April. The current speed was as high as 10 cm $s^{-1}$ before ice freeze-up and reduced to 2–3 cm $s^{-1}$ during the ice-covered period. The solid ice cover caused a significant reduction in vertical mixing of the water under the ice, with little momentum transfer from the wind to the water body. The main forcing of the circulation is derived from the two inlets and some major rivers, and heating of the water body by the heat flux from the bottom sediments and solar heating become relevant as soon as the snow has melted.

The under-ice boundary layer can be characterized by three types of flow regimes: a laminar flow regime close to the ice-ocean boundary, a transitional flow regime from laminar to turbulent flow, and a turbulent flow regime furthest away from the boundary. The Reynolds number is commonly used to distinguish between fully turbulent and laminar or transitional flow, and the roughness

Reynolds number is used to determine whether the flow is hydrodynamically rough or smooth. The Reynolds number, Re, and the roughness Reynolds number, Re*, are defined as (e.g., Shirasawa and Ingram 1991a):

$$Re = \frac{UD}{\nu} \quad \text{(Equation 3.7.16)},$$

$$Re^* = \frac{30 u_* z_0}{\nu} \quad \text{(Equation 3.7.17)},$$

where $D$ is the distance from the ice bottom and $\nu$ is the kinematic viscosity of seawater. The roughness length can be estimated as $z_0 = 30k_s$, where $k_s$ is the mean height of roughness elements. $k_s$ can be taken to be the standard deviation of ice thickness and $z_0$ was found to be 0.3–0.4 mm for smooth landfast sea ice sheet (Shirasawa et al. 1997). In the turbulent regime, the eddy correlation technique is applied to estimate the turbulent fluxes (see Equations 3.7.7–3.7.9).

## 3.7.6 ESTIMATION OF OCEANIC HEAT FLUX

The oceanic heat flux from the water body to the ice bottom can be obtained from various methods. The eddy correlation technique requires high-frequency three-dimensional current and temperature data in the OBL beneath sea ice. In addition to the eddy correlation technique, three other methods—the bulk formula approach, the temperature profile method, and the residual method—are explained with observed data below.

An example of the oceanic heat flux $Q_w$ calculated by Equation 3.7.8 is shown in Figure 3.7.9. An average of 0.77 W m$^{-2}$ with the standard deviation of 10.1 W m$^{-2}$ was obtained for the period between February 10 and March 11, 1999. Re* indicated

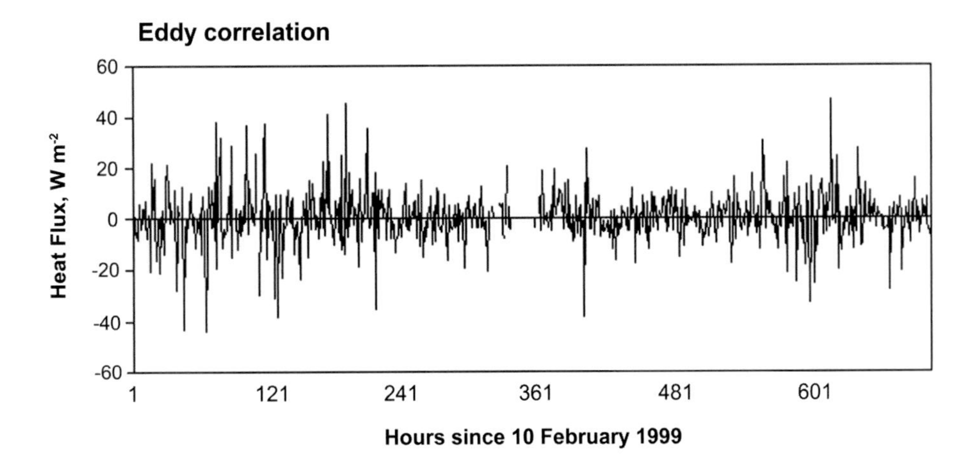


**Figure 3.7.9.** The oceanic heat flux at the ice-ocean interface calculated by the eddy correlation technique for the period from 10 February through 11 March 1999 at Saroma-ko Lagoon

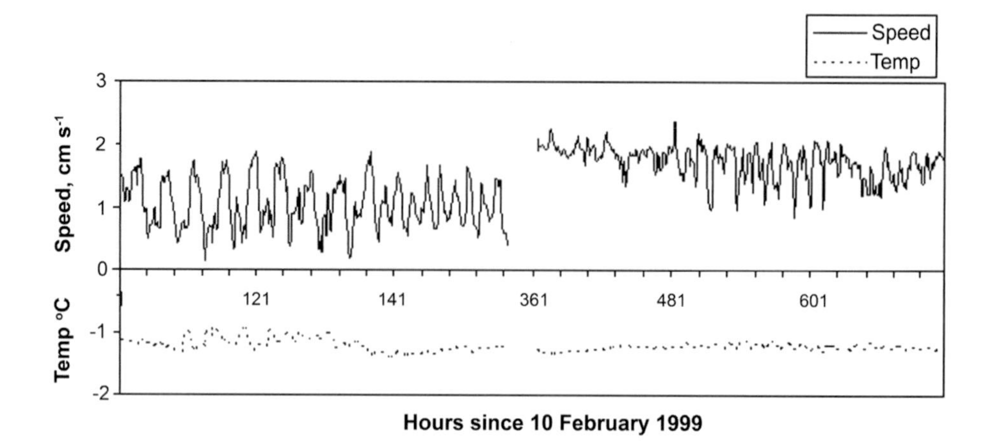


**Figure 3.7.10.** The current speeds and water temperatures obtained at the depth of 0.5 m below the ice-ocean interface during the period from 10 February through 11 March 1999 at Saroma-ko Lagoon. The gaps in the time series of current speed and temperature show redeployment of the current meter.

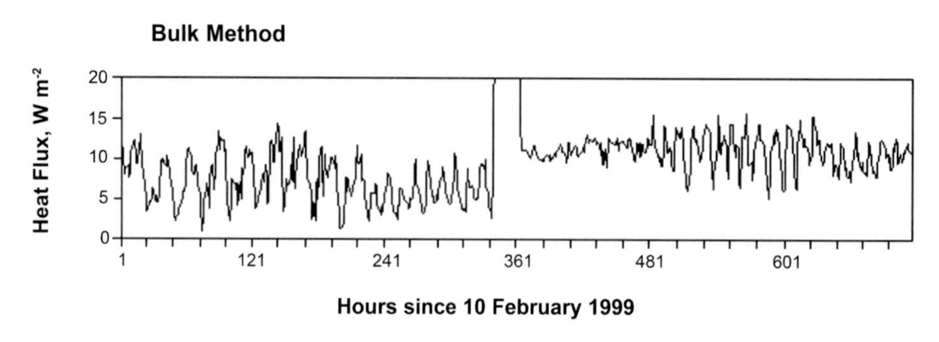


**Figure 3.7.11.** The oceanic heat flux at the ice-ocean interface estimated by the bulk method for the period from 10 February to 11 March 1999 at Saroma-ko Lagoon. The gap in the time series shows redeployment of the current meter. Heat transfer coefficient $C_H = 3{\cdot}10^{-4}$ (Shirasawa et al. 2006) was used in Eq. 3.7.12.

that the measurements were made with hydrodynamically smooth or in a transition from smooth to rough flow, and the flow regime was laminar and/or in transition from laminar to turbulent flow. Thus, the results obtained from the eddy correlation technique are not entirely relevant since the under-ice currents have not fully developed turbulent flow.

The data set of current speeds and temperatures was obtained at a depth of 0.5 m below the ice-ocean interface during the period from February 10 through March 11, 1999 (Figure 3.7.10). The temperature difference $T - T_0$ was 0.3–0.7°C and the current velocity $U_w$ was 0.5–2 cm s$^{-1}$. The estimated oceanic heat flux $Q_w$ from the bulk method was 5–15 W m$^{-2}$ (Figure 3.7.11).

## 3.7.7 INFLUENCE OF OCEANIC HEAT FLUX ON ICE THICKNESS

The influence of the oceanic heat flux on sea ice thickness can be examined through mathematical modeling. Ice in natural waters grows and decays as forced by the fluxes through the upper and lower boundaries. Analytic models can be used to examine initial growth and equilibrium thickness under fixed external conditions, and with perturbation techniques for cyclic oceanic forcing, too. In the cyclic case low frequencies (0.1 cycles per day or less) and low relaxation constants (i.e., fast response times for ice) can lead to remarkable ice thickness cycles. An important feature of the oceanic heat flux is that it influences not only the total ice thickness but also the stratification of the ice sheet, since it melts ice at the bottom and thus provides the potential for more snow-ice formation.

### 3.7.7.1 Analytic Modeling

Simple analytic models provide good results for the climatology of sea ice thickness (Leppäranta 1993). A generalized form including the oceanic heat flux and air-ice interaction reads in differential form:

$$\frac{dh}{dt} = a\frac{T_f - T_a}{h+d} - \frac{Q_w}{\rho_i L} \quad (T_a \leq T_f) \quad \text{(Equation 3.7.18)},$$

where $h$ is ice thickness, $t$ is time, $a = \kappa_t/(\rho_t L) \approx 5.5\ \text{cm}^2/(°\text{C}\cdot\text{day})$ is the freezing-degree-day coefficient, $\kappa_i = 2.1\ \text{W}/(\text{m}\ °\text{C})$ is the thermal conductivity of ice, $\rho_i$ is the ice density, $L$ is the latent heat of freezing, $T_f$ is the freezing point temperature, $T_a$ is the air temperature, and $d \approx 10$ cm is the effective insulating thickness of the surface snow layer. The Stefan solution $h^2 = 2aS$, where $S = \int_o^t \max(0, T_f - T_a)ds$, comes from $d = Q_w = 0$ and can be taken as an upper bound for favorable ice growth conditions.

If the oceanic heat flux is a nonzero constant and air temperature is $T_a$ = constant, the analytic model gives the steady-state solution:

$$h + d = \kappa_t(T_f - T_a)/Q_w \quad \text{(Equation 3.7.19)}.$$

Since $h \geq 0$ and $d \geq 0$, it is seen that the heat flux of $Q_w = \kappa_t(T_f - T_a)/d$ is sufficient to prevent ice formation. If the heat flux is less than that, the steady-state ice thickness becomes $h^* = \kappa_t(T_f - T_a)/Q_w - d$. With $T_a$ = constant, the time evolution is obtained from Equation 3.7.18 as:

$$\frac{h}{h^*} + \log\left[1 - \frac{h}{h^*}\right] = -\frac{\kappa_i(T_f - T_a)}{h^{*2}}\, t \quad \text{(Equation 3.7.20)}.$$

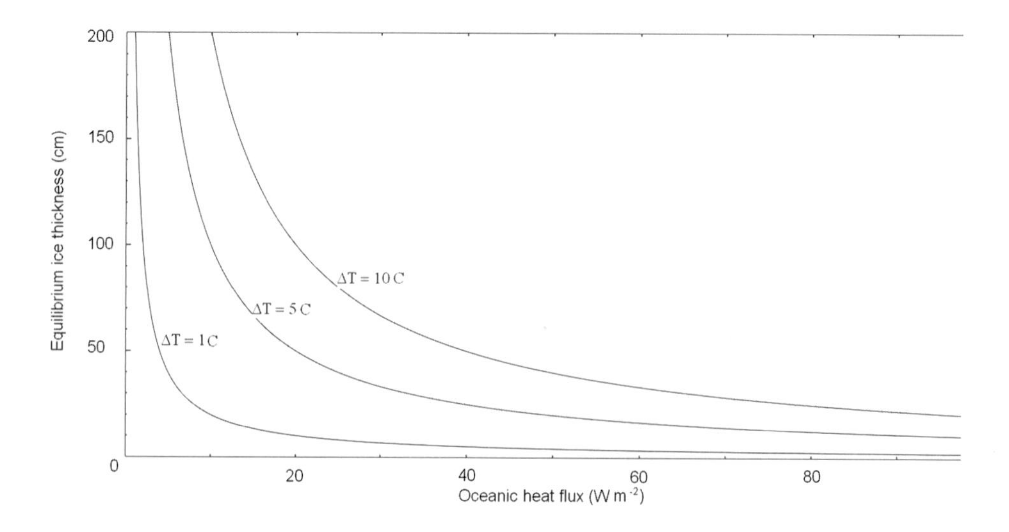


**Figure 3.7.12.** The equilibrium ice thickness as a function of oceanic heat flux. ΔT=1C, 5C and 10C indicate the differences between air temperature and freezing point in degree centigrade.

In the beginning ($h << h^*$) the ice thickens proportional to the square root of time (Figure 3.7.12).

To understand the role of the variable oceanic heat flux, perturbation techniques can be utilized. Writing the ice thickness as $h = h^* + \eta$, and assuming $\eta << h^*$, the ice growth equation can be approximated as:

$$\frac{d\eta}{dt} = a\frac{T_f - T_a}{h^*}\left[1 - \frac{\eta}{h^*}\right] - \frac{Q_w}{\rho L} \quad \text{(Equation 3.7.21).}$$

For cyclic oceanic forcing $Q_w/\rho L = q_o + q_1\sin(\omega t)$, the thickness disturbance has a persistent cyclic part:

$$\eta = \frac{q_1}{\sqrt{\beta^2 + \omega^2}}\sin(\omega t + \phi) \quad \text{(Equation 3.7.22),}$$

where $\beta = -a(T_f - T_a)/h^{*2}$ is a relaxation constant for ice growth and $\phi$ is the phase shift. When both relaxation constant and the frequency of oceanic heat flux are small, considerable thickness variations result. The amplitude $q_1$ could be of the order of 1 cm $d^{-1}$, and therefore with $\beta$, $\omega << 1\ d^{-1}$ the thickness variations would be 10 cm or more. Consequently, tidal heat transport may show up in the fortnightly cycle but not significantly in diurnal or semidiurnal cycles.

### 3.7.7.2 Numerical Modeling

In numerical modeling the temperature profile and resulting heat flow are solved in a dense grid across the ice sheet (Maykut and Untersteiner 1971; Shirasawa et al. 2005). The salinity of the ice is prescribed and together with the temperature determines the brine volume. Compared with analytical models, more realistic boundary conditions can be applied and the thermal inertia of the heat flow through the ice can be included. The heat conduction equation is:

$$\rho_i c_i \frac{\partial T}{\partial t} = \kappa_i \frac{\partial^2 T}{\partial z^2} - q(z) \quad \text{(Equation 3.7.23)},$$

where $c_i$ is the specific heat of ice, $z$ is vertical coordinate, and $q$ is the absorption of solar radiation by the ice. The boundaries of the system move according to changes in ice thickness driven by the heat fluxes, at the bottom $\kappa_i \partial T/\partial z = \rho_i L dh/dt + Q_w$ and at the top similarly except that for growth liquid water or slush must be available.

For a numerical modeling effort, a three-layer (snow/snow-ice/congelation ice) thermodynamic sea ice model (Shirasawa et al. 2005) was used to examine the evolution of ice thickness. Simulations were performed with zero, nonzero constant, and cyclic oceanic heat flux (Figure 3.7.13). The constant was taken as 10 W m$^{-2}$ corresponding to a typical level for subarctic seas and the Southern Ocean, while in the cyclic case the mean was the same and the amplitude also the same, that is, the heat flux varied between 0 and 20 W m$^{-2}$. It is clear that significant ocean heat transfer has a strong influence on the ice thickness cycle. During vigorous ice growth the differences are small, but when the growth rates abate the ocean heat starts to melt the ice, and ice breakup can set in earlier by as much as three weeks. Remarkably, the cyclic oceanic heat flux case did not differ much from the constant heat flux case as long as the averages are equal. With an amplitude of 5 W m$^{-2}$ or for higher frequencies the differences were even smaller.

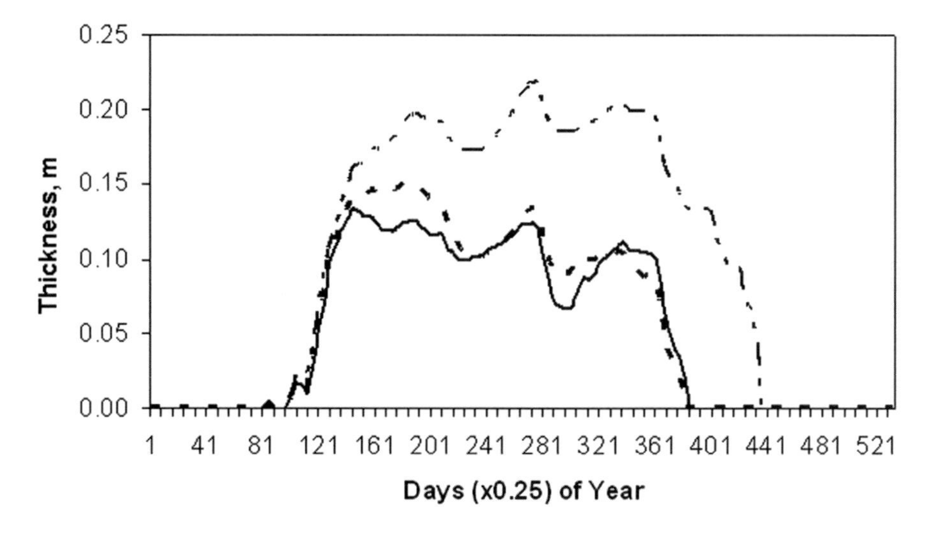


**Figure 3.7.13.** Simulations of the evolution of ice thickness with the oceanic heat flux of zero (top dashed curve), constant 10 W m$^{-2}$ (solid curve), and cyclic 10 W m$^{-2}$ (dotted curve)

We can turn around the question and examine the oceanic heat flux on the basis of ice measurements. If ice thickness and temperature are known, the oceanic heat flux is obtained as the residual:

$$Q_w = \kappa_i \left.\frac{\partial T}{\partial z}\right|_{z=0+} - \rho_i L \frac{dh}{dt} \quad \text{(Equation 3.7.24).}$$

This residual method requires high-resolution measurements of ice growth to obtain a time series of the oceanic heat flux. The ice thickness evolution in winter 1999 is shown in Figure 3.7.14. The ice thickness was measured three times (on February 8–11, February 18–20, and March 2–3, 1999) at several different sampling sites during the whole winter. The average ice growth was 2 mm $d^{-1}$, which corresponds to a release of latent heat of 7 W $m^{-2}$, where $\rho_\iota$ is 910 kg $m^{-3}$ and $L$ is 333.5 kJ $kg^{-1}$. The time series of ice temperatures at depths of 5, 20, and 40 cm from the surface are shown in Figure 3.7.15. The temperature gradient in the ice sheet

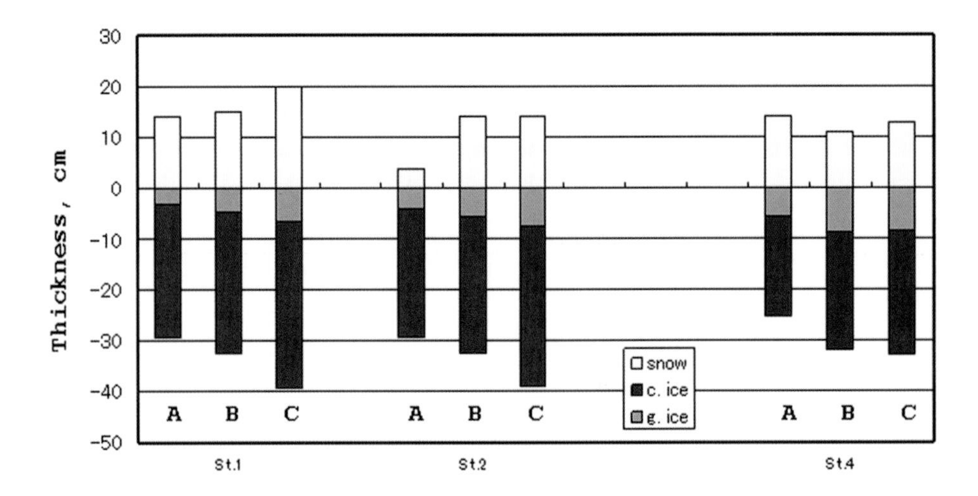


**Figure 3.7.14.** The snow/ice thickness evolution at Saroma-ko Lagoon on 8–11 February 1999 (A), 18–20 February 1999 (B), and 2–3 March 1999 (C), at Sts. 1, 2, and 4 on the survey line from the river mouth towards the second inlet (in Fig. 3.7.6). "Snow," "c. ice" and "g. ice" indicate the thicknesses of snow layer, columnar ice, and granular ice, respectively. Reproduced from Kawamura et al. (2004).

is 1–1.5°C/0.2 m between 20 cm and 40 cm depth (close to the ice bottom) during cold periods. Therefore, the first term on the right-hand side of Equation 3.7.24 is over 15 W $m^{-2}$ at maximum (Figure 3.7.16). The heat flux $Q_w$ is then given as the difference between the heat flux from heat conduction and the latent heat release during ice growth or melt, and consequently both terms must be determined to a high degree of accuracy in order to estimate the heat flux from the ocean.

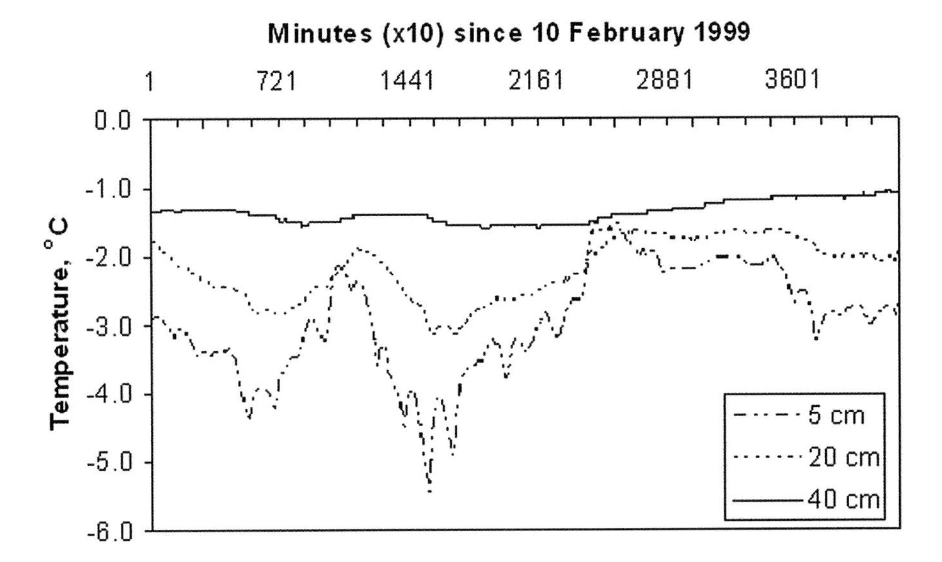


**Figure 3.7.15.** The ice temperatures measured from a thermistor string installed in the ice sheet, at the depths of 5, 20, and 40 cm (close to the ice bottom), at a sampling interval of ten minutes during the period from 10 February through 1 March 1999 at Saroma-ko Lagoon

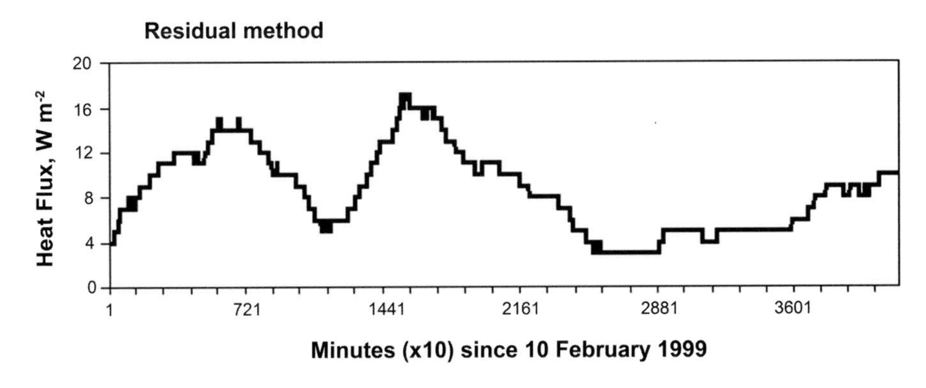


**Figure 3.7.16.** The oceanic heat flux at the ice-ocean interface calculated as the residual term of Eq. 3.14.24, for the period from 10 February through 11 March 1999 at Saroma-ko Lagoon

### 3.7.8 SUMMARY

Ice-ocean interaction plays an important role in physical not only but also biogeochemical processes in the oceanic boundary layer beneath sea ice in seasonally and perennially ice-covered waters. The general theoretical framework with application to the oceanic boundary layer was presented, and methods to estimate turbulent fluxes of momentum, heat, and salt at the ice-water interface were discussed. Then, field measurement technologies and equipment necessary for flux determination were described. Various methods, including eddy correlation technique, bulk formula, temperature profile, and residual methods to estimate the oceanic heat flux, were demonstrated by measurements made in undeformed level ice in a frozen lagoon. The results from those measurements demonstrated that the under-ice currents had not fully developed turbulent flows, but they also indicated that such methods can be applied to the oceanic boundary layer in dynamic ice-covered waters as well. Analytic and mathematical models were used to examine the influence of the oceanic heat flux on the ice thickness evolution. Seasonal and perennial sea ice covers are currently undergoing significant changes, such as the decline of arctic summer ice extent and decreases in ice thickness. One of the important factors to cause those changes is most likely the increase of the oceanic heat flux in ice-covered waters. The approach of concurrent measurements and modeling can be useful to grapple with such problems.

### Acknowledgments

Field experiments in this work are part of the project "Ice Climatology of the Okhotsk and Baltic Seas" supported by the Japan Society for the Promotion of Science; the Japanese Ministry of Education, Culture, Sports, Science and Technology; and the Academy of Finland.

## REFERENCES

Andreas, E. L. (1998), The Atmospheric Boundary Layer over Polar Marine Surfaces, in *Physics of Ice-Covered Seas,* vol. II, edited by M. Leppäranta, pp. 715–773, Helsinki University Press, Helsinki.

Andreas, E. L., R. E. Jordan, and A. P. Makshtas (2005), Parameterizing turbulent exchange over sea ice: the Ice Station Weddell results, *Bound.-Layer Meteorol., 114,* 439–460.

Arrigo, K. R. (2003), Primary Production in Sea Ice, in *Sea Ice: An Introduction to its Physics, Chemistry, Biology, and Geology*, edited by D. N. Thomas and G. S. Dieckmann, pp. 143–183, Blackwell Science, London.

Brennecke, W. (1921), Die ozeanographischen Arbeiten der Deutschen Antarktischen Expedition 1911–1912, *Archiv der Deutschen Seewarte, XXXIX (Nr. 1).*

Brown, R. A. (1980), Boundary-Layer Modeling for AIDJEX, in *Sea Ice Processes and Models*, edited by R. S. Pritchard, pp. 387–401, University of Washington Press, Seattle.

Cushman-Roisin, B. (1994), *Introduction to Geophysical Fluid Dynamics*, Prentice Hall, Englewood Cliffs, NJ.

Eicken, H. (2003), From the Microscopic, to the Macroscopic, to the Regional Scale: Growth, Microstructure and Properties of Sea Ice, in *Sea Ice: An Introduction to Its Physics, Chemistry, Biology and Geology*, edited by D. N. Thomas and G. S. Dieckmann, pp. 22–81, Blackwell Science, London.

Ekman, V. W. (1904), On Dead Water, in *The Norwegian North Polar Expedition 1893–1896: Scientific Results,* vol. V, edited by F. Nansen Copp, pp. 1–152, Clark, Mississauga, Ontario.

Guest, P. S., and K. L. Davidson (1991), The aerodynamic roughness of different types of sea ice, *J. Geophys. Res., 96*(C3), 4709–4721.

Hudier, E. J.-J., R. G. Ingram, and K. Shirasawa (1995), Upward flushing of sea water through first year ice, *Atmosphere-Ocean, 33*(3), 569–580.

Joffre, S. M. (1984), *The Atmospheric Boundary Layer over the Bothnian Bay: A Review of Work on Momentum Transfer and Wind Structure*, Report No. 40, Winter Navigation Research Board, Helsinki.

Kawamura, T., K. Shirasawa, M. Ishikawa, T. Takatsuka, T. Daibou, and M. Leppäranta (2004), On the annual variation of characteristics of snow and ice in Lake Saroma, *Proceedings of the 17th IAHR International Symposium on Ice, Saint Petersburg, Russia, 21–25 June 2004*, pp. 212–220.

Leppäranta, M. (1990), Observations of free ice drift and currents in the Bay of Bothnia, *Acta Regiae Societatis Scientiarum et Litterarum Gothoburgensis, Geophysica, 3*, 84–98.

Leppäranta, M. (1993), A review of analytical sea ice growth models, *Atmosphere-Ocean, 31*(1), 123–138.

Leppäranta, M. (2005), *The Drift of Sea Ice*, Praxis, Chichester, UK.

Leppäranta, M., and A. Omstedt (1990), Dynamic coupling of sea ice and water for an ice field with free boundaries, *Tellus, 42A*, 482–495.

Leppäranta, M., and K. Shirasawa (2007), Influence of the ice ocean heat flux on the ice thickness evolution in Saroma-ko Lagoon, Hokkaido, Japan, *Proceedings of the 16th Northern Research Basins Symposium and Workshop, Petrozavodsk, Russia, 27 August–2 September 2007*, pp. 83–88.

Li, W. H., and S. H. Lam (1964), *Principles of Fluid Mechanics*, Addison-Wesley, Reading, MA.

Lytle, V. I., and S. F. Ackley (1996), Heat flux through sea ice in the western Weddell Sea: Convective and conductive transfer processes, *J. Geophys. Res., 101*(C4), 8853–8868.

Martinson, D. G., and C. Wamser (1990), Ice drift and momentum exchange in winter Antarctic pack ice, *J. Geophys. Res.*, *95*(C2), 1741–1755.

Maykut, G. A., and N. Untersteiner (1971), Some results from a time-dependent, thermodynamic model of sea ice, *J. Geophys. Res.*, *76*, 1550–1575.

McPhee, M. G. (1982), *Sea Ice Drag Laws and Simple Boundary Layer Concepts, Including Application to Rapid Melting*, Report 82-4, U.S. Army Cold Regions Research and Engineering Laboratory, Hanover, NH.

McPhee, M. G. (1986), The upper ocean, in *Geophysics of Sea Ice*, edited by N. Untersteiner, pp. 339–394, Plenum Press, New York.

McPhee, M. G. (1990), Small-scale processes, in *Polar Oceanography, Part A: Physical Science*, edited by W. O. Smith, Jr., pp. 287–334, Academic Press, San Diego.

McPhee, M. G. (2008), *Air-Ice-Ocean Interaction: Turbulent Oceanic Boundary Layer Exchange Processes*, Springer, Heidelberg, Germany.

McPhee, M. G., S. F. Ackley, P. Guest, B. A. Huber, D. G. Martinson, J. H. Morison, R. D. Muench, L. Padman, and T. P. Stanton (1996), The Antarctic zone flux experiment, *Bulletin of the American Meteorological Society, 77*, 1221–1232.

Nansen, F. (1902), The oceanography of the North Polar Basin, *Norwegian North Polar Expedition 1893–1896: Scientific Results,* vol. 3, No. 9, Longman Green & Co., Kristiania, Norway.

Omstedt, A. (1998), Freezing estuaries and semi-enclosed basins, in *Physics of Ice-Covered Seas,* vol. II, edited by M. Leppäranta, pp. 483–516, Helsinki University Press, Helsinki.

Omstedt, A., L. Nyberg, and M. Leppäranta (1996), The role of ice inertia in ice-ocean dynamics, *Tellus, 48A*, 593–606.

Omstedt, A., and J. S. Wettlaufer (1992), Ice growth and oceanic heat flux: Models and measurements, *J. Geophys. Res.*, *97*(C6), 9383–9390.

Overland, J. E. (1985), Atmospheric boundary layer structure and drag coefficients over sea ice, *J. Geophys. Res., 90*(C5), 9029–9049.

Perovich, D., T. Grenfell, J. Richter-Menge, B. Light, W. Tucker III, and H. Eicken (2003), Thin and thicker: Sea ice mass balance measurements during SHEBA, *J. Geophys. Res., 108*(C3), 8050, doi:10.1029/2001JC001079.

Rossby, C. G., and R. G. Montgomery (1935), Frictional influence in wind and ocean currents, *Papers in Physical Oceanography and Meteorology*, vol. 3, no. 3, Massachusetts Institute of Technology and Woods Hole Oceanographic Institute, Cambridge, MA.

Shirasawa, K. (1986), Water stress and ocean current measurements under first-year sea ice in the Canadian Arctic, *J. Geophys. Res., 91*(C12), 14,305–14,316.

Shirasawa, K., and M. Aota (1991), Atmospheric boundary layer measurements over sea ice in the Sea of Okhotsk, *J. Marine Systems, 2*(1), 63–79.

Shirasawa, K., and G. Ingram (1991a), Characteristics of the turbulent oceanic boundary layer under sea ice. Part 1: A review of the ice-ocean boundary layer, *J. Marine Systems, 2*(1), 153–160.

Shirasawa, K., and G. Ingram (1991b), Characteristics of the turbulent oceanic boundary layer under sea ice. Part 2: Measurements in southeast Hudson Bay, *J. Marine Systems, 2*(1), 161–169.

Shirasawa, K., and G. Ingram (1997), Currents and turbulent fluxes under the first-year sea ice in Resolute Passage, Northwest Territories, Canada, *J. Marine Systems, 11*, 21–32.

Shirasawa, K., G. Ingram, and E. J.-J. Hudier (1997), Oceanic heat fluxes under thin sea ice in Saroma-ko Lagoon, Hokkaido, Japan, *J. Marine Systems, 11*, 9–19.

Shirasawa, K., and M. Leppäranta (2003), Hydrometeorological and sea ice conditions at Saroma-ko lagoon, Hokkaido, Japan, *Proc. The seminar "Sea Ice Climate and Marine Environments in the Okhotsk and Baltic Seas—The Present*

*Status and Prospects," Seili, Finland, 10–13 September 2001, Report Series in Geophysics, Division of Geophysics, University of Helsinki, No. 46*, pp. 161–168.

Shirasawa, K., M. Leppäranta, T. Kawamura, M. Ishikawa, and T. Takatsuka (2006), Measurements and modeling of the water-ice heat flux in natural waters, *Proceedings of the 18th IAHR International Symposium on Ice, Sapporo, Japan, 28 August–1 September 2006, vol. 1*, pp. 85–91.

Shirasawa, K., M. Leppäranta, T. Saloranta, T. Kawamura, A. Polomoshnov, and G. Surkov (2005), The thickness of coastal fast ice in the Sea of Okhotsk, *Cold Regions Science and Technology, 42*, 25–40.

Shuleikin, V. V. (1938), Drift of ice fields, *Doklady Academy of Sciences USSR, 19*(8), 589–594.

Sverdrup, H. U. (1928), The wind-drift of the ice on the Northern Siberian Shelf: The Norwegian north polar expedition with the "Maud" 1918–1925, *Scientific Results, 4*(1), 1–46.

Tennekes, H., and J. L. Lumley (1972), *A First Course in Turbulence*, MIT Press, Cambridge, MA.

Uusikivi, J., J. Ehn, and M. A. Granskog (2006), Direct measurements of turbulent momentum, heat, and salt fluxes under landfast ice in the Baltic Sea, *Annals of Glaciology, 44*, 42–46.

Waters, J. K., and M. S. Bruno (1995), Internal wave generation by ice floes moving in stratified water: Results from a laboratory study, *J. Geophys. Res., 100*(C7), 13,635–13,639.

Wettlaufer, J. S. (1991), Heat flux at the ice-ocean interface, *J. Geophys. Res., 96*(C4), 7215–7236.

## APPENDIX: NOMENCLATURE

| Parameter | Symbol | Value | Unit | Appearance | Source |
|---|---|---|---|---|---|
| Ice density | $\rho_i$ | 910 | $kgm^{-3}$ | Eq. 3.7.18 | Crocker and Wadhams (1989) |
| Molecular thermal conductivity | $\kappa$ | 0.6 | $Wm^{-1o}C^{-1}$ | Eq. 3.7.15 | |
| Thermal conductivity of ice | $\kappa_i$ | 2.1 | $Wm^{-1o}C^{-1}$ | Eq. 3.7.18 | Yen (1981) |
| Specific heat of seawater | $c_w$ | | $kJkg^{-1o}C^{-1}$ | Eq. 3.7.8 | |
| Specific heat of ice | $c_i$ | 2.114 | $kJkg^{-1o}C^{-1}$ | Eq. 3.7.23 | Yen (1981) |
| Latent heat of freezing | $L$ | 333.5 | $kJkg^{-1}$ | Eq. 3.7.18 | Yen (1981) |
| Molecular kinematic viscosity | $\nu$ | $1.8{\cdot}10^{-6}$ | $m^2s^{-1}$ | Eq. 3.7.16 | |
| Kinematic eddy viscosity | $K$ | | | Eq. 3.7.1 | |
| Gravity acceleration | $g$ | 9.8 | $ms^{-2}$ | Eq. 3.7.1 | |
| Coriolis parameter | $f$ | $1.4{\cdot}10^{-4}$ | $s^{-1}$ | Eq. 3.7.1 | |
| von Karman constant | $\kappa$ | 0.4 | | Eq. 3.7.2 | |
| Freezing-degree-day coefficient | $a$ | 5.5 | $cm^{2o}C^{-1}day^{-1}$ | Eq. 3.7.18 | |
| Air temperature | $T_a$ | | $^oC$ | Eq. 3.7.18 | |
| Water temperature fluctuation | $T'$ | | $^oC$ | Eq. 3.7.8 | |
| Freezing point temperature | $T_f$ | | $^oC$ | Eq. 3.7.18 | |
| Ice bottom temperature | $T_0$ | | $^oC$ | Eq. 3.7.13 | |
| Water salinity fluctuation | $S'$ | | psu [practical salinity unit] | Eq. 3.7.9 | |
| Water salinity at the ice bottom | $S_0$ | | psu [practical salinity unit] | Eq. 3.7.13 | |
| Time | $t$ | | | Eq. 3.7.1 | |
| Wind/Current velocity | $U$ | | $ms^{-1}$ | Eq. 3.7.1 | |
| Down-flow fluctuation | $u'$ | | $ms^{-1}$ | Eq. 3.7.7 | |
| Cross-flow fluctuation | $v'$ | | $ms^{-1}$ | Eq. 3.7.7 | |
| Vertical flow fluctuation | $w'$ | | $ms^{-1}$ | Eq.e 3.7.7 | |
| Water curent velocity | $U_w$ | | $ms^{-1}$ | Eq. 3.7.12 | |
| Geostrophic velocity | $U_g$ | | $ms^{-1}$ | Eq. 3.7.4 | |
| Friction velocity | $u_*$ | | $cms^{-1}$ | Eq. 3.7.2 | |
| Ice thickness | $h$ | | m | Eq. 3.7.18 | |
| Steady state ice thickness | $h^*$ | | m | Eq. 3.7.20 | |
| Roughness length. | $z_o$ | | cm | Eq. 3.7.3 | |
| Mean height of roughness elements | $k_s$ | | cm | Eq. 3.7.17 | |
| Monin–Obukhov length | $L_{MO}$ | | m | Eq. 3.7.6 | |
| Reynolds number | Re | | | Eq. 3.7.16 | |
| Roughness Reynolds number | $Re^*$ | | | Eq. 3.7.17 | |
| Vertical shear stress | $\tau$ | | $Nm^{-2}$ | Eq. 3.7.1 | |
| Oceanic heat flux | $Q_w$ | | $Wm^{-2}$ | Eq. 3.7.8 | |
| Salt flux | $S_w$ | | $ms^{-1}$ | Eq. 3.7.9 | |
| Ice-ocean drag coefficient | $C_w$ | | | Eq. 3.7.11 & Table 3.7.1 | |
| Heat transfer coefficient | $C_H$ | $3{\cdot}10^{-4}$ | | Eq. 3.7.12 | Shirasawa et al. (2006) |

# Chapter 3.8

# Biogeochemical Properties of Sea Ice

*Andrew McMinn, Rolf Gradinger, and Daiki Nomura*

## 3.8.1 INTRODUCTION

Sea ice harbors an entire microcosm of organisms (Chapter 3.9), that substantially contribute to the biogeochemical cycles of polar waters (Legendre et al. 1992). Organisms are not evenly distributed throughout the ice but are layered either in the ice interior or attached to the bottom. This chapter outlines some of the most common variables assessed and different approaches to assess them. This chapter cannot cover all processes occurring in sea ice and thus focuses on the contribution of sea ice to the carbon cycle, while other significant aspects (e.g., contributions to the sulfur cycle by dimethylsulphide production) are not considered.

## 3.8.2 GENERAL INTRODUCTION TO FIELD SAMPLING FOR BIOGEOCHEMICAL PROPERTIES

Artifacts due to improper sampling can easily be introduced when collecting sea ice–related information. Dirty clothing and shoes can alter the chemistry and sediment load of the surface sea ice while walking on the ice has a similar effect to smoking. Therefore, great care has to be taken to secure the dedicated field sampling site from unnecessary human activities, which can be a difficult task when working close to a human settlement or research station or when working jointly with other ice research teams. Each set of variables has its own specific sampling needs, which will be discussed further below. However, some general best practices should be followed to reduce any contamination.

Washing hands prior to any sampling with clean water and wearing clean clothing are important prerequisites to minimize contamination. In addition minimize any foot traffic through your sampling site, which could alter properties such as surface sediment load, snow compaction, and therefore also light regimes. If needed, sampling sites can be marked with flags or bands to minimize the risk of contamination. It also helps to mark sampling sites close to human settlements

**Figure 3.8.1.** Ice corer connected to a portable generator allows work reliably even at temperatures of –35°C. Note that sampling site is not impacted by shadow of any equipment and clear of any foot traffic.

with a display explaining the purpose of this study and contact information for the scientists. Under such circumstances a public presentation outlining the research effort to the people living in this area is often considered a valuable approach not only for outreach but also to ensure the long-term success of the project.

When selecting a sampling site, consideration should be given to the aspect of the site, as artificial shading by sleds, research gear, icebreakers, etc. will affect light levels and algal production estimates (Figure 3.8.1). Ideally, you should minimize the disturbance of the snow cover to keep conditions as natural as possible at least before measuring light in areas where algal growth will be assessed. Also, work should not be conducted downwind of the ship or machinery as their emissions could impact snow and ice properties.

### 3.8.2.1 Ice Coring

Ice coring with CRREL-type corers is the main method of collecting sea ice samples for further processing; ice-coring techniques are described in detail in Chapter 3.3. For biogeochemical purposes, contamination issues need to be considered much more seriously than for factors such as salinity as concentrations of most biochemical variables are measured in μmol/l of melted sea ice.

Ice corers, which are quite expensive, are available from specialist vendors with various diameters and designs. Ice corers should be free of any contamination by organic or inorganic substances (e.g., gasoline, stone salt used to keep ship decks free of ice), which might originate from storage and transport to the field. One to two holes are often cored into the ice to clean the ice corer before starting to work on the real sampling program. The use of an electric power drill connected to a portable power generator additionally minimizes not only the noise pollution but potential contamination with condensed chemicals from the exhaust of the gas-powered ice corer (e.g., iron compounds, oil, organic substances). After collecting, the core needs to be processed as quickly as possible to minimize loss of dissolved and particulate material due to brine drainage. A well-trained team of two people can section a 1 m long ice core into 10 cm segments in less than five minutes. The collected ice segments can be placed into solid plastic containers or disposable, resealable zipper storage bags (a 10 cm segment fits into one-gallon bags). While using zipper storage bags minimizes the volume required for transport to the field, they are prone to being punctured—double bagging appears to be a useful solution.

Great care needs to be taken to label all bags with a predefined numbering scheme, as any one sampling event can easily produce more than fifty samples depending on the vertical resolution and number of replicates. Specifically, under very cold conditions it is recommended that the bags be labeled with a black permanent marker before going into the field. During the process of taking the core, handling it, and sectioning the core, it is important to wear lab gloves to avoid contamination.

The vertical resolution used for sectioning the ice cores will largely depend on the research interest. In both the Arctic and Antarctic, biological activity mostly occurs in densely populated bands either at the surface, within the interior, or at the bottom of the ice (Figure 3.8.2). Studies to resolve these vertical gradients typically

**Figure 3.8.2.** (a) Imaging an ice core immediately after collection documents layering in the core, occurrence of sediment, and algal layers; (b) A sheet of black plastic strongly enhances the contrast. Note that such pictures should be taken on a core not used for any further processing to avoid unnecessary loss of brine or damage to algal photosynthetic performance by exposure to bright daylight.

use sampling intervals of 1–5 cm, while in other parts of the core segments of 10–20 cm lengths are sufficient to resolve biogeochemical gradients.

How the ice is melted has an important impact on many biological and chemical parameters. If the core is to be used for measuring active biological processes, the ice must be melted into filtered seawater. If the dissolved silica is to be measured, the rate of melting is also important.

### 3.8.2.2 Determination of Photosynthetically Active Radiation (PAR)

Accurate light measurements are not only needed to assess the effect of sea ice on the heat balance of the Arctic, but are also crucial to assess primary productivity patterns within and below the sea ice. Currently we have no means to measure directly the light availability within the sea ice in the field. However, we can make reliable assumptions based on known attenuation coefficients for snow, ice, and algal pigments in addition to sediment layers. Several such models have been published that provide estimates for such coefficients (e.g., Smith et al. 1988). Surface albedo and ice optical properties are discussed in detail in Chapter 3.6. From the biological perspective, spectral absorbance, which for algal pigments falls in the range from 400 to 700 nm and is called photosynthetically active radiation (PAR), is of most interest. PAR is typically measured as photon flux density (in μmol photons $m^{-2}$ $s^{-1}$) and not in energy units (in W $m^{-2}$), as the number of available photons in the PAR range is more important than their energy from the perspective of algal growth. Extensive foot traffic on the ice or the presence of holes through the ice will negatively impact the accuracy and quality of the light data, so light measurements are among the first activities at any ice station and should be taken after drilling a single hole. PAR sensors are deployed through the hole (either spherical [4π] or planar [2π] sensors have been used) into the ice-water interface region, ideally with an expandable L-shaped arm to distance the light sensor from the drilling hole (Figure 3.8.3). In addition the hole should be covered to minimize light penetration towards the sensor. Simultaneous readings of the under-ice and the incoming radiation (achieved with a second light sensor at the surface) are taken or recorded with data loggers over longer time periods depending on the research question. While the amount of light will control algal photosynthesis, the amount of algae in the ice will also determine how much light penetrates the ice. Ice algae are adapted to live at low PAR intensities (1 to 50 μmol photons $m^{-2}$ $s^{-1}$) but they themselves alter the light in both its quantity and quality through alterations of the spectrum by absorption with their pigments. It has therefore been proposed in both arctic and antarctic studies that under-ice spectral light composition can be used as a remote sensing tool to determine ice algal biomass.

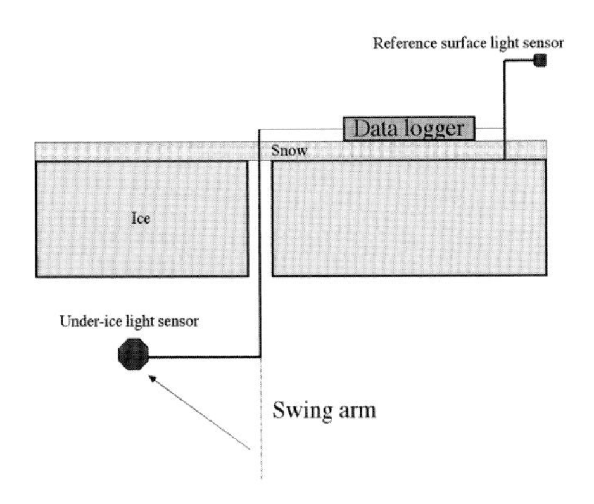


**Figure 3.8.3.** Schematic diagram illustrating the deployment of light sensors under undisturbed parts of ice floes

### 3.8.2.3 Determination of Inorganic Nutrients

Inorganic macronutrients (mainly various nitrogen sources, silicate, phosphate) are required for microalgal growth within the sea ice. These nutrients are required in a certain ratio, which we often refer to as the Redfield ratio (C:N:P = 106:16:1 as molar ratio). Knowing the nutrient concentrations within the ice is relevant not only to determine which inorganic nutrient might be limiting the algal growth (Liebig's law of the minimum) but also to estimate and model ice algal growth. As for all algae, algal growth can be expressed as a function of nutrient availability and is often parameterized according to the Michaelis-Menton kinetics:

$$\mu = \mu_{max} N/(N + Ks) \quad \text{(Equation 3.8.1)},$$

with $\mu$ growth rate (typically $d^{-1}$) $\mu_{max}$ = maximum growth rate (see above), $N$ = concentration of inorganic nutrient (typically in µmol/l), and $Ks$ = nutrient concentration where $\mu$ equals 0.5 $\mu_{max}$.

The example provided in Figure 3.8.4 shows an initial linear relationship and an asymptotical approach towards µmax had high concentrations.

In most aquatic environments, carbon is not considered a limiting resource. While our studies focused on the relation of photosynthetic rate to nitrogen and phosphorous, we also realized over the last decade that inorganic carbon can be depleted within the ice system due to the immense C-uptake during ice algal growth. Therefore estimation of the availability of $CO_2$ for photosynthesis should be included in any study focusing on the growth aspects of ice algae. The following paragraphs outline general considerations of various approaches to assess

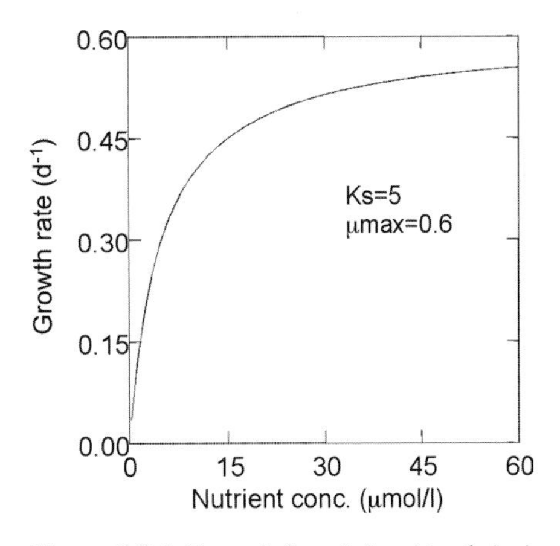


**Figure 3.8.4.** Example for relationship of algal growth rate versus nutrient concentrations

nutrient concentrations which are unique to the ice environment and which need to be considered when determining and reporting nutrient concentration.

## Bulk vs. Brine Concentrations

Two fundamentally different nutrient concentrations can be determined in sea ice, both of which have their specific application. Care needs to be taken to avoid contamination of the samples, as nutrient concentrations are typically very low (in the µmolar range—see general recommendation at the beginning of the chapter). Similar to bulk salinity estimates, bulk nutrient concentrations are determined from completely melted ice core segments. Immediately after coring, ice cores are cut into segments of various length (depending on research question and budget). To resolve the gradients occurring within sea ice, it is recommended that a vertical resolution of at least 10 cm be used. In some instances, such as in bottom communities of arctic sea ice, a < 1 cm vertical resolution is suggested. Ice core segments are placed into labeled and clean leakproof containers made out of inert materials like zipper bags and placed in the dark. Melting needs to be rapid to avoid any changes in concentrations resulting from continuing biological processes. Cores are usually melted in the dark at temperatures between 4 and 15°C and complete melting is achieved in less than six hours. After complete melting has occurred, 25 to 50 ml subsamples are taken for nutrient analysis. These volumes are sufficient to analyze for nitrate, nitrite, silicate, and phosphate using standard techniques (Grasshoff et al. 1999). Salinity also needs to be measured at the same time to correct the measured values. Also note that high concentrations of particulates (could be sediment load or algal biomass) need to be removed prior to measurements; this can be achieved by prefiltration with a syringe filter. The example in Figure 3.8.5 demonstrates the strong vertical gradients present in the bulk nutrient concentrations but also points towards a weakness of this approach. Simply melting the ice exposes the cells to substantial osmotic stress and can cause cell rupture and release of internal pools of inorganic nutrients, thus artificially increasing the bulk nutrient value. This has been observed at several locations in the Arctic and is also likely to be the cause in the example shown in Figure 3.8.5. Here the algal pigment concentrations exceeded 1000 µg/l in the bottom layers of the ice at this station.

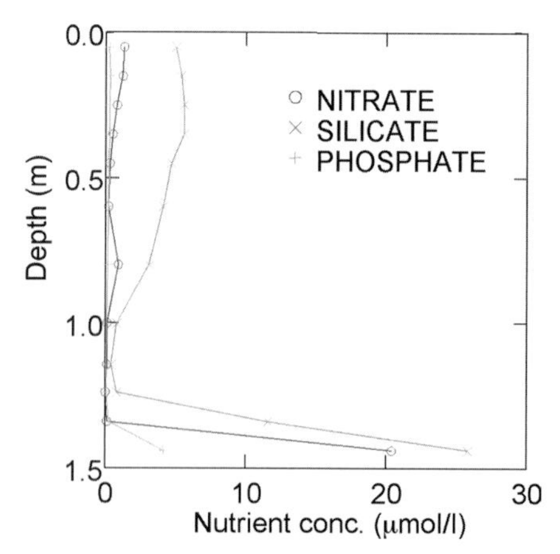


**Figure 3.8.5.** Example for vertical nutrient gradient in arctic pack ice collected on May 14, 2002, at 71.9°N and 166.0°W. Data points are in the middle of the individual segments. Total ice thickness: 1.44m.

While the bulk nutrient concentration provides valuable information about the nutrient status of the ice and its impact on water chemistry during ice melt, it does not provide the relevant information needed to assess algal growth. For this an estimate of the brine nutrient concentration that the algae are exposed to in their natural environment is needed. This can be achieved in one of two ways: either through a calculation based on bulk nutrient concentration and knowledge of total brine volume, or through actual measurement of properties of collected brine. In the first approach a combination of bulk nutrient concentration divided by the fraction of brine per core segment (estimation based on ice temperature and bulk salinity [Chapter 3.3]) gives a reliable estimate for brine concentrations. To directly collect the brine, two different approaches have been successfully used in the field. Brine can be collected in so-called sackholes, holes produced by only partially drilling through the ice. Brine drains out of open brine channels and pockets into the sackhole for thirty to sixty minutes and can then be collected with syringes or pumps. The holes are covered with an insulator to avoid exposure to ambient atmospheric conditions. While this method can produce large quantities of brine over short time periods (depending on ice temperature and bulk salinity), it does not give any insights into the vertical gradients of brine properties on small scales. It is also limited to periods when no melt puddles form on the sea ice surface. During spring and summer melting, the sackholes will be immediately flooded by runoff from the melt ponds and thus no longer represent true brine values. Under such conditions, and to assess the small-scale gradients in brine properties, centrifugation techniques have been successfully applied. Ice core segments 2 to 5 cm thick are taken by standard techniques and placed in a cooled centrifuge and spun at about 1400 rpm for five to fifteen minutes (Figure 3.8.6). Brine will be drained out of the segments and can be analyzed for nutrient concentrations—but not for biomass estimates, as a large fraction of the biomass remains within the core segments, likely attached to the

**Figure 3.8.6.** Sea ice core segments placed into the buckets of a cooled centrifuge

walls or as aggregates. The centrifugation technique was developed by Weissenberger et al. (1992) to study the brine channel morphology in sea ice. Brine nutrient concentrations are the measurements needed to relate in situ living conditions of sea ice algae to their growth rate and/or productivity.

### 3.8.2.4 Sea Ice Algal Pigment Composition and Concentration

Phytoplankton and ice algal cells contain a large number of pigments, which are primarily used for photosynthesis and photoprotection. Chlorophyll *a* is the primary pigment used in photosynthesis by all plants. Because it is ubiquitous, it has been widely used as a measure of phytoplankton biomass. While chlorophyll *a* content per cell varies both with both species and light environment, this pigment is primarily responsible for harvesting the sun's energy and so provides information on the capacity of the ecosystem to trap and utilize solar energy.

Sea ice contains several different microalgal communities, including surface, brine, intermediate, interstitial, hanging, and platelet communities. Most of them can be sampled by coring. Some hanging communities are so loosely attached that any coring will detach them. The only way to sample this type is from underneath by a diver.

*Methods*

Care must be taken at all times to avoid exposing the ice sample to be measured to any direct sunlight. Ice algae are typically growing under very low light levels and exposure to high light will bleach the pigments and damage the photosystems. For this reason once the ice corer containing the ice core has been raised to the surface, the bottom of the corer is immediately placed in a thick dark plastic bag for transport back to the laboratory if possible. The cores can be stored in the dark in the freezer for short periods of time (no more than one or two hours). If processing in the lab cannot be achieved, processing needs to be achieved in the shade on the ice—tents and tarp can provide shade under these circumstances.

To measure pigments the algae has to be extracted from the ice and this requires melting the ice. The speed of melting is a tradeoff. Rapid removal of cells from the ice is desirable because algal cells can change their cellular chlorophyll concentration in response to changed light by up to five times over the course of three to five days. This means that the concentration in the ice core starts to change as soon as it is taken (storage at −20°C delays the response). However, overrapid melting causes cell rupture and loss of chlorophyll.

The following processing is suitable for all biological process studies. The ice cores should be sectioned into appropriate lengths for melting. In most studies ice segment length is between 5 and 20 cm; for appropriate resolution of dense biomass maxima higher resolution (< 1 cm) is recommended to reflect the naturally occurring gradients. The best saws to use are wide-toothed garden saws with individual teeth up to 1 cm wide. Each core segment should be placed into an equal volume of filtered (< 0.2 to 1 μm) seawater. This minimizes membrane rupture and osmotic shock and also speeds up melting. Crushing or breaking the ice up into smaller pieces also increases the speed of melting. Melting should be completed within six hours and observed regularly. As soon as the ice has melted, the total volume of seawater should be measured in a graded cylinder and then, depending on biomass concentration, filtered in total or in a defined subsample onto a glass fiber filter (GF/F). The filter can then be stored frozen. Storage at –80°C or below will allow storage for a period of days to weeks. Storage at –20°C is suitable for only a few days.

There are three main types of equipment used to measure algal pigments: the fluorometer, the spectrophotometer, and the HPLC.

## The Fluorometer

Chlorophyll *a* in all plants naturally produces a low red fluorescence when stimulated with blue light. This fluorescence is proportional to the amount of chlorophyll present and so can be used to measure chlorophyll biomass. The fluorometer was designed to use this characteristic and is now the most widely used method in marine science. It is a very sensitive method and can detect biomass down to 0.01 μg chl a $l^{-1}$.

In practice the frozen filter with algal cells is placed in a test tube with 10 ml of an organic solvent. The most common solvents used are methanol, acetone, or N, N-dimethylformamide (DMF). Each solvent has advantages and disadvantages, but the fluorometer you will use has been calibrated to only one of them and the same solvent should be used until the instrument is recalibrated with pure standard solutions of chlorophyll *a*. Each sample should be extracted in the dark and cold (a refrigerator) for a minimum of eight hours. The extraction time should be constant for all samples. Chlorophyll in solution is far less stable than in a cell, and an attempt should be made to make the measurements quickly and in dim light.

The chlorophyll in solution is transferred to a fluorometer tube and placed in the dark to adjust to room temperature. After placing the sample vial into the instrument it takes twenty to thirty seconds to stabilize before the reading can be taken. After writing down the value three drops of 0.1 M HCl are added, the sample inverted to mix, and then read again. The acid changes the chlorophyll into phaeophytin with a lower fluorescence and the difference between the two values is equivalent to the chlorophyll concentration. To do this chlorophyll *a* = reading 1 (chlorophyll *a* + phaeophytin) – reading 2 (phaeophytin).

This value then needs to be multiplied by a machine-specific acid factor (usually around 0.9) and the chlorophyll *a* specific fluorescence (fluorescence per unit chlorophyll) to give the real chlorophyll *a* concentration in μg chl*a* $l^{-1}$.

This value often has to be converted to an areal concentration. To do this Chl*a* (mg chl*a* $m^{-2}$) = V/v × 1/(3.1459×$R^2$) /1000 /100.

Where V = total volume of ice melt plus filtered seawater
v = volume of melt filtered,
R = radius of ice core in meters,
/1000 to convert µg to mg,
/100 to compensate for extracting 10 ml of solvent not a liter.

## Spectrophotometer

The spectrophotometer is an older method than the fluorometer but is still widely used. While it is an order of magnitude less sensitive than the fluorometer, it has the advantage that it can simultaneously measure Chl*a*, Chl*b,* and Chl*c*. Unlike the fluorometer there is not a problem with interference from Chl*b*.

Samples are extracted in a solvent as described for the fluorometer. After extraction the sample is placed in the spectrophotometer and the absorption measured at wavelengths of 663, 645, and 630 nm (Parsons and Strickland 1963).

The concentration of Chl*a* is calculated by

$$\text{Chl}a\text{ (ug/L)} = [11.64 \times(\text{abs } 663) - 2.16 \times(\text{abs } 645) +0.10 - (\text{abs } 630)] \times E(F)$$
$$V(L). \quad \text{(Equation 3.8.2)},$$

Where F = Dilution factor,
E = Volume of solvent used,
V = Volume of sample filtered,
L = The cell path length (usually 1 cm), (abs XXX) = absorption at XXX nm.

## HPLC

The most accurate method of pigment concentration determination is by high performance liquid chromatography (HPLC). This instrument can accurately measure up to one hundred different algal pigments. However, this is not an instrument that is routinely used in the field. Samples should be collected in the same way as for the fluorometer and the frozen filters returned to the laboratory for later analysis. Samples for HPLC pigment analysis need to be stored at –80°C or colder.

### *Other Pigments*

Sea ice algae and phytoplankton also contain additional accessory pigments, some of which are used for expanding the wavelength of light capture (e.g., fucoxanthin) and others for photoprotection. Photoprotection pigments include the microsporine-like amino acids (MAAs), which are used for UV protection, and the xanthophyll pigments (diadinoxanthin is converted to diatoxanthin), which are used to protect against excess energy absorption. For most of these pigments it is necessary to use an HPLC.

### 3.8.2.5 Primary Production and Photosynthesis

Carbon fixation by marine plants, i.e., primary production, determines the amount of energy available to all trophic levels in a marine environment. Consequently, it is one of the most fundamental measurements that can be made in any ecological study.

Sea ice poses special problems for obtaining primary production measurements. Traditionally, the micro algae have been extracted from the ice by melting and the cells then treated like phytoplankton. However, in sea ice micro algae live in substrate-bound communities that are usually associated with steep gradients of light, nutrients, and dissolved gases. Melting the ice destroys these gradients and creates an artificial homogeneous environment. In situ studies, though, are extremely challenging because accessing the algae through the hard ice substrate is very difficult.

Primary production and photosynthesis has usually been assessed by measuring either the amount of $CO_2$ fixed or the amount of $O_2$ evolved. For both these measurements the algae needs to be extracted from the ice by melting either before or after the measurement. Particular care needs to be taken with all physiological measurements. Melting the ice potentially exposes the cells to unnatural elevated temperatures and low salinities, and so the ice must be melted into an equal quantity of cold-filtered seawater to ensure minimal temperature and osmotic shock. Cell physiology starts to change as soon as the algae are removed from their natural environment and so all measurements must be made as soon as possible after sampling. These samples cannot be stored and used later. Care must also be taken to avoid exposing the cells to unnatural light levels so most of this work needs to be done in the dark.

Once the samples have been prepared, standard phytoplankton procedures for $^{14}C$ or $O_2$ incubations can be followed.

#### In Situ Methods

The most successful in situ methods of sea ice primary productivity measurement have been based on micro sensor techniques. Typically an oxygen electrode or the more recently developed optodes are used. The method measures the flux of oxygen from the underside of the ice and interprets this in terms of productivity. Measurements have been made with either a diver-assisted lander or by extended arms through ice holes.

Beneath the ice there is a thin diffusive boundary layer, approximately 0.1–2.0 mm thick. All gases and nutrients must pass through this boundary layer. As oxygen is one of the gases involved in photosynthesis, its measurement enables productivity to be calculated. The electrode needs to be stepped through this boundary layer at as small an interval as possible, preferably at no more than 10 μm spacing. From these measurements the thickness of the boundary layer and the oxygen flux is estimated. The oxygen diffusion is calculated as:

$$J = D_0 \frac{d[O_2]}{dx} \quad \text{(Equation 3.8.3),}$$

where $D_0$ is the diffusivity constant, which varies with temperature, salinity, and pressure. At –1.88 and 34 psu it equals 1.11 x $10^{-5}$ $cm^2 s^{-1}$.

The electrodes produce a signal in either milliamps or volts (or fluorescence units if optodes are used) and these need to be converted to oxygen concentration. This is most easily done with a two-point calibration. A small aquarium pump is used to bubble seawater at ambient temperature (~1.8°C) for twenty minutes. This provides air-saturated water and the concentration of oxygen at any given temperature and salinity is known. The concentration is measured with the electrode and the value recorded. The second point is provided by deoxygenating the same water with sodium sulphite. The sodium sulphite rapidly strips the oxygen from the water providing a 0 percent oxygen concentration value. This water is then also measured and the value recorded.

This method has produced very good results from fast and pack ice in both the Arctic and Antarctic (McMinn et al. 2000). Its strength is that it produces rapid, real-time measurements of productivity at intervals of less than two minutes. As it never actually has to touch the ice, there are minimal handling problems. The disadvantages are that it can only be used reliably on bottom communities, the equipment is not commercially available, and it is difficult to construct and deploy.

An alternative approach has been to use an in situ incubation of ice core segments within the original ice core. A sea ice core is taken, sectioned into 1 to –5 cm thick sections placed into incubation chambers (Mock and Gradinger 1999). A tracer such as $^{14}C$ is added and the ice is returned to the core hole. This is then left for up to twenty-four hours, after which it is retrieved. The samples are melted and filtered, and production of fixed $^{14}C$ is measured by standard methods. This method has the advantage of being able to measure productivity at all levels within the ice and exposing the biota to natural light and temperature regimes. However, it also has the disadvantages of handling, tracer absorption, and incubation chambers affecting light transmission and bottle effects.

### *Fluorescence (PAM)*

Active fluorescence is now widely used to measure photosynthesis and photosynthetic stress in marine algal communities. As outlined before all plants contain chlorophyll *a*, which produces red fluorescence. The chlorophyll molecule absorbs light and either passes the energy on as photosynthesis of re emits it as fluorescence, heat or the chlorophyll molecule is damaged. The changes of chlorophyll *a* fluorescence at different actinic light levels can be used to investigate photosynthetic physiology. If a cell has been in the dark, all its photochemical reaction centers will be "open" and ready to receive a photon. Any incoming photons will be absorbed and passed on for photosynthesis, thus fluorescence will be at a minimum. The process of passing on the energy (as electrons) takes a small but finite length of time. If another photon were to strike the same reaction center microseconds after the first, the reaction centre would effectively be "busy." It wouldn't be able to process the energy and so fluorescence would increase with an increase of actinic light and saturation of the reaction centers. Thus the change in fluorescence is proportional to photosynthesis and can be used to measure it.

Another important measurement that can be made is the level of stress on photosynthesis. If a cell is stressed through lack of nutrients, excess light, UV, toxins, etc., it will close reaction centers and its ability to photosynthesize will decrease. This stress level is notated in the stress parameter Fv/Fm (variable fluorescence/maximum fluorescence). Healthy phytoplankton typically have Fv/Fm values of 0.65. Nutrient-limited phytoplankton can have values below 0.2. This ratio applies not only to sea ice algal cells but to any photosynthetic organisms that contain photosystem II. Sea ice samples rarely have values greater than 0.5, although in culture they also approach 0.65.

The change of photosynthetic performance over the naturally occurring light range can be assessed in rapid light curves (RLC) (Figure 3.8.7), which are another valuable data source from PAM fluorometers. An RLC is the fluorescence equivalent of the Productivity vs. Irradiance (P vs. E) curve that is used widely in biological oceanography. The sample is taken through a series of increasing irradiances, and the resulting photosynthesis (measured as electron transport rate [ETR]) recorded. From these curves it is possible to calculate the photosynthetic parameters rETRmax (maximum photosynthetic rate, ~Pmax), $\alpha$ (photosynthetic efficiency), and $E_k$ (photoadaptive index) using curve fits similar to what is used in traditional P vs. E curves.

P vs. E data from both traditional incubations and RLCs are normally fitted to an exponential function (Jassby and Platt 1976) using a nonlinear multiple regression to estimate the photosynthetic parameters:

$$P = Pmax\ (1-e^{(-\alpha E/Pmax)}) \quad \text{(Equation 3.8.4)},$$

where $P$ (ETR) = productivity, $Pmax$ (ETRmax) = maximum productivity, $\alpha$ = photosynthetic efficiency, and $E$ = irradiance.

We have used these instruments to examine the effects of photo inhibition, high and low salinity, and low temperatures on sea ice algal communities (Ralph et al. 2005). They have also been used to examine photosynthesis in both the Arctic (McMinn et al. 2008) and the Antarctic (McMinn et al. 2003).

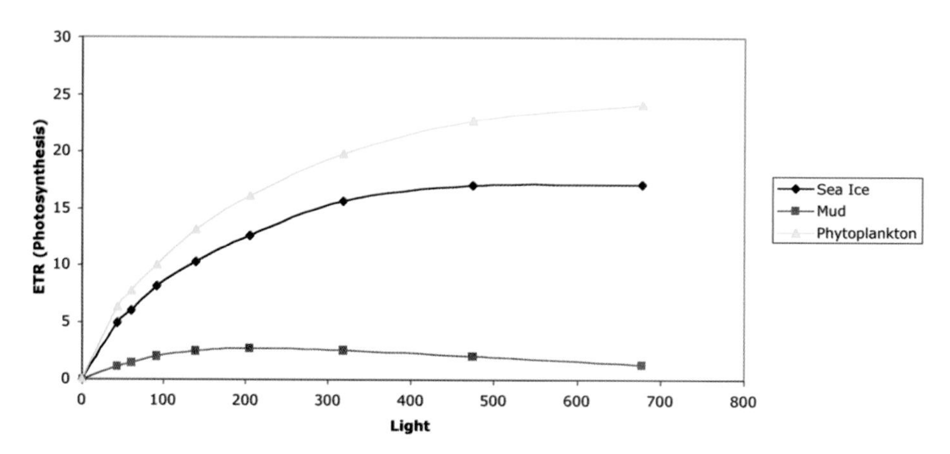


**Figure 3.8.7.** RLCs from ice-covered Saromo-ko Lagoon, northern Hokkaido. Curves were generated from ice, water, and sediment-surface samples taken at the same time.

## Measurement of Sea Ice Algal Photophysiological Parameters in Variable Fluorescence Fluorometers

Great care must be taken when sampling to avoid all exposure to light. As ice cores are being extracted, both ends of the corer need to be covered and the core placed in darkness. Samples can be measured while on the ice if care is taken. Ice shavings are taken and put in the quartz cuvette and placed in a specialized fluorometer (Figure 3.8.8) with sufficient sensitivity like the WaterPAM (Waltz). Fv/Fm values are obtained by hitting the start button. Alternatively, an RLC can be run. The data are saved to the control unit and downloaded later. These instruments can be run from laptop control, but this is unwise in the field where cold and saltwater are likely to disable the computer.

Each new measurement requires a new sample. For many types of analysis the samples need to be dark-adapted. This involves placing the sample in complete dark for fifteen to thirty minutes prior to measuring at temperatures close to in situ values. Samples can be transferred immediately to the cuvette and left in the instrument for this time if this practical. Dark adaptation gives a measure of maximum potential photosynthesis while samples measured immediately give a snapshot of

**Figure 3.8.8.** A WaterPAM (Walz, Germany) set up for use on sea ice.

the photosystem at the time of sampling. Both sets of data are valuable and it is important to clearly document the approach used for each set of measurements.

RLC data is subsequently uploaded from the control unit and saved. The file is then imported to a statistics program and a multiple nonlinear regression done on the light and ETR data to generate the photosynthetic parameters.

### 3.8.2.6 Estimation of Sea Ice Heterotrophic Activity

While several studies have directly measured sea ice primary production, no detailed measurements exist from neither the Arctic nor Antarctica regarding the grazing of ice organic matter by sea ice meiofauna. The few estimates that exist are based solely on the application of allometric equations established for a wide range of marine invertebrates. These equations relate the biomass of an individual animal to its metabolic rate, which includes respiration, excretion, and ingestion based on either nitrogen or carbon units. A common set of equations used in several arctic and antarctic sea ice studies was published by Moloney and Field (1989). When interested in the sea ice carbon flow, the potential ingestion rates of the ice meiofauna can be calculated using the allometric mass-specific ingestion rate and by assuming an ice temperature of –1°C and a $Q_{10}$ value of 2 (typical for plankton metazoans):

$$I_{max} = 63 \times M^{-0.25} \times 0.23326 \quad \text{(Equation 3.8.5)},$$

with $I_{max}$ ($d^{-1}$) as daily mass-specific maximum potential ingestion rate and M (pg C) as body mass of one organism. Estimating the community grazing by sea ice herbivores requires knowledge about their abundances and biomass, both of which can be assessed based on information from ice core samples (see Chapter 3.9). Table 3.8.1 shows that grazing by sea ice meiofauna is removing less than 10 percent of the algal daily primary production during the spring–summer period.

The second major consumers of sea ice–derived organic matter are various amphipod species living either seasonally or permanently in association with sea ice in the Arctic and Antarctic. Estimates on their grazing pressure and carbon consumption are based on observation of their abundances (see diversity chapter) and bottle experiments. A typical experimental design follows a modified Frost grazing experimental scheme (e.g., Werner 1997), where amphipods feed on Chl*a*-containing sea ice for a set time. At the end of the experiment, the decrease in Chl*a* is determined and compared to a control of sea ice without amphipods present. This approach works well for herbivorous species like *Apherusa glacialis* or juveniles of *Gammarus wilkitzkii*, but will underestimate the carbon consumption of omnivorous or carnivorous taxa (e.g., adult *G. wilkitzkii*) that also feed on, for example, pelagic food items.

*Bacterial Activity in Sea Ice*

Bacterial biomass within the sea ice can exceed 50 percent of the total at times. Several observations have shown that bacterial activity can actually exceed primary productivity for certain periods and can occur at temperatures below –20°C. Nevertheless, we know very little about the functional role of the bacterial communities in polar seas, which makes it an exciting field of research for the next decades. The determination of bacterial activity in sea ice largely follows the scheme outlined in the primary productivity section. Ice cores are melted and various tracers have been used to assess their activity in lab experiments where the incorporation of, for example, [3H]leucine into proteins to [3H]thymidine incorporation into DNA are determined. In addition, tracer additions have been conducted to thin ice segments with incubations within the ice core, again providing a more realistic environmental regime for the activity measurement, specifically for the ice interior parts.

**Table 3.8.1.** Estimated daily (mgC $m^{-2}$ $d^{-1}$) and yearly (gC $m^{-2}$ $y^{-1}$) maximum ingestion rates of arctic and antarctic pack ice meiofauna in comparison to sympagic primary production rates (gC $m^{-2}$ $y^{-1}$). Modified after Gradinger (1999).

| Season/author | Duration (days) | Arctic | Weddell Sea |
|---|---|---|---|
| Spring (per day) | 92 | 0.27 | 0.96 |
| Summer (per day) | 91 | 0.27 | 8.15 |
| Autumn (per day) | 91 | 0.21 | 1.91 |
| Winter (per day) | 91 | 0.21 | 1.45 |
| **Yearly ingestion rate** | | 0.1 | 1.1 |
| *Yearly primary production rates* | | | |
| Kirst and Wiencke (1995) | | 0.7 | 1.9 |
| Arrigo et al. (1997) | | | 6.7 (for 7 months) |
| Gosselin et al. (1997) | | 4.0 | |

### 3.8.2.7 Measurement of the Air–Sea Ice $CO_2$ Flux and Dissolved Organic Carbon in Sea Ice

While carbon dioxide ($CO_2$) is the primary gas involved in the exchange of carbon between ocean and atmosphere, sea ice has been thought to prevent this exchange. Consequently, carbon cycle models have typically not included $CO_2$ exchange between the two through sea ice. However, recently a number of studies have reported the possibility of $CO_2$ exchange through the sea ice (Nomura et al. 2006; Zemmelink et al. 2006). These results suggest that the air–sea ice $CO_2$ flux can strongly influence the annual global carbon budget (Delille 2006).

The main components of dissolved organic matter (DOM) are dissolved inorganic carbon (DOC), dissolved organic nitrogen (DON), and dissolved organic phosphorus (DOP). DOM is defined as the material that can pass through a filter, commonly 0.45 μm. Here, we will focus on the DOC in sea ice. DOC concentrations in sea ice are generally higher than those from under the ice due to the selective retention of DOC during ice formation and organisms living in sea ice. These results suggest that the contribution of the sea ice DOC to carbon cycling in the ocean is significant.

Here, we explain how to measure the $CO_2$ flux between the atmosphere and sea ice in the field. We will also show how to measure the gas exchange through the sea ice by measuring physical properties of sea ice, the snow deposited on top of the sea ice, the $pCO_2$ in the brine and the under-ice water, and the DOC in sea ice.

## *Air–Sea Ice $CO_2$ Flux Measurements*

There are two kinds of method to measure the air–sea ice $CO_2$ flux: chamber techniques (e.g., Delille 2006) and eddy correlation technique (e.g., Zemmelink et al. 2006). Here, we will demonstrate the chamber technique as conducted in the Arctic Ocean and the southern part of the Okhotsk Sea.

The $CO_2$ measuring system to examine the air–sea ice $CO_2$ flux consists of a stainless-steel chamber, a nondispersive infrared gas (NDIR) analyzer (GMP343 Carbon dioxide probe, Vaisala Ojy, Finland), a chemical desiccant column ($Mg[ClO_4]_2$, CDC), a diaphragm pump, a mass flow controller (MFC), and a data logger (DL) (Figures 3.8.9 and 3.8.10). A stainless-steel chamber, 50 cm diameter and 30 cm in height, which has a saw at the bottom of the chamber to minimize gas leak through the chamber-ice interface, is used. The chamber is attached to the data logger (DL) with Teflon tube connectors that introduce sample air into the NDIR. Temperature in the chamber is measured during each estimation with a temperature sensor. The air sample in the chamber is circulated at a flow rate of 0.7 L $min^{-1}$. These devices are installed in the portable adiabatic box (40 × 25 × 30 cm). The electricity is provided with an electric generator (EG) or AC source. The $CO_2$ concentration in the chamber is measured every ten seconds during the forty

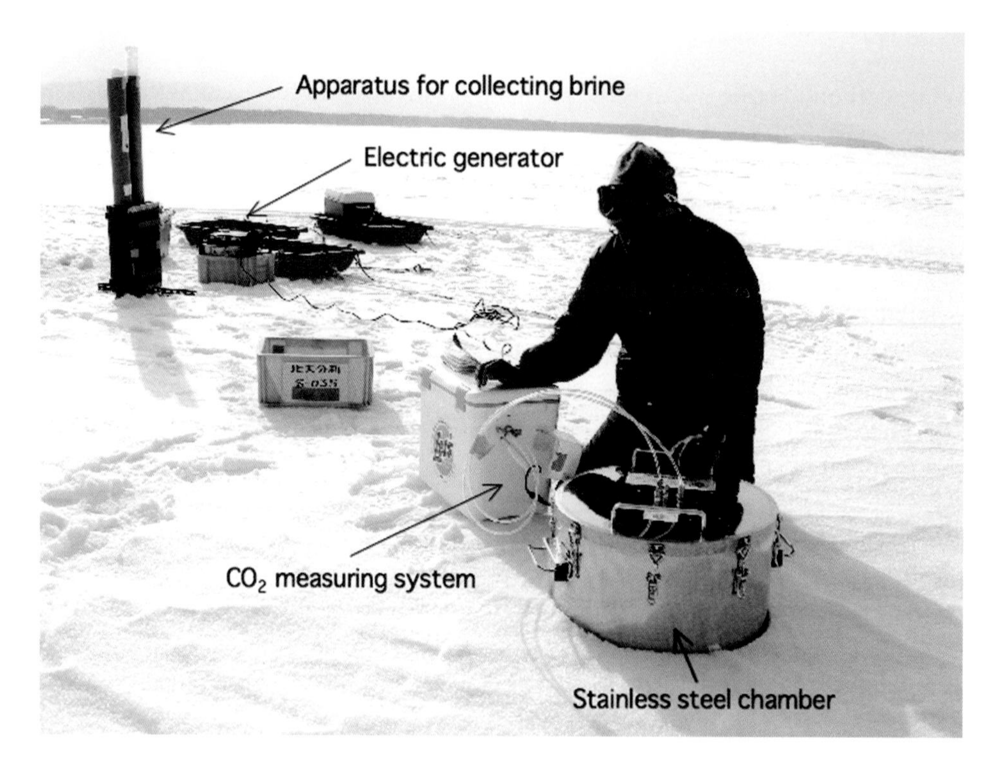


**Figure 3.8.9.** Technique for measuring the $CO_2$ flux using a chamber at Saroma-ko lagoon, connecting to the Sea of Okhotsk, on the northern coast of Hokkaido, Japan, on March 2006

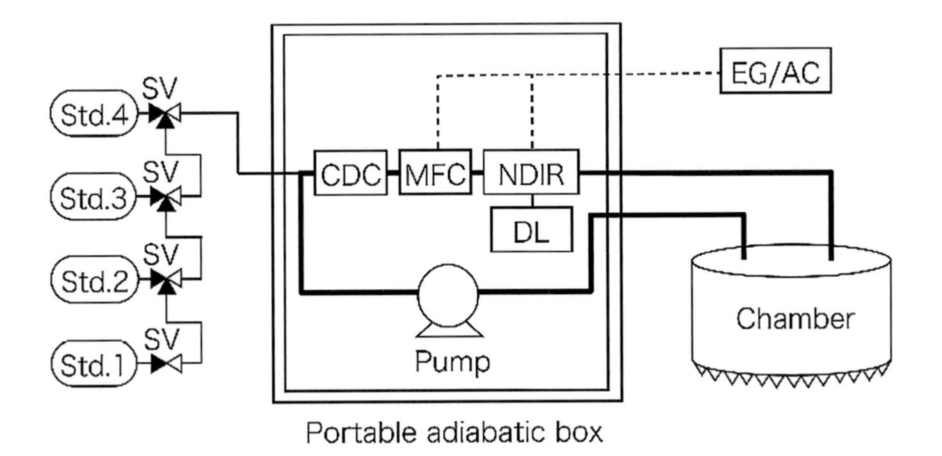


**Figure 3.8.10.** Schematic diagrams of the $CO_2$ measuring system and a stainless-steel chamber. Solid thick line indicates the closed circuit that circulates air to measure the air $CO_2$ concentration in the chamber.

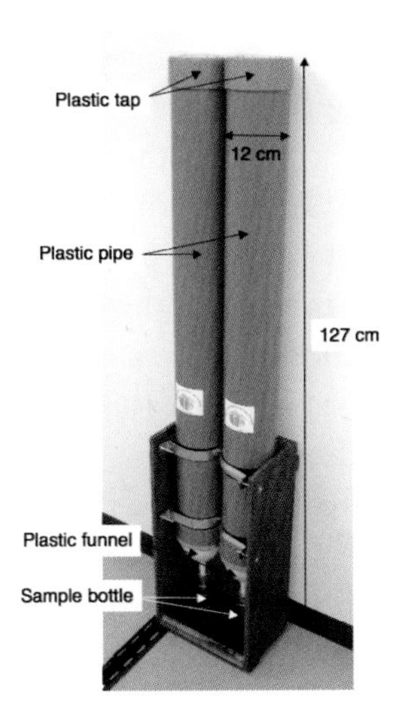


**Figure 3.8.11.** Apparatus for collecting brine samples in sea ice by the gravity drainage method. From Nomura et al. (2009).

minute experiment. Four standards (typically 324 ppm, 341 ppm, 363 ppm, and 406 ppm $CO_2$ in natural air) are used to calibrate the $CO_2$ measuring system prior to the field measurements. The $CO_2$ flux is calculated from the changes in $CO_2$ concentration in chamber against to the elapsed time.

## Sampling Methods for Sea Ice, Brine in Sea Ice, and Under-Ice Water

Samples of sea ice core, brine in sea ice, and under-ice water are collected to measure the physical properties of sea ice, DOC concentration in sea ice, and the $pCO_2$ in the brine and under-ice water.

Sea ice cores are collected with an ice-core sampler (see Chapter 3.3). Immediately after the ice core is collected, ice thickness and temperature are measured by scale and inserting a needle-like thermometer sensor into a drilled hole of the core, respectively. Thereafter, the ice cores are kept in a polythene bag and in the freezing room. For DOC samples, in particular, we have to take care not to contaminate ice cores as shown in the general introduction of this chapter.

Sea ice brine from thicker ice is collected by the sackhole method (see above) in polar seas and during laboratory experiments. The seeped brine at the bottom of the sackhole in sea ice is carefully sampled using a plastic syringe. The brine extracted from the ice core for measuring brine salinity is placed into a 10 ml glass vial and into a 120 ml glass vial for measuring dissolved inorganic carbon (DIC) and brine total alkalinity (TA). 50 μl mercuric chloride ($HgCl_2$) is added to the DIC sample to stop biological activity.

During the ice-melting season, however, the sea ice is relatively thinner and temperatures and brine volume fractions are higher. At this time, under-ice water frequently penetrates up into the sackhole through the well-developed network of brine channels. Use of the sackhole method is therefore inappropriate during this season. As an alternative, the gravity drainage method can be used for brine collection (Nomura et al. 2009). A schematic diagram of the apparatus is shown in Figure 3.8.11. Immediately after the ice cores are collected, they are stored in plastic pipes, and the brine in the sea ice is extracted through brine channels by gravity into a 200 ml sample bottle for about ten minutes. The ice cores are kept in plastic pipes under ambient conditions for about ten minutes without exposing them to the atmosphere, where the temperature difference between air and seawater could cause freezing or melting of the ice core. The brine samples are stored in the same manner as those collected by the sackhole method.

The under-ice water sample is collected from 1 m below the surface of the sea ice with a 500 ml Teflon water sampler through an ice core hole. The water samples are collected approximately fifteen minutes after coring to avoid the disturbance caused by the drilling. The under-ice water samples are kept in the same manner as that for the brine samples for further analysis.

### Sea Ice Sample Preparation

To estimate the brine volume fraction, an ice-core sample is cut to a 4.5 × 2.5 cm block by a band saw and then into 3 cm thick slices in a cold room at –15°C. The bulk volume and mass of the 3 cm thick sea ice section is measured to calculate the density of sea ice. The sea ice sections are then melted to measure salinity, and finally the brine volume fraction of sea ice is calculated using the equations by Eicken (2003).

The procedure for the measurement of the DOC concentration in sea ice is that of Thomas et al. (1995). First, a 1 to 2 cm outer layer of the ice core is cut off with a band saw to remove contamination; it is then cut into a 4.5 × 2.5 cm block by a band saw and then sliced into a 3 cm thick section in a cold room at –15°C. These sections are also trimmed with an acid-washed ceramic knife and placed into an acid-washed Teflon container and melted at 4°C. The melted water samples are filtered through Whatman GF/F (glass fiber filters) into glass vials. Both filter and glass vial are precombusted for five hours at 450°C.

### Sample Analysis

Salinity of the brine and under-ice water is measured using a salt analyzer (SAT-210, Toa Electronics Ltd., Japan). DIC is determined by the coulometry technique (Johnson et al. 1985). DIC measurements are calibrated with working seawater standards traceable to the Certified Reference Material. The precision of DIC analyses from duplicate determinations is within ±0.1 percent (Wakita et al. 2005). TA is analyzed by the improved single-point titration method (Culberson et al. 1970) and measured by using a glass electrode calibrated with Tris buffer and 2-aminopyridine buffer (DOE 1994). The precision of the TA analysis from duplicate determinations is within ±0.2 percent (Wakita et al. 2005). By measuring two variables describing the carbonate system in seawater, we can calculate the $pCO_2$ in the brine and under-ice water. The $CO_2$ solubility is given as a function of temperature and salinity (Weiss 1974). Equilibrium constants of carbonic acids are also given as a function of temperature and salinity. The equilibrium constants reported earlier as functions of temperature and salinity may have lead to results that deviated from typical sea ice conditions. DOC concentration in sea ice is measured by high-temperature combustion oxidation as described in Granskog et al. (2004).

## 3.8.3 CONCLUSION AND OUTLOOK

Sea ice ecosystems are critical to all life in polar seas. Contributing on average 25 percent of the primary production of ice-covered seas, they are critical to overwintering zooplankton communities, provide a major pulsed input to underlying benthic invertebrate communities, and seed springtime ice-edge blooms. Sea ice ecosystems are highly structured and dynamic. Steep physical and biochemical gradients have profound effects on the entrained biological communities. Nutrient concentrations are strongly influenced by distance from the ice-water interface and biological activity. $CO_2$ concentration, although never limiting in the ocean, can be severely limiting in the narrow confines of brine channels. Biological activity within the ice is affected by light transmittance, temperature, brine salinity, and nutrient concentration. This activity itself also strongly modifies some of these same parameters.

Measurements of many biochemical and biological parameters within sea ice have long presented challenges. In so many cases the ice has to be melted to make the measurements and the melting process itself often radically affects the integrity of the measurement. Finding ways to measure key gradients and parameters without having to sample and/or melt the ice will continue to be sought. In areas of biomass and productivity there has been some success using emerging fluorescent and microsensor technologies, but these still have some unresolved issues. Unfortunately, unlike the open ocean where many physical and biological parameters can be measured remotely, snow cover on the surface of ice largely prevents these measurements being possible. A recent approach showing some promise, however, has been to measure biomass and other properties from underneath the ice using semiautonomous vehicles.

The sea ice zone in both the Arctic and the Antarctic is undergoing unprecedented change. Understanding the biochemical and biological processes occurring now will enable us to better anticipate likely changes in the future.

## REFERENCES

Arrigo, K. R., M. P. Lizotte, D. L. Worthen, P. Dixon, and G. Dieckmann (1997), Primary production in Antarctic sea ice, *Science*, *276*, 394–397.

Culberson, C., R. M. Pytkowicz, and J. E. Hawley (1970), Seawater alkalinity determination by the pH method, *J. Mar. Res*,. *28*, 15–21.

Delille, B. (2006), Inorganic carbon dynamics and air-ice-sea $CO_2$ fluxes in the open and coastal waters of the Southern Ocean, Ph.D. thesis, University of Liége, Belgium.

DOE (1994), *Handbook of Methods for the Analysis of the Various Parameters of the Carbon Dioxide System in Sea Water*, Version 2, edited by A. G. Dickson and C. Goyet, ORNL/CDIAC-74, Oak Ridge National Laboratory, Oak Ridge, TN.

Eicken, H. (2003), From the Microscopic to the Macroscopic to the Regional Scale: Growth, Microstructure, and Properties of Sea Ice, in *Sea Ice: An Introduction to Its Physics, Chemistry, Biology, and Geology*, edited by D. N. Thomas and G. S. Dieckmann, pp. 22–81, Blackwell Science, London.

Gosselin, M., M. Levasseur, P. A. Wheeler, R. A. Horner, and B. C. Booth (1997), New measurements of phytoplankton and ice algal production in the Arctic Ocean, *Deep-Sea Res.*, *44*, 1623–1644.

Gradinger, R. (1999), Integrated abundance and biomass of sympagic meiofauna in Arctic and Antarctic pack ice, *Polar Biol.*, *22*, 169–177.

Granskog, M. A., K. Virkkunen, D. N. Thomas, J. Ehn, H. Kola, and T. Martma (2004), Chemical properties of brackish water ice in the Bothnian Bay, the Baltic Sea, *J. Glaciol.*, *169*, 292–302.

Grasshoff, K., M. Ehrhardt, K. Kremling, and L. G. Anderson (1999), *Methods of Seawater Analysis*, Wiley-VCH, New York.

Jassby, A. D., and T. Platt (1976), Mathematical formulation of the relationship between photosynthesis and light for phytoplankton, *Limnol. Oceanog., 21,* 540–547.

Johnson, K. M., A. E. King, and J. M. Sieburth (1985), Coulometric $TCO_2$ analyses for marine studies: An introduction, *Mar. Chem., 16,* 61–82.

Kirst, G. O., and C. Wiencke (1995), Ecophysiology of polar algae, *J. Phycol., 31,* 181–199.

Legendre, L. S. F. Ackley, G. S. Dieckmann, B. Gulliksen, R. Horner, T. Hoshiai, I. A. Melnikov, W. S. Reeburgh, C. Spindler, and C. W. Sullivan (1992), Ecology of sea ice biota 2. Global Significance, *Polar Biol., 12,* 429–444.

McMinn, A., C. Ashworth, and K. Ryan (2000), In situ net primary productivity of an Antarctic fast ice bottom algal community, *Aquat. Microbial. Ecol., 21,* 177–185.

McMinn, A., H. Hattori, T. Hirawake, and A. Iwamoto (2008), Preliminary investigation of Okhotsk Sea ice algae; Taxonomic composition and photosynthetic activity, *Polar Biol., 31,* 1011–1015.

McMinn, A., K. Ryan, and R. Gademann (2003), Photoacclimation of Antarctic fast ice algal communities determined by pulse amplitude modulation (PAM) fluorometry, *Marine Biol., 143,* 359–367.

Mock, T., and R. Gradinger (1999), Changes in photosynthetic carbon allocation in algal assemblages of Arctic sea ice with decreasing nutrient concentrations and irradiance, *Mar. Ecol. Prog. Ser., 202,* 1–11.

Moloney, C. L., and J. G. Field (1989), General allometric equations for rates of nutrient-uptake, ingestion, and respiration in plankton organisms, *Limnol. Oceanog., 34,* 1290–1299.

Nomura, D., Y. H. Inoue, and T. Toyota (2006), The effect of sea ice growth on air-sea $CO_2$ flux in a tank experiment, *Tellus, 58B,* 418–426.

Nomura, D., T. Takatsuka, M. Ishikawa, T. Kawamura, K. Shirasawa, and H. Y. Inoue (2009), Transport of chemical components in sea ice and under-ice water during melting in the seasonally ice-covered Saroma-ko Lagoon, Hokkaido, Japan, *Estuarine, Coastal and Shelf Science, 81,* 201–209.

Parsons, T. R., and J. D. H. Strickland (1963), Discussion of spectrophotometric determinations of marine plant pigments, with revised equations for ascertaining chlorophylls and carotenoids, *J. Mar. Res., 21,* 155–163.

Ralph, P. J., A. McMinn, K. G. Ryan, and C. Ashworth (2005), Effect of salinity and temperature on the photokinetics of brine channel algae, *J. Phycol., 41,* 763–769.

Smith, R. E. H., J. Anning, P. Clement, and G. Cota (1988), Abundance and production of ice algae in Resolute Passage, Canadian Arctic, *Mar. Ecol. Prog. Ser., 48,* 251–263.

Thomas, D. N., R. J. Lara, H. Eicken, G. Kattner, and A. Skoog (1995), Dissolved organic matter in Arctic multi-year sea ice during winter: major components and relationship to ice characteristics, *Polar Biol., 15,* 477–483.

Wakita, M., S. Watanabe, Y. W. Watanabe, T. Ono, N. Tsurushima, and S. Tsunogai (2005), Temporal change of dissolved inorganic carbon in the subsurface water at station KNOT (44°N, 155°N) in the western North Pacific subpolar region, *J. Oceanog., 61,* 129–139.

Weiss, R. F. (1974), Carbon dioxide in water and seawater: The solubility of a non-ideal gas, *Mar. Chem., 2,* 203–215.

Weissenberger, J., G. Dieckmann, and R. Gradinger (1992), Sea ice—A cast technique to examine and analyze brine pockets and channel structure, *Limnol Oceanog., 37,* 179–183.

Werner, I. (1997), Grazing of Arctic under-ice amphipods on sea ice algae, *Mar. Ecol. Prog. Ser., 160,* 93–99.

Zemmelink, H. J., B. Delille, J. L. Tison, E. J. Hintsa, L. Houghton, and H. Decey (2006), $CO_2$ deposition over the multi-year ice of the western Weddell Sea, *Geophys. Res. Lett., 33,* doi:10.1029/2006GL026320.

**Chapter 3.9**

# Assessment of the Abundance and Diversity of Sea Ice Biota

*Rolf Gradinger and Bodil Bluhm*

## 3.9.1 INTRODUCTION

While the substantial contribution of sea ice to polar marine primary productivity and its relevance for the marine food web, including charismatic megafauna such as polar bears and seals (Chapter 3.10), is well established, much less is known of the hidden life within the brine channel system and at the bottom of the ice. The close association of living organisms with sea ice, however, has been documented for both polar seas for more than a century (Horner 1985). Today, several hundred species of viruses, bacteria, fungi, algae, and uni- and multicellular animals have been described as living inside, on top of, or attached to the bottom of the sea ice (see Figure 3.9.1 for examples). However, this inventory is far from complete and new species of bacteria, primary producers, and animals are regularly described from the ice environment. For example, one study in Antarctica revealed a total of forty-six species of ciliates in sea ice, with seventeen of these completely new to science (Petz et al. 1995). Very recently the first-ever record of hydrozoans, in the form of a new genus and species, living within arctic sea ice was published (Bluhm et al. 2007). Thus, new discoveries can be expected in the next decade.

In this chapter, we focus on appropriate field methodology to assess the abundance and diversity of sea ice flora and fauna. The chapter will deal with aspects unique to sea ice, with some comments regarding biological sampling design, which is complicated in its own right and covered in available textbooks (such as Bakus 2007). Most of the sea ice biota living inside the ice are of microscopic nature given the small sizes of available spaces with brine channel diameters of typically well below 500 μm (Krembs et al. 2000). Microscopic techniques have therefore been the traditional tools in sea ice diversity studies, but molecular tools focusing on genetic information obtained from ribonucleic acid (RNA) and desoxyribonucleic acid (DNA) sequence analyses are of increasing relevance (e.g., Kaartokallio et al. 2008). The fauna and flora using the bottom of sea ice as a hard substrate can be considerably larger, and specifically macrofauna (dominated by amphipods,

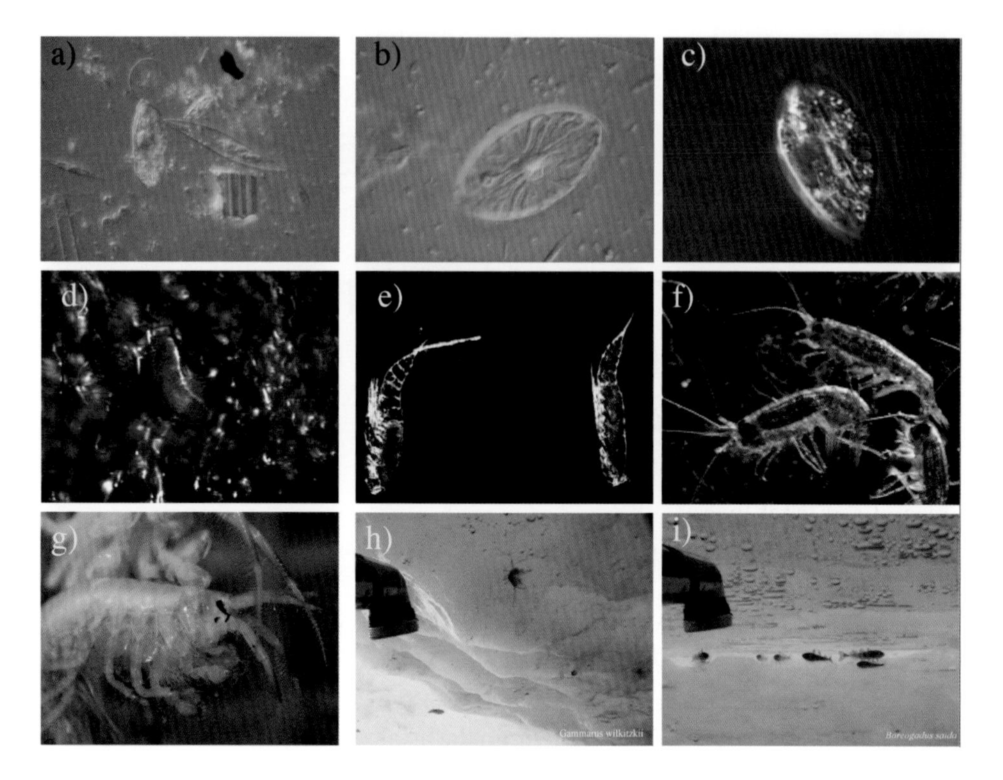


**Figure 3.9.1.** Diversity of biota occurring in sea ice. All examples are from arctic studies. (a) Overview shows diatoms, flagellates, and various aggregates; (b) pennate diatom (60μm long); (c) heterotrophic flagellate (Anisonema sp.) with ingested pennated diatom; (d) bright red acoel turbellarian (400μm long) inside brine channel matrix; (e) harpacticoid copepod (450μm long); (f) amphipod, Apherusa glacialis (5mm); (g) amphipod, Gammarus wilkitzkii (female with juveniles in brood pouch; 4cm); (h) under-ice picture of G. wilkitzkii; and (i) under-ice picture of arctic cod, Boreogadus saida.

euphausiids, and fish) can be detected and counted without any microscopic tools. Therefore, we split this contribution into two parts—in-ice flora and fauna and sub-ice flora and fauna—as these two groups of organisms require different approaches for sampling and analysis. The emphasis is on the fieldwork component (including sampling considerations, field collection, and initial sample processing), as the analysis of samples itself by microscopy or other tools does not differ from the analysis in any other marine environment (e.g., Kemp et al. 1993). This contribution ends with some recommendations for data processing and our perspective on future directions in sea ice–related diversity research.

### 3.9.1.1 Sample Size and Variability with Regard to Sea Ice Biota

One of the most basic and important questions to cope with prior to any field study (in addition to safety concerns) is to estimate the number of stations and samples needed to address the scientific question or hypothesis. Biological fieldwork is very labor- and time-intensive, and time and access to sampling on the ice are often limited during expeditions. Thus, the number of sampling events should be determined prior to the fieldwork based on the best available estimates for a given set of variables. One useful statistical tool is an a priori power analysis, available in most statistical packages and discussed in standard textbooks (e.g., Barker Bausell and Li 2002). The power analysis provides insights on the probability of meaningful statistical results (i.e., rejecting a false null hypothesis) based on a given sample size using estimated means and variability and can be performed for, say, analysis of variance, t-tests, z-tests, etc. The few studies targeting the spatial and/or temporal variability in sea ice biological data revealed relatively small biological variability (often less than one order of magnitude) within a site with homogeneous environmental setting, as defined by key variables such as snow depth, ice thickness, porosity, and sediment load in ice (Stoecker et al. 2000; Gradinger and Bluhm 2005), while the spatial and temporal variability specifically of species abundances may exceed four orders of magnitude between stations in the same ocean basin (Smith et al. 1988; Gosselin et al. 1997; Haecky and Andersson 1999; Horner 1980; Michel et al. 2002; Steffens et al. 2006; Gradinger 2008).

In the following we will provide an example for the manifold applications of power analysis for sea ice biota sampling. In our example, we design a field experiment to test the hypothesis that the abundance of ice algal species is significantly different in winter (sampling event 1) compared to spring (sampling event 2). Based on published data we estimate that the difference between the mean abundances at these two sampling events might be five times larger than the standard deviation, which can be expressed as a standardized effect size of 5. Furthermore we set the Type 1 error level $\alpha$ (which indicates the likelihood to erroneously reject a correct null hypothesis that predicted no difference between the two sampling events) to 0.05. A power analysis (performed with the program SYSTAT 11) based on these conservative estimates reveals that a minimum of three true replicates per sampling event would be sufficient to detect significant differences between these locations using a two-sample t-test with a power (= probability to reject a false null hypothesis) of 0.9. True replicates are completely independent samples, so splitting a single ice core into three subsamples that are individually analyzed would not count as replicates but be considered pseudoreplication.

In general three true replicates for all biological parameters should be the minimum number of samples per site and variable to be determined. Based on personal experience, collecting triplicate ice core samples for measuring physical, chemical, and biological bulk parameters and collecting field measurements relevant for biota abundance and distribution can be achieved in less than five hours with a team of three people even in arctic pack ice (excluding pressure ridges). However, if environmental conditions are challenging or if thicker parts of ice floes (thickness > 3m) are targeted, one might have to adjust the sampling scheme to what is realistically achievable.

### 3.9.1.2 Assessing the Diversity of In-Ice Biota

The patchy distribution of sea ice biota in terms of abundance, biomass, and diversity can largely be attributed to the physical conditions (bottom-up control). Light availability, ice temperature, brine salinity, and nutrient concentrations or food availability (Chapter 3.8) have been characterized as main driving forces to explain the horizontal and vertical patchiness of ice biota. Any biodiversity study should thus aim to provide auxiliary information at each sampling location for the aforementioned variables and is thus automatically interdisciplinary in nature. At a minimum, snow depth and ice thickness (both determining light availability to ice algae), ice temperature as well as bulk salinity of melted samples should be determined. The methods for these measurements are described in Chapters 3.2 and 3.3. The measurement approaches for light, nutrients, and bulk biomass parameters (algal pigments, particulate organic matter) are outlined in Chapter 3.8.

*Field Sampling*

The primary tool for collecting samples for in-ice biota diversity assessment is an ice corer and explanations on how to use this tool are given in detail in Chapter 3.3 and in the video footage on the accompanying DVD. Additional basic gear needed for field sampling includes a saw, sample bags, or containers and coolers (Figure 3.9.2).

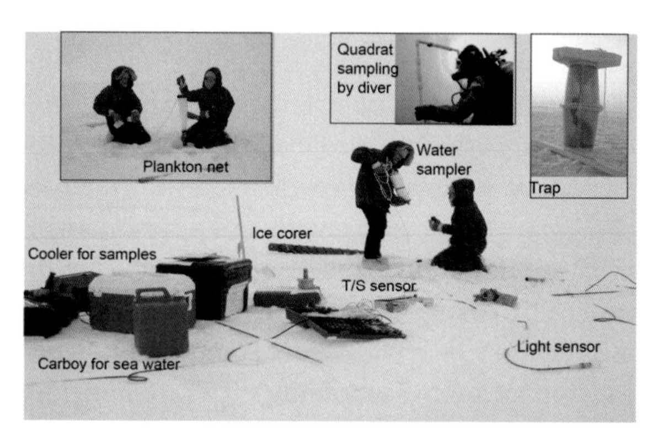


**Figure 3.9.2.** Example for equipment used in interdisciplinary field sampling. Picture taken in April 2002 on landfast ice near Barrow, Alaska. Quadrat sampling by diver. Picture: S. Harper.

Sea ice biota is not evenly distributed throughout the vertical extent of sea ice and different taxa and assemblages occur in layers at different depths (Figure 3.9.3) (v. Quillfeldt et al. 2003). In both the Arctic and Antarctic, biomass maxima occur in dense, only a few centimeter thick layers at the bottom of the ice (Horner 1985). In the Antarctic, a surface assemblage layer is also common (e.g., Stoecker et al. 2000). Ideally community compositions should be investigated over the entire core length with a vertical resolution similar to that used for bulk chemical, physical, and other biological parameters, often with high resolution near the bottom and lower resolution (e.g., 10 or 20 cm) higher up in the core. However, for practical reasons, most arctic and antarctic biological studies focus on the few centimeter thick dense ice algal layers (as illustrated in Chapter 3.8), which also typically contain the highest concentrations of sea ice animals (e.g., Nozais et al. 2001). Focusing on these dense layers might be an acceptable assumption prior to a study, but the biotic distribution should be verified for the region of interest during the first field visit, because high biotic concentrations in the ice interior have also been observed in the Arctic (Gradinger 1999). From the perspective of organism diversity, it should be noted that considerable diversity with unique physiological characteristics might be hidden in the less frequently studied low biomass regions (for examples see below), which often are exposed to the most extreme temperature and brine salinity regime. The upper parts of ice floes exhibit considerably higher environmental stress on biota due to the tremendously varying temperature and brine salinities ranging from very cold and saline in winter to relatively warm and fresh in summer. In antarctic fast ice, the dinoflagellates species *Polarella glacialis* excysts and grows at temperatures of about −3°C and salinities of 60 (Stoecker et al. 1997). An even more extreme example are arctic fast ice bacteria, which were active at even lower ice temperatures down to −20°C (Junge et al. 2004). Consequently, studies aiming at describing the regional diversity associated with sea ice should include collections over the entire ice thickness. This was nicely illustrated by v. Quillfeldt et al. (2003), who observed 135 algal species in the bottom 20 cm of one ice core in Chukchi Sea ice, but 237 vertically stratified species in a nearby second ice core that had been analyzed over its entire thickness.

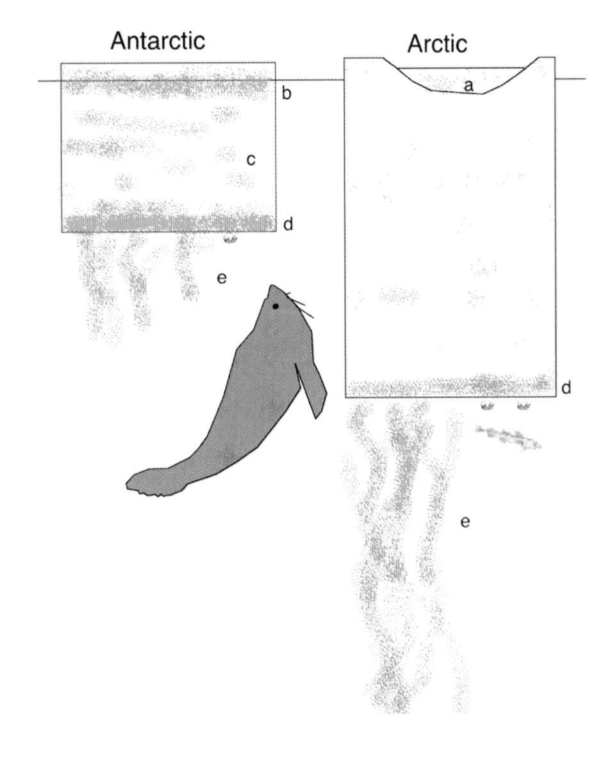


**Figure 3.9.3.** Vertical distribution of sea ice communities within arctic and antarctic sea ice: (a) melt puddle communities, (b) surface/ infiltration communities, (c) internal/ interior communities, (d) bottom communities, and (e) sub-ice communities. Green coloration indicates biomass concentration. In the sub-ice layer amphipods, fish, and seal are presented.

Once you have successfully collected a core, two considerations are of immediate urgency: (1) to get the core samples into a sealed container (to minimize brine loss), and (2) to maintain in situ temperature (to minimize osmotic stress). Immediately after coring, the core is placed horizontally on the ice or on a cutting board and sawed into 1 to 10 or 20 cm long segments, which are placed into clean (for microbiological studies: acid-cleaned) containers (see Chapter 3.8 for details). One-gallon-sized freezer-quality, resealable zipper storage bags available in most supermarkets have proven to be reliable for transporting samples back to the lab even during long snowmobile rides at ambient temperatures below –30°C. All bags need to be clearly labeled with location, core, and segment identification using permanent markers and stored in a dark, insulated cooler to prevent further cooling or warming of the sample. Temperature change can cause substantial changes in brine salinity and thus osmotic stress for biota. Alternatively, sections can be placed into solid plastic containers with lids.

*Melting of Ice Samples*

A major discrepancy exists in the literature with respect to the approach taken in further processing of the ice core segments for ice biota studies. Most arctic and antarctic working groups recommend adding between 0.1 and 0.5 l of 0.2 μm to 1 μm filtered seawater per 1 cm of ice core length to avoid osmotic stress for the biota (Garrison and Buck 1986). Pore size of the filters used for seawater filtration will depend on the targeted group. For bacterial studies 0.2 μm prefiltered seawater is required for the melting procedure, while for most other purposes seawater filtered through glass microfiber filters (such as Whatman GF/F filters typically used for algal pigment analysis; see Chapter 3.8 for details) is adequate. Some groups do not add any filtered seawater to the melting ice segments (e.g., for ice bacterial work in the Baltic Sea: Kartokallio et al. 2008). The adequate amount of seawater to be added should be determined for the region of interest and for the specific taxonomic group being targeted, and pre-experiments should verify that the chosen approach is adequate. The difference between the final sample volume and the volume of added seawater, all determined with a graded cylinder, provides the volume of melted sea ice which is needed to determine organism abundances.

*Fixation and Determination of Abundance of Prokaryotes and Protists*

For analysis of bacterial and protist (unicellular eukaryotes, either single celled or colony forming) abundances, a defined subsample, typically 100 to 250ml of the melted ice, is taken after gently mixing the sample for a few minutes. The subsample is then fixed with a suitable fixative (e.g., formalin, glutaraldehyde, or lugol) and stored in a dark glass bottle. Microscopic analyses of these samples can include epifluorescence counts of bacteria (Meiners et al. 2008) and viruses (Maranger et al.

1994) or abundance determination of ice algae by inverted microscopy (Table 3.9.1; see also video footage on the accompanying DVD). A broad array of fixatives, fluorescence stains, and microscopes have been successfully used in such studies and the approach should thus be adjusted based on (a) the taxonomic group targeted and (b) the availability of equipment in the lab.

**Table 3.9.1.** Methodological approach for abundance determination and range of abundances for dominant taxonomic group within and underneath the sea ice

| Taxonomic group | Typical lab methodology | Typical abundance range | Examples |
|---|---|---|---|
| Bacteria and viruses | Epifluorescence microscopy | $10^5$–$10^9$ ml$^{-1}$ | Meiners et al. 2008 |
| Flagellates (photo- and heterotrophic) | Epifluorescence and inverted microscopy; life microscopy | $10^2$–$10^9$ l$^{-1}$ | Ikävalko and Gradinger 1997; Gradinger 1999 |
| Diatoms and other protists | Inverted microscopy | $10^2$–$10^9$ l$^{-1}$ | Garrison and Buck 1986 |
| Meiofauna | Microscopy of preserved samples, life microscopy, stereomicroscope | $<1$–$10^3$ l$^{-1}$ | Gradinger et al. 1999; Petz et al. 1995 |
| Macrofauna | Under-ice camera, quadrat sampling (SCUBA divers); suction pumps; and see Table 3.9.2 | $<1$–$10^3$ m$^{-2}$ | Hop et al. 2000; Werner and Gradinger 2002; Gradinger and Bluhm 2004; Werner 2006 |

Abundance of organisms in melted ice core samples can be determined per volume (as cells/l melted sea ice) as well as per unit area (as cells/m$^2$; see Chapter 3.8 for details). A more detailed analysis of the taxonomic diversity of certain groups requires more specialized microscopic approaches because either these may not preserve well or the common fixatives listed above may not work. Alternative methods include life microscopy (for flagellates, e.g., Ikävalko and Gradinger 1997), protagol staining (for ciliates, e.g., Petz et al. 1995), or permanent slides of cleaned frustules for diatoms (e.g., Poulin 1993; v. Quillfeldt et al. 2003).

The multitude of approaches to study bacterial diversity (e.g., Kaartokallio et al. 2008) is not described as this book concentrates on the field collection part and new laboratory analysis tools are developed at a rapid rate (e.g., Sogin et al. 2006). In general readers should consult lab manuals for aquatic habitats for further reading on these and other methods (e.g., *Handbook of Methods in Aquatic Microbial Ecology* by Kemp et al. 1993). It is important to realize that no single approach will allow one to study all taxa occurring within the ice, and thus any study needs to be adjusted with respect to both sampling and processing according to the scientific target (Table 3.9.1).

*Determination of Abundance of Ice Meiofauna*

Sea ice meiofauna includes mainly multicellular life-forms with a typical size range of 50 to 1000 μm in body length. Some studies include foraminifera and ciliates in meiofauna due to overlap in size ranges with the metazoan meiofauna. In arctic and antarctic sea ice, meiofauna comprises a broad range of taxonomic groups like hydroids, nematodes, various copepod taxa, turbellarians, gastropods, and rotifers. At coastal sites the inventory is augmented by larvae and juveniles of benthic gastropods and polychaetes that occur seasonally in the ice. Meiofauna abundances can vary greatly with season. For example, we observed a one order of magnitude increase between February and May 2004 in the landfast ice close to Barrow, Alaska, closely related to the increase in the algal food availability (Figure 3.9.4) (Gradinger and Bluhm 2005).

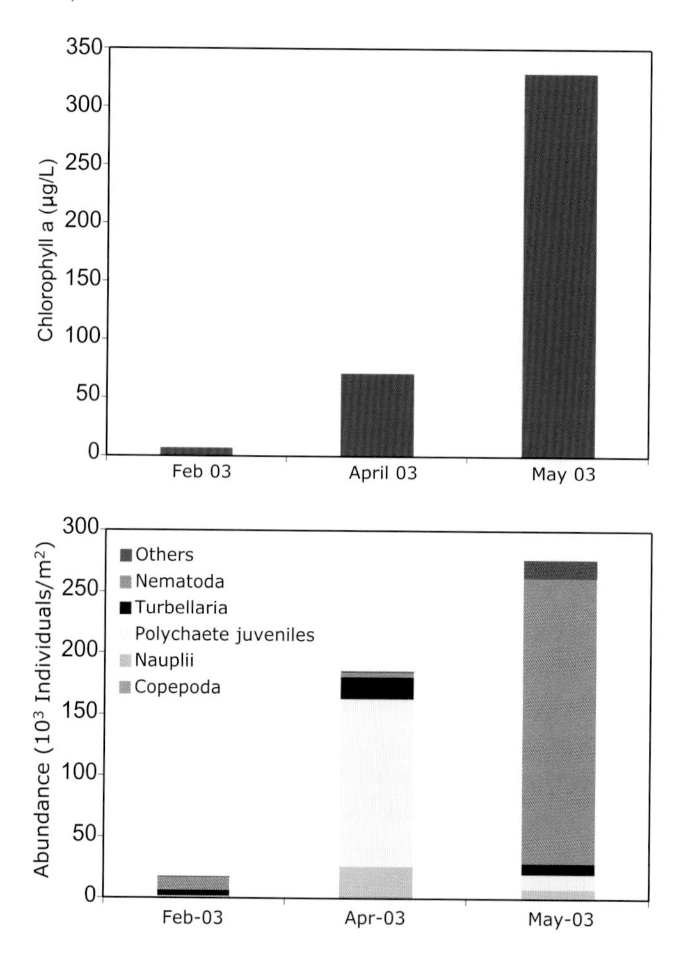


**Figure 3.9.4.** Seasonal increase in ice algal biomass **(a)** and meiofauna abundance **(b)** in the bottom segments of Chukchi Sea fast ice (modified after Gradinger and Bluhm 2005)

The determination of abundance and composition of sea ice meiofauna is again based on melted ice core segments. However, larger sample volumes are needed compared to bacteria and algae, as meiofauna typically occurs in far lower abundances (Table 3.9.1). We normally use the entire volume of a melted core segment and concentrate all animals living in this segment over 20 μm gauze before further processing. Meiofauna will be retained on the sieve and can be washed into a counting chamber with a squeeze bottle containing cooled filtered seawater. We typically use a Bogorov chamber, although a simple petri dish will also do. Counting and sorting of the samples occurs under a stereomicroscope ideally in a temperature-controlled room at 0°C or with intermediate or constant cooling of the counted sample in a normal lab environment (Figure 3.9.5). Alternatively, preserved samples can be sorted with or without previous staining, although some taxa such as hydroids and ciliates do not preserve well. After sorting, samples are fixed for further taxonomic analysis as needed. Typical fixatives include formalin for morphological study and long-term storage and ethanol for molecular study. Precautions for working with hazardous materials need to be taken to protect sample integrity but also the health of the investigators.

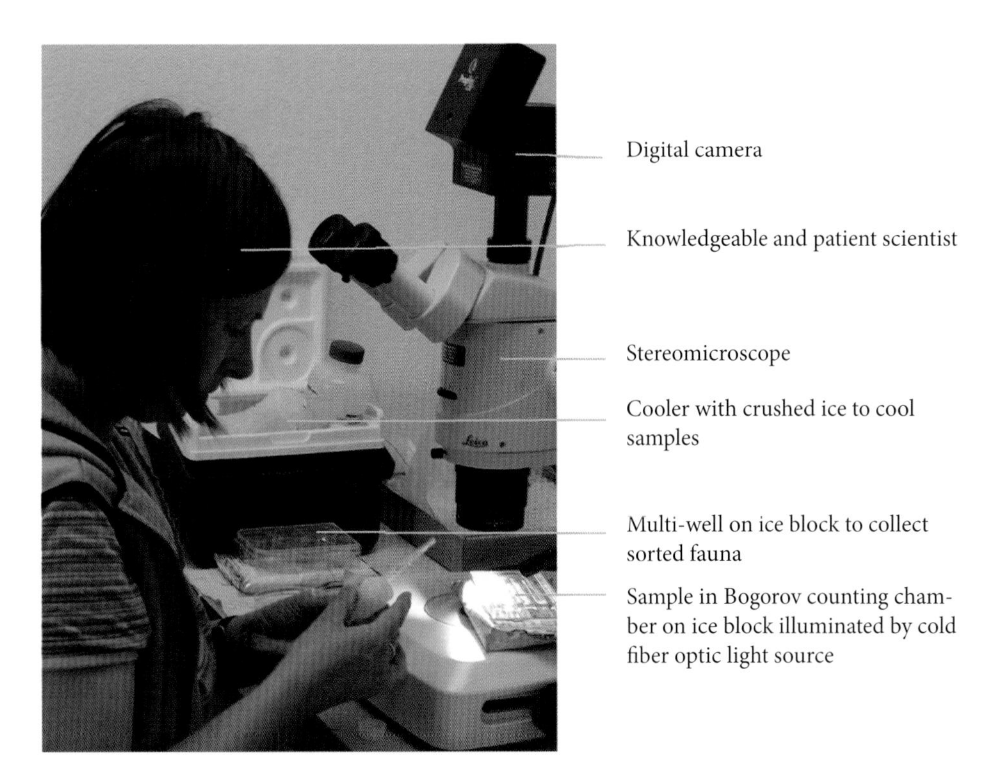


**Figure 3.9.5.** Laboratory setting for live microscopy of meiofauna

### 3.9.1.3 Assessing the Diversity of Under-Ice Biota

While brine channel size sets an upper limit to the size of the in-ice fauna, considerably larger animals and algal aggregates occur at the bottom of the ice in the sub-ice layer. Typical algal representatives observed in these layers include microalgal mats attached to the bottom of the ice, mainly of the arctic diatom species *Melosira arctica,* and unattached aggregates drifting along the bottom with the current (e.g., Melnikov 1997; Ambrose et al. 2005). The fauna is composed of a mixture of sea ice in-fauna (specifically when released during periods of ice melt), surface water phyto- and zooplankton and under-ice specific macrofauna, mainly crustaceans and fish, some of which are unique (endemic) to the interface region (e.g., Bluhm and Gradinger 2008).

Collection tools for the interface regions are water samplers, pump systems, baited traps, nets, and optical tools (Table 3.9.2; Figure 3.9.2).

*Water samplers* are lowered into the interface region to collect water samples for analysis of the phyto- and zooplankton composition directly below the ice. We use small hand-deployed Kemmerer-type water samplers (Figure 3.9.2). Holes big enough for water samplers can be made with ice augers in diameters up to 25 cm, which are available in any ice-fishing supplies store. Holes should be cleaned from any ice shavings and newly forming ice at cold temperatures with a heavy-duty ice skimmer prior to each deployment of the water sampler. Water samplers are not suitable to estimate attached aggregates or macrofauna abundances.

Different types of *pump systems* have been designed to concentrate the biota from large volumes of under-ice water through filtration through sieves of typically 20 to 200 µm. Some pump systems like the NIPR net (Tanimura et al. 1996) and diver-operated hand pumps (Lønne 1988) have been designed to operate under the sea ice. Other systems use a pump and filter system on top of the ice with connected hoses leading through holes in the sea ice to defined water depths. The possibly most sophisticated under-ice water sampler is ADONIS (Dieckmann et al. 1992), which uses vacuum bottles to collect water samples and resolves small-scale vertical gradients under sea ice. It is, however, not designed to collect large amounts of water and therefore not suitable for macrofauna collection. None of these systems are commercially available but they can be built relatively easily with parts such as garden hoses, plumbing tools, and home water treatment systems available in hardware stores. Any system should be tested at low temperatures prior to field deployments.

*Baited traps* have been deployed to collect animals for diversity studies and experimental work (e.g., Kaufman et al. 2008). Commercially available minnow traps (Figure 3.9.2), lined with mosquito screen or gauze to reduce the mesh size, are deployed through auger holes and have collected up to thousands of amphipods in a twenty-four hour period in the nearshore Arctic, with success depending on bait, time of year, and location. Occasionally, traps will catch juvenile arctic cod. Traps should be deployed immediately under the ice in a horizontal fashion, leaving both entrances open. The ice holes should be covered with insulating material to reduce refreezing and allow for easier recovery. The line attached to the trap should go through the middle of the insulating material, not along the edge of the ice hole where it refreezes easily at low temperatures. No general recommendation can be given for bait, but cat food, trout paste, salmon eggs, canned fish, etc. have worked for the arctic amphipod *Onisimus litoralis*. A broad range of tests may be needed to determine suitable bait for specific target species.

In freshwater systems, under-ice gill *nets* have been deployed successfully with various techniques, for example, by stringing a line under the ice which is then used to set a gill net over distances that can exceed 70 m (Hill et al. 1996). We are not aware of the use of these approaches in sea ice, where the milky appearance might make these approaches more difficult compared to the more transparent freshwater ice. For quantitative collections of antarctic under-ice macrofauna, the promising Surface and Under Ice Trawl (SUIT) (Franeker et al. 2009) has recently been designed to glide immediately under the sea ice at an angle to the vessel at 1.5–2 knots trawling speed, using wheels and floats with bars guiding ice out of the 15 m long net. A weight attached to the towing cable avoids interference of the cable with ice floes.

*Optical tools* are likely the best approach to get reliable estimates of the abundances of microalgal mats and sea ice macrofauna. The logistically easiest approach is lowering camera systems on a stick into the ice-water interface through a corer or auger hole. Ideally, the camera should be connected to an arm that can be controlled remotely to alter the orientation of the camera (Werner and Gradinger 2002). Underwater camera systems are commercially available and range from a $150 grayshade camera developed for sports fishing to a $3,000 color high-definition television system. The camera is connected to a video recording unit on top of the ice and small ice areas surrounding the drilled hole can be scanned to assess macrofauna abundance (e.g., Werner and Gradinger 2002). The resolution of the camera will determine whether macrofauna can be determined to the genus and some instances species level. Ice area imaged can be quantified based on the known distance between the beams of two lasers (mounted on the camera system) in the recorded image. For true sample replication, the camera system needs to be lowered through several holes to avoid pseudoreplication. In addition to the small bottom area covered and potential pseudoreplication problems, the need to drill through the ice generally excludes investigation of thick pressure ridges with this technique.

Greater thickness ranges and bottom areas of sea ice can be covered by mobile optical systems. *Remotely operated vehicles* (ROVs) have successfully mapped the occurrence of microalgal mats (Ambrose et al. 2005) and reported the occurrence of sea ice macrofauna in the Arctic and Antarctica (Marschall 1988). ROVs can cover large areas while recording the camera feed. Ice-related work is ideally done with a small and portable ROV system that is battery powered and lowered through several overlapping auger holes.

Scuba divers can also cover relatively large areas, although smaller than ROV coverage, and depth ranges including, for example, studies of the occurrence of macrofauna along pressure ridges (Hop et al. 2000). Divers can directly count under-ice amphipods in replicate quadrats (typically 0.25 $m^2$) along line transects while also recording ice structural features and depth. Quadrats get placed at the under-ice surface while divers hold their breath to avoid displacing fauna with air bubbles during counts (Gradinger et al. submitted). It is helpful for the quadrats to be positively buoyant, because the divers do not need to hold them in place in that case. Plastic tubes with connector pieces containing O-rings work well. Notes get recorded with pencil on waterproof paper held in place by a clipboard.

**Table 3.9.2.** Tools to study under-ice fauna and flora

| Tool | Sophistication/ costs | Logistics | Area covered | Depth range covered |
|---|---|---|---|---|
| ROV | High/high | Moderate to high | Large (> 1000$m^2$) | 0–> 100m (typical for small ROVs) |
| Surface and under-ice trawl | High/high | Moderate to high | Large (> 1000$m^2$) | 0–2m |
| Scuba diver | High/moderate | High | Medium (1 to 100$m^2$) | 0–30m (depending on national rules) |
| On-ice or diver-operated pump | Low/low | Low | Small (stationary)/ medium (diver-operated) | Small (stationary)/ medium (diver-operated) |
| Baited trap | Low/very low | Low | Not quantitative | Unknown |
| Camera through ice hole | Low/low | Low | Small (stationary) | Small (stationary) |
| Water sampler | Low to medium/ low | Low | Small (stationary) | Small (stationary) |

Alternatively, scaled video recordings made by scuba divers have been used to estimate abundances of ice-associated macrofauna (Gradinger and Bluhm 2004). Divers can also collect specimens and water samples during their dives, which is normally not possible with small-sized ROVs. Specimen collections are recommended for proper species identification. The major drawbacks for using divers in polar regions are safety concerns and the substantial additional logistical needs. Diving in waters at the freezing point and under a solid ice cover is very challenging and potentially hazardous, and therefore certain standards regarding diver certification, supervision protocols, and gear suitability need to be adhered to for scientific under-ice diving (Lang and Sayer 2007). Rules and regulations differ between countries and research platforms, and researchers need to contact their national funding agency, logistics providers, and scientific diving centers well in advance of a field expedition to ensure success and compliance.

### 3.9.2 DATA ANALYSIS AND SHARING

Given the occurrence of several hundred species of eukaryotic and prokaryotic life-forms in association with sea ice, no single study will be able to analyze the complete diversity with one set of samples. Consultation of taxonomic experts is often needed because of the multitude of species and the potential for encountering undescribed species. After completion of a species-station table based on either presence/absence or abundance or biomass estimates, a large range of indices can be calculated to detect and describe patterns in the data set as outlined in various textbooks (e.g., Magurran 2004). In addition, multivariate statistics (e.g., factor and principal component analyses, analysis of similarities) are useful tools to explore complex species occurrence and their relation to environmental data sets (Bakus 2007). The most complete software package offering the above and other analyses is PRIMER (Clarke and Gorley 2006).

In the field of marine ecology, two open-access databases, the Ocean Biogeographic Information System OBIS (http://www.iobis.org) and the Global Biodiversity Information Facility GBIF (http://www.gbif.org), are widely used to archive data and will likely be maintained over the next decades. Currently, information on sea ice biota posted in these databases is still very limited. In the case of arctic ice-endemic macrofauna, Figure 3.9.6 illustrates that only 99 records for *Gammarus wilkitzkii* and 414 records for *Apherusa glacialis* were available through OBIS on

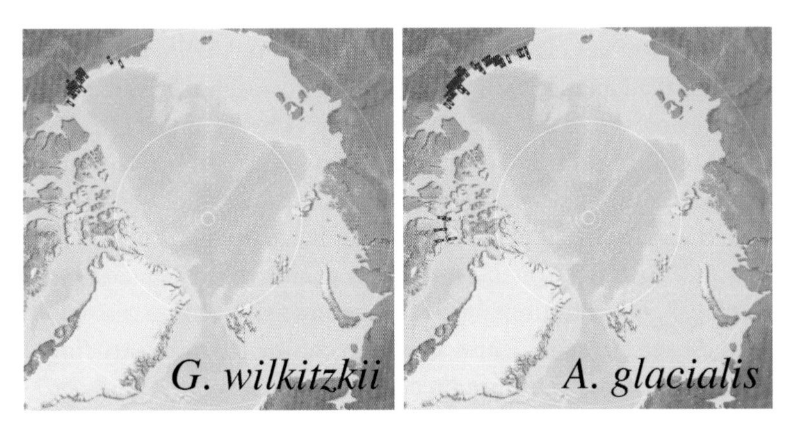


**Figure 3.9.6.** Occurrences of two arctic ice endemic amphipods (*Apherusa glacialis*, *Gammarus wilkitzkii*) in the Arctic based on information provided in the Ocean Biogeographic Information System (http://www.iobis.org). Data were downloaded on October 22, 2008. The complete lack of the species in the central Arctic and on most shelves does not reflect a biological pattern, but the lack of data within OBIS. Both species are known to occur arctic-wide in the seasonal and multiyear ice zones.

October 22, 2008, with nearly all of the information coming from the nearshore waters of the Beaufort Sea. Reasons for the low availability of data are mostly: (a) the reluctance of researchers to release their data and (b) the lack of diversity-focused research in many polar regions. Both hurdles need to be overcome and highlight the need for including diversity aspects in interdisciplinary sea ice field research. Research in benthic and pelagic marine habitats has demonstrated the strong tie between ecosystem structure, diversity, and functioning in polar waters, an approach that would be of benefit to many ongoing and future ice investigations.

### 3.9.2.1 Outlook

Understanding the diversity of sea ice biota is essential to identify the biota's functional role in the arctic system from many different aspects. The tight relation between species composition and entire food-web structure was recently documented for an antarctic food web, where a climate-induced shift in phytoplankton species composition from large diatoms to small cryptophytes contributed to a shift from krill- to salp-dominated zooplankton communities (Moline et al. 2004). We currently can only speculate whether similar ice algal species changes or shifts of ice algae to pelagic algae might occur related to changes in the arctic sea ice regime, and it is worth investigating whether the current dominance of large diatoms in bottom-ice algal communities is critical for the nutrition of sea ice meio- and macrofauna. On a regional scale, sea ice–derived organic matter provides a valuable prey field for ice-associated higher trophic levels. Recent studies suggest that benthic bivalves prefer settling ice algae over phytoplankton due to their elevated levels of essential fatty acids (McMahon et al. 2006), suggesting that ice algae may be a critical food source for sea ice, pelagic, and even benthic biota during periods of ice cover (Bluhm and Gradinger 2008).

Diversity of ice biota is also of interest outside the biological domain as specific physiological adaptations vary from species to species. We are just beginning to understand how species composition might directly impact sea ice physical and chemical properties during ice formation, periods of ice cover, and decay. For example, there is some evidence that ice bacterial species might be able to avoid ice formation in their immediate proximity maybe due to unique cell surface properties (Junge et al. 2008). Also, the often-high concentrations of translucent exopolymeric substance (TEP) concentrations in sea ice produced by some diatoms, prymnesiophytes, and bacteria might cause alterations of the ice microstructure (Krembs and Deming 2008). As a third example, organic sulfur components produced by some sea ice algal taxa appear to be relevant in the context of cloud formation and contribute to the regulation of the earth's climate system (Trevena and Jones 2006). The unique physiological properties of cold-adapted sea ice species might also be of interest for biotechnological screening (Nichols et al. 1999; Cavicchioli et al. 2002). All these processes and adaptations depend on the functioning of individual species within the ice and contribute to the relevance of ice biota in an interdisciplinary context. Emerging molecular tools like 454-based tag sequencing might help to allow fast and reliable estimations of relative abundance and diversity of both dominant and rare members of the microbial community (Sogin et al. 2006).

In the under-ice environments, autonomous underwater vehicles will open up the possibility for regional biological remote sensing, which map abundances of under-ice fauna in relation to other ice physical and biological properties on scales of hundreds of kilometers (Dowdeswell et al. 2008). On a small local scale, under-ice webcams connected to satellite phones could be used to get species occurrence information with higher temporal resolution as currently available. Hopefully, these developments can be implemented on a sufficiently fast time scale to allow a proper documentation of current sea ice biodiversity needed as baseline to document future changes.

## REFERENCES

Ambrose, W., C. v. Quillfeldt, L. Clough, P. Tilney, and T. Tucker (2005), The sub-ice algal community in the Chukchi sea: Large- and small-scale patterns of abundance based on images from a remotely operated vehicle, *Polar Biol., 28,* 784–795.

Bakus, G. J. (2007), *Quantitative Analysis of Marine Biological Communities,* Wiley-Interscience, Hoboken, NJ.

Barker Bausell, R., and Y. F. Li (2002), *Power Analysis for Experimental Research: A Practical Guide for the Biological, Medical, and Social Sciences*, Cambridge University Press, New York.

Bluhm, B. A., R. Gradinger, and S. Piraino (2007), First record of sympagic hydroids (Hydrozoa, Cnidaria) in Arctic coastal fast ice, *Polar Biol., 30,* 1557–1563.

Bluhm, B., and R. Gradinger (2008), Regional variability in food availability for Arctic marine mammals, *Ecolog. Applic., 18,* S77–S96.

Cavicchioli, R., K. S. Siddiqui, D. Andrews, and K. R. Sowers (2002), Low-temperature extremophiles and their applications, *Curr. Opin. Biotechnol., 13,* 253–261.

Clarke, K. R., and R. N. Gorley (2006), *PRIMER v6: User manual/tutorial.* PRIMER-E, Plymouth, MA.

Dieckmann, G. S., K. Arrigo, and C. W. Sullivan (1992), A high-resolution sampler for nutrient and chlorophyll a profiles of the sea ice platelet layer and underlying water column below fast ice in polar oceans: Preliminary results, *Mar. Ecol. Prog. Ser., 80,* 291–300.

Dowdeswell, J. A., J. Evans, R. Mugford, G. Griffiths, S. D. McPhail, N. Millard, P. Stevenson, M. A. Brandon, C. Banks, K. J. Heywood, M. R. Price, P. A. Dodd, A. Jenkins, K. W. Nicholls, D. Hayes, E. P. Abrahamsen, P. A. Tyler, B. J. Bett, D. O. B. Jones, P. Wadhams, J. P. Wilkinson, K. Stansfield, and S. Ackley (2008),

Autonomous underwater vehicles (AUVs) and investigations of the ice-ocean interface: Deploying the Autosub AUV in Antarctic and Arctic waters, *J. Glaciol.*, *54*, 661–672.

Franeker, J. A. van, Flores, H., and Dorssen, M. van (2009), The surface and under-ice trawl (SUIT). In: Flores, H. Frozen desert alive: the role of sea ice for pelagic macrofauna and its predators: implications for the Antarctic pack-ice food web. Chapter 9. PhD dissertation, University of Groningen, pp. 181–187.

Garrison, D. L., and K. R. Buck (1986), Organism losses during ice melting: A serious bias in sea ice community studies, *Polar Biol.*, *6*, 237–239.

Gosselin, M., M. Levasseur, P. A. Wheeler, R. A. Horner, and B. C. Booth (1997), New measurements of phytoplankton and ice algal production in the Arctic Ocean, *Deep-Sea Res. II*, *44*, 1623–1644.

Gradinger, R. (1999), Vertical fine structure of algal biomass and composition in Arctic pack ice, *Mar. Biol.*, *133*, 745–754.

Gradinger, R. (2008), Sea ice algae: Major contributors to primary production and algal biomass in the Chukchi and Beaufort Sea during May/June 2002, *Deep-Sea Res. II*, doi:10.1016/j.dsr2.2008.10.016.

Gradinger, R., C. Friedrich, and M. Spindler (1999), Abundance, biomass and composition of the sea ice biota of the Greenland Sea pack ice, *Deep-Sea Res. I*, *46*, 1457–1472.

Gradinger, R. R., and B. A. Bluhm (2004), *In situ* observations on the distribution and behavior of amphipods and Arctic cod (*Boreogadus saida*) under the sea ice of the high Arctic Canadian Basin, *Polar Biol.*, *27*, 595–603.

Gradinger, R., and B. Bluhm (2005), Susceptibility of sea ice biota to disturbances in the shallow Beaufort Sea: Phase 1: Biological coupling of sea ice with the pelagic and benthic realms, *Final Report. OCS Study MMS 2005-062*, University Alaska Fairbanks, Coastal Marine Institute.

Gradinger, R., B. A. Bluhm, and K. Iken (submitted), Arctic sea ice ridges—safe havens for sea ice fauna during periods of extreme ice melt?, *Deep-Sea Res. II*.

Haecky, P., and A. Andersson (1999), Primary and bacterial production in sea ice in the northern Baltic Sea, *Aquat. Microb. Ecol.*, *20*, 107–118.

Hill, T. D., S. T. Lynott, S. D. Bryan, and W. G. Duffy (1996), An efficient method for setting gill nets under ice, *North American J. Fish. Managem.*, *16*, 960–962.

Hop, H., M. Poltermann, O. J. Lonne, S. Falk-Petersen, R. Korsnes, and W. P. Budgell (2000), Ice amphipod distribution relative to ice density and under-ice topography in the northern Barents Sea, *Polar Biol.*, *23*, 357–367.

Horner, R., (1980), Ecology and productivity of Arctic sea ice diatoms, in *Proc. 6th Symp. Recent Fossil Diatoms, Budapest. Otto Koeltz, Königstein,* edited by R. Ross, pp. 359–369.

Horner, R. (1985), *Sea Ice Biota.* CRC Press, Boca Raton, FL.

Ikävalko, J., and R. Gradinger (1997), Flagellates and heliozoans in the Greenland Sea ice studied alive using light microscopy. *Polar Biol.*, *17*, 473–481.

Junge, K. and B. D. Swanson (2008), High-resolution ice nucleation spectra of sea ice bacteria: Implications for cloud formation and life in frozen environments, *Biogeosciences, 5,* 865–873.

Junge, K., H. Eicken, and J. W. Deming (2004), Bacterial activity at –2 to –20°C in Arctic wintertime sea ice, *Appl. Envir. Microb., 70,* 550–555.

Kaartokallio, H., J. Tuomainen, H. Kuosa, J. Kuparinen, P. Martikainen, and K. Servomaa (2008), Succession of sea ice bacterial communities in the Baltic Sea fast ice, *Polar Biol., 31,* 783–793.

Kaufman, M., R. Gradinger, B. Bluhm, and D. O'Brien (2008), Using stable isotopes to assess carbon and nitrogen turnover in the Arctic sympagic amphipod *Onisimus litoralis, Oecol., 16,* 11–22.

Kemp, P. F., B. F. Sherr, E. B. Sherr, and J. J. Cole (1993), *Handbook of Methods in Aquatic Microbial Ecology,* Lewis Publications, Boca Raton, FL.

Krembs, C., and J. W. Deming (2008), The role of exopolymers in microbial adaptation to sea ice, in *Psychrophiles: From Biodiversity to Biotechnology,* edited by R. Margesin, F. Schinner, J. C. Marx, and C. Gerday, pp. 247–264, Springer, Heidelberg.

Krembs, C., R. Gradinger, and M. Spindler (2000), Implications of brine channel geometry and surface area for the interaction of sympagic organisms in Arctic sea ice, *J. Exp. Mar. Biol., Ecol., 243,* 55–80.

Lang, M. A., and M. D. J. Sayer (eds.) (2007), *Proceedings of the International Polar Diving Workshop,* Svalbard, Smithsonian Institution, Washington, DC.

Lønne, O. J. (1988), A diver-operated electric suction sampler for sympagic (=under-ice) invertebrates. *Polar Res., 6,* 135–136.

Magurran, A. E. (2004), *Measuring Biological Diversity,* Blackwell Pub., London

Maranger, R., D. F. Bird, and S. K. Juniper (1994), Viral and bacterial dynamics in Arctic sea ice during the spring algal bloom near Resolute, N.W.T., Canada. *Mar. Ecol. Prog. Ser., 111,* 121–127.

Marschall, H. P. (1988), The overwintering strategy of Antarctic krill under the pack-ice of the Weddell Sea. *Polar Biol., 9,* 129–135.

McMahon, K. W., W. G. Ambrose, B. J. Johnson, M.-Y. Sun, G. R. Lopez, L. M. Clough, and M. L. Carroll (2006), Benthic Community Response to Ice Algae and Phytoplankton in Ny Ålesund, Svalbard. *Mar. Ecol. Prog. Ser. 310,* pp. 1–14.

Meiners, K., C. Krembs, and R. Gradinger (2008). Exopolymer particles: Microbial hotspots of enhanced bacterial activity in Arctic fast ice (Chukchi Sea), *Aquatic Microb. Ecol., 52,* 195–207.

Melnikov, I. A. (1997), *The Arctic Sea Ice Ecosystem,* Gordon Breach Science Publications, Amsterdam.

Michel, C., T. G. Nielsen, C. Nozais, and M. Gosselin (2002), Significance of sedimentation and grazing by ice micro- and meiofauna for carbon cycling in annual sea ice (northern Baffin Bay), *Aquat. Microb. Ecol., 30,* 57–68.

Moline, M. A., H. Claustre, T. K. Frazer, O. Schofield, and M. Vernet (2004), Alteration of the food web along the Antarctic Peninsula in response to a regional warming trend, *Global Change Biol., 10*, 1973–1980.

Nichols, D., J. Bowman, K. Sanderson, C. M. Nichols, T. Lewis, T. A. McMeekin, and P. D. Nichols (1999), Developments with Antarctic microorganisms: Culture collections, bioactivity screening, taxonomy, PUFA production, and cold-adapted enzymes, *Curr. Opin. Biotechnol., 10,* 240–246.

Nozais, C., M. Gosselin, C. Michel, and G. Tita (2001), Abundance, biomass, composition, and grazing impact of the sea ice meiofauna in the North Water, northern Baffin Bay, *Mar. Ecol. Prog. Ser., 217*, 235–250.

Petz, W., W. Song, and N. Wilbert (1995), Taxonomy and ecology of the ciliate fauna (protozoa, ciliophora) in the endopagial and pelagial of the Weddell Sea, Antarctica, *Stapfia, 40,* 1–223.

Poulin, M. (1993), *Craspedopleura* (Bacillariophyta), a new diatom genus of arctic sea ice assemblages, *Phycol., 32,* 223–233.

Smith, R. E. H., J. Anning, P. Clement, and G. Cota (1988), Abundance and production of ice algae in Resolute Passage, Canadian Arctic, *Mar. Ecol. Prog. Ser., 48*, 251–263.

Sogin, M. L., H. G. Morrison, J. A. Huber, D. M. Welch, S. M. Huse, P. R. Neal, J. M. Arrieta, and G. J. Herndl (2006), Microbial diversity in the deep sea and the underexplored rare biosphere, *PNAS, 103,* 12,115–12,120.

Steffens, M., M. A. Granskog, H. Kaartokallio, H. Kuosa, K. Luodekari, S. Papadimitriou, and D. N. Thomas (2006). Spatial variation of biogeochemical properties of landfast sea ice in the Gulf of Bothnia (Baltic Sea), *Annals Glaciol., 44,* 80–87.

Stoecker, D. K., D. E. Gustafson, C. T. Baier, and M. M. D. Black (2000), Primary production in the upper sea ice, *Aquat. Microb. Ecol., 21*, 275–287.

Stoecker, D. K., D. E. Gustafson, J. R. Merrell, M. M. D. Black, and C. T. Baier (1997), Excystment and growth of chrysophytes and dinoflagellates at low temperatures and high salinities in Antarctic sea ice, *J. Phycol., 33*, 585–595.

Tanimura, A., T. Hoshiai, and M. Fukuchi (1996), The life cycle strategy of the ice-associated copepod, *Paralabidocera antarctica* (Calanoida, Copepoda), at Syowa station, Antarctica, *Antarct. Sci., 8,* 257–266.

Trevena, A. J., and G. B. Jones (2006), Dimethylsulphide and dimethylsulphoniopropionate in Antarctic sea ice and their release during sea ice melting, *Marine Chem., 98*, 210–222.

v. Quillfeldt, C. V., W. Ambrose, and L. Clough (2003), High number of diatom species in first year ice from the Chukchi Sea, *Polar Biol., 12*, 806–818.

Werner, I. (2006), Seasonal dynamics of sub-ice fauna below pack ice in the Arctic (Fram Strait), *Deep Sea Res. I, 53,* 294–309.

Werner, I., and R. Gradinger (2002), Under-ice amphipods in the Greenland Sea and Fram Strait (Arctic): Environmental controls and seasonal patterns below the pack ice, *Mar. Biol., 140,* 317–326.

## Chapter 3.10

# Studying Seals in Their Sea Ice Habitat

## Application of Traditional and Scientific Methods

*Brendan P. Kelly*

### 3.10.1 INTRODUCTION

Arctic sea ice has influenced the evolution and ecology of marine mammals by serving as a barrier to dispersal, enhancing primary productivity, creating refuges from predation, and providing a substrate for reproduction and other important life history events.

As a barrier to dispersal, sea ice has contributed to divergence and speciation among cetaceans (whales, dolphins, and porpoises) and pinnipeds (seals, sea lions, and walruses) (Davies 1958; Fay 1982; Deméré et al. 1993; Dyke et al. 1996; Harington 2008). For example, North Atlantic and North Pacific populations of humpback whales (*Megaptera novaeangliae*), gray whales (*Eschrictius robustus*), harbor porpoises (*Phocoena phocoena*), and harbor seals (*Phoca vitulina*) diverged from one another, because gene flow was thwarted by Arctic Ocean ice cover. Even among species more adapted to occupying ice-covered waters, sea ice limited gene flow leading to the establishment of distinct subspecies of walruses (*Odobenus rosmarus*) and stocks of bowhead whales (*Balaena mysticetus*), beluga whales (*Delphinapterus leucas*), and probably other species.

The productivity of the Arctic Ocean and its marginal seas is enhanced by sea ice dynamics (Buckley et al. 1979; Alexander and Niebauer 1981; Legendre et al. 1992), allowing those waters to support a large biomass of vertebrate predators including blue whales (*Balaenoptera musculus*), fin whales (*B. physalus*), sei whales (*B. borealis*), minke whales (*B. acutorostrata*), humpback whales, killer whales (*Orcinus orca*), northern right whales (*Eubalaena glacialis*), gray whales, bowhead whales, beluga whales, narwhals (*Monodon monoceros*), walruses, and ringed (*Phoca hispida*), spotted (*P. largha*), ribbon (*P. fasciata*), bearded (*Erignathus barbatus*), harp (*P. groenlandica*), hooded (*Cystophora cristata*), and gray seals (*Haliochoerus grypus*) (Fay 1974; Lavigne and Kovacs 1988).

The degree to which these species have adapted to sea ice varies widely, with some species benefitting from ice-associated productivity while mostly avoiding the ice themselves (e.g., humpback whales and harbor seals) and others penetrating far into the ice by creating their own openings (e.g., bowhead whales, walruses, and ringed seals). The ice creates refuges from predation for these more ice-adapted species as few terrestrial or marine predators have colonized sea ice environments. Killer whales do hunt along the edge of the pack ice, and the more ice-adapted species find refuge from those predators by moving farther into the ice. The refuge is not complete, however, as Greenland sharks (*Somniosus microcephalus*) and polar bears (*Ursus maritimus*) penetrate even the heaviest ice. Still, those predators occur in low density under and on the vast extent of arctic sea ice, which remains an important partial refuge from predation.

The role of ice as a refuge from predation has been especially important in the evolution of pinnipeds (Stirling 1977; Kelly 2001). Pinnipeds diverged from terrestrial carnivores twenty to twenty-five million years ago (Arnason et al. 2006) and evolved as marine predators while retaining their ties to terrestrial existence for the purposes of giving birth, nursing their young, and molting. That amphibious lifestyle entailed selective tradeoffs, and the morphological adaptations favoring marine foraging (fusiform body shape and limbs modified as flippers) make locomotion out of water slow and awkward. Thus, when giving birth and nursing their young, pinnipeds are vulnerable to predation. That vulnerability has favored rearing young on substrates minimally accessible to predators, and the most expansive such substrate has been the ice cover of polar seas.

The difficulty of accessing marine mammals in their natural habitat has meant that knowledge of their basic biology has come slowly. For example, the echolocation abilities of toothed whales and dolphins were unknown before the 1950s (Kellogg et al. 1953; Norris 1968), and the physiological adaptations to thermoregulating and diving in the ocean were not well understood until the 1970s (Øritsland 1970; Kooyman 1973; Blix et al. 1975; Elsner et al. 1977; Kooyman et al. 1981). For species inhabiting sea ice, access is even more difficult, and the importance of sea ice in their evolution and ecology was long underappreciated (Burns et al. 1981; Stirling 1997; Kelly 2001). In recent decades, however, our scientific understanding of the biology and ecology of ice-associated marine mammals has progressed considerably. That progress has been greatly enhanced by insights gained from Native hunters and by increasingly sophisticated technology. Monographic works on ice-associated species (e.g., McLaren 1958; Fay 1982; Smith 1987; Burns et al. 1993) acknowledged considerable intellectual debt to Native teachers and collaborators. Traditional knowledge amongst hunters is especially well developed with regards to seasonal availability and behavior of the animals at local scales.

Many marine mammals migrate over vast distances and spend much of their lives beneath the surface. Instruments that allow tracking of seasonal movements and subsurface activities have greatly enhanced our understanding of marine mammal ecology at regional and global scales. The use of sea ice as habitat by some species presents additional complications and opportunities. On the one hand, the extreme cold of polar environments contributes to materials failures and increases the power demands for electronic instruments. On the other hand, the sea ice provides a platform allowing researchers access to marine habitats that otherwise would require the expensive deployment of ships. For example, researchers in Alaska have long used the shorefast ice at Point Barrow as a platform from which to count and study migrating bowhead whales (Clark et al. 1986; Krogman et al. 1989; George et al. 2004). Visual counts of migrating whales were made from the top of ice ridges several meters high at the lead edge, and the researchers camped on the ice adjacent to the lead, emulating Iñupiat methods of travel and camping. Those counts were supplemented with estimates based on whales tracked acoustically using hydrophone and recording stations established on the ice. Hydroacoustic investigations of bowhead prey likewise have been conducted from floating ice camps (Crawford and Jorgenson 1990).

Pioneering studies of ringed seal life history by Chapskii (1940) and McLaren (1958) and Native collaborators took place in camps established on the shorefast ice. Subsequent ice-based studies have increased our understanding of ringed seal reproductive biology (Smith and Stirling 1975; 1978; Lydersen and Gjertz 1986; Hammill et al. 1991; Lydersen and Hammill 1993a), foraging (Gjertz and Lydersen 1986; Weslawski et al. 1994; Simpkins et al. 2001a; Wathne et al. 2000), predation (Smith 1976; Lydersen and Gjertz 1986; Lydersen and Smith 1989; Hammill and Smith 1991; Smith et al. 1991; Furgal et al. 1996), diving behavior (Lydersen et al. 1992; Lydersen and Hammill 1993b; Kelly and Wartzok 1996; Simpkins et al. 2001b,c), thermoregulation (Taugbøl 1984), vulnerability to oil contamination (Smith and Geraci 1975; Geraci and Smith 1976; Engelhardt et al. 1977), and habitat use (McLaren 1958; Lukin and Potelov 1978; Hammill and Smith 1989, 1990; Kingsley et al. 1990; Kunnasranta 2001). Seals harvested from the sea ice have been important in studies of nutritional value and contamination with heavy metals and other pollutants (Addison et al. 1986; Addison and Smith 1974; 1998; Hyvärinen et al. 1998; Cooper et al. 2000; Bang et al. 2001; Woshner et al. 2001; Kucklick et al. 2002, 2006; Dehn et al. 2005).

Sea ice has also been an important platform from which telemetry devices, including time-depth recorders and tracking devices (radio, satellite, and acoustic tags), have been attached to walruses as well as ringed, bearded, ribbon, harp, and hooded seals (Kelly and Quakenbush 1990; Folkow et al. 1996; Kelly and Wartzok 1996; Stenson and Sjare 1997; Gjertz et al. 2000; Jay and Gardner 2002; Born et al. 2005; Bornhold et al. 2005; Simpkins et al. 2001a,b,c; Cameron 2005).

In this chapter, I describe methods learned from Iñupiat and Inuit hunters, others adapted from scientific colleagues, as well as novel methods used to investigate the ecology of ringed seals. While each of the methods has been briefly described elsewhere, it seems useful to elaborate on each in a single publication. Some of these methods are specific to ringed seals, but most have been or could be adapted to other ice-associated seals.

After providing background information on the biology and ecology of ice-associated marine mammals, I provide details of methods of locating subnivean seal holes; live-capturing seals in their breathing holes and when resting on the ice; telemetric methods for locating seals in their lairs, tracking local movements under the ice, and tracking long-distance movements; and collecting tissue samples for the investigation of molecular markers. I conclude with a brief discussion of permitting authorities and requirements and some thoughts on the future of research on ice-associated marine mammals.

## 3.10.2 BACKGROUND

Seven species of pinnipeds depend strongly on arctic sea ice. Three of those species—ringed seals, bearded seals, and walruses—are circumpolar or nearly so. Four species—harp seals, hooded seals, spotted seals, and ribbon seals—are restricted to only a portion of the arctic basin.

Ringed seals are circumpolar and found in all seasonally ice-covered seas of the Northern Hemisphere as well as in some freshwater lakes. Abrading and maintaining breathing holes in the ice with thick claws on their foreflippers allows ringed seals to remain in areas of otherwise continuous ice cover including both shorefast- and pack-ice areas (Chapskii 1940; McLaren 1958). They also use those claws to excavate lairs in snowdrifts above a subset of their breathing holes. The seals rest in the lairs during the winter months and give birth there in April and May. The pups nurse in the subnivean lairs for the first six to eight weeks of their lives (Burns 1970; Smith 1987). The snow cover insulates the small pups (3–5 kg at birth) from extreme cold and conceals them from predators. Ringed seals forage on a wide assortment of fish and zooplankton; important prey species include arctic cod (*Boreogadus saida*), krill (*Euphausia* sp.), and gamarid amphipods.

Bearded seals also are nearly circumpolar but restricted by benthic feeding to continental shelves (Chapskii 1938; Mansfield 1967; Potelov 1975; Braham et al. 1984; Kingsley et al. 1985; Kelly 1988). They also are restricted by and large to active ice in which natural openings are available for surfacing. Occasionally, however, they appear in areas of continuous ice cover where they use breathing holes and, rarely, subnivean lairs. More typically, their young are born and nursed in the open on pack-ice floes with ready access to the water (Kovacs et al. 1996; Simpkins et al. 2003; Bengtson et al. 2005).

Walruses were once very nearly circumpolar, but they have not returned to some areas from which they were extirpated by commercial hunters (Fay et al. 1990). Like bearded seals, walruses are restricted to continental shelf waters by a diet of benthic and epibenthic prey, primarily bivalve and gastropod molluscs (Fay 1982; Fay and Burns 1988). Walruses come ashore—typically in dense aggregations—on shorelines and sea ice, and the latter is the primary habitat for mating and nursing young. The large size of walruses (up to 1500 kg) and their tendency to cluster in dense herds restrict them to the thicker ice floes that predominate 50 or more km north of the outer ice front (Fay 1974). They often open holes in ice as thick as 22 cm by breaking the ice with their massive skulls.

Harp seals inhabit the pack ice of the North Atlantic Ocean and breed during February, March, and April in concentrations in the Gulf of St. Lawrence to the coast of Labrador, eastern Greenland, and the White Sea (Lavigne and Kovacs 1988). They molt on that ice in April and May and move northward with the ice in summer (Stenson and Sjare 1997). Harp seals consume zooplankton as well as pelagic and benthic fishes. Their importance as consumers of Atlantic cod (*Gadus morhua*) and other commercially important species (Stenson et al. 1997; Hammill and Stenson 2000; Wathne et al. 2000) and as the objects of commercial harvests (Lavigne 2006) drives much of the research interests in the species. The harvests have provided considerable data on population biology. More recently, that knowledge has been supplemented by the deployment on the ice of telemetry devices (Boness and Bowen 1996; Hammill and Stenson 2000).

Hooded seals also are broadly distributed in the pack ice of the North Atlantic Ocean, but they tend to occupy deeper areas than do harp seals (Sergeant 1976; Lavigne and Kovacs 1988). They breed on the ice in three distinct stocks, all of which congregate in Denmark Strait for molting during summer months (Sergeant 1974; King 1983). Hooded seals have the shortest lactation of any of the seals, averaging four days (Bowen et al. 1985; Kovacs and Lavigne 1992). That brief lactation apparently is adaptive for breeding in rapidly moving ice where a prolonged mother-pup bond cannot be assured. Hooded seals eat a wide variety of fish and squid with important energy contributions from Greenland halibut (*Reinhardtius hippoglossoides*), redfish (*Sebastes* sp.), arctic cod, and Atlantic herring (*Clupea harengus*).

Spotted seals are restricted to the Okhotsk, Bering, Chukchi, and Beaufort seas where they inhabit moving ice in winter and spring and coastal haulouts in summer and fall. Spotted seals primarily feed pelagically but also benthically on fish, crustaceans, and octopus (Bukhtiyarov et al. 1984). They give birth in April and May on the edge of the pack ice where they nurse a single pup for two to four weeks (Burns 1970).

Ribbon seals have a similar distribution as the spotted seal during the spring breeding period, pupping and nursing their young in April and May on the ice edge (Burns 1970; Braham et al. 1984). They do not, however, use terrestrial haulouts in the Bering and Chukchi seas during summer and fall when their distribution spans from south of the Aleutians (Cameron 2005) to the ice edge of the Chukchi Sea (Kelly, unpublished). Ribbon seals feed on crustaceans as juveniles and increasingly on cephalopods and pollock (*Theragra chalcogramma*) as they get older (Shustov 1965; Frost and Lowry 1980; Dehn et al. 2007).

The diversification of pinnipeds on arctic sea ice opened a niche for higher-level predators, and members of the Ursidae (bear family) and Canidae (dog family) adapted forms to exploit that opportunity. Polar bears split apart from brown bears (*Ursus arctos*) and developed dentition, claws, and a white coat specialized for preying on ice-associated seals. Molecular divergence from brown bears dates back approximately 250,000 years (Cronin et al. 1991; Talbot and Shields 1996; Waits et al. 1998), but the oldest specimens showing the specialized dentition of modern polar bears are less than 40,000 years old (Stanley 1979; Talbot and Shields 1996).

Other carnivores—including wolves (*Canis lupus*), wolverines (*Gulo gulo*), brown bears, and red foxes (*Vulpes vulpes*)—make occasional forays onto the ice to prey on ringed seals, but only the arctic fox (*Alopex lagopus*) frequents that habitat. On the ice, arctic foxes prey on newborn ringed seals and scavenge at sites where polar bears have killed pinnipeds or whales. The small foxes are capable of killing only the smallest ringed seal pups, which they attack after tunneling in to the seal's birth lair. When fox populations are high, they can take substantial proportions of local seal production (Smith 1976).

Efforts to enumerate ice-associated marine mammals have been considerable but in many cases unsuccessful. A recent analysis concluded that current survey methods have no chance of detecting a 50 percent decline in populations of ice-associated seals within fifteen years in contrast to a 95 percent probability of detecting such a decline among pinnipeds on terrestrial haulouts (Taylor et al. 2007).

Even the systematics of marine mammals are not secure. As molecular evidence of gene flow increasingly contravenes classifications based on morphological assessments, it has been estimated that proper identification of cryptic species could increase the number of extant mammals by as much as 50 percent (Baker and Bradley 2006). The systematics of marine mammals are further complicated by high levels of interspecific and intergeneric hybridization. Thus, intergeneric hybrids have been documented between harp and hooded seal (Kovacs et al. 1997) as well as among fur seals (Lancaster et al. 2006) and whales (Spilliaert et al. 1991; Bérubé and Aquilar 1998; Willis et al. 2004). Such hybridization quite likely occurs also between spotted seals and ringed seals, and spotted seals and harbor seals (*Phoca vitulina*).

For the past fifteen years, the National Marine Fisheries Service and U.S. Fish and Wildlife Service researchers and managers have struggled to define even finer taxa, namely independent stocks within marine mammal species to satisfy stock-specific management mandates (Dizon et al. 1992; Wade and Angliss 1997). Progress has been slow, especially for ice-associated species (Swanson et al. 2006; O'Corry-Crowe 2008; Davis et al. 2008). It seems likely that patterns of gene flow will change dramatically as the continent-sized sea ice habitat rapidly diminishes (Overland and Wang 2007; Serreze et al. 2007; Comiso et al. 2008), and those changes may well outpace our ability to describe the previous patterns (Kelly et al. in review).

### 3.10.3 USING DOGS TO LOCATE SUBNIVEAN SEAL HOLES

Dogs are the oldest example of domestication (Clutton-Brock 1995), having originated from Eurasian gray wolves—most likely in East Asia about fifteen thousand years ago (Savolainen et al. 2002). When the first humans crossed the Bering land bridge, ten thousand to fifteen thousand years ago, they brought domesticated dogs with them (Arnold 1979; Leonard et al. 2002). Nonetheless, dogs may have been sparse among arctic people until the emergence of Thule culture about a thousand years ago (Morey and Aaris-Sorensen 2002). The expense of maintaining dogs was likely compensated by their contribution to the harvest of marine mammals. Dogs could pack and pull sleds containing gear and meat, and they helped hunters locate the breathing holes of ringed seals hidden beneath the snow. Thus, when Europeans penetrated the High Arctic, the use of dogs to locate subnivean seal holes was prevalent (Hall 1866).

Researchers began using dogs to locate seal holes only in the past fifty years in the Canadian Arctic (Smith and Stirling 1975), the White Sea (Lukin and Potelov 1978), Svalbard (Taugbøl 1984), Alaska (Kelly and Quakenbush 1987), and the Baltic Sea (Kunnasranta pers. comm.). In 1981, I was introduced to the use of dogs to locate subnivean seal holes by Thomas G. Smith (then of the Canadian Department of Fisheries and Oceans) and Jimmy Memorana, an Inuit hunter in Holman Island, Northwest Territories, Canada. While Eskimo hunters used sled dogs to locate seal holes, researchers have favored more tractable working breeds, especially Labrador retrievers (Smith and Stirling 1975; Kelly and Quakenbush 1987). Tractability has been selected for in domesticated dogs (Hare et al. 2002) and is especially well developed in retrievers and other working breeds. People commonly assume that the dog's olfactory sensitivity is the most important trait but, in my experience, it seldom is limiting to the dog's performance. That is, virtually all dogs have highly sensitive noses, and their abilities to successfully and consistently locate seal holes varies more with their tractability. Male or female dogs can be trained successfully; I have tended to prefer females as they seem to have greater stamina for long searches.

Young dogs well trained in basic obedience and in taking direction in the form of verbal and hand signals are the best students for seal-sniffing school. Training them to locate seal holes entails teaching them to:

- associate a voice command with the odor of ringed seals;
- traverse the ice in directions specified by the handler;
- follow scent plumes to their sources; and
- dig in snow or otherwise alert at a seal hole.

Refinements include training the dog to ignore the temptations of other odors and to appropriately double back when scent plumes are disrupted by, for example, complex ice topography.

The preferred approach to training the dog to associate the odor of ringed seals with a voice command involves taking her to actual seal holes or other sites where seal odor is present and consistently praising her when she is sniffing the site. It is important that a handler, who knows well the particular dog's behavior in many contexts, observes the dog's behavior at the seal hole and gives the greatest reinforcement when she is actively using her nose to investigate the seal hole and suspends reinforcement when the dog engages in other behaviors.

I prefer to expose the trainees to seal odor by allowing them to follow experienced dogs to subnivean seal holes, an approach that keeps the desired behavior in context and—with the right social relationship between the dogs—benefits from social facilitation.

In the absence of an experienced dog, I have trained novice dogs by directing them to run across ice in the springtime, after seals have abandoned their subnivean lairs and while they are visible resting next to breathing holes on the ice surface. Because I could see the seals on the ice from distances greater than could the dogs, I was able to direct the dogs on a path that I knew would take them downwind of the seals. Often, the seals would escape through their breathing holes and into the water without the dogs seeing them, but by keeping track of where the seal had been and the wind direction, I could direct the dogs into the scent plume. Taking the trainee to basking sites located visually has the threefold advantage that the handler has greater control over the frequency of reinforcement, the dog's motivation is enhanced by sometimes seeing the seals, and the handler can observe when the trainee begins to respond appropriately to seal odor.

I have also used skin and blubber samples to teach dogs to associate seal odor with a search command, but that method is less desirable as it is out of context and might be perceived quite differently by the dog than would be the odor of live seals. The approach does have the advantage, however, that it can be employed even at times when it is impractical to take the dog into seal habitat.

To effectively detect airborne seal odor, the dog should work in front of the handler where she can be easily observed and where she is not subjected to the handler's odor. At the same time, she needs to frequently look back to take direction from the handler. Dogs can be trained to traverse in specific directions in response to voice commands and hand signals. The latter typically are more practical in the Arctic where the noise of wind and snowmobiles can interfere with vocal communication. I train puppies to take hand signals using the standard "baseball" game used to train retrievers (Wolters 1964).

Teaching the dog to run on command on a given course does not depend on being in seal habitat and, in fact, is best done in a variety of terrains and circumstances. Most retrievers are inclined by nature and training to work close to their handler, and they are especially inclined to do so when they are in unfamiliar environments. Before taking a trainee to the Arctic, I introduce them to the command "go ahead" while following them from a snowmobile or other vehicle. Whenever the dog starts to advance in front of the vehicle, I issue the command and praise the behavior. If the dog drops back, I cease the command and stop the vehicle. I avoid chastising the dog and wait until she again advances in front of the vehicle, at which point I reissue the command and words of praise.

A dog's ability to follow an airborne scent to its source is largely innate, but her proficiency in doing so improves greatly with practice. Because the scent is invisible to the handler, it is necessary to have an independent means of judging when a dog in training is downwind of a seal hole. The ideal method is to let the inexperienced dog run with an experienced dog. The novice dog typically will chase after the experienced dog and, thereby, be led to a seal hole while hearing and seeing the handler's commands. Thus, many of the necessary behaviors will develop even before the novice has developed the ability to find holes on her own. Letting her follow an experienced dog is especially valuable for developing efficient quartering behavior.

The dogs turn in to the wind when they detect an odor and run generally upwind until the odor fades at which time they turn back quartering upwind. Each time the strength of the scent diminishes, the dogs again turn back to something less than 180 degrees of their previous path (Figure 3.10.1). Experienced dogs work

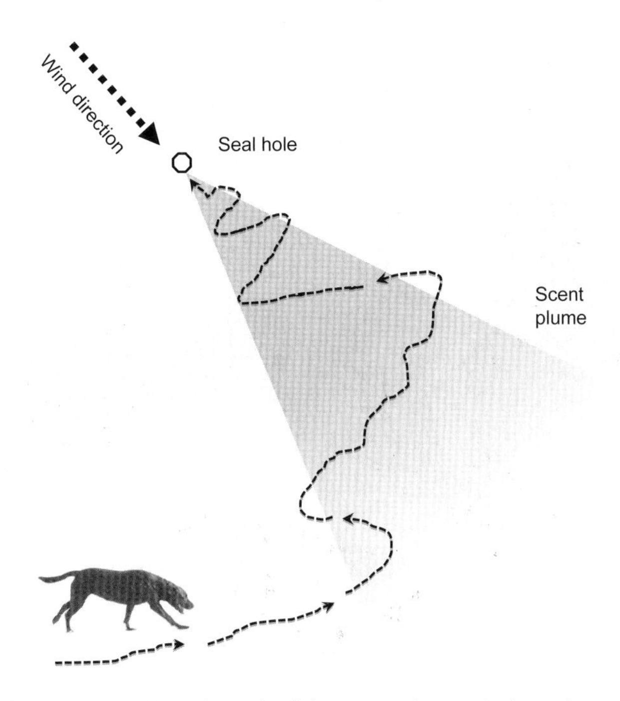


**Figure 3.10.1.** When the trained dog enters the scent plume downwind of a seal hole, she turns upwind. Each time she detects the odor fading, she turns back and upwind. The resulting zigzags become more frequent as she approaches the source of odor.

the scent plume efficiently, turning abruptly as soon as they detect a diminution of the odor. Inexperienced dogs tend to overshoot the scent plume and, once they are beyond the plume, they lack clues of which direction to travel.

Most often, a dog's natural inclination when detecting odor buried beneath snow or any other surface is to attempt to excavate to the source (see video on accompanying DVD). Reinforcing the digging behavior each time a dog digs at a known source of seal odor encourages the dog in an obvious alert. Individual dogs, however, vary widely in their inclination to dig, and each dog is inclined to give its own cues when it has located a source of seal odor. I had one dog who tended to stott on the snow above seal holes and another whose alert was detectable only as a rapid back-and-forth movement of her head while standing in one spot. Learning the alerting and other behaviors of an individual dog is crucial to their use in finding seals.

Each time a dog successfully locates a seal hole, I reinforce the behavior with enthusiastic praise. I withhold the praise, however, until I am certain that the dog has located seal odor. Most often, confirming the find requires probing the snow with an avalanche probe until a distinct ice layer in the snow—indicative of a lair—or open water—indicative of a breathing hole—is felt (Figure 3.10.2, and Video 3.10.1 on the accompanying DVD). If a lair or breathing hole is not detected by probing within a couple of minutes of arriving where the dog has alerted, I press my nose to the snow at the holes made by the probe and confirm the presence of seal by its odor.

**Figure 3.10.2.** Cooper, a trained seal dog, alerts to seal odor by digging in snow above a breathing hole. The author uses an avalanche probe to locate the subnivean hole. Joe Prouk guards against polar bear attack. Beaufort Sea, May 2008.

Repetition and consistency, of course, are critical in training dogs to locate seals by scent. Traversing long miles in search of seal odor, however, can fatigue the dogs, and their enthusiasm can wane, especially in habitats with low densities of seals. The dogs and their handlers tend to perform better in several short bouts of searching than when the same amount of search time occurs in a single bout. The dog's drive also is enhanced by occasional exposure to live seals.

Quantification of the search effort using dogs is difficult because factors such as wind speed that affect their performance cannot be controlled. Hammill and Smith (1990) used a modification of the removal sampling method to estimate the total number of seal holes in an area based on a sample found by trained dogs. We have used measures of distance and time spent searching in comparing success rates of searches. GPS tracking of working dogs can provide records of their search paths.

A great deal of ringed seal behavior and natural history can be inferred from subnivean breathing holes and lairs located by trained dogs. Evidence of birthing, molting, and depredations by predators, as well as indications of the seals' diets, often can be seen in the snow (Smith and Stirling 1975; Kelly 1988). The distribution of breathing holes and lairs reveal the features of ice and snow that are important in the seals' ecology (Hammill and Smith 1989; Kelly and Quakenbush 1990; Sipila 1990; Furgal et al. 1996; Kunnasranta et al. 2001). For example, the distribution of breathing holes often traces the networks of cracks that opened and refroze in the ice (Figure 3.10.3). When those cracks open, seals use them as breathing sites, and

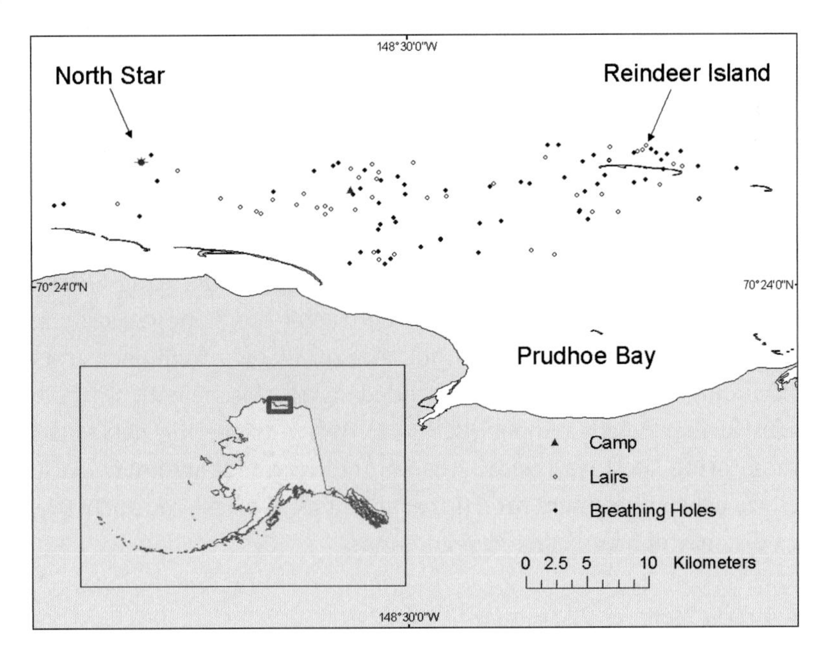


**Figure 3.10.3.** Ringed seal subnivean breathing holes and lairs located by trained dogs in the Beaufort Sea in 2000 (after Kelly et al. 2006). Linear sequences of breathing holes and lairs reflect refrozen cracks and pressure ridges, respectively.

they maintain breathing holes there as the water in the cracks refreezes. Similarly, the distribution of lairs typically reflects the distribution of pressure ridges where drifting increases snow depth. We also measure a standard suite of physical features that characterize each breathing hole or lair (Appendix 3.10.1).

## 3.10.4 LIVE CAPTURING RINGED SEALS AT BREATHING HOLES

While the distribution of breathing holes and lairs and evidence in the snow at those sites can reveal the seals' interactions with the biotic and abiotic environment, investigations of many aspects of their physiology, behavior, and ecology, however, require capture and handling. Previously, I described a method for live-capturing ringed seals in their breathing holes (Kelly 1996). The method has evolved significantly since then, and I use this opportunity to update my description of the method.

Netting seals in their breathing holes (Inuktitut *aglu*, Iñupiaq *alluvak* or *allu*) takes advantage of their need to frequently visit those holes to breathe and maintain them. Throughout the winter and spring, the seals must frequently abrade the ice to keep their breathing holes from freezing. Ice formation can also impede the functioning of the "*aglu* net," and I minimized those effects by modifying the net so as to be remotely triggered by the researchers (as opposed to by way of a seal-activated trigger) and by shortening the net and lowering its position in the breathing hole.

In late spring and early summer, ringed seals rest on top of the snow next to their breathing holes. At that time, they can be captured in nets that we refer to as *basking nets* and that are set on the surface of the ice.

Our remotely operated triggering system was designed and constructed by John Benevento, electronics shop supervisor, Geophysical Institute, University of Alaska Fairbanks. The system consists of a programmable VHF radio transmitter (Bendix King GPH-5102XP) with a touch-pad signal encoder and a "trigger box" housing a companion VHF radio receiver (also a Bendix King GPH-5102XP), a series of relays, and a signal decoder. The trigger box is powered by a 12-volt dc lead-acid battery. Each trigger box operates on its own frequency and activates a 12-volt circuit when it receives an encoded signal. A schematic of the trigger box provides further details (Appendix 3.10.2). When triggering nets within 2 km of our monitoring sites, we typically use omnidirectional antennas. At greater distances, we use multielement directional antennas. We have routinely triggered nets from a distance of 5 km using Yagi antennas.

### 3.10.4.1 *Aglu* Nets

Ringed seals have been captured in ice-free waters using tangle nets (Smith et al. 1973), and that method is effective in capturing large numbers of seals in summer or early fall. Studying seals during the breeding season when they occupy ice-covered waters (Kelly and Quakenbush 1990; Wartzok et al. 1992a,b; Kelly and Wartzok 1996; Kelly et al. 1998; Kelly 2001; 2004; 2008; Simpkins et al. 2001a,b,c; Harwood et al. 2007; Kelly et al. 2008), however, required a different capture method. I developed the *aglu* net to capture seals in breathing holes in the ice (Kelly 1996). The net lines the lower portion of the breathing hole. A lower curtain of netting is dropped and pursed after a seal passes through the open net to surface and breathe.

The *aglu* nets are made of 10 cm stretched mesh knotted nylon (#48 twine) fastened to two flexible hoops of galvanized-steel wire rope (13 cm—7/19) held in circular form with compression sleeves. The upper hoop measures 46 cm in diameter, and the lower hoop measures 70 cm in diameter. The two hoops are connected to one another by 23 cm of netting tied to the circumference of each hoop at 5–10 cm intervals (Figure 3.10.4). An additional 75 cm of netting hangs from the lower hoop, and a steel wolf snare is threaded through the mesh to serve as a purse line at the bottom of the net. The snares are made of 9.5 mm diameter, seven-stranded aircraft cable passing through a camlock lashed to the bottom of the net. I crimp a "deer stop" onto the snare 10–15 cm from the camlock to keep the snare from closing completely and injuring the captive seals. Successful use of the nets depends on

**Figure 3.10.4.** John Moran (right) supervises as Ori Badajos (left) and the author (center) rig an *aglu* net suspended from the hut's ceiling. Anchor lines are visible as coiled (white) lines hanging from the upper hoop. The bottom of the net has been pulled up at one point (just left of center) and tied with cotton thread to the lower hoop. When fully rigged, the bottom will be fastened to the lower hoop at sixteen points. Running through the bottom of the mesh is the snare/purse line.

careful attention to detail in preparing and setting the nets, and I provide step-by-step instructions of those procedures in Appendices 3.10.3 and 3.10.4.

The *aglu* nets are monitored by way of audio signals transmitted to a listening station on the ice. The sound of a seal breathing indicates that it is positioned such that the anterior portion of its body is above the net. We then send a radio signal to the trigger box, which releases a lead weight attached to the purse line under the ice. The falling weight purses the net beneath the seal, preventing it from leaving the *aglu*. Details of triggering *aglu* nets and recovering seals are provided in Appendix 3.10.5.

#### 3.10.4.2 Basking Nets

Basking nets permit capturing seals in late spring or early summer when they rest on the ice surface but no longer are concealed beneath the snow. The basking nets are made of a single wire rope hoop (70–100 cm diameter) to which is tied the same stretched mesh nylon netting used for *aglu* nets. The netting creates a shallow bag (~30 cm deep) but has no opening.

One advantage of basking nets is that they are easier to prepare than are *aglu* nets. There is no falling portion of the net requiring fastening with threads and there is no snare/purse line. The only preparation necessary before setting a basking net is lashing a wooden anchor (50 x 100 x 600 cm) to one point on the net's hoop. Methods of setting, triggering, and recovering seals from basking nets are detailed in Appendices 3.10.6 and 3.10.7.

### 3.10.5 TELEMETRICALLY RECORDING RINGED SEAL MOVEMENTS AND DIVES

To understand habitat use by ringed seals, we have investigated their at-sea and on-ice behavior and movements. We have relied primarily on:

- radio tracking via VHF transmitters to investigate on-ice resting locations in the subnivean and basking periods—primarily April, May, and June;
- acoustic tracking via ultrasonic transmitters to investigate under-ice movements primarily April, May, and June; and
- tracking via satellite-linked transmitters to investigate on-ice resting locations throughout the year.

We also have used underwater cameras deployed manually and on remotely operated vehicles under the ice and attached to a seal to visually explore the under-ice habitat (Kelly 2008).

### 3.10.5.1 Radio Tracking

In the early 1980s, we adopted and modified methods developed by Fedak et al. (1983) and others to glue VHF transmitters to the hair of ringed seals (Kelly et al. 1988; Kelly and Quakenbush 1990). We have successfully tracked ringed seals on the ice by way of VHF transmitters glued to the hair on the dorsum at a point midway between the point of maximal girth and the tail (Figure 3.10.5). We chose that attachment point so as to not restrict the animals' movement within breathing holes.

**Figure 3.10.5.** A ringed seal dives into its breathing hole carrying a VHF radio transmitter. The transmitter is glued to the hair on the seal's back.

We use a thermosetting epoxide polymer (Devcon 5-Minute Epoxy) mixed into the hair to attach the transmitters. We find it useful to use a short section of plastic pipe (7.5 cm diameter) as a dam to contain the polymer in one location while it is setting. It is important to keep the layer of polymer thin (≤ 2 cm) to minimize the seal's exposure to heat from the exothermic reaction of the polymer and its catalyst. Packing snow around the curing polymer further helps draw away the heat. In many of our early deployments, we included a nylon mesh between the hair and the instrument, but based on the experience of colleagues (John Bengston and Peter Boveng), we found that attachments without the mesh held every bit as well.

We have attached transmitters made by Telonics, Inc. (model L2B5, 30 g) and by Advanced Telemetry Systems (model MM170, 26 g) to the backs of ringed seals. Each tag was identified by a unique frequency. Most of our receiving stations were on the ice, and we received signals from distances of 5–10 km when using eight-element Yagi antennas on 11 m high rotating masts.

During the molt, the tags are shed, leaving no permanent marking on the seals. To track seals during and beyond the annual molt, we have attached Advanced Telemetry Systems VHF transmitters (model MM420, 23 g) to one hindflipper of seals by way of a cattle ear tag (Temple Tag, Inc., model 73200). A leather punch is used to create two holes (≤ 1 cm diameter) in the webbing of the hindflipper, and the ear tag is woven through the two holes (Figure 3.10.6). Brass or stainless steel screws are used to reinforce the closure of the tag.

Most often, we monitor the tags from an on-ice monitoring station (most typically an insulated hut). Typically, we monitor the frequencies of deployed tags hourly while rotating the antenna through 360 degrees and recording the direction from which each signal is received. The specific location of tagged seals can be

**Figure 3.10.6.** A plastic cattle ear tag (yellow) is woven through a pair of holes punched in the hind flipper of a ringed seal. A satellite-linked transmitter is glued to the opposite side of the tag, and its distal end and antenna are visible extending to the left past the end of the flipper. This attachment method is designed for longer deployments than are possible when instruments are attached to the hair, and we have tracked seals for as long as fourteen months by this method.

determined using a mobile receiver and handheld directional antenna array. The directional antenna array consists of two Yagi antennas communicating with the acoustic receiver by way of a null combiner. The bearing from the array to a transmitter is indicated by a null surrounded by high amplitude signals. Telonics, Inc. provides instructions for building and using a portable, direction antenna array (http://www.telonics.com/products/vhfAntennas/RA-NS_index.php). To locate radio-tagged ringed seals in their subnivean lairs, we circle the signal on skis or foot and take five or more bearings (accuracy of approximately ±3°) from points surrounding the signal. The seal's position is plotted as the intersection of those bearings. Multiple locations used by a seal can be plotted using MapSource (Garmin Corporation), transferred to an ArcView shape file, and the home range delineated as minimum convex polygons (Worton 1987).

### 3.10.5.2 Acoustic Tracking

In winter and spring, ringed seals spend most of their time under the ice (Kelly 2004), where radio tracking is of no facility. To track their under-ice movements, we attach ultrasonic transmitters (60–73 kHz, Model V3P-5HI or V3/V4P-8HI-

CPU, VEMCO Ltd.) to the hair on the seals' dorsums. A short stretch (~15 cm) of nylon line (3 mm diameter) is glued to the seal's hair and tied to the ultrasonic transmitter. Experience has shown that signal reception is enhanced by having the transmitter trail just off of the seal's body. The transmitters pulse at one-to-two-second intervals with the interpulse interval modulated by a pressure transducer.

Signals from the ultrasonic transmitters are monitored via an array of four hydrophones precisely placed under the ice in a diamond-shape configuration with a long axis of 800 m and a short axis of 80 m. Theodolites and steel measuring tapes are used to survey the hydrophone locations. Power or hand augers are used to drill holes (20 cm diameter) through the ice. Hydrophones are passed through a plastic pipe (4 cm diameter, 3 m length) housed in a second plastic pipe (6 cm diameter, 3 m length). The two pipes are separated from one another at the top and at the bottom by rubber O-rings. The pipes are placed in the holes in the ice such that 50 cm or more extend above and below the ice. When the water in the holes refreezes, the outer pipe becomes fixed in the ice, but the hydrophones can be recovered by pulling the inner pipe out of the outer pipe.

Data output from acoustic receivers (Model VR-15, VEMCO, Ltd.) are cabled to a purpose-built digital processor that precisely records the times of each pulse's arrival at the four hydrophones. The difference in arrival times is used to calculate the location in the x-y plane, while depth is determined from the pulse interval. The location data are plotted in real time to a computer screen as well as recorded on digital media. Further details of the tracking system are available in Wartzok et al. (1992a). Further details also are available on the use of three-dimensional, under-ice tracks in the analysis of navigation (Elsner et al. 1989; Wartzok et al. 1992b), diving behavior (Kelly and Wartzok 1996; Simpkins et al. 2001a), and prey searching (Simpkins et al. 2001b,c).

### 3.10.5.3 Satellite Tracking

The small home ranges occupied by ringed seals throughout the ice-bound months (Kelly 2004) are conducive to following their movements with VHF radio and ultrasonic acoustic tracking. During the open-water months, however, ringed seals range over much greater distances, and satellite tracking is a more effective approach to following those long-distance movements (Born et al. 2002; Harwood et al. 2007). The duration of tracking, however, is limited when the satellite-linked transmitters are glued to the seals' hair. Those transmitters are shed during the annual molt.

We are interested in interannual fidelity to breeding sites, so we pioneered using very small ARGOS satellite-linked transmitters (Wildlife Computer, SPOT4 and SPOT5) attached to the hindflippers by way of cattle ear tags as described previously for the attachment of VHF radio transmitters (Video 3.10.1). Battery life is prolonged by programming the tags to transmit only for a few days per week and by use of a conductivity switch that turns the transmitter off when in seawater. These tags have provided the first yearlong tracks of ringed seals (Kelly et al. 2008) and now are being deployed on other species.

## 3.10.6 GENETIC SAMPLING IN RINGED SEAL BREEDING HABITAT

The telemetric tracking of ringed seals (see video on accompanying DVD) demonstrated that at least adult seals are faithful to the same small home ranges in successive breeding seasons despite long-distance movements in between (Kelly et al. 2008). One of the implications of that finding is the possibility that ringed seals are breeding in their own natal sites. Alternatively, juvenile seals may disperse and, as adults, develop fidelity to particular breeding sites independent of their birth sites. The implications for the population structure and conservation of ringed seals are substantial, yet telemetric tracking is ill suited to distinguishing between these hypotheses in these long-lived animals. Therefore, we are seeking evidence for or against natal philopatry in the heterogeneity of microsatellite and mitochrondrial DNA markers.

Tissue samples from seals collected in the Native harvest of ringed seals have been good sources of DNA (Davis et al. 2008), but the question of stock structure hinges on knowing the extent of genetic exchange between breeding sites. The fact that ringed seals wander large distances between breeding seasons means that samples taken outside of the breeding season bear little relationship to breeding stocks.

Skin samples from seals live-captured in *aglu* nets in their breeding sites are valuable for assessing the philopatry hypothesis, but that approach is labor intensive and somewhat invasive. For the past three years, we have been obtaining DNA from shed skin samples collected from the ice around ringed seal breathing holes (Swanson et al. 2006). Flakes of skin are picked from the ice using sterilized forceps, placed in to a paper envelope, and dried (Video 3.10.1). The location of each collection site is recorded using a handheld GPS device. Collection time has been decreased by a suggestion from Kevin Bakker (University of Georgia) to shovel snow laden with shed skin onto a paper filter and then melt away the snow. We still are comparing the yield of DNA extracted from skin picked dry off of the ice versus that filtered from wet snow.

Because multiple seals sometimes shed skin next to the same breathing hole, care must be taken to separate samples that—based on melt patterns in the snow—appear to be from different individuals. Details of analytic methods are presented in Swanson et al. (2006).

## 3.10.7 PERMITTING ISSUES

Research activities involving marine mammals are regulated by local, state, and national laws, some of which require specific permits. Most funding agencies (e.g., the National Science Foundation) require documentation of permits in order to process research proposals. The time required to obtain necessary permits can be considerable, and more than a few projects have been delayed or canceled because permits were denied or delayed. Lack of compliance has resulted in substantial interruptions of research programs (Dalton 2006; Morell 2008) and fines to investigators (http://www.fakr.noaa.gov/newsreleases/2006/steller040406.htm).

In the Arctic, local governments may require authorization for research activities on their lands (e.g., the North Slope Borough, Alaska). In Canada, local governments in Nunavut, the Northwest Territories, and the northern provinces have varying regulations (http://www.ipy-api.ca/english/documents/ipy_reslic_updated_0705_e.pdf).

In the United States, the Animal Welfare Act and the Public Health Service Policy require that federally funded research on vertebrate animals be reviewed and approved by an institutional animal-use oversight committee (IACUC). Most universities maintain an IACUC (e.g., http://www.uaf.edu/iacuc/training/module_1/1_uaf_iacuc.html).

Research on marine mammals in U.S. waters requires a permit under the Marine Mammal Protection Act of 1972 as amended (http://www.nmfs.noaa.gov/pr/pdfs/laws/mmpa.pdf). The permits are issued either by the U.S. Fish and Wildlife Service (http://www.fws.gov/forms/3-200-43.pdf) in the case of manatees, sea otters, walruses, or polar bears or by the National Marine Fisheries Service (http://www.nmfs.noaa.gov/pr/laws/mmpa/) in the case of all other marine mammal species. Under the broad definitions in the Marine Mammal Protection Act, even research that only involves observations requires permitting.

### 3.10.8 FUTURE DIRECTIONS

Scientific interest in ice-associated seals has increased dramatically with recent rapid reductions in arctic sea ice cover (Kelly 2001; Stirling and Smith 2004; Ferguson et al. 2005; Freitas 2008). Their dependence on sea ice and its snow cover make them particularly vulnerable to the projected decreases in ice extent and its changing seasonality. Current knowledge of stock structure, resource selection functions, population biology, and the scope for adaptive change is inadequate for assessing the likely impacts to these species. For example, the most dramatic changes involve summer sea ice, which may be absent in less than one hundred years. Summer ice is important to ice-associated seals as a platform for molting as well as for its impact on ocean productivity. The costs of losing that substrate in terms of the energetics of molting or in terms of trophic relationships are poorly known.

There is reason for hope, however, in that scientific researchers are improving rapidly in their ability to learn from traditional knowledge and in their inventive application of new technologies. Some of the methods described here may be useful to teams of scientists and hunters working to improve our knowledge of ice-associated animals. Extending our knowledge of this ecosystem will require new innovations, and I hope that our adaptation of traditional knowledge and modern technologies to specific research problems will have heuristic value as an approach to solving research problems as much as providing specific tools and methods.

### Acknowledgments

Many friends and colleagues in the Native and scientific communities provided ideas, inspiration, and tremendous companionship on the ice. Lew Shapiro introduced me to sea ice fieldwork. John J. Burns provided my first opportunity to work with ringed seals, and he has been of great help as an employer, a collaborator, and a friend. Tom Smith taught me much about ringed seals and the use of dogs in locating them, and he continues to be a valued colleague. I was honored to work in Prince Albert Sound with Tom's longtime friend and teacher Jimmy Memorana. In recent years, Jimmy's son Roger has captured seals with us in Canada and Alaska. Mike Halstead of Research Nets in Seattle, Washington, was a wealth of ideas during the development of the net and came up with the idea of wire rope hoops to create flexible supports needed for the *aglu* nets. Many people participated in the long development of the *aglu* and basking nets, including Holly Cleator, Laura Higgins, Sue Hills, Stuart Innes, Gordon Jarrell, Karen Kelly, Lori Quakenbush, Mike Simpkins, William Stortz, Brian Taras, and Doug Wartzok. Doug was the brains behind the under-ice tracking system, coadviser on my Ph.D. committee, and a great camp companion. Peter Waser—also a coadviser on my Ph.D. committee—asked many excellent questions, was a wealth of fresh ideas, provided competent help on the ice at Resolute, and proved that it indeed is possible to get bit by a ringed seal. Henry Stone played an essential role in the design and construction of the digital processor used to measure time delays for the acoustic tracking.

Many other scientific colleagues and students participated in field excursions and provided their ideas and energy, including Steve Amstrup, Larry Aumiller, Edan Badajos, Ori Badajos, Kevin Bakker, John Bengtson, Brita Bishop, Karen Blejwas, Peter Boveng, Michelle Cronin, Ned Conway, Shannon Crowley, Raychelle Daniel, Craig Dorman, Jessie Dunton, George Durner, Lara Dzinich, Bob Elsner, Kristine Faloon, Rowena Flinn, Scott Foster, Mike Hammill, Lois Harwood, Kyndall Hildebrant, Eran Hood, Zak Hoyt, Rachel Ingram, Amy Kamerainen, Kevin Kelly, Michael Kingsley, Mervi Kunnasranta, Josh London, Dusty McDonald, John Moran, Bob Nelson, Julie Nielsen, David O'Leary, Micaela Ponce, Stephanie Sell, Lew Shapiro, Philip Singer, Alan Steffert, Gretchen Steiger, Brad Swanson, Daniella Swenton, Carrie Talus, Sam Thalman, J. J. Vollenweider, Lynn Waterhouse, and Jamie Womble. Peter Amariluk and Terry Manik provided excellent help on the ice near Resolute Bay, as did Taylor and Jimmy Moto in Deering; Jimmy Jones and Rex Snyder in Barrow; Raymond Ettagiak and Roger Memorana in Inuvik; and Ross Schaeffer and Alex Whiting in Kotzebue. Financial and logistic support was provided by DEW line site personnel at Port Clarence and Oliktok Point, the Polar Continental Shelf Project (Canada), the Resolute Hunters' and Trappers' Association, the Department of Wildlife Management (North Slope Borough), the Nanuuq Commission, the Ice Seal Committee, the Native Village of Kotzebue, the U.S. Fish and Wildlife Service, the National Marine Mammal Laboratory (NOAA), the Department of Fisheries and Oceans (Canada), BP Alaska, ConocoPhillips, the Minerals Management Service, the Office of Naval Research, the National Science Foundation, the Natural Resources Fund and the Coastal Marine Institute (University of Alaska Fairbanks), the North Pacific Research Board, and the National Geographic Society. My son, Corwin, has become increasingly adept at on-ice work and tolerated my backseat driving while he put in many miles on the snowmobile.

## REFERENCES

Addison, R. F., and T. G. Smith (1974), Organochlorine residue levels in Arctic ringed seals: Variations with age and sex, *Oikos, 25,* 335–337.

Addison, R. F., and T. G. Smith (1998), Trends in organochlorine residue concentrations in ringed seal (*Phoca hispida*) from Holman, Northwest Territories, 1972–1991, *Arctic, 51,* 253–261.

Addison, R. F., M. E. Zinck, and T. G. Smith (1986), PCBs have declined more than DDT-group residues in Arctic ringed seals (*Phoca hispida*) between 1972 and 1981, *Environ. Sci. Technol., 20,* 253–256.

Alexander, V., and H. J. Niebauer (1981), Oceanography of the eastern Bering Sea ice-edge zone in spring, *Limnol. Oceanogr., 26,* 1111–1125.

Arnason, U., A. Gulberg, A. Janke, M. Kullberg, N. Lehman, E. A Petrov, and R. Väinölä (2006), Pinniped phylogeny and a new hypothesis for their origin and dispersal, *Mol. Phylogenet. Evol., 41,* 345–354.

Arnold, C. D. (1979), Possible evidence of domestic dog in a paleoeskimo context, *Arctic, 32*(2), 263–265.

Baker, R. J., and R. D. Bradley (2006), Speciation in mammals and the genetic species concept, *J. Mamm., 87,* 643–662.

Bang, K., B. M. Jenssen, C. Lydersen, and J. U. Skaare (2001), Organochlorine burdens in blood of ringed and bearded seals from north-western Svalbard, *Chemosphere, 44,* 193–203.

Bengtson, J. L., L. M. Hiruki-Raring, M. A. Simpkins, and P. L. Boveng (2005), Ringed and bearded seal densities in the eastern Chukchi Sea, 1999–2000, *Polar Biol., 28,* 833–845.

Bérubé, M., and A. Aguilar (1998), A new hybrid between a blue whale, *Balaenoptera musculus*, and a fin whale, *B. physalus*: Frequency and implications of hybridization, *Mar. Mamm. Sci., 14,* 82–98.

Blix, A. S., H. J. Grav, and K. Ronald (1975), Brown adipose tissue and the significance of venous plexes in pinnipeds, *Acta Physiol., 236,* 133–135.

Boness, D. J., and W. D. Bowen (1996), The evolution of maternal care in pinnipeds, *Biosci., 46,* 645–654.

Born, E. W., M. Acquarone, L. Ø. Knutsen, and L. Toudal (2005), Homing behaviour in an Atlantic Walrus (*Odobenus rosmarus rosmarus*), *Aquat. Mamm., 31,* 23–33.

Born, E. W., J. Teilman, and F. Riget (2002), Haul-out activity of ringed seals (*Phoca hispida*) determined from satellite telemetry, *Mar. Mamm. Sci., 18,* 167–181.

Bornhold, B. D., C. V. Jay, R. McConnaughey, G. Rathwell, K. Rhynas, and W. Collins (2005), Walrus foraging marks on the seafloor in Bristol Bay, Alaska: A reconnaissance survey, *Geo-Mar. Ltrs., 25,* 293–299.

Bowen, W. D., O. T. Oftedal, and D. J. Boness (1985), Birth to weaning in four days: Remarkable growth in the hooded seal, *Cystophora cristata, Can. J. Zool., 63,* 2841–2846.

Braham, H. W., J. J. Burns, G. A. Fedoseev, and B. D. Krogman (1984), Habitat partitioning by ice-associated pinnipeds: Distribution and density of seals and walruses in the Bering Sea, April 1976, U.S. Department of Commerce, *NOAA Tech. Rep. NMFS, 12,* 25–48.

Buckley, J. R., T. Gammelsrød, A. Johannessen, 0. M. Johannessen, and L. P. Røed (1979), Upwelling: Oceanic structure at the edge of the arctic ice pack in winter, *Science, 203,* 165–167.

Bukhtiyarov, Y. A., K. J. Frost, and L. F. Lowry (1984), New information on the foods of the spotted seal, *Phoca largha*, in the Bering Sea in spring, in *Soviet-American Cooperative Research on Marine Mammals*, edited by F. H. Fay and G. A. Fedoseev, vol. I, pp. 55–59, Tech. Rep. 12. NOAA, Washington, DC.

Burns, J. J. (1970), Remarks on the distribution and natural history of pagophilic pinnipeds in the Bering and Chukchi seas, *J. Mamm., 51,* 445–454.

Burns, J. J., J. J. Montague, and C. J. Cowles (eds.) (1993), The bowhead whale, *Special Publication No. 2,* The Society for Marine Mammalogy, Lawrence, KS.

Burns, J. J., L. H Shapiro, and F. H. Fay (1981), Ice as Marine Mammal Habitat in the Bering Sea, in *The Eastern Bering Sea Shelf: Oceanography and Resources,* Vol. 2, edited by D. W. Hood and J. A. Calder, pp. 781–797, University of Washington Press, Seattle.

Cameron, M. (2005), Seasonal movements of ribbon seals in the Bering Sea and North Pacific, National Marine Mammal Laboratory, NMFS, Polar Ecosystems Program,

Chapskii, K. K. (1938), The bearded seal (*Erignathus barbatus* Fabr.) of the Kara and Barents seas, *Game mammals of the Barents and Kara Seas, 124,* 7–70, Fisheries and Marine Service, *Translation Series No. 3162.*

Chapskii, K. K. (1940), The ringed seal of western seas of the Soviet Arctic (The morphological characteristic, biology, and hunting production), *Tr. Vses. Arkt. Inst. (Leningrad), 145,* 1–72. Translated from Russian by Fish. Res. Board Can., 1971, *Translation Ser., 1665.*

Clark, C., W. Ellison, and K. Beeman (1986), Acoustic tracking of migrating bowhead whales, *Oceans, 18,* 341–346.

Clutton-Brock, J. (1995), Origins of the dog: domestication and early history. In *The Domestic Dog: Its Evolution, Behaviour, and Interactions with People,* edited by J. Serpell, pp. 7–20, Cambridge University Press, Cambridge.

Comiso, J. C., C. L. Parkinson, R. Gersten, and L. Stock (2008), Accelerated decline in the Arctic sea ice cover, *Geophys. Res. Lett., 35,* L01703, doi:10.1029/2007GL031972.

Cooper, L. W., I. L. Larsen, T. M. O'Hara, S. Dolvin, V. Woshner, and G. F. Cota (2000), Radionuclide contaminant burdens in arctic marine mammals harvested during subsistence hunting, *Arctic, 53*, 174–182.

Crawford, R., and J. Jorgenson (1990), Density distribution of fish in the presence of whales at the Admiralty Inlet landfast ice edge, *Arctic, 43*, 215–222.

Cronin, M. A., S. C. Amstrup, G. W. Garner, and E. R. Vyse (1991), Interspecific and intraspecific mitochondrial DNA variation in North American bears (*Ursus*), *Can. J. Zool., 69*, 2985–2992.

Dalton, R. (2006), Sea-lion studies come to a halt after court judgement, *Nature, 441*, 677.

Davies, J. L. (1958), Pleistocene geography and the distribution of the northern pinnipeds, *Ecology, 39,* 97–113.

Davis, C. S., I. Stirling, C. Strobeck, and D. W. Coltman (2008), Population structure of ice-breeding seals, *Mol. Ecol., 17,* 3078–3094.

Dehn, L. A., G. G. Sheffield, E. H. Follmann, L. K. Duffy, D. L. Thomas, G. R. Bratton, R. J. Taylor, and T. M. O'Hara (2005), Trace elements in tissues of phocid seals harvested in Alaskan and Canadian Arctic: Influence of age and feeding ecology, *Can. J. Zool., 83,* 726–746.

Dehn, L. A., G. G. Sheffield, E. H. Follmann, L. K. Duffy, D. L. Thomas, and T. M. O'Hara (2007), Feeding ecology of phocid seals and some walrus in the Alaskan and Canadian Arctic as determined by stomach contents and stable isotope analysis, *Polar Biol., 30,* 167–181.

Deméré, T. A., A. Berta, and P. J. Adam (2003), Pinnipedo-morph evolutionary biogeography, *Bulletin of the American Museum of Natural History, 279,* 32–76.

Dizon, A. E., C. Lockyer, W. F. Perrin, D. P. DeMaster, and J. Sisson (1992), Rethinking the stock concept: A phylogenetic approach, *Conserv. Biol., 6,* 24–36.

Dyke, A. S., J. Hooper, and J. M. Savelle (1996), A history of sea ice in the Canadian Arctic Archipelago based on postglacial remains of the bowhead whale (*Balaena mysticetus*), *Arctic, 49,* 235–255.

Elsner, R., D. D. Hammond, D. M. Dension, and R. Wyburn (1977), Temperature regulation in newborn Weddell seal *Leptonychotes weddelli*, in *Proceedings of the 3rd SCAR Symposium on Antarctic Biology*, edited by G. A. Llano, pp. 531–540, Smithsonian Institution, Washington, DC.

Elsner, R., D. Wartzok, N. B. Sonafrank, and B. P. Kelly (1989), Behavioral and physiological reactions of arctic seals during under-ice pilotage, *Can. J. Zool., 67,* 2506–2513.

Engelhardt, F. R., J. R. Geraci, and T. G. Smith (1977), Uptake and clearance of petroleum hydrocarbon in the ringed seal, *Phoca hispida, J. Fish. Res. Bd. Can., 34,* 1143–1147.

Fay, F. H. (1974), The role of ice in the ecology of marine mammals of the Bering Sea, in *Oceanography of the Bering Sea,* edited by D. W. Hood and E. J. Kelley, pp. 383–399, Institute of Marine Science, University of Alaska Fairbanks.

Fay, F. H. (1982), Ecology and biology of the Pacific walrus, Odobenus rosmarus divergens Illiger, U.S. Department of the Interior, Fish and Wildlife Service, *North American Fauna, 74,* 1–279.

Fay, F. H., and J. J. Burns (1988), Maximal feeding depth of walruses, *Arctic, 41,* 239–240.

Fay, F. H., B. P. Kelly, and B. A. Fay (Eds.) (1990), *The Ecology and Management of Walrus Populations: Report of an International Workshop*, 26–30 March 1990, Seattle, Washington. U.S. Marine Mammal Commission, Washington, DC.

Folkow, L. P., P. Mårtensson, and A. S. Blix (1996), Annual distribution of hooded seals (*Cystophora cristata*) in the Greenland and Norwegian seas, *Polar Biol., 16,*179–189.

Freitas, C. (2008), Habitat selection by Arctic pinnipeds: patterns, tactics and predictions in a dynamic environment, Ph.D. dissertation, University of Tromsø, Norway.

Frost, K. J., and L. F. Lowry (1980), Feeding of ribbon seals (*Phoca fasciata*) in the Bering Sea in spring, *Can. J. Zool., 58,* 1601–1607.

Furgal, C. M., S. Innes, and K. M. Kovacs (1996), Characteristics of ringed seal, *Phoca hispida,* subnivean structures and breeding habitat and their effects on predation, *Can. J. Zool., 74,* 858–874.

George, J. C., J. Zeh, R. Suydam, and C. Clark (2004), Abundance and population trend (1978–2001) of western Arctic bowhead whales surveyed near Barrow, Alaska, *Mar. Mamm. Sci., 20,* 755–773.

Geraci, J. R., and T. G. Smith (1976), Direct and indirect effects of oil on ringed seal (*Phoca hispida*) of the Beaufort Sea, *J. Fish. Res. Bd. Can., 33,* 1976–984.

Gjertz, I., K. M. Kovacs, C. Lydersen, and Ø. Wiig (2000), Movements and diving of bearded seal (*Erignathus barbatus*) mothers and pups during lactation and post-weaning, *Polar Biol., 23,* 559–566.

Gjertz, I., and C. Lydersen (1986), The ringed seal (*Phoca hispida*) spring diet in northwestern Spitzbergen, Svalbard, *Polar Research, 4,* 53–56.

Hall, C. F. (1866), *Arctic Researches and Life among the Esquimaux: Being the Narrative of an Expedition in Search of Sir John Franklin, in the Years 1860, 1861, and 1862.* Harper and Brothers, New York.

Hammill, M. O., C. Lydersen, M. Ryg, and T. G. Smith (1991), Lactation in the ringed seal (*Phoca hispida*), *Can. J. Fish. Aquat. Sci., 48*, 2471–2476.

Hammill, M. O., and T. G. Smith (1989), Factors affecting the distribution and abundance of ringed seal structures in Barrow Strait, Northwest Territories, *Can. J. Zool., 67*, 2212–2219.

Hammill, M. O., and T. G. Smith (1990), Application of removal sampling to estimate the density of ringed seals (*Phoca hispida*) in Barrow Strait, Northwest Territories, *Can. J. Fish. Aquat. Sci., 47*, 244–250.

Hammill, M. O., and T. G. Smith (1991), The role of predation in the ecology of the ringed seals in Barrow Strait, Northwest Territories, Canada, *Mar. Mamm. Sci., 7*, 123–135.

Hammill, M. O., and G. B. Stenson (2000), Estimated prey consumption by harp seals (*Phoca groenlandica*), hooded seals (*Cystophora cristata*), grey seals (*Halichoerus grypus*), and harbour seals (*Phoca vitulina*) in Atlantic Canada, *J. Northw. Atl. Fish. Sci., 26*, 1–23.

Hare, B., M. Brown, C. Williamson, and M. Tomasello (2002), The domestication of social cognition in dogs, *Science, 298*, 1634–1636.

Harington, C. R. (2008), The evolution of Arctic marine mammals, *Ecological Applications, 18*(2), S23–S40, Supplement.

Hartley, J. W., and D. Pitcher (2002), Seal finger—tetracycline is first line, *J. Infection, 45*(2), 71–75.

Harwood, L., T. G. Smith, and H. Melling (2007), Assessing the potential effects of near shore hydrocarbon exploration on ringed seals in the Beaufort Sea region 2003–2006, *Environmental Studies Research Funds Report No. 162,* Fisheries and Oceans Canada. Yellowknife, Northwest Territories, Canada.

Hunt, T. D., M. H. Ziccardi, F. M. D. Gulland, P. K. Yochem, D. W. Hird, T. Rowles, and J. A. K. Mazet (2008), Health risks for marine mammal workers, *Disease Aquat. Org., 81*, 81–92.

Hyvärinen, H., T. Sipilä, M. Kunnasranta, and J. T. Koskela (1998), Mercury pollution and the Saimaa ringed seal (*Phoca hispida saimensis*), *Mar. Poll. Bull., 36*, 76–81.

Jay, C. V., and G. W. Garner (2002), Performance of a satellite-linked GPS on Pacific walruses (*Odobenus rosmarus divergens*), *Polar Biol., 25*, 235–237.

Kellogg, W. N., R. Kohler, and H. N. Morris (1953), Porpoise sounds as sonar signals, *Science, 117*, 239–243.

Kelly, B. P. (1988), Bearded Seal (*Erignathus barbatus*), In *Selected Marine Mammals of Alaska: Species Accounts with Research and Management Recommendations,* edited by J. W. Lentfer, pp. 77–94, Marine Mammal Commission, Washington, DC.

Kelly, B. P. (1996), Live capture of ringed seals in ice-covered waters, *J. Wildl. Man., 60*, 678–684.

Kelly, B. P. (2001), Climate Change and Ice Breeding Pinnipeds, in *"Fingerprints" of Climate Change: Adapted Behaviour and Shifting Species' Ranges*, edited by G.-R. Walther, C. A. Burga, and P. J. Edwards, pp. 43–55, Kluwer Academic/ Plenum Publishers, New York and London.

Kelly, B. P. (2004), Correction factor for ringed seal surveys in northern Alaska. Final Reprt to the Coastal Marine Institute, University of Alaska Fairbanks.

Kelly, B. P. (2008), Ringed seals: An instrumentation challenge, in *Proceedings of the 2007 Animal-borne Imaging Symposium,* Oct. 10–13, 2007, edited by G. Marshall, pp. 145–148, National Geographic Society, Washington, DC.

Kelly, B. P., P. Boveng, and B. R. Swanson (2008), Ice seal movements and stock structure in a changing cryosphere, *North Pacific Research Board Final Report 515,* accessed electronically at http://doc.nprb.org/web/05_prjs/515_Final_report.pdf.

Kelly, B. P., J. J. Burns, and L. T. Quakenbush (1988), Responses of ringed seals (*Phoca hispida*) to noise disturbance, in *Port and Ocean Engineering under Arctic Conditions,* vol. II, edited by W. M. Sackinger, M. O. Jeffries, J. L. Imm, and S. D. Treacy, pp. 27–38, Symposium on Noise and Marine Mammals in Ice-Covered Waters. Geophysical Institute, University of Alaska Fairbanks.

Kelly, B. P., and L. T. Quakenbush (1987), Trained dogs and wild seals, *Whalewatcher, Journal of the American Cetacean Society*, *21*, 8–11.

Kelly, B. P., and L. T. Quakenbush (1990), Spatiotemporal use of lairs by ringed seals (*Phoca hispida*), *Can. J. Zool., 68,* 2503–2512.

Kelly, B. P., and D. Wartzok (1996), Ringed seal diving behavior in the breeding season, *Can. J. Zool., 74,* 1547–1555.

Kelly, B. P., A. R. Whiteley, and D. A. Tallmon (in review), Diminished sea ice cover and the potential for extinctions by introgressive hybridization, *Evol. Appl.*

King, J. E. (1983), *Seals of the World.* Cornell University Press, Ithaca, NY.

Kingsley, M. C. S., M. O. Hammill, and B. P. Kelly (1990), Infrared sensing of the under-snow lairs of the ringed seal, *Mar. Mamm. Sci., 6,* 339–347.

Kingsley, M. C. S., I. Stirling, and W. Calvert (1985), The distribution and abundance of seals in the Canadian high Arctic, 1980–1982, *Can. J. Fish. Aquat. Sci., 42,* 1189–1210.

Kooyman, G. L. (1973), Respiratory adaptations in marine mammals, *Am. Zool., 13,* 457–468.

Kooyman, G. L., M. A. Castellini, and R. W. Davis (1981), Physiology of diving in marine mammals, *Annu. Rev. Physiol., 43,* 343–356.

Kovacs, K. M., and D. M. Lavigne (1992), Mass-transfer efficiency between hooded seal (*Cystophora cristata*) mothers and their pups in the Gulf of St. Lawrence, *Can. J. Zool., 70,* 1315–1320.

Kovacs, K. M., C. Lydersen, and I. Gjertz (1996), Birth-site characteristics and prenatal molting in bearded seals (*Erignathus barbatus*), *J. Mamm., 77,* 1085–1091.

Kovacs, K. M., C. Lydersen, M. O. Hammill, B. N. White, P. J. Wilson, and S. Malik (1997), A harp seal X hooded seal hybrid, *Mar. Mamm. Sci., 13,* 460–468.

Krogman, B., D. Rugh, R. Sonntag, J. Zeh, and K. Daijin (1989), Ice-based census of bowhead whales migrating past Point Barrow, Alaska, 1978–1983, *Mar. Mamm. Sci., 5,* 116–138.

Kucklick, J. R., M. M. Krahn, P. R. Becker, B. J. Porter, M. M. Schantz, G. S. York, T. M. O'Hara, and S. A. Wise (2006), Persistent organic pollutants in Alaskan ringed seal (*Phoca hispida*) and walrus (*Odobenus rosmarus*) blubber, *J. Environ. Monitoring, 8,* 848–854.

Kucklick, J. R., W. D. J. Struntz, P. R. Becker, G. W. York, T. M. O'Hara, and J. E. Bohonowych (2002), Persistent organochlorine pollutants in ringed seals and polar bears collected from northern Alaska, *The Sci. Total Environ., 287,* 45–59.

Kunnasranta, M. (2001), Behavioural biology of two ringed seal (*Phoca hispida*) subspecies in the large European lakes Saimaa and Ladoga, PhD. dissertation, University of Joensuu, Finland.

Kunnasranta, M., H. Hyvärinen, T. Sipila, and M. Medvedev (2001), Breeding habitat and lair structure of the ringed seal (*Phoca hispida ladogensis*) in northern Lake Ladoga in Russia, *Polar Biol., 24,* 171–174.

Landcaster, M. L., N. J. Gemmell, S. Negro, S. Goldsworthy, and P. Stunnucks (2006), Ménage á troison Macquarie Island: Hybridization among three species of fur seals (*Arctocephalus* spp.) following historical population extinction, *Mol. Ecol., 15,* 3681–3692.

Lavigne, D. M., and K. M. Kovacs (1988), *Harps and Hoods: Ice-Breeding Seals of the Northwest Atlantic.* University of Waterloo Press, Ontario, Canada.

Legendre, L., S. F. Ackley, G. S. Dieckmann, B. Gulliksen, R. Horner, T. Hoshiai, I. A. Melnikov, W. S. Reeburgh, M. Spindler, and C. W. Sullivan (1992), Ecology of sea ice biota, *Polar Biol., 12,* 429–444.

Leonard, J. A., R. K. Wayne, J. Wheeler, R. Valadez, S. Guillén, and C. Vilà (2002), Ancient DNA evidence for Old World origin of New World dogs, *Science, 298,* 1613–1616.

Lewin, M. R., P. Knott, and M. Lo (2004), Seal finger, *Lancet, 364,* 448.

Lukin, L. R., and V. A. Potelov (1978), Living conditions and distributions of ringed seals in the White Sea in winter, *Biol. Morya (Vladivost.), 3,* 62–69.

Lydersen, C., and I. Gjertz (1986), Studies of the ringed seal (*Phoca hispida,* Schreber, 1775) in its breeding habitat in Kongsyorden, Svalbard, *Polar Research, 4,* 57–63.

Lydersen, C., and M. O. Hammill (1993a), Activity, milk intake, and energy consumption in free-living ringed seal (*Phoca hispida*) pups during the nursing period, *Can. J. Zool., 71,* 991–996.

Lydersen, C., and M. O. Hammill (1993b), Diving in ringed seal (*Phoca hispida*) pups, *J. Comp. Physiol. B, 163,* 433–438.

Lydersen, C., M. S. Ryg, M. O. Hammill, and P. J. O'Brien (1992), Oxygen stores and aerobic dive limit of ringed seals (*Phoca hispida*). *Can. J. Zool., 70,* 458–461.

Lydersen, C., and T. G. Smith (1989), Avian predation on ringed seal *Phoca hispida* pups, *Polar Biol., 9*, 489–490.

Mansfield, A. W. (1967), Seals of arctic and eastern Canada, *Fish. Res. Bd. Can. Bull., 137, 1–35*.

McLaren, I. A. (1958), The biology of the ringed seal (*Phoca hispida Schreber*) in the eastern Canadian Arctic, *Fish. Res. Board Can. Bull., 118*.

Morell, V. (2008), Puzzling over a Steller whodunit, *Science, 320,* 44–45.

Morey, D. F., and K. Aaris-Sørensen (2002), Paleoeskimo dogs of the eastern Arctic, *Arctic, 55,* 44–56.

Norris, K. S. (1968), The Evolution of Acoustic Mechanisms in Odontocete Cetaceans, in *Evolution and Environment*, edited by E. T. Drake, pp. 297–324, Yale University Press, New Haven.

O'Corry-Crowe, G. (2008), Climate change and the molecular ecology of Arctic marine mammals, *Ecol. Appl., 18,* S56–S76.

Overland, J. E., and M. Wang (2007), Future regional Arctic sea ice declines, *Geophys. Res. Lett., 34,* L17705, doi:10.1029/2007GL030808.

Øristland, N. A. (1970), Energetic significance of absorption of solar radiation in polar homeotherms, in *Antarctic Ecology*, edited by M. W. Holgate, vol. 1, pp. 464–470, Academic Press, London.

Potelov, V. A. (1975), Biological background for determining the abundance of beard seals (*Erignathus barbatus*) and ringed seals (*Pusa hispida*), *Rapp. P.-v. Reun. Cons. int. Explor. Mer., 169*, 553.

Savolainen, P., Y. Zhang, J. Luo, J. Lundeberg, and T. Leitner (2002), Genetic evidence for an East Asian origin of domestic dogs, *Science, 298,* 1610–1613.

Sergeant, D. E. (1974), A rediscovered whelping population of hooded seals *Cystophora cristata* Erxleben and its possible relationship to other populations, *Polarforschung, 44,* 1–7.

Sergeant, D. E. (1976), History and present status of populations of harp and hooded seals, *Biol. Conserv., 10,* 95–117.

Serreze, M. C., M. M. Holland, and J. Stroeve (2007), Perspectives on the Arctic's shrinking sea ice cover, *Science, 315,* 1533–1536.

Shustov, A. P. (1965), The food of ribbon seal in the Bering Sea, *Izv. TINRO, 59,* 178–183.

Simpkins, M. A. (2001a), Three-dimensional analysis of search behaviour by ringed seals, *Anim. Behav., 62,* 67–72.

Simpkins, M. A., B. P. Kelly, and D. Wartzok (2001b), Three-dimensional diving behaviors of ringed seals (*Phoca hispida*), *Marine Mammal Science, 17,* 909–925.

Simpkins, M. A., B. P. Kelly, and D. Wartzok (2001c), Three-dimensional movements within individual dives by ringed seals (*Phoca hispida*), *Can. J. Zool., 79,* 1455–1464.

Simpkins, M. A., L. M. Hiruki-Raring, G. Sheffield, J. M. Grebmeier, and J. L. Bengtson (2003), Habitat selection by ice-associated pinnipeds near St. Lawrence Island, Alaska in March 2001, *Polar Biol., 26,* 577–586.

Sipila, T. (1990), Lair structure and breeding habitat of the Saimaa ringed seal (*Phoca hispida saimensis Nordq.*) in Finland. *Finnish Game Res., 47,* 11–20.

Smith, T. G. (1976), Predation on ringed seal pups (*Phoca hispida*) by the arctic fox (*Alopex lagopus*), *Can. J. Zool, 54,* 1610–1616.

Smith, T. G. (1987), The ringed seal, *Phoca hispida*, of the Canadian western Arctic. *Can. Bull. Fish and Aquat. Sci., 216.*

Smith, T. G. (1980), Polar bear predation of ringed and bearded seals in the land-fast sea ice habitat, *Can. J. Zool., 58,* 2201–2209.

Smith, T. G., B. Beck, and G. A. Sleno (1973), Capture, handling, and branding of ringed seals, *J. Wildlf. Manage., 37,* 579–583.

Smith, T. G., and J. R. Geraci (1975), The effect of contact and ingestion of crude oil on ringed seals of the Beaufort Sea, *Tech. Rep. Beaufort Sea Proj. No. 5.*

Smith, T. G., M. O. Hammill, and G. Taugbøl (1991), A review of the developmental, behavioural and physiological adaptations of the ringed seal, *Phoca hispida*, to life in the arctic winter, *Arctic, 44,* 124–131.

Smith, T. G., and I. Stirling (1975), The breeding habitat of the ringed seal (*Phoca hispida*). The birth lair and associated structures, *Can. J. Zool., 53,* 1297–1305.

Smith, T. G., and I. Stirling (1978), Variation in the density of ringed seal (*Phoca hispida*) birth lairs in the Admunsen Gulf, Northwest Territories, *Can. J. Zool., 56,* 1066–1071.

Spilliaert, R., G. Vikingsson, U. Arnason, A. Palsdottir, J. Sigurjonsson, and A. Arnason (1991), Species hybridization between a female blue whale (*Balaenoptera musculus*) and a male fin whale (*B. physalus*): Molecular and morphological documentation, *J. Hered., 82,* 269–274.

Stanley, S. M. (1979), *Macroevolution, Pattern, and Process.* W. H. Freeman, San Francisco.

Stenson, G. B., M. O. Hammill, and J. W. Lawson (1997), Predation by harp seals in Atlantic Canada: Preliminary consumption estimates for Arctic cod, capelin, and Atlantic cod, *J. Northw. Fish. Sci., 22,* 137–154.

Stenson G. B., and B. Sjare (1997), Seasonal distribution of harp seals, *Phoca groenlandica,* in the Northwest Atlantic, *Int. Counc. Explor. Sea Comm. Fleet 1997/CC, 10.*

Stirling, I. (1977), Adaptations of Weddell and Ringed Seals to Exploit the Polar Fast Ice Habitat in the Absence or Presence of Surface Predators, in *Adaptations within Antarctic Ecosystems,* edited by G. A. Llano, pp. 741–748, *Proc. 3rd Symp. Antarc. Biol.* Gulf Publishing Co., Houston.

Stirling, I. (1997), The importance of polynyas, ice edges, and leads to marine mammals and birds, *J. Mar. Sys., 10,* 9–21.

Stirling, I., and T. G. Smith (2004), Implications of warm temperatures and an unusual rain event for the survival of ringed seals on the coast of southeastern Baffin Island, *Arctic*, *57*, 59–67.

Swanson, B. J., B. P. Kelly, C. K. Maddox, and J. R. Moran (2006), Shed skin as a source of DNA for genotyping seals, *Mol. Ecol. Notes*, *6*, 1006–1009.

Talbot, S. L., and G. G. Shields (1996), A phylogeny of the bears (*Ursidae*) inferred from complete sequences of three mitochondrial genes, *Mol. Phylogen. Evol.*, *5*, 567–575.

Taugbøl, G. (1984), Ringed seal thermoregulation, energy balance and development in early life, a study of *Pusa hispida* in Kongsfj., Svalbard, thesis. University of Oslo, Norway.

Taylor, B. L., M. Martinez, T. Gerrodette, J. Barlow, and Y. N. Hrovat (2007), Lessons from monitoring trends in abundance of marine mammals, *Mar. Mamm. Sci.*, *23*, 157–175.

Wade, P. R., and R. Angliss (1997), *Guidelines for Assessing Marine Mammal Stocks: Report of the GAMMS Workshop April 3–5, 1996*, Seattle, Washington, NOAA Tech. Memo. NMFS-OPR-12.

Waits, L. P., S. L. Talbot, R. H. Ward, and G. F. Shields (1998), Mitochondrial DNA phylogeography of the North American brown bear and implications for conservation, *Conserv. Biol.*, *12*, 408–417.

Wartzok, D., S. Sayegh, H. Stone, J. Barchak, and W. Barnes (1992a), Acoustic tracking system for monitoring under-ice movements of polar seals, *J. Acoustic Soc. Am.*, *92*, 682–687.

Wartzok, D., R. Elsner, H. Stone, B. P. Kelly, and R. W. Davis (1992b), Under-ice movements and sensory basis of hole finding by ringed and Weddell seals, *Can. J. Zool.*, *70*, 1712–1722.

Wathne, J. A., T. Haug, and C. Lydersen (2000), Prey preference and niche overlap of ringed seals *Phoca hispida* and harp seals *P. groenlandica* in the Barents Sea, *Mar. Ecol. Prog. Ser.*, *194*, 233–239.

Weslaswki, J. M., M. Ryg, T. G. Smith, and N. A. Øritsland (1994), Diet of ringed seals (*Phoca hispida*) in a fjord of West Svalbard, *Arctic*, *47*, 109–114.

Willis, P. M., B. J. Crespi, L. M. Dill, R. W. Baird, and M. B. Hanson (2004), Natural hybridization between Dall's porpoises (*Phocoenoides dalli*) and harbour porpoises (*Phocoena phocoena*), *Can. J. Zool.*, *82*, 828–834.

Wolters, R. A. (1964), *Water Dog*. E. P. Dutton, New York.

Worton, B. J. (1987), A review of models of home range for animal movement, *Ecol. Model.*, *38*, 277–298.

Woshner, V. M., T. M. O'Hara, G. R. Bratton, R. S. Suydam, and V. R. Beasley (2001), Concentrations and interactions of selected essential and non-essential elements in ringed seals and polar bears of Arctic Alaska, *J. Wildl. Dis.*, *37*, 711–721.

## APPENDIX 3.10.1: STANDARD DATA RECORDED AT RINGED SEAL HOLES

Each hole located is assigned an accession number that incorporates the year and the type of hole:

B = Breathing hole, a hole maintained through the ice by seals without an associated lair

H = "Haulout" lair, a breathing hole above which is a chamber excavated by a seal in the snow or a natural cavity within deformed ice (typically a pressure ridge) showing evidence of seal occupation

P = "Pupping" lair, as with a haulout lair but also containing evidence of a occupation by a newborn seal

Also recorded are the location (GPS coordinates), date, and time. The status of each hole is recorded as:

O = "Open" if the breathing hole shows no sign of refreezing at the surface

P = "Partially frozen" if the breathing hole has begun to refreeze but still has an opening large enough for a seal's nose

F = "Frozen" if the surface of the breathing hole is completely refrozen

I = "Inactive" refers to lairs only when ice buildup around the periphery of the breathing hole or collapsing snow above indicate that the site is in recent use as a breathing hole that the seal is no longer coming out on to the surface of the ice

We also record the method of examination of the hole as:

D = Sniffed by a "dog" standing above the hole

P = "Probed" with an avalanche probe

O = "Opened" if a portion of the snow cover was removed by the investigator

A = "Altered" if a net or other device was placed in the hole

The odor of rutting males at the hole is recorded as:

0 = None detected

1 = Faint

2 = Moderate

3 = Strong

Evidence of predator activity at a hole is recorded as:

FM = Marked with urine or feces by a fox
FE = Lair entered by a fox as evidenced by an open or drifted-over tunnel
FK = Remains of a seal pup killed by foxes
FT = Fox tracks around the hole
BM = Marked with urine or feces by a bear
BE = Lair or breathing hole uncovered by a bear
BK = Remains of a seal killed by a bear
BT = Bear tracks around a hole
N = No evidence of predator activity

Measurements taken at breathing holes include the diameter (cm) of the breathing hole at the ice surface, the water depth (m), snow depth (cm) measured with a graduated avalanche probe, and percentage of the ice surface within 200 m of the hole that is deformed (hummocks and/or ridges).

Additional measurements taken at lairs include the height (cm) of the ice hummock or ridge adjacent to the snowdrift in which the lair is excavated, the compass bearing of snowdrift (from high to low point), the longest dimension (cm) of the lair, the floor-to-ceiling depth (cm) of the lair at its highest point, and the maximal depth (cm) of the portion of the snowdrift containing the lair.

For those lairs opened and inspected, we record whether there was evidence of a seal pup:

L = Lanugo (natal hair)
T = Small, pup-sized tunnels in the snow
B = Blood-stained birth site
P = Live pup
K = Dead pup or a kill site

## APPENDIX 3.10.2: SCHEMATIC DRAWING OF TRIGGER BOX

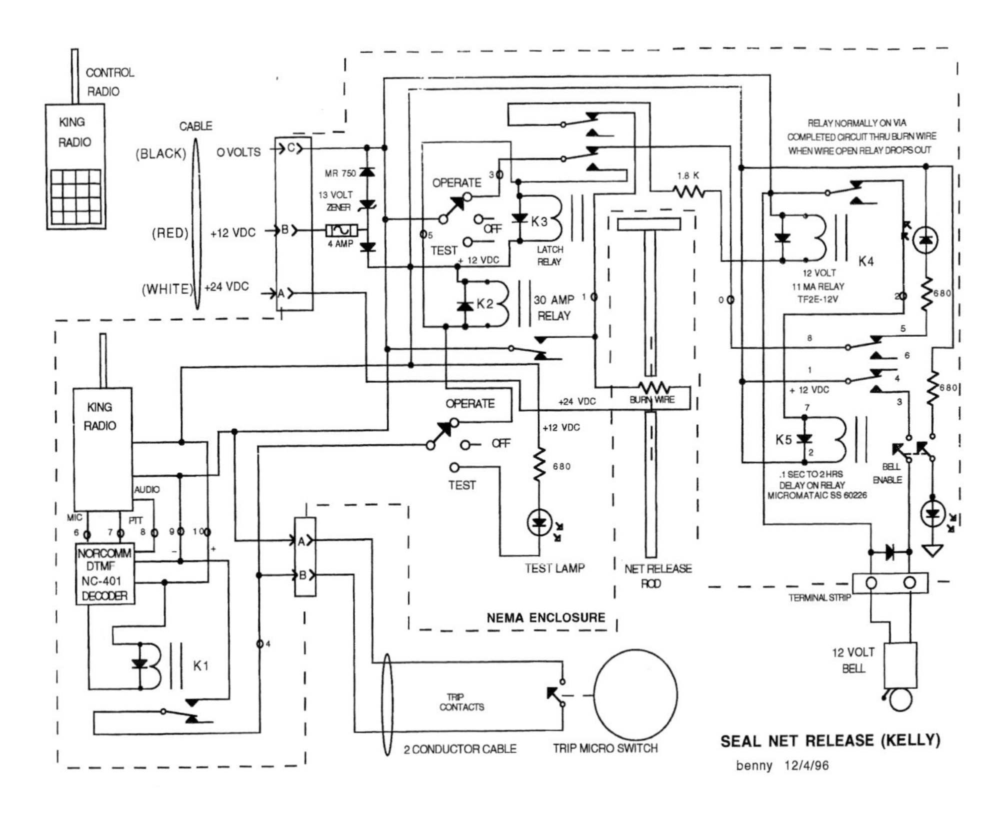


## APPENDIX 3.10.3: *AGLU* NET PREPARATION INSTRUCTIONS

Preparing the *aglu* nets is best accomplished with two pairs of hands.

Suspend the net at a comfortable height.

Tie four anchor lines (9.5 mm diameter x 2 m long) to the upper hoop at equally spaced points.

Pull the snare/purse line and the bottom of the net up to the lower hoop, and fasten them to the hoop at sixteen equally spaced points using cotton thread. The netting should be equally distributed around the hoop and the threads tied securely to minimize the bulk of net hanging from the hoop.

Tie a 1–2 m length of 3 mm diameter nylon line to a loop in the free end of the snare/purse line and temporarily coil it and attach it to the upper hoop. The latter line will be used to attach a weight that will close the net, and the attachment line must be long enough to put the weight at least 1 m below the undersurface of the ice when the net is set but not so long as to allow the weight to land on the seafloor before it closes the net. The combined length of the net, the extended snare/purse line, and the weight attachment line limit the net to water depths of 3 m or greater.

Prepare a burn wire using insulated copper wire (20 gauge, seven-stranded). Cut a piece of wire to a length of 25 cm and strip the insulation from a 5 cm stretch in the middle and from 1 cm at each end of the wire. Unwind the strands in the middle of the wire, cut away three or four of the seven, and rewind the remaining strands.

## APPENDIX 3.10.4: *AGLU* NET SETTING INSTRUCTIONS

We set *aglu* nets in breathing holes that do not have a lair above them or in the breathing holes of lairs where evidence in the snow indicates that the seals are still using the site for breathing but not for coming out of the water. When seals are using such sites only for breathing, the maintained hole is smaller in diameter and a rim of ice forms around it as a result of water being pushed upward when seals surface.

Setting an *aglu* net is best accomplished with two or three people. Ideally, a breathing hole will be uncovered, the net set, and the hole covered again before the seal attempts to visit the site. The likelihood of the seal surfacing in a net is diminished if they attempt to visit while the hole is uncovered. Thus, it is preferable to set nets efficiently and minimize time at the site. We have found that a checklist of tools and equipment minimizes delays (Table 3.10.4.1).

**Appendix Table 3.10.4.1.** Net Setting Equipment Checklist

- ❑ Avalanche probe
- ❑ Shovels (square-bladed spade)
- ❑ Ice strainer (plastic mesh/Vexar stretched across a metal frame, 20 cm diameter)
- ❑ Spikes (or wooden anchors)
- ❑ Hammer for driving spikes (2.5 pound)
- ❑ Ice chisel (socket slick mounted on 2 m wooden handle)
- ❑ Cotton thread (in case some securing the netting broke in transport)
- ❑ Pushing pole (1 × 5 × 150 cm, one end notched to fit over net hoop)
- ❑ Lead weight (4 kg, "cannon ball" style salmon fishing weight)
- ❑ Monofilament line (27 kg test)
- ❑ Burn wire (and extras)
- ❑ Trigger box
- ❑ Battery for trigger box (12 volt)
- ❑ Antenna for triggering receiver (VHF, omni or directional depending on range)

- ❑ Microphone
- ❑ Audio transmitter (Telonics TS200 or TS300)
- ❑ Battery for audio transmitter (12 volt)
- ❑ Antenna for audio transmitter (VHF, omni or directional depending on range)

Carefully remove the snow from above the breathing hole. If possible, remove hard, metamorphosed snow in a single block.

If the breathing hole is partially refrozen, use an ice chisel to open it to its original full diameter. A woodworker's socket slick on a long handle (1.5–2 m) makes an effective ice chisel.

Use the ice strainer to clear the *aglu* surface of ice and snow. Examine the shape of the hole—particularly at the lower surface of the ice—and determine where the monofilament line supporting the weight should enter the hole. Most seal holes are asymmetric and wear patterns indicate biases in the direction from which seals enter the holes. Close examination will reveal the best location for the weight-supporting line such that it hangs as far to the periphery of the hole as possible. Less desirable locations would have that line hanging near the center of the hole at the lower surface of the ice.

Establish four evenly spaced attachment points around the hole either by driving 30 cm long spikes into the ice or by burying wooden anchors (50 × 100 × 600 cm) in the snow. Most often, we use spikes placed 30–50 cm from the *aglu*. The wooden anchors are useful in warm conditions when the ice is too soft to securely hold spikes.

Examine the net carefully to ensure that all of the threads fastening the lower portion to the lower hoop are intact, that the camlock is not hooked to the webbing, and that the anchor lines are securely fastened.

Lay the steel hoops of the net one upon another, compress them into an oval, and push them down into the breathing hole (Figure 3.10.7).

As one or two assistants lower the net by way of the anchor lines, observe its progress underwater and signal when it is completely below the ice. Rotate the net so that the snare's camlock is in line with the point where the monofilament line will support the weight.

Observe the net as the assistant(s) pull the anchor lines upward until the upper hoop of the net is wedged in the lower portion of the breathing hole and the lower hoop is against the undersurface of the ice (Figure 3.10.8).

**Figure 3.10.7.** The hoops of an *aglu* net are compressed as it is inserted into an *aglu*. The four anchor lines (white) are being held by assistants (out of the frame). After the net is in place at the bottom of the *aglu*, the anchor lines will be tied to the spikes driven into the ice.

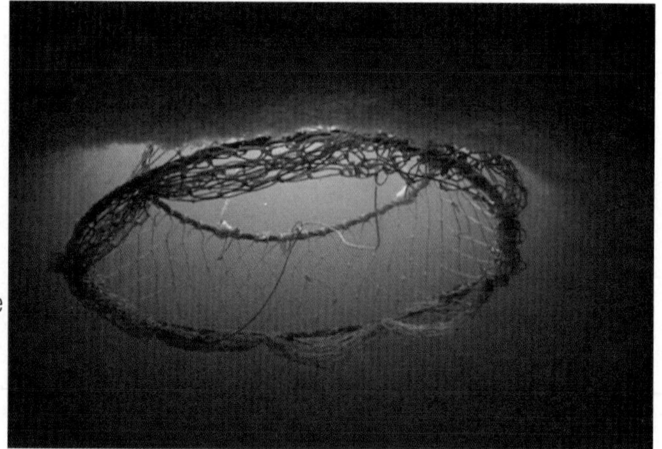

**Figure 3.10.8.** An *aglu* net being deployed under the ice. The lower curtain of netting is tied to the lower hoop. The net is being pulled up by the anchor lines until the upper hoop is within the *aglu* and the lower hoop against or nearly against the under surface of the ice.

Tie the anchor lines to the spikes or wooden anchors.

Tie the weight attachment line and the monofilament line to the lead weight.

Lower the weight slowly on the monofilament line until it is 1 m or more below the underside of the ice depending on the water depth and the length of the nylon line connecting the weight to the snare/purse line. At least a few centimeters of slack must remain in the latter line. Note that the choice of monofilament is deliberate as it will pull freely through any ice that forms around it whereas lines with even a small amount of texture tend to become stuck in such ice.

Tie the monofilament line to an anchor next to the breathing hole. We typically use the battery powering the triggering circuit as the anchor point.

Wrap the bared middle portion of the burn wire tightly around the monofilament line between its anchor point and the breathing hole. Successive wraps of the burn wire should touch one another so that the heat will be concentrated on the monofilament line (Figure 3.10.9). Be careful to position the burn wire on the breathing hole side of the knot fastening the monofilament to the anchor. A knot on the traveling end of the severed monofilament can catch on ice or other surfaces and prevent the net from closing.

**Figure 3.10.9.** Remotely operated triggering system for *aglu* nets. The blue crate houses a 12-volt battery and the trigger box (gray). The monofilament line supporting the net closing weight is visible passing through the round hole on the front right of the crate. A loop in the end of the monofilament line rests on a carriage bolt connected to the inside wall of the crate. The burn wire (red) is wrapped around the monofilament line between the loop and the round hole in the crate and terminates at the trigger box. The yellow cable delivers power from the battery to the trigger box. The black coaxial cable connected to the trigger box leads to an antenna (out of frame). The yellow device attached to the inside of the crate (and seen supporting one end of a large clevis pin) is a radio transmitter and part of an experiment to detect inadvertent breakage of the monofilament line. The Leatherman tool (left) has gone missing. Let us know if you find it.

Attach the ends of the burn wire to terminals of the trigger box. Arm the trigger box by setting the power toggle switch to the "on" position.

Carefully cover the breathing hole with snow using—whenever possible—the metamorphosed snow that originally covered it (Figure 3.10.10). Bore a small hole (~1 cm diameter) in the snow over the breathing hole and insert a small microphone such that it is roughly centered over and about 20 cm above the breathing hole. Backfill the original excavation filling voids with loose snow taking care to

**Figure 3.10.10.** Mervi Kunnasranta prepares to re-cover an *aglu* in the Arctic Ocean. A dome carved in the underside of the snow block will be situated above the *aglu*. Faintly visible in the apex of the dome is a hole bored to accommodate a microphone. The net's anchor lines are seen tied to spikes driven in to the ice around the *aglu*. Behind the *aglu*, the trigger box's antenna is seen mounted on a wooden mast.

re-create—to the extent possible—the original snow cover. While it is not possible to exactly duplicate that cover, it is important to insulate the hole from the cold air and, more importantly, to approximate the original light penetration to the hole. Ringed seals are very sensitive to the light from above breathing holes and will avoid holes if the light penetration has substantially increased or decreased. Mark the perimeter of the excavated area with wooden stakes or other markers. Drifting snow often obliterates the outline of the excavation, and the stakes will expedite locating and uncovering the hole when a seal is captured.

Secure the audio transmitter, and connect the microphone, antenna, and power supply. We use Telonics TS200 and TS300 audio transmitters. Test the audio transmissions and the communication between the triggering transmitter and receiver (the system includes a testing code) before leaving the site.

## APPENDIX 3.10.5: TRIGGERING THE *AGLU* NET AND RECOVERING THE SEAL

At the monitoring station, assign a unique channel to the FM receiver (we use an ICOM R7000, Bellevue, Washington) corresponding to the frequency of the audio transmitter. Monitor the signal continuously or at 5–15 second intervals if scanning multiple channels corresponding to multiple nets.

When a seal approaches a breathing hole, the first audible indications typically are the sound of water surging in the breathing hole or bubbles (exhaled by the seal) bursting at the surface. Whenever such sounds are received, lock the receiver on the corresponding channel, power up and set the channel on the triggering radio, and listen carefully for breathing, which can vary from explosive to extremely soft. Transmit the trigger code only when the seal is breathing as that signals that she is sufficiently encircled to be captured when the net closes. We typically wait until we hear three breaths in a row before triggering. Skittish seals sometimes will take one or two breaths and then abruptly dive, and triggering in such instances often results in missing the seal. On the other hand, when a seal takes three breaths in sequence, it usually indicates she is not about to dive abruptly.

The coded transmission will be decoded in the trigger box and close the circuit containing the burn wire. The midsection of the compromised wire will heat up and melt through the monofilament allowing the weight to fall. As the weight falls, it pulls the snare/purse line, breaking the threads holding the lower curtain of net to the lower hoop. The weight continues to pull the snare/purse line downward, pursing the net. Once closed, the camlock on the snare/purse line prevents the net from being reopened. The captive seal can surface and breathe but upon diving is contained within the net (Figure 3.10.11).

Once the trigger code has been transmitted, listen carefully to the seal's response. Most often, they detect the net closing and dive a second or two later, audible as sharp splash. Subsequent sounds of water surging and splashing are good indications that the seal did not escape.

Return to the net site, remove the snow cover, and, if needed, use the ice chisel to open the breathing hole to its full diameter. The seal can be pulled from the hole by grabbing either a hind or a foreflipper. Sometimes, the hindflipper is presented at

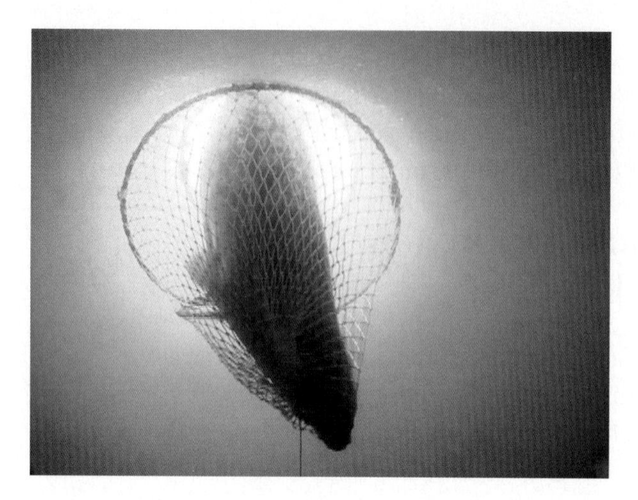

**Figure 3.10.11.** A seal searches for an escape route in a pursed *aglu* net. The snare/purse line is seen hanging beneath the net. The lead weight (out of frame) is at the lower end of that line.

the surface; other times, it is necessary to reach in to the breathing hole to grasp the hindflipper. When the seal surfaces to breathe, it is possible to reach down and take hold of their foreflippers. Most often, the seal will only present its nares at the surface, and she will not be able to see what is above because of ice and snow floating on the surface. Reach into the breathing hole next to the seal, taking care not to touch the sensitive vibrissae. Slide a hand down along the wall of the breathing hole until you feel a foreflipper. The claws of the foreflipper will be pressed into the wall of the breathing hole. Grab the flipper firmly and pull the seal upward until a coworker can grab the second foreflipper. Gain a good footing and pull the seal from the hole. We have capturedmore than two hundred seals in this manner and have never been bit while extracting one from a breathing hole. Most ringed seals remain docile after removal from the breathing hole. Young seals are especially inclined to be aggressive and to bite after being removed from the breathing hole.

It is advisable to wear rubber gloves while capturing and handling seals, although a barehanded grasp is much surer during captures. Regardless of the grasp, however, it is highly inadvisable to directly contact seals if you have any open wounds. Even very small skin breaks are sufficient to admit *Mycoplasma* spp. or *Erysipelothrix rhusiopathiae,* bacteria that can cause serious infections. "Seal finger" infections are painful and lead readily to septicemia (Lewin et al. 2004; Hunt et al. 2008) if not promptly treated with a prolonged course of tetracycline (Hartley and Pitcher 2002).

## APPENDIX 3.10.6: BASKING NET SETTING INSTRUCTIONS

We set basking nets next to breathing holes showing evidence that one or more seals have recently rested on the ice next to the hole. Typical indications include melted depressions where the seals laid, claw marks in the snow, and shed epidermis on the ice. An additional advantage of the basking nets is that they can be easily set by one person.

Determine from wear and melt matters around the breathing hole where the seal(s) exited the hole and where the seal(s) rested when out of the water (see video on accompanying DVD).

Excavate a channel in the snow (shovel) or ice (chisel) that will receive the wooden anchor. The channel should be situated such that the net is centered over the breathing hole and the anchor is not in the seal's path from the water to its resting site (Figure 3.10.12). The anchor need only be recessed sufficiently to keep the net in position when it is compressed. If necessary, the anchor can be fastened to the ice with spikes.

**Figure 3.10.12.** A basking net is laid over a basking hole. Note melted depression in foreground where seal had been lying.

Tie a monofilament line to the net's hoop at a point 180 degrees from where the hoop is lashed to the anchor, and pull that point across the breathing hole until it makes contact with the opposite side of the hoop (Figure 3.10.13).

**Figure 3.10.13.** The basking net is held in a compressed configuration by a line leading to a burn wire attachment to an anchor (out of frame).

Run the free end of the monofilament between the hoop and the anchor and tie it to a secure anchor. We typically attach to the battery that will power the trigger box.

Attach a burn wire to the monofilament line and the trigger box as described for the *aglu* nets.

Use blocks of ice or snow to conceal the battery and trigger box from a seal's-eye view. Cover the compressed hoop with a white fabric (Figure 3.10.14). Woven polypropylene sacks work well for disguising basking nets.

**Figure 3.10.14.** The basking net is disguised under white fabric. To the left of the author, two snow blocks conceal the battery and trigger box.

## APPENDIX 3.10.7: TRIGGERING THE BASKING NET AND RECOVERING THE SEAL

Periodically observe the basking site from 1–2 km distance using binoculars or a spotting scope.

When a seal(s) is visible resting next to the hole, transmit the trigger code. The coded transmission will be decoded in the trigger box and close the circuit containing the burn wire. The midsection of the compromised wire will heat up and melt through the monofilament, releasing the compressed hoop. The hoop will expand abruptly to its circular shape, covering the breathing hole with netting. Any attempts by the seal(s) to dive will be thwarted by the netting (Figure 3.10.15). Often, they do not attempt to dive until approached closely by the researchers.

**Figure 3.10.15.** A ringed seal is thwarted from diving by a basking net spanning its hole. The tip of the antenna on the trigger box is barely visible protruding above the snow block in the upper left.

# Chapter 3.11

# Community-Based Observation Programs and Indigenous and Local Sea Ice Knowledge

*Henry P. Huntington, Shari Gearheard, Matthew Druckenmiller, and Andy Mahoney*

## 3.11.1 INTRODUCTION

For much of the Arctic, sea ice is a defining characteristic of coastal environments and a major factor shaping human activities (Nelson 1969; Stonehouse 1989). For arctic residents, sea ice provides several essential regulatory, provisioning, cultural, and supporting services (e.g., Eicken et al. 2009). On a large scale, sea ice helps regulate climate, producing the conditions that arctic residents are used to. Sea ice allows travel by snowmobile and dog team while constraining access by boat. It is habitat for the marine mammals that have supported indigenous communities for millennia (Huntington and Moore 2008). Sea ice is a platform for fishing and hunting; multiyear ice is a source of drinking water; and when present sea ice protects susceptible coasts from storm waves. But sea ice is also a dynamic environment, intimate knowledge of which is necessary to reduce risks of falling through thin ice or open water, being carried out to sea on floes that have broken off from the shorefast ice, or being crushed or losing gear to ice overrides and the formation of pressure ridges (e.g., George et al. 2004; Gearheard et al. 2006).

Arctic residents therefore have accumulated a great deal of knowledge about sea ice. Some of this knowledge is generally applicable to sea ice environments, whereas a great deal is highly specific to a particular location and its currents, winds, topography, bathymetry, and ecology (Aporta 2002; Gearheard et al. 2006; Laidler and Elee 2008). Intricacies in the differences between the local knowledge possessed by different communities are often important as they may indicate phenomena overlooked by physical scientists, who typically seek generally applicable knowledge rather than particular localized knowledge. For hunters and fishermen, two factors are paramount: the ability to find and produce food, and the ability to return home safely (Nelson 1969). These imperatives often conflict, for example when hunters need to search farther to find seals or be out late in the sea ice season to hunt for whales (George et al. 2004). Sea ice knowledge is highly valued, shared among hunters, and passed down through the generations (Oozeva et al. 2004). Hunters and fishermen recognize key indicators in ice and weather, weigh the risks and rewards of going out on a particular day, and balance many factors in making a decision one way or another.

The observations and understanding of arctic residents overlap with scientific observations and understanding, but not perfectly (Gearheard et al. 2006; Laidler 2006). The concern for safety often manifests itself in great attention to unusual events, the type that may be regarded as "outliers" in scientific studies. Hunters regard some events as having spiritual origins or resulting from the intentions of animals or other parts of the environment, interpretations that have no counterpart in scientific worldviews (e.g., Huntington et al. 2006). More prosaically, hunters are often concerned with localized, short-term phenomena (what is happening here, today) rather than broader trends, and may rely on indicators such as clouds that scientists do not track (e.g., Oozeva et al. 2004). (Clouds show wind conditions up high and thus may indicate the approach of a storm; for this reason, cloudless days can be regarded as dangerous because a key indicator is absent.)

There are several examples of productive interactions between scientific and traditional knowledge. One of the most prominent concerns counting bowhead whales near Barrow, Alaska (see Huntington 2000a for further discussion). In 1977, the International Whaling Commission voted to ban the bowhead whale hunt, based on concern that the population was too low to sustain the traditional Alaska Eskimo hunt. The Iñupiat and Yupik whalers disputed the population figures and questioned the methods used for the whale count. Acting on the whalers' advice, Tom Albert (former senior scientist with the North Slope Borough Department of Wildlife Management) and other scientists developed new methods to supplement the visual count conducted at the sea ice edge. Aerial surveys confirmed that bowheads migrated farther offshore than could be seen from the ice edge, and acoustic surveys showed that many whales migrated past without being seen, either under the ice or otherwise unseen by the visual counters. These results helped show that the bowhead population was larger than had been realized, confirming the whalers' statements by using their knowledge to design better studies.

Scientists and arctic residents can learn a great deal from one another about sea ice and the human–sea ice environment. In some communities, arctic residents use satellite imagery to learn more about ice patterns in their location. Scientists have relied on experienced hunters to guide them on the ice, to explain ice types and dynamics, and to explain how ice affects human activity. With a strong shared interest in sea ice, many hunters and scientists have greatly enjoyed discussing their experiences on sea ice and their understanding of it. For example, to understand the potential magnitude of an *ivu*, the Iñupiaq term for an ice-push event that deposits ice on the beach, Lou Shapiro, a sea ice geophysicist, and Kenneth Toovak Sr., a Barrow whaling captain, interviewed Barrow elders in the Iñupiaq language about past experiences with these potentially dangerous events. Toovak both translated these interviews into English and assisted Shapiro in interpreting these accounts from a physical science perspective (Shapiro 1979). This chapter describes various ways in which hunters and scientists can collaborate to share what they know and generate new information.

Collaborative research on sea ice can take many forms, from interviews with locally recognized experts to monitoring with measuring instruments, from indoor workshops to going out onto the ice together. Such research can enhance understanding of ice, by identifying key features and local processes, by documenting local perceptions of and perspectives on ice, and by characterizing the range of services provided by sea ice. Additionally, improved mutual understanding can help improve the ability of scientists to provide useful information to hunters and arctic communities, if they better understand what local residents look for and how they use their observations and other information.

As with any other type of research, an essential starting point is a clear set of goals and expectations. These may change while a project is underway, but it is nonetheless crucial to know what one intends to gain from collaborative research. Otherwise, those involved in the project may be confused as to their roles and whether they are accomplishing what is needed from them. In this context, miscommunication and misunderstanding can take many forms, from failure to understand what is possible or desirable from either point of view, to misinterpretation of objectives or results ending in research products of little or no use to anyone. A further area for potential misunderstanding is that of compensation for local participants, a point to which we will return in the Discussion (Section 3.11.3).

The chapter begins with a description of various methods for collaborative research on sea ice. The methods discussed are neither exhaustive nor mutually exclusive. Instead, they provide a range of options for researchers to consider when designing a project. Next, we discuss the types of results that collaborative research has produced and can produce. Here, too, our intent is to be illustrative rather than comprehensive. We conclude with discussion of collaborative research, its benefits and pitfalls, and key elements of successful collaborations.

## 3.11.2 METHODS

Many of the methods for collaborative research come from the social sciences, where research about or with people is the norm. These methods may not be as familiar to physical and biological scientists, but they can be learned. It is important to remember that the interaction is a social one, involving not just the information being discussed but also the personalities, perspectives, and histories of those involved. In other words, research of this kind consists of forming a relationship, which may be brief or lasting, with all the give and take that interpersonal relationships typically involve.

With that in mind, the researcher needs to be attuned to the potential for misunderstandings, both innocent and otherwise. Differences in culture, language, experience, and other factors may lead one or more participants astray in their attempts to understand others (e.g., Huntington et al. 2006). For example, local terminology may use common terms in ways unfamiliar to the visitor (e.g., "north" may mean "northeast" or "up the coast" rather than due north). Less obviously, many indigenous cultures place high value on harmonious relations between people, so that public contradiction and conflict may be avoided even at the expense of accurate information. Thus, a statement that many know to be incorrect may pass without comment to avoid embarrassing the speaker. For the newcomer, however, subtle clues that indicate such an event may be difficult to spot, resulting in misunderstanding. Less benignly, there are times when local residents may express frustration or annoyance by providing incomplete or misleading answers.

There are no infallible ways to avoid such complications, but attention to nonverbal communications (expression, body position, etc.) and interpersonal relationships (showing respect, allowing for silence, etc.) can help (cf., Briggs 1986). Cultural practices and norms vary between regions and communities, so there is no all-encompassing set of guidelines. Instead, seeking advice from or involvement of someone familiar with the community is useful. It is also important to listen to local styles of conversation, which often involve speaking more slowly and pausing for what seem to be long silences before continuing (e.g., Huntington et al. 2006). Interrupting someone who has merely paused to gather his thoughts is likely to be considered rude, though perhaps not unexpected in visitors. Taking the time to learn the rhythms of the community and local speech patterns is time well spent. It can also be valuable to take time to allow everyone involved to get to know one another, so that everyone has the sense of working with a whole person rather than just with someone filling a particular role. Thus, time together but separate from the direct work of the project can be valuable.

Upon entering a community for the first time, researchers may consider the use of a liaison to assist with introductions and communication. A liaison may be someone in or well acquainted with the community who has established trustworthy relationships. Typically such a person will have a history of working at the interface between visiting researchers and the community, which is oftentimes made up of their friends and family members. Given this trust, a liaison is often accepting a significant responsibility by introducing an "outsider." Researchers should acknowledge this responsibility and freely share the details of their intended work as well as present themselves on a personal level. Researchers who openly discuss their intentions will likely have a more positive experience and help to maintain the liaison's trust in the community. A researcher may also consider a liaison as a source of information on how best to approach specific groups of people, such as elders, local governments, or hunters, who may use different unwritten cultural protocols for interacting with visitors. A liaison may provide invaluable guidance on seeking the appropriate permissions or adjusting research methods so that the community may receive the researcher more favorably.

Research involving community members also entails questions of ethics, which can have both formal and informal implications. Funding from the U.S. government, and from many other sources carries an obligation to protect human subjects involved in any research project (see, for example, the National Science Foundation's website, http://www.nsf.gov/bfa/dias/policy/human.jsp). The ultimate goal is to ensure that someone participating in a project is not harmed, physically or otherwise, by the research or the dissemination of data and results. Those who are involved in a project must give their informed consent, meaning that they are aware of the purpose of the research, the potential costs and benefits, and so on. Collaborative research on a topic such as sea ice may not seem to involve any persons as "subjects" of the research, but it is nonetheless better to err on the side of caution. Universities and some other research institutions have committees or boards to review research proposals and determine if they follow human subjects rules. Often, funding will not be approved until such a committee or board has approved research plans.

### 3.11.2.1 Key Informants

A common approach to gathering local information is the use of key informants, or individuals recognized in their community as particularly knowledgeable or experienced on a particular topic. This approach contrasts with censuses or surveys in which a researcher seeks a representative sample of the community. With key informants, one seeks a select few to gain the "best" information available on a particular topic, rather than "average" or "representative" information. Deciding on the appropriate approach depends, of course, on the goals of the study. If one seeks to know how commonly multiyear sea ice is used for drinking water in a community, the census or survey is appropriate. If one seeks to know how to identify multiyear ice and where to find it, asking key informants is appropriate.

Perhaps the most common method for engaging key informants is the *interview.* Interviews can be conducted with an individual or with a group, with benefits and drawbacks on both sides. The interview is generally most suitable when the researcher seeks to document or learn about a specific topic. The researcher may be able to share information that s/he has or results from other methods (e.g., satellite imagery or data from a monitoring station), but the primary goal of the interview is for the researcher to gather information from the key informant.

Interviews can use questionnaires or follow more open-ended approaches, as is done in the semi-directive interview (Huntington 1998). In that method, the researcher may start with a list of topics to cover, but the order in which they are discussed is fluid, allowing the key informant to make connections that s/he sees or follow the associations that s/he sees between different aspects of sea ice. An

**Figure 3.11.1.** Andy Mahoney and Warren Matumeak discussing satellite imagery during an interview in January 2008 in Barrow, Alaska. Photo by Henry Huntington.

advantage of this approach is that it allows the key informant to add new information that the researcher might not have anticipated in writing a questionnaire. A drawback is that the information in the interview cannot be compiled as easily as responses to a fixed questionnaire, thus requiring greater knowledge and time from the researcher.

As noted earlier, the interview is a form of social interaction. Especially with an open-ended discussion, it is important that the researcher has some knowledge of the topic to be able to keep up his/her end of the conversation. Most key informants will quickly tire of responding to someone who does not understand their replies and who therefore is unlikely to be able to interpret correctly what they are saying. The key informant interview is therefore not an ideal method for a researcher just beginning to study sea ice. Instead, participant observation (discussed below under Collaborative Fieldwork) allows the beginning researcher to learn by doing and build up sufficient knowledge to sustain an intelligent conversation with a key informant. (Note that the interviewer need not know about local sea ice use, just that s/he should know enough about sea ice to give the key informant confidence that the key informant is being understood.)

In cases where a two-way flow of information is important, the *workshop* is a variation on the key informant approach. Workshops can take a variety of forms and can have different titles (conference, symposium, seminar, etc.), but the central feature is that they invite active participation from many individuals and offer a

**Figure 3.11.2.** Members of the Siku-Inuit-Hila research group during a workshop in Barrow, Alaska, in May 2007. Photo by Henry Huntington.

forum where information can be exchanged. In this sense, they differ from lectures, where one or more individuals impart their knowledge to a largely passive audience.

A workshop can have many of the same challenges as interviews, compounded (or perhaps alleviated) by the number of people involved and thus the number of interpersonal interactions that take place (Huntington et al. 2002). Planning and management of a workshop are thus critical, both for determining who should participate and for deciding how to organize discussions. In February 2000, the Marine Mammal Commission, a small federal agency, organized a workshop on sea ice near Anchorage, Alaska. There were about two dozen participants, more or less equally divided between academic researchers and arctic coastal residents. The organizers arranged several introductory talks, setting the stage for the discussion sessions. These were followed by several parallel discussion groups addressing specific topics and returning to the full group with lists of observations, conclusions, and recommendations, which were compiled in a report (Huntington 2000b).

In November of the same year, the Barrow Symposium on Sea Ice was held in Barrow, a coastal village in northern Alaska where participants could visit sea ice during a field trip that was part of the event. For this event, a series of case studies was prepared using key informant interviews and other methods. These case studies were presented to the group and formed the basis for discussion. Some of the case studies were published as stand-alone papers (George et al. 2004; Norton and Gaylord 2004), as was a review of the symposium as a whole (Huntington et al. 2001).

Both events were useful and productive, stimulating interactions among groups who may not previously have met in such settings. Both led indirectly to follow-up activities, but the inability to take action more quickly on ideas developed at the workshops was a weakness of both events. Workshops in other settings that have allowed participants to determine further research programs or other actions have arguably been more successful in creating long-term collaboration because they have created their own follow-up activities. Not all workshops have that luxury, however, but that is no reason not to plan a workshop. Instead, it merely means that organizers should have goals that are compatible with the constraints of their particular event.

Interviews and workshops are retrospective research, in that they seek information from past events or experiences. Information can also be gathered prospectively through journals and similar approaches. One outcome of the Marine Mammal Commission workshop was a small project on St. Lawrence Island, Alaska, in which local Yupik whalers kept a record of their observations of sea ice through a winter. The results were published in a book, *Watching Ice and Weather Our Way* (Oozeva et al. 2004), which also includes material from archival sources, a list of Yupik sea ice terms with drawings and explanations, and reflections on the sea ice winter and the project. Ironically, one of the difficulties in starting the project was convincing the participants that their own observations of sea ice, in their own terms, was the goal, rather than their taking measurements with a thermometer or depth gauge in lieu of scientists being able to do so at the same locations.

In Wales and Barrow, Alaska, researchers Hajo Eicken and Matthew Druckenmiller have also collaborated with local Iñupiat hunters who kept near-daily journals on ice conditions. Similar to the previously mentioned project, these local observers kept a record on how the local ice cover evolved throughout the year and were encouraged to document how the ice conditions influenced community activities on or near the ice, such as launching boats from the ice to hunt walrus or simply chipping holes through the shorefast ice to fish for cod. While these observers were expected to freely document any aspects of the environment they felt were important to the ice cover, the researchers also clearly explained their interest, which was the timing of key events in the annual evolution of the ice cover, such as the appearance of the first ice, the time at which the shorefast ice became stable enough for human travel, and the onset of breakup. In this project the journal observations served as a rich data set that was analyzed alongside the project's other forms of geophysical-based observations, which included satellite images, ice coring, thickness measurements, and coastal radar data. At various times throughout the project interviews were held with the local observers to clarify and expand upon their written observations, especially as they related to the project's objectives and how the geophysical-based sampling plans could be adapted to best complement the types of observations the local observers felt were most important.

The journal approach places a great deal of initiative and responsibility with the local resident, which can allow deeper insights into local perspectives than might come across in a relatively brief interview or workshop. It also provides a record of changing ice over the course of a winter (or other time period), rather than a reconstruction after the fact. In the case of the St. Lawrence Island project, one of the participating elders supplemented his observations with descriptions of learning about sea ice, the importance of various phenomena such as tides and clouds, and other details that he thought important and which might not have been committed to paper otherwise.

### 3.11.2.2 Collaborative Fieldwork

Key informant approaches generally take place indoors. Collaborative fieldwork, on the other hand, is as the name implies a means of getting out on the sea ice to learn about it in situ. This entails considerations of safety, logistics, and working with local guides or partners. The last of these includes the aspects of interpersonal relationships discussed earlier, with the additional wrinkle of traveling and living together in potentially harsh conditions in a risky environment. A local partner may be concerned about safety and comfort of the visiting researcher, placing an additional burden on him/her. A visiting researcher may be concerned about the same things, possibly distracting him/her from paying attention to sea ice and other subjects of research. While potentially highly rewarding, collaborative fieldwork is not to be entered into lightly.

A standard social science method is *participant observation*, in which the researcher joins in local activities to learn by doing and observing, rather than by verbal description or instruction as occurs in an interview or workshop. This is a powerful method, providing a shared set of experiences as the basis for further interactions between visitor and host. Richard Nelson used this method in Wainwright, Alaska, in the late 1960s to produce his landmark study, *Hunters of the Northern Ice* (Nelson 1969). He lived in Wainwright for a year, observing the full seasonal round, and documenting in detail the local patterns and understanding of the environment and its use.

Many researchers today also spend time in arctic communities, visiting at different times of the year, interviewing residents, and traveling with Inuit on the sea ice. Claudio Aporta, James Ford, Martin Tremblay, Chris Furgal, and Gita Laidler are Canadian researchers who have used participant observation in studies of sea ice in several communities in Nunavut and Nunavik (northern Quebec).

**Figure 3.11.3.** Traveling on sea ice near Clyde River, Nunavut, in April 2008. Photo by Henry Huntington.

Participant observation has helped these researchers to document Inuit wayfinding skills and travel routes (Aporta 2002), understand community vulnerability and adaptation to sea ice change (Ford et al. 2008), understand how ice changes improve or constrain access to resources (Tremblay et al. 2006), and document how Inuit characterize the sea ice environment (Laidler and Elee 2008). Anne Henshaw is a U.S. researcher who has spent many years in Inuit communities and traveled with Inuit in part to document the relationship between Inuit toponymy (placenames) and changing environments (Henshaw 2006). Scott Heyes, an Australian, has worked and traveled with residents of Nunavik communities over several years to create a detailed ethnography of traditional knowledge of coastal landscapes (Heyes and Jacobs 2008).

A variation on the participant observation approach is to combine it with physical science research so that journeys onto the ice examine the ice from scientific as well as local perspectives. Matthew Druckenmiller has done research on Barrow's spring ice trails that are built by the whaling crews to access the edge of the shorefast ice during the bowhead whale hunt (see Chapter 3.2). This project used differential GPS to map the trail locations and to assess ice topography and employed an electromagnetic induction device to derive ice thickness along the trails (see Figure 3.2.30 in Chapter 3.2). Given that this work involved a great deal of time on the ice, often spent in the company of whaling crew members, opportunities existed to discuss diverse perspectives on ice stability, potential hazards, and ice features of importance. Furthermore, this collaboration between the whaling

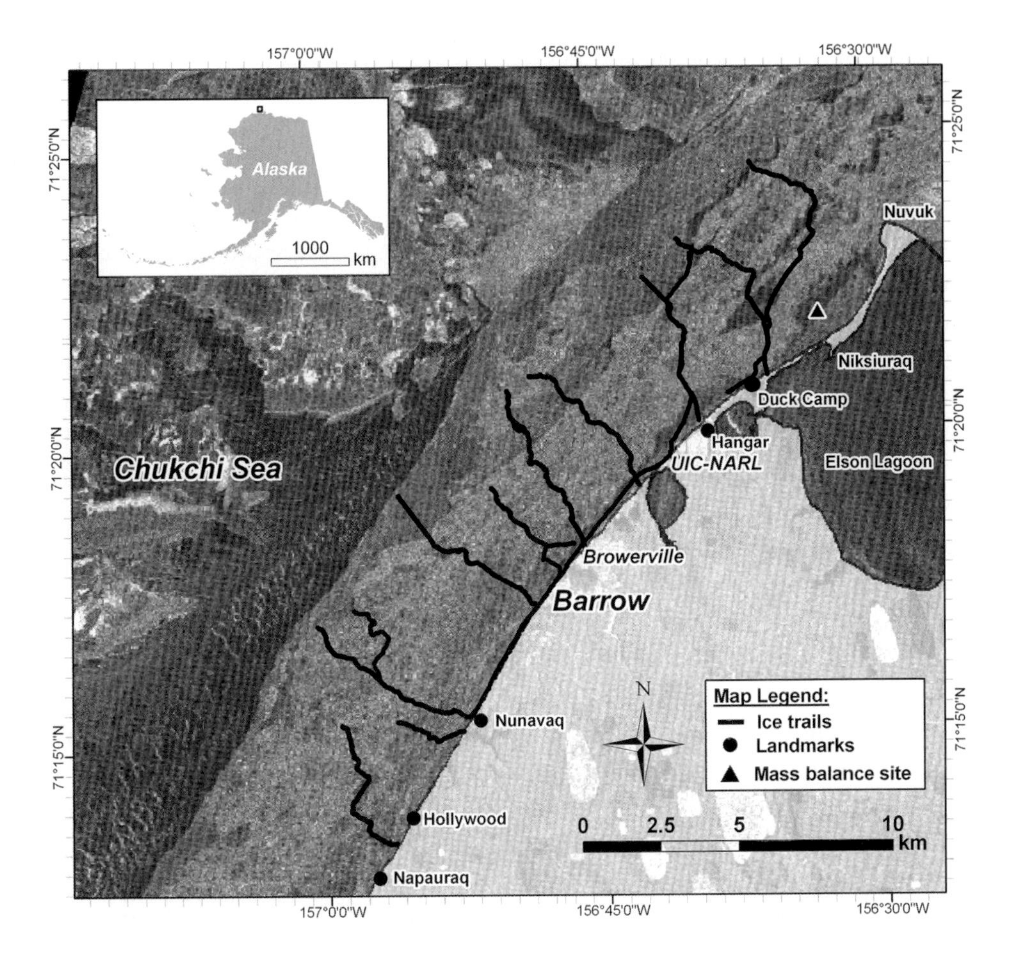


**Figure 3.11.4.** Map of ice trails on the shorefast ice off Barrow, Alaska, in spring 2008. These trails are made by whaling crews so that they may set up hunting camps at the edge of the shorefast ice and along the flaw lead through which the bowhead whales migrate. The background of this map is a synthetic aperture radar (SAR) satellite image from March 22, 2008. While in this image many of the trails appear not to reach the shorefast ice edge, they do in satellite images later in the season as the shorefast ice extent is reduced through different minor calving events at the edge. A version of this map was provided to the whaling community and local search and rescue office during the spring 2008 whaling season. See Figure 3.2.30 in Chapter 3.2 for ice thickness data along these trails. The black triangle represents a mass balance site operated by Eicken et al. from the University of Alaska Fairbanks (see Chapter 3.17).

community and physical scientists led to products like whaling trail maps (Figure 3.11.4) that benefited the Barrow community by providing a resource for navigation and the local search and rescue office. These maps also served as a tool for conducting interviews with whaling captains and crew members about the sea ice conditions they encountered throughout the season. Ultimately this work is an example of how community sea ice use and local sea ice knowledge may guide research projects in order to obtain information relevant to all involved. In this example, the community use of the ice actually determined where geophysical-based measurements were taken.

A team of Inuit, Iñupiat, and academic researchers studied sea ice in Barrow, Alaska, and Clyde River, Nunavut, in 2004 (Gearheard et al. 2006). One of the participants, Jim Maslanik, is a sea ice physicist conducting various research projects in the Arctic. He was able to share his understanding of ice and data from satellite imagery and other sources to make the fieldwork in both locations a collaborative undertaking and sharing of information, rather than simply a means of educating visitors about local sea ice patterns and uses. An advantage of this approach is that researchers who do not specialize in sea ice may not fully understand the phenomena that local residents describe. A sea ice scientist, on the other hand, may be able to better make sense of the causes and dynamics of local observations, or place those observations in a broader context. This is not to dismiss the role of the social scientist, but to emphasize the cross-disciplinary nature of collaborative research in just about any form.

### 3.11.2.3 Combinations of Methods

As mentioned in the Introduction, the methods described here are not mutually exclusive choices, but a range of possibilities that can be combined in various ways. An example is the Siku-Inuit-Hila (Sea Ice-People-Weather) project currently underway, which builds on the Barrow–Clyde River project just mentioned with the addition of the community of Qaanaaq, Greenland. This project has three major components that complement each other in documenting and analyzing information. The first component includes a series of "sea ice knowledge exchanges," visits by all participants (residents of all three communities plus the visiting researchers) to each of the study locations for participant observation (the hosts of each visit also become "participant observers" in the other locations they visit). During these trips, the emphasis is on traveling the sea ice together. The sea ice itself acts as the common denominator for the participating hunters and elders from different communities, and scientists from different disciplines, to share their knowledge and perspectives. The host community leads the sea ice travel and activities during the visits, allowing the visiting team members to experience local hunting and travel techniques and to exchange knowledge about diverse issues such as tools, clothing, food, and navigation.

The second component of the project involves regular (monthly) meetings of sea ice experts in each community. Led by local team members of Siku-Inuit-Hila, these working groups provide an opportunity to assess current sea ice conditions throughout the sea ice season and document local knowledge of sea ice, ranging widely from traditional stories and mythology of sea ice, to sea ice terminology, to extreme events, to strategies for hunting and traveling in different sea ice environments. The working groups record their meetings through digital audio and photographs and often use maps.

**Figure 3.11.5.** Joe Leavitt listing Iñupiaq sea ice terms during a monthly experts' group meeting in Barrow, January 2008. Photo by Henry Huntington.

The last component of Siku-Inuit-Hila involves the establishment of a sea ice monitoring network in the three communities. Trained by the project's sea ice physicist and supported by a handbook created especially for the local monitors (Mahoney and Gearheard 2008), local technicians measure physical properties of sea ice and snow on a weekly basis at three to four stations installed at each community. Local sea ice experts chose the location of the stations according to key areas of importance for sea ice use. Results from the measurements are continually updated and displayed in the communities (e.g., information about ice thickness and snow thickness). In combination with local historical records, available climate data, and local knowledge, the data from the observing network provides detailed information about local and regional sea ice growth and melt processes (Mahoney et al. submitted).

The different components of the project are tied together in a number of ways. For example, the expert working groups review the latest data (visualized graphically) collected in the sea ice monitoring program, sparking discussion about current conditions and progression of the sea ice season. In turn, local knowledge about currents, winds, snow accumulation, and other environmental factors affecting sea ice characteristics and processes documented during expert meetings are used to help understand the sea ice station data. For example, local knowledge of changing ocean currents near Qaanaaq have helped to explain early and dramatic thinning of sea ice during spring in recent years (Mahoney 2009). The team visits to each community and the ongoing expert meetings provide the face-to-face and

**Figure 3.11.6.** Taking measurements at a monitoring station near Clyde River, Nunavut, in April 2008. Photo by Andy Mahoney.

discussion time needed to continually link different components of the project and work toward multilayered research products in the form of journal articles, maps, and a book.

Because the project is still underway, it is too early to report more specific results. Nonetheless, the research team has displayed great enthusiasm, particularly for the community visits and the chance to learn about new areas. For the local residents, some aspects of sea ice in the other locations are familiar, but some details are new and at times surprising. However, the logistics and management of the project require a considerable investment of time and funding, not least in communication to keep all participants on the same page.

## 3.11.3 RESULTS

As with any other type of research, the test of the utility of community-based programs is in the results they produce. Involving arctic residents in sea ice research can occur at various stages of a project, from planning and preparatory research through fieldwork, analysis, and interpretation. As noted earlier, it is essential to give careful thought to the goals and organization of community-based activities, both so that local participants know what is expected of them and so that the community effort is focused and productive rather than vague and disconnected. Engaging local residents at the outset of a project is ideal, so that they can help the visiting researchers set appropriate goals and design activities that are likely to produce the desired results. Involving local residents later in a project reduces the range of available options, but can still be beneficial, especially in terms of sharing the results of scientific studies.

Early discussions with local residents can help frame hypotheses and other elements of study design. In the Siku-Inuit-Hila project, local knowledge was essential in selecting sites for instrumental measurements to make sure the sites captured local variability, were of local interest so that results would be seen as relevant, and were connected to the other activities of the project such as the local discussion groups tracking their own observations of ice conditions during the winter. As an example, one hypothesis that grew out of discussions with Qaanaaq hunters was that under-ice currents played an important role in determining the thickness of ice and the onset of melting. Accordingly, the ice measuring stations were distributed in a transect across the fjord. The ice thickness data supported the hypothesis, demonstrating that there is substantial heat flux from the ocean, retarding ice growth. A large proportion of the spring ice melt occurred at the bottom of the ice, in contrast to sites in the other communities. Information from the measuring stations also helped provide information back to the hunters. Toku Oshima, who was in charge of the measuring effort, had expected almost all of the melt to occur at the bottom of the ice, and was surprised to see how much melted at the top.

Gathering and interpretation of data can also benefit from local input. The instrumental sites in the Siku-Inuit-Hila project require a local resident to take measurements throughout the sea ice season, when the visiting researchers are not present. The record of measurements is thus far more complete and regular than would be possible otherwise, particularly when several communities are involved (Mahoney and Gearheard 2008). Interpretation also benefits from local insight. In the Barrow Symposium on Sea Ice, one case study focused on an event in which the shorefast ice had shattered, nearly leading to disaster for several whaling crews on the ice at the time (Huntington et al. 2001). Discussions during the symposium revealed that the shorefast ice had had a much higher than usual percentage of multiyear ice, which was more likely to shatter than to deform like first-year ice. This crucial aspect of the conditions during the event in question had not come out in the interviews on which the case study was based, and helped the group better evaluate what had occurred and the likelihood that such an anomalous event might happen again.

To complement in-depth details of local conditions, broad assessments of sea ice across many communities can identify broad patterns and concerns. The Marine Mammal Commission's workshop in 2000 involved participants from villages along much of the northern and western coasts of Alaska. All participants noted changes in sea ice conditions in recent years, and all expressed concern about the likely impacts that such changes would have on hunting, coastal erosion, and other aspects of their lives. Recognizing the widespread nature of the problem was important both for scientists and for coastal residents, who saw that they had much in common with people from other parts of the state. Approaching sea ice from the community perspective also highlighted gaps in scientific research, within and between disciplines, pointing to the potential for new studies and new approaches to involving communities and their knowledge (Huntington 2000b).

Workshops, research, and other interactions also afford the opportunity for scientific researchers to share their findings with local residents. The Barrow Symposium on Sea Ice made this a formal goal and included one case study on ice monitoring and prediction to show how the National Weather Service provides information to the public on sea ice in Alaska. Other researchers at the symposium showed animations of shorefast ice breakup from radar imagery and discussed the measurement of ice and internal stresses in ice floes, which were instructive and enjoyable for all participants (Huntington et al. 2001). Active involvement in a study appears to help when it comes to sharing information, in part by establishing a common basis for communication by shared experience and observation, and in part by creating interpersonal relationships that improve communication and understanding while also providing additional social incentive to interact. While some benefits are intangible, many researchers who have worked with local residents have gotten a great deal of satisfaction from their collaboration, a sentiment shared by their local partners.

Finally, the goal of most research is to produce results for publication. Community-based research is no different, though it may have other goals such as local dissemination of information (for which academic publication is unlikely to be suitable). A challenge of publishing the results of a community-based study is that of including the wide range of information that may be gathered or discussed, not all of which may fit into a concise and linear narrative of hypothesis, result, and discussion. Nonetheless, publication is an opportunity to capture and share observations and concerns that may illustrate the significance of sea ice for arctic residents and communities, helping introduce other researchers to potential applications of their studies. Publication also helps raise the status of community ideas, observations, knowledge, and concerns by recognizing them as legitimate topics of research and writing rather than sideline curiosities ancillary to standard research. Collaborative projects raise an additional issue of joint authorship or other acknowledgment of community research partners, which researchers should consider early in a project (see Huntington 2006 for further discussion of authorship on such studies).

## 3.11.4 DISCUSSION

From full partners in research to an interested audience for scientific findings, arctic community residents can add a great deal to sea ice research in many ways. The material presented here is simply a set of examples or illustrations of this point. Researchers and community members can be creative in developing ways to work together for mutual benefit, to meet the needs of a particular study and the needs and interests of a particular community at a particular time. The important thing is to start the discussion.

Working with communities is not, however, always straightforward (Gearheard and Shirley 2007 provide several case studies). Community residents, particularly those who are active on the ice, often have many commitments, some of which will inevitably conflict with the schedules of visiting researchers. Personalities may mesh or clash, communication may be difficult (or worse, may seem smooth but turn out not to be). Weather and other factors beyond anyone's control can add complications. As with most fieldwork in the Arctic, actual events often do not follow carefully laid plans, requiring a degree of flexibility and innovation. Above all, researchers need to be cautious in interpreting what they have seen and heard. A few hours, days, or even weeks with arctic residents is not sufficient to gain full understanding of local practices, knowledge, cultures, values, or communication styles. The potential for misinterpretation is thus high, requiring a healthy degree of skepticism of one's own depth of understanding.

As mentioned in the Introduction, the question of compensation can further complicate relations between local participants and researchers. There are no uniform standards across the Arctic, nor even within a single community. Typically, we have paid people for their time for many activities, such as participating in key-informant interviews, taking measurements, recording observations, or traveling to other communities. In other instances, such as broader workshops, we have not paid people merely for attending an event. Researchers are best advised to find out about local practices and expectations and to discuss the topic with local participants in advance, even though it can be awkward, to avoid larger problems later on. When payments are to be made, promptness is appreciated and reliability is essential. Overall, it is important to remember that researchers are rarely volunteering their time (though they may volunteer to do additional tasks such as school or community talks during the course of a project), and that expecting local residents to contribute extensively of their time or knowledge without compensation may be seen as unjust.

Setting clear goals is critical to community-based research, as for any type of research. The goals may be developed collaboratively, or may change over the course of the project as visitors and locals learn more about the project and one another. The potential for changing goals and expectations is one more reason for clear communication among all those involved. The project *Watching Ice and Weather Our Way* grew as it progressed, thanks to the enthusiasm of the participants and the willingness of the Marine Mammal Commission to provide additional funding to support new activities. That said, there were still occasions when various participants became confused as to what was taking place, when certain activities were to be completed, and so on. Long-standing relationships between the local and visiting members of the project team helped overcome most of the problems.

Looking to the future, we can anticipate greater involvement of local residents in all aspects of research, including more arctic residents seeking academic qualifications. This trend responds both to local interest in participating in and contributing to research and to an increasing awareness in the scientific community of what can be gained from working with local residents. One consequence of collaboration is an increasing sophistication on the part of sea ice users with regard to scientific research, satellite imagery, and so on. Just as the use of GPS has become common in much of the Arctic, the use of sea ice images and other products may become more widespread as hunters and travelers find those materials available and useful.

In summary, community-based research and other involvement by local residents can be a valuable addition to sea ice studies. Putting these approaches into practice requires preparation, skill, experience, and most importantly a willingness to be open to new ideas in the hopes of enhancing one's own understanding of the sea ice environment, the nature of research, and the role of science in the lives of arctic residents today. For arctic residents as well as scientists, there are many benefits from collaboration and communication.

## REFERENCES

Aporta, C. (2002), Life on the ice: Understanding the codes of a changing environment, *Polar Record, 38*(207), 341–354.

Briggs, C. L. (1986), *Learning How to Ask: A Sociolinguistic Appraisal of the Role of the Interview in Social Science Research.* Cambridge University Press, Cambridge.

Eicken, H., A. L. Lovecraft, and M. L. Druckenmiller (2009), Sea ice system services: A framework to help identify and meet information needs relevant for arctic observing networks, *Arctic,* 62(2), 119–136.

Ford, J., B. Smit, J. Wandel, M. Allurut, K. Shappa, H. Ittusarjuat, and K. Qrunnut (2008), Climate change in the Arctic: Current and future vulnerability in two Inuit communities in Canada, *Geographical Journal, 174*(1), 45–62.

Gearheard, S., W. Matumeak, I. Angutikjuaq, J. Maslanik, H. P. Huntington, J. Leavitt, D. Matumeak Kagak, G. Tigullaraq, and R. G. Barry (2006), "It's not that simple": A collaborative comparison of sea ice environments, their uses, observed changes, and adaptations in Barrow, Alaska, USA, and Clyde River, Nunavut, Canada, *Ambio, 35*(4), 203–211.

Gearheard, S., and J. Shirley (2007), Challenges in community-research relationships: Learning from natural science in Nunavut, *Arctic, 60*(1), 62–74.

George, J. C., H. P. Huntington, K. Brewster, H. Eicken, D. W. Norton, and R. Glenn (2004), Observations on shorefast ice failures in Arctic Alaska and the responses of the Inupiat hunting community, *Arctic, 57*(4), 363–374.

Henshaw, A. (2006), Pausing along the journey: Learning landscapes, environmental change, and toponymy amongst the Sikusilarmiut, *Arctic Anthropology, 43*(1), 52–66.

Heyes, S., and P. Jacobs (2008), Losing Place: Diminishing Traditional Knowledge of the Arctic Coastal Landscape, in *Making Sense of Place: Exploring Concepts and Expressions of Place through Different Lenses and Senses*, edited by F. Vanclay, M. Higgins, and A. Blackshaw, pp. 135–154. National Museum of Australia, Canberra.

Huntington, H. P. (1998), Observations on the utility of the semi-directive interview for documenting traditional ecological knowledge, *Arctic*, *51*(3), 237–242.

Huntington, H. P. (2000a), Using traditional ecological knowledge in science: Methods and applications, *Ecological Applications*, *10*(5), 1270–1274.

Huntington, H. P. (ed.) (2000b), *Impacts of Changes in Sea Ice and Other Environmental Parameters in the Arctic*, Report of the Marine Mammal Commission Workshop, 15–17 February 2000, Girdwood, Alaska, Marine Mammal Commission, Bethesda, MD.

Huntington, H. P. (2006), Who are the "authors" when traditional knowledge is documented? Commentary, *Arctic*, *59*(3), iii–iv.

Huntington, H. P., H. Brower, Jr., and D. W. Norton (2001), The Barrow symposium on sea ice, 2000: Evaluation of one means of exchanging information between subsistence whalers and scientists, *Arctic*, *54*(2), 201–204.

Huntington, H. P., P. K. Brown-Schwalenberg, M. E. Fernandez-Gimenez, K. J. Frost, D. W. Norton, and D. H. Rosenberg (2002), Observations on the workshop as a means of improving communication between holders of traditional and scientific knowledge, *Environmental Management*, *30*(6), 778–792.

Huntington, H. P., and S. E. Moore (eds.) (2008), The implications of climate change for Arctic marine mammals, *Ecological Applications*, *18*(2), Supplement.

Huntington, H. P., S. F. Trainor, D. C. Natcher, O. Huntington, L. DeWilde, and F. S. Chapin III (2006), The significance of context in community-based research: Understanding discussions about wildfire in Huslia, Alaska, *Ecology and Society*, *11*(1), 40, available at http://www.ecologyandsociety.org/vol11/iss1/art40/.

Laidler, G. J. (2006), Inuit and scientific perspectives on the relationship between sea ice and climate change: The ideal complement?, *Climatic Change*, *78*, 407–444.

Laidler G. J., and P. Elee (2008), Human geographies of sea ice: Freeze/thaw processes around Cape Dorset, Nunavut, Canada, *Polar Record*, *44*(228), 51–76.

Mahoney, A., and S. Gearheard (2008), Handbook for Community-Based Sea Ice Monitoring, NSIDC Special Report 14, National Snow and Ice Data Center, Boulder, CO, available at http://nsidc.org/pubs/special/nsidc_special_report_14.pdf.

Mahoney, A., S. Gearheard, T. Oshima, and T. Qillaq (2009), Sea ice thickness measurements from a community-based observing network, Bulletin of the American Meteorological Society, 90(3), 370–377.

Nelson, R. (1969), *Hunters of the Northern Ice*. University of Chicago Press, Chicago.

Norton, D. W., and A. G. Gaylord (2004), Drift velocities of ice floes in Alaska's northern Chukchi Sea flaw zone: Determination of success by spring subsistence whalers in 2000 and 2001, *Arctic*, *57*(4), 347–362.

Oozeva, C., C. Noongwook, G. Noongwook, C. Alowa, and I. Krupnik. (2004), *Watching Ice and Weather Our Way.* Arctic Studies Center, Smithsonian Institution, Washington, DC.

Shapiro, L. (1979), *Historical References to Ice Conditions along the Beaufort Sea Coast of Alaska.* Geophysical Institute, University of Alaska Fairbanks, Fairbanks.

Stonehouse, B. (1989), *Polar Ecology.* Blackie, London.

Tremblay, M., C. Furgal, V. Lafortune, C. Larrivee, J.-P. Savard, M. Barrett, T. Annanack, N. Enish, P. Tookalook, and B. Etidloie (2006), Communities and Ice: Linking Traditional and Scientific Knowledge, in *Climate Change: Linking Traditional and Scientific Knowledge*, edited by R. Riewe and J. Oakes, pp. 185–201, Aboriginal Issues Press, Winnipeg, Manitoba, Canada.

# Chapter 3.12

# Ship-Based Ice Observation Programs

*Anthony P. Worby and Hajo Eicken*

## 3.12.1 INTRODUCTION

Ship-based observations of sea ice conditions are perhaps one of the easiest data sets to collect in polar regions, requiring nothing more than a keen set of eyes and a good knowledge of sea ice nomenclature. Observations taken from ships can provide a very useful overview of the synoptic sea ice conditions along a ship's track that greatly complement other high-precision in situ measurements as well as remote-sensing studies (e.g., Jeffries et al. 2001; Eicken et al. 2005), and they are an important source of historic ice-condition data (e.g., Vinje 2001; Ackley et al. 2003). Since the days of the earliest explorers, ship logs have recorded encounters with sea ice. Earliest sightings recorded the location of the ice edge, and as vessels have become more ice-capable other variables such as ice concentration, thickness, floe size, and snow-cover characteristics have also been recorded.

In the framework of services delivered by the sea ice system (see Chapter 2), ship-based observations figure prominently as a measurement approach that directly ties the variable observed (e.g., ice concentration, ice thickness, and ice type) to the service, in this case navigational hazard. This linkage is perhaps best illustrated by studies of ice conditions on icebreakers instrumented to record the mechanical impact of ice on the vessel's hull, allowing direct conclusions about the mechanics and energetics of ice breaking in relation to ice conditions (see Chapter 3.5). Ship-based ice observations can also contribute greatly to the study of sea ice provisioning and supporting services. For example, some studies combine ice observations with marine mammal or seabird counts to investigate linkages between ice type and species distributions (van Franeker 1992; Gelatt and Sinniff 1999).

### 3.12.1.1 Standardization of Ice Observations—Antarctic

Historically, one of the biggest problems with ship-based observations has been the lack of a standardized technique for observing and recording ice conditions.

Most vessels operating in the pack ice have maintained some form of ice log, but invariably these have recorded different sea ice variables using different codes or definitions and in varying levels of detail. For example, some logs have recorded only ice type (with no indication of thickness) or have recorded ice and snow thickness as a total value without differentiating between them. Consequently, it has been extremely difficult (and in some cases impossible) to marry such data sets into a larger archive that makes full value of the data that have been collected. For example, data collected along frequently repeated shipping routes to antarctic coastal stations provides an opportunity to obtain data that may identify seasonal and, possibly, interannual changes in ice conditions. Similarly, observations from voyages to different regions, but at a similar time of year, can help identify regional differences in ice conditions.

In 1997, the Scientific Committee on Antarctic Research (SCAR) established the Antarctic Sea Ice Processes and Climate (ASPeCt) program, which initiated a system for making standardized sea ice observations from ships in Antarctica. This was based on work already being done by a number of national programs and was the culmination of significant international effort. The basis of the observing scheme is the World Meteorological Organization (WMO) Sea Ice Nomenclature, which is used to define the ice types present in any given location (electronic versions of the WMO nomenclature are included on the accompanying DVD). For each ice type the thickness, floe size, topography, snow cover type, and snow cover thickness are recorded. Observations are recorded hourly from the ship's bridge using a defined set of codes that makes it possible to record a full twenty-four hours of observations onto a single page. The data can then be digitized using purpose-built software that facilitates basic quality control as the data are entered.

Figure 3.12.1 is an aerial photograph over antarctic sea ice showing the range of ice conditions that can occur over an area of 1 km$^2$, which is the approximate coverage of a ship-based ASPeCt ice observation (from Worby et al. 2008). The scene represents the net effect of dynamic and thermodynamic processes that have influenced the evolution of the ice, resulting in a broad range of ice types and thicknesses. The ASPeCt observing scheme cannot capture all the complexity of this scene, but does provide the capability to record the characteristics of the three dominant ice classes and their topography, floe size, and snow cover characteristics, as outlined below. A training CD-ROM has also been produced that provides a step-by-step tutorial for making ice observations, so that the ice conditions in scenes such as this can be accurately and consistently recorded. This material is also accessible in its entirety on the accompanying DVD.

### 3.12.1.2 Standardization of Ice Observations—Arctic

It is important to note that the ASPeCt ice observing protocols have been specifically designed for antarctic conditions, and while the technique has been used by

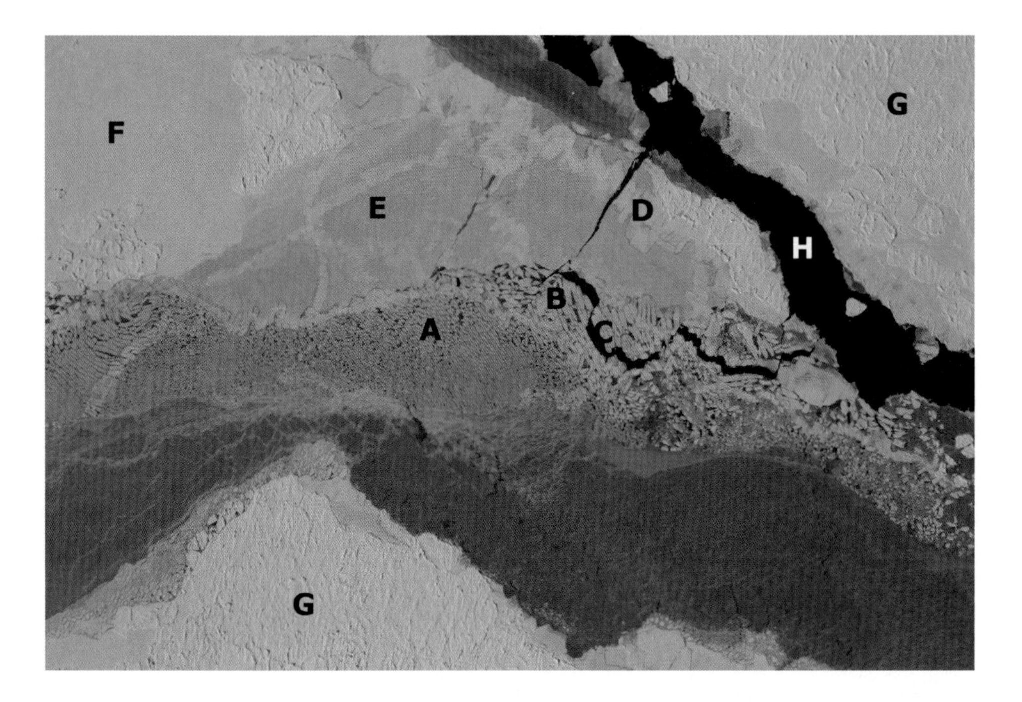


**Figure 3.12.1.** Aerial photograph over east antarctic sea ice showing the range of ice types that can occur within a small area (after Worby et al., 2008). The image is approximately 1200 x 800 m in size which is the approximate coverage of an ASPeCt ice observation. The scene represents the net effect of dynamic and thermodynamic processes that have influenced the evolution of the ice. The thin ice at (A) has been broken into tiny floes by wave action, some of which have subsequently refrozen together (B) before being split apart again by the formation of a new lead (C). The effects of finger rafting can be seen in the gray-white nilas (D), while steady thermodynamic growth has formed new gray-white (E) and first-year (F) ice. Older, thicker floes (G) have clearly undergone significant deformation and have a variable thickness snow cover, while new grease ice is forming in the leads (H).

different groups in the Arctic, there has not been the same coordinated effort to collect arctic data. Paradoxically, this situation may in part be due to the higher degree and long history of shipping and other maritime activities in the Arctic, which has resulted in a number of parallel observation approaches that are employed by different sea ice user groups or different agencies charged with overseeing ice conditions. In recent years, however, attempts at standardization have received increasing support, e.g., by the International Ice Charting Working Group or by a nascent sea ice working group under the auspices of the Climate and the Cryosphere (CliC) Project. An excellent resource for arctic ice observations is the Canadian Ice Service's *Manual of Standard Procedures for Observing and Reporting Ice Conditions* (Environment Canada 2005), which is included in its entirety on the accompanying DVD. In recent years, the number of icebreaker research cruises into the Arctic has steadily increased. This increase in activity and opportunity for ship-based observations has resulted in the development of a Web-based ice observation system, originally implemented on the U.S. Coast Guard icebreaker *Healy* (Eicken et al. 2005).

The observation protocol is based on an adaptation of the ASPeCt protocol to arctic conditions. Arctic sea ice is generally thicker and occurs at higher concentration, and also has different surface characteristics (particularly melt ponds) that vary seasonally. While this chapter and the accompanying DVD address both antarctic and arctic observations, the focus is on the former, in particular since the observation protocol for the Arctic is still evolving, for example through a planned review by the CliC Arctic Sea Ice Working Group. It is hoped that this book may also contribute to broader exchange and development of a common protocol, and readers interested in contributing are encouraged to contact the authors of this chapter.

## 3.12.2 OBSERVATION TECHNIQUES—ANTARCTIC

A standard set of observations are made hourly within a radius of approximately 1 km of the ship. The observations are made for up to three ice thickness categories (there may be less than three), which are defined as those with the greatest concentration. The thickest of these is defined as the primary ice type, followed by secondary and tertiary categories. The observations are entered on log sheets using a standard set of codes based on the World Meteorological Organization (1970) sea ice nomenclature. A set of blank forms is accessible on the accompanying DVD. Below, we present the antarctic observation protocol, with observations of arctic sea ice discussed in the subsequent section.

### 3.12.2.1 Ice Concentration (c)

The total ice concentration is an estimate of the total area covered by all types of ice, expressed in tenths and entered as an integer between 0 and 10. In regions of very high ice concentration (95–99 percent) where only very small cracks are present, the recorded value should be 10 and the open-water classification should be 1 (small cracks). Regions of complete ice cover (100 percent) will be distinguished by recording an open-water classification of 0 (no openings). An estimate of the concentration of each of the three dominant ice thickness categories is also made. These values are also expressed in tenths and should sum to the value of the total ice concentration. It is sometimes difficult to divide the pack into three distinct categories, and it may be necessary to group some categories of similar thickness together to ensure their representation.

### 3.12.2.2 Ice Type (ty)

The different ice categories, together with the codes used to record the observations, are shown in Table 3.12.1. The ice categories are based on the WMO (1970) sea ice classifications. First-year ice greater than approximately 0.1 m thick is classified by its thickness (e.g., young gray ice 0.1–0.15 m; first-year ice 0.7–1.2 m), while thinner ice is generally classified by type (e.g., frazil, shuga, grease, and nilas). A single category is defined for multiyear ice. There is also a category for brash, which is common between floes in areas affected by swell and where pressure ridging has collapsed. Books by, for example, Armstrong et al. (1973), Steffen (1986) or the Environment Canada MANICE (2005) handbook included on the accompanying DVD provide illustrated examples of different sea ice types. The ASPeCt CD-ROM (Worby 1999) and the ASPeCt online image library at http://www.aspect.aq also provide extensive image libraries of sea ice types.

**Table 3.12.1.** Ice thickness classifications used for ship-based observations

| Ice type classification | Ice thickness, m | Code |
|---|---|---|
| New ice: | <0.1 | |
| Frazil | | 10 |
| Shuga | | 11 |
| Grease | | 12 |
| Nilas | | 20 |
| Pancakes | <0.2* | 30 |
| Young gray ice | 0.10–0.15 | 40 |
| Young gray-white ice | 0.15–0.3 | 50 |
| First-year ice | 0.3–0.7 | 60 |
| First-year ice | 0.7–1.2 | 70 |
| First-year ice | >1.2 | 80 |
| Multiyear ice | <20* | 85 |
| Brash | <0.5* | 90 |
| Fast ice | <3* | 95 |

**Range is a guide only and may be exceeded.*

### 3.12.2.3 Ice Thickness (z)

Ice thickness is estimated for each of the three ice types. On some ships, a purpose-built rule, with 20 cm graduations marked on it, is extended from the side of the ship to gauge the thickness of the turned blocks. Alternatively, it is helpful to the observer to suspend an inflatable buoy of known diameter over the side of the ship, approximately 1 m above the ice, to provide a scaled reference against which floe thickness can be estimated. The ice thickness can then be determined quite accurately as floes are broken and turned sideways along the ship's hull. A further refinement of this approach is the acquisition of photographs or video footage of ice blocks turned on their side by the ship. While more labor intensive to process, such imagery can be evaluated quantitatively to yield information about the level ice thickness distribution (Tunik 1994). Moreover, images of floes turned on their side often also contain substantial information on ice stratigraphy. Hence, they can provide insight into ice formation and deformation processes or the distribution of ice algae and sediments (see example shown in Figure 3.12.2 or Eicken et al. 2005) and other information relevant for on-ice studies, such as the degree of potential lateral heterogeneity of importance in ice-core sampling (see Chapter 3.3).

It is extremely important to note that only the thickness of level floes, or the level ice between ridges, is estimated. This is because ridges tend to break apart into their component blocks when hit by the ship, making it impossible to estimate their thickness. In order to determine the thickness of ridged floes, observations of the areal extent and mean sail height of the ridges are made (see below) and combined with the level ice thickness data into a simple model to generate total thickness (see Section 3.12.3). This is a more reliable way of determining total ice thickness, rather than simply guessing at the thickness of heavily ridged floes that do not turn sideways when hit by the ship.

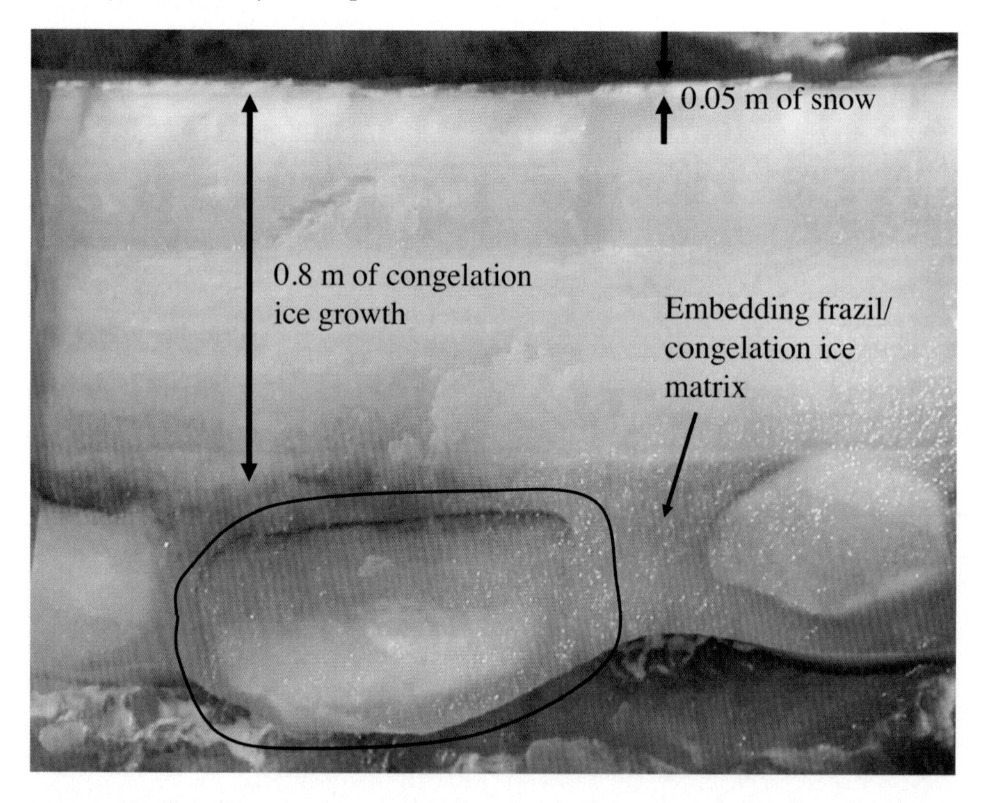


**Figure 3.12.2.** Cross-section through arctic sea ice floe as photographed down the side of the icebreaker *Healy* during a cruise to the Chukchi Sea in May of 2002. The figure illustrates the type of information that can be obtained from such photographs, including an estimate of total level ice thickness and an assessment of ice stratigraphy. With respect to the latter, the photo provides information on the thickness of level ice, accumulation of deformed ice blocks, such as the one outlined in black or frazil ice underneath a pre-existing ice cover, presence of sediment inclusions visible as brown layers on top of the highlighted ice block, or the nature of the ice-water interface.

Thinner, snow-free, ice thickness categories can be reliably classified by a trained observer from their apparent albedo, while the thickness of very thick floes may be estimated by their freeboard. The accuracy of careful observations will be within 10 to 20 percent of the actual thickness, and a large sample of observations can be expected to provide a good statistical description of the characteristics of the pack. This is particularly true at the thin end of the thickness distribution where changes are most important for both radiant and turbulent heat transfer (e.g., Worby and Allison 1991).

On dedicated scientific voyages, it is usually possible to make regular in situ measurements of ice and snow thickness, both on level ice and across ridges, to "calibrate" the ship-based observations. Worby et al. (1996) demonstrated a technique for combining in situ and ship-based observations to estimate the ice thickness distribution in the Bellingshausen Sea. Dedicated scientific voyages also usually provide the opportunity to follow specific routes to optimize data quality, which may be compromised or biased towards thin ice if the ship follows the most easily navigable routes. It is the observer's responsibility to clearly indicate on the observation sheet when the ship is preferentially following leads so that this may be considered during data processing.

### 3.12.2.4 Floe Size (f)

Floe size can be difficult to determine because it is not always clear where the boundary of a floe is located. Cracks and leads delineate floe boundaries whereas ridges do not. Where smaller floes have been cemented together to form larger floes, the larger dimension is recorded, but usually with a comment to indicate that smaller floes are visible. Where two floes have converged and ridged, the floe size is taken as the combined size of the two. A good rule of thumb is: If you could walk from point A to point B, then both points are on the same floe. This guide can be helpful when trying to determine floe size. The length of the ship (about 100 m for most icebreakers) can act as a good guide for estimating floe size. The ship's radar can be useful for determining the size of very large floes.

Floe size is recorded using a code between 100 and 700. New sheet ice (code 200) is normally used for nilas. This code does not specify a floe size, but is a descriptor for refrozen leads and polynyas. It is often used in conjunction with topography codes 100 (level ice) and 400 (finger rafting).

### 3.12.2.5 Topography (t)

As discussed above, the ice thickness estimates are only made of the level ice in a floe. This is because the thickness of ridges cannot reliably be estimated from a ship, since they tend to break up into their component blocks when hit by the ship, rather than turning sideways so that their thickness can be estimated. However, drilled transects across ridged ice floes indicate that the mass of ice in ridges is a major contributor to the total ice mass of the pack, hence it is important to quantify the extent of ridging within the pack. To do this, the areal extent and mean sail height of ridges (the portion of the ridge jutting above the level ice surface) is recorded for each ice type within the pack. The extent of surface ridging is estimated to the nearest 10 percent. It is very important that observers not look too far from the ship when estimating the areal extent of ridges, otherwise only the ridge peaks are seen and not the level ice between them. This gives a false impression of more heavily deformed ice than is actually present and may lead to an overestimate of the extent of ridging. The mean sail height is estimated to the nearest half meter below 2 m, and to the nearest meter above 2 m. It is important to remember that it is the mean sail height that is recorded. This can be difficult to estimate, particularly in flat light when the sky is overcast. Our experience has shown that ridge sail height is generally underestimated due to the vertical perspective from the bridge, so it is important for observers to 'calibrate' their eye with in situ measurements, or by observing from lower levels of the ship.

Ridges are classified using a three-digit code between 500 and 897. The first digit (5–8) is a description of the type of ridge, which may be unconsolidated, consolidated, or weathered. This is determined from the appearance of the ridge and is useful for estimating ridge sail density. The second digit (0–9) describes the areal coverage of ridges, and the third digit (0–7) records the mean sail height to the nearest half meter. These observations are probably the most subjective of those made from the ship, and it is particularly important to standardize them between observers.

The observations of surface ridging are input to a model formulation that is described in detail in Worby et al. (2008), to calculate total ice thickness and the thickness distribution of the ice.

### 3.12.2.6 Snow Type (s)

The snow type classification is a descriptor for the state of the snow cover on sea ice floes and uses a code between 1 and 10. The snow cover properties are important for estimating the area-averaged albedo of the pack as discussed in Section 3.12.3, estimating precipitation over the sea ice and the impacts on the Southern Ocean freshwater balance, and for understanding snow-ice formation processes. The snow cover on sea ice can be highly complex, hence the snow-cover classification describes only the surface snow. For example, in a case where fresh snow has fallen over older wind-packed snow, the classification code should describe the freshly fallen snow cover. However, it is very important that the total snow-cover thickness is still recorded.

### 3.12.2.7 Snow Thickness (sz)

An estimate of the snow-cover thickness is made for each of the three dominant ice thickness categories. Snow thickness is relatively straightforward to estimate for floes turned sideways along the ship's hull, although at times the ice-snow interface is difficult to distinguish, particularly when the base of the snow layer has been flooded and snow-ice has formed.

### 3.12.2.8 Open Water (o/w)

The codes for open water are descriptors for the size of the cracks or leads between floes, not a concentration value (in tenths). As discussed above, the length and breadth of the ship can act as a useful guide when estimating lead dimensions. The ship's radar can also be useful, particularly at night.

### 3.12.2.9 Meteorological Observations

Instantaneous conditions are usually recorded hourly, but this may be reduced to three-hourly. The standard set of observations includes water temperature, air temperature, true wind speed and direction, total cloud cover, visibility, and current weather. On most research vessels, water temperature, air temperature, and wind speed and direction will be displayed on the bridge, and may even be logged for the duration of the voyage. Cloud cover can be estimated by the observer in eighths, and visibility is estimated in kilometers from the ship. Wind speed is recorded in m $s^{-1}$ and wind direction relative to north (°T). The current weather is recorded using the Australian meteorological observer's two-digit codes, which that are provided on the accompanying DVD. Only a subset of these codes, pertinent to antarctic conditions, has been included in the software.

### 3.12.2.10 Photographic Records

During daylight hours a photographic record of ice conditions can be kept. Photographs are usually taken from the bridge at the time of each observation, and the log book has a column for recording frame numbers (and film numbers, if photography is with film rather than digital cameras). There is also scope for recording the frame number for a time-lapse video recorder which the authors have mounted on the ship's rail. This captures a single video frame every eight seconds, providing a comprehensive visual record of ice conditions on a single videotape for each thirty-day period. This photographic archive is not generally used for quantitative analyses, but provides an excellent reference that can be used in conjunction with the ship-based observations. At night the camera is angled closer to the ship to view an area that can be adequately lit by floodlights mounted on the ship's rail.

#### 3.12.2.11 Comments

In addition to the hourly observations entered by code, there is scope for additional comments to be recorded. These usually include a brief description of the characteristics of the pack, in particular features that are not covered by the observation codes, such as frost flowers on dark nilas or swell penetrating the pack. Brief details of sampling sites, buoy deployments, or other on-ice activities may also be recorded and, if necessary, a comment on how typical the ice along the ship's route is of the surrounding region. Lengthy comments can be recorded on a separate sheet and cross-referenced by placing a number in the "reference number" column.

## 3.12.3 OBSERVATION TECHNIQUES—ARCTIC

While there is no internationally recognized standard observational protocol for arctic sea ice, most protocols currently in use (such as that described in the MANICE handbook, Environment Canada 2005; see accompanying DVD) follow similar approaches. Here we discuss a protocol that has been employed on U.S. Coast Guard icebreaker voyages, both in the Alaskan Arctic (Eicken et al. 2005) and on cruises into the central Arctic (Perovich et al. 2009). The protocol has been implemented as a web-based interface that automatically ingests a number of ship and weather parameters from the vessel's automated data acquisition system and allows the observer to enter data into online forms and download digital photographs of ice conditions. In addition to ease of entry, including the use of a predetermined set of options for different categories that can be selected from a pull-down menu, this approach allows immediate access to the ice-observation data set by other cruise participants or even researchers onshore if data are transferred from the vessel to an onshore server. An example of such archived data sets, in this case for a 2007 cruise into the Bering Sea, can be accessed through the Bering Sea Ecosystem Study (BEST) Field Catalog that is hosted by the Earth Observing Laboratory (EOL) at the National Center for Atmospheric Research;* retrieved on 25 October 2008). Examples of completed data sheets and linked photographs are also included on the accompanying DVD.

The protocol discussed here (see also Eicken et al. 2005) builds mostly on the ASPeCt protocol, with a number of differences that are mostly the result of different ice conditions found in the Arctic. Table 3.12.2 lists the variables that are captured and includes the different categories and options for all of these. One of the major differences relative to the antarctic protocol is the description of surface melt features, such as melt ponds, which are of great importance both for the surface energy budget (see Chapter 3.6) and for ice strength, stability, and its role in

* National Center for Atmospheric Research (http://catalog.eol.ucar.edu/cgibin/best/research/prod_browse?platform=ice_observations&prod=report&howmany=All&start=Start+Date&end=End+Date&submit=Get+Data)

polar ecosystems. In the Arctic, surface melt is much more pronounced than in the Antarctic, and the different stages of melt listed in Table 3.12.2 under the category "Surface Melting" reflect the complexity of this process. Similarly, arctic sea ice is often discolored by high concentrations of sediment, entrained over water depths of less than 30 m or so and subsequently dispersed throughout the Arctic by ice drift (see discussion in Eicken et al. 2005). The observation protocol addresses this aspect of ice particulate loadings, along with observations of colonization of sea ice by microalgae which can lead to strong discoloration (see Chapters 3.8 and 3.9). Finally, the protocol also allows for observations of the number of marine mammals, birds, and larger fish and crustaceans to be recorded.

**Table 3.12.2.** Summary of categories and options for ship-based ice observations in the Arctic (see Eicken et al. 2005 for details)

**1. Basic information**
Name of observer:
Date:
Time: (start-stop time of observation, ideally 10-min. period)
Position (Latitude, Longitude): (at start and stop time)

**2. Navigation**
Ship speed (kt):
Ship heading (°):
Ship progress (code): (on station; traveling outside of ice edge; traveling in lead or other open water within pack ice; proceeding through loose ice; continuous icebreaking; ramming; stopped due to ice)

**3. Meteorological and hydrographic variables**
Air temperature (°C):
Air pressure (mbar/hPa):
Wind speed (m/s):
Wind direction (°T):
Visibility (km): (instrument data or visual estimate)
Cloud cover (octa): (instrument data or visual estimate)
Surface water temperature (°C):
Surface water salinity (psu):
Water depth (m):

**4. Ice Conditions**
Total ice concentration (%):
For up to three prevailing ice types (primary, secondary, tertiary) indicate the following (incl. additional comments on each ice type):

**A. Ice**

Ice type (code): frazil; shuga; grease; brash ice; dark nilas; light nilas; pancakes; young gray ice; young gray-white ice; first-year white ice; multiyear white ice; fast ice; other (specify in comments)

Ice concentration (%):

Floe size (code): (brash (<10 m); individual pancakes (<10 m); ice cakes (<20 m); small floes (<100 m); medium floes (<1000 m); large/giant floes (>1000 m); vast (>10 km)

Thickness of level ice (m):

Ridging (type): new, unconsolidated ridges; consolidated ridges; weathered ridges; rounded, highly weathered hummocks; rubble-field

Areal fraction of surface covered by ridges (%):

Typical height of ridge sail (m):

Rafting: areal fraction of surface affected by rafting (%)

**B. Snow**

Snow type (code): patches of snow; continuous snow cover; sastrugi and snow dunes; flooded snow (but not due to surface melting!)

Snow depth (m):

**C. Surface melting**

General stage of surface melt (code): no melt; grayish melting snow; meltwater pooling in snow; complete removal of snow but lack of ponds; meltwater pooling on bare ice (with remnants of snow cover on level ice); meltwater pooling on bare ice (with no remnants of snow cover on level ice); meltwater pooling with few ponds melted all the way through ice; meltwater pooling with large number of ponds melted through; meltwater pooling with vast majority of ponds melted through; rotten ice

Current stage of surface melt (code): dusting of new snow; ponds frozen over; both

If present, areal fraction of melt ponds (%):

Width and length of melt ponds (m):

If observable, depth of melt ponds (m):

**D. Sediment-laden sea ice**

Type of sediment inclusions/distribution (code): surface accumulations; layers in ice interior; pockets or patches in ice interior

Fraction of sediment-laden ice (%):

Degree of ice discoloration by sediment inclusions: faint, light, medium, strong

**E. Sea ice algae**

Type of algae: (semiattached strands and patches of algae at bottom and sides of floes; bottom algal layer; interior algal layer; surface algal layer; melt-pond community)

Fraction of ice colonized (%):

Degree of ice discoloration by algae: (faint, light, medium, strong)

**5. Open water**

Type of open water (code): (no open water; small cracks (<1 m wide); fractures (<10 m wide); lead (>10 m); polynya; patches of open water between individual floes; open water with scattered floes; open sea)

If applicable, width of open water (m):

If applicable, length of crack, fracture, lead (m):

Sea state: (calm, swell, swell with breaking waves)

**6. Fauna**

Type of organism: (whale, walrus, seal, polar bear, gull, sea-duck, alcid, cormorant, arctic cod, amphipod/shrimp)

Number:

**7. General Comments**

**8. Digital photographs**

A photograph of ice conditions representative of observation interval should be taken to both port and starboard (wide-angle setting, pointing away perpendicular from the ship's length, with ice/ocean surface filling most of the picture and the horizon in the uppermost fifth of the frame; additionally, a photo down the ship's side showing broken ice floes tilted on their side to indicate ice stratigraphy).

As discussed above and evident from Figure 3.12.2, photographs of ice conditions, including the stratigraphic cross-section of floe fragments turned on their side by the ship, can be of great value in linking on-ice sampling activities to larger-scale ice conditions and ice growth history. Hence, the observation protocol includes at least three such photographs for each observation (see Table 3.12.2 for further specifications). Since these photographs can then be linked to the individual observations in the context of a Web-based database (such as the one referred to above), they also provide a good visual summary of ice and weather conditions.

## 3.12.4 RESULTS FROM PAST OBSERVATIONAL PROGRAMS

Many valuable insights into the character of antarctic sea ice have been derived from ship-based observations, including those that predate the introduction of the ASPeCt observing protocols. For example, Wadhams et al. (1987) published a detailed description of the conditions in the Weddell Sea; Jeffries et al. (2001) and Jeffries and Weeks (1992) have published details of ice conditions in the Bellingshausen, Ross, and Amundsen seas, while Worby et al. (1996; 1998) describe ice conditions in the Bellingshausen Sea and East Antarctica, respectively.

In addition to establishing the ice observing protocols, the ASPeCt program has also undertaken a comprehensive assessment of all the available ship observations since 1980 and collated as many as possible into a single archive. This effort focused primarily on the Australian, German, U.S., and Russian national antarctic programs, which were known to have dozens of data sets containing information on the concentration, thickness, and snow-cover characteristics of the antarctic sea ice zone. Many of the data sets used were still in the form of old analog charts that had never been published, while others, even if digitized, were stored across multiple software applications using different formats, codes, and abbreviations, highlighting the value of a standardized system. The results of this work have been published by Worby at al. (2008) and include an analysis of more than twenty thousand individual observations from eighty-three voyages. The results have been used to create a circumpolar climatology of sea ice and snow-cover thickness, and their regional and seasonal variability around Antarctica. This is the first climatology ever published based on observational data, thus highlighting the value of underway measurements from icebreakers. The data are publicly available via the ASPeCt Web site (http://www.aspect.aq).

In the Arctic, such a comprehensive effort at pulling together sea ice observation data is lacking to date. However, individual studies have demonstrated the great value to be derived from ship-based observations, be it to assess ice thickness along the—relatively unbiased—track of nuclear icebreakers traveling to the North Pole (Tunik 1994), determining the distribution and entrainment history of sediments into sea ice (Eicken et al. 2005), or providing a means of areally integrating measurements of ice albedo and heat exchange in a transarctic survey (Perovich et al. 2009). The latter study is also of interest in conjunction with an earlier, detailed analysis of ice thickness, concentration, ice type, and ice-entrained sediment distribution across the Arctic along a comparable ship track in 1994 (Tucker et al. 1999). A nascent Arctic Sea Ice Working Group under the auspices of the CliC Project may help us to arrive at a more concerted, coordinated approach in recording and disseminating ice observation data.

## 3.12.5 OUTLOOK

With an increase in a range of activities in the polar regions (see Chapter 2), the opportunities for ship-based ice observations are likely to increase substantially. For example, destinational shipping has increased substantially along some routes in the Arctic, as has the number of research cruises into the Arctic Basin. These voyages could provide valuable scientific data to complement remote-sensing studies and other measurement programs. While a common protocol as outlined above is important in ensuring acquisition of data of consistent quality and allowing for intercomparison, the availability of trained observers is also a critical prerequisite. Easing the process of data entry and guiding observations by providing computer-based entry forms, which ideally ingest a lot of environmental data automatically, goes a long way towards being able to attract competent volunteers (such as ship officers or scientists traveling on research cruises) to participate in observation programs. It may furthermore help to emphasize the value of ship-based observations in particular to those most affected by sea ice system services, such as the shipping industry and others, since it can help provide data that are much more closely tied to specific ice uses and associated information needs than generic products.

At the same time, automation of ice observations may also help in building a broader observation database. Recent efforts by Japanese researchers, summarized in Kubo et al. (2000) and Toyota et al. (2006), have demonstrated that automated acquisition, perspective correction, and mosaicking of digital imagery obtained automatically from a ship's bridge or camera mounted in the superstructure can be of great value, for example, in the study of floe-size distributions. The highly sophisticated approach has been implemented by Weissling et al. (2009), who developed an automated ship-based image acquisition and processing system that allows extraction of quantitative parameters based on unsupervised classification of imagery. While observations by a skilled observer are difficult to supplant, there is significant benefit to be derived from an integrated approach that combines imagery of the local ice conditions, photographs of overturned floes, automated surface weather observations, and automated measurements of ice thickness from an electromagnetic induction device (see Chapter 3.2). With a trend towards fewer and fewer personnel on commercial vessels, such approaches may be the only viable ones in the mid- to long term if ship traffic into the Arctic were to increase.

## REFERENCES

Ackley, S., P. Wadhams, J. C. Comiso, and A. P. Worby (2003), Decadal decrease of Antarctic sea ice extent inferred from whaling records revisited on the basis of historical and modern sea ice records, *Polar Res., 22,* 19–25.

Armstrong, T., B. Roberts, and C. Swithinbank (1973), *Illustrated Glossary of Snow and Ice*, Special Publication, 4, Scott Polar Research Institute, Cambridge.

Eicken, H., R. Gradinger, A. Graves, A. Mahoney, I. Rigor, and H. Melling (2005), Sediment transport by sea ice in the Chukchi and Beaufort Seas: Increasing importance due to changing ice conditions?, *Deep-Sea Res. II, 52*, 3281–3302.

Environment Canada (2005), *MANICE—Manual of Standard Procedures for Observing and Reporting Ice Conditions,* revised 9th edition, Canadian Ice Service—Environment Canada, Ottawa (electronic document, available on accompanying DVD).

Gelatt, T. S., and D. B. Sinnif (1999), Line transect survey of crabeater seals in the Amundsen-Bellingshausen Seas, 1994, *Wildlife Soc. Bull., 27,* 330–336.

Jeffries, M. O., K. Morris, T. Maksym, N. Kozlenko, and T. Tin (2001), Autumn sea ice thickness, ridging and heat flux variability in and adjacent to Terra Nova Bay, Ross Sea, Antarctica, *J. Geophys. Res., 106,* 4437–4448.

Jeffries, M. O., and W. F. Weeks (1992), Structural characteristics and development of sea ice in the western Ross Sea, *Antarctic Science, 5*(1), 63–75.

Kubo, M., D. Kutsuwada, and K.-I. Muramoto (2000), Sea ice video image processing using geometric transformation and template matching, IGARSS 2000, *IEEE 2000 Int. Geosci. Remote Sens. Symp., 2*, 475–477.

Perovich, D. K., T. C. Grenfell, B. Light, B. C. Elder, J. Harbeck, C. Polashenski, W. B. Tucker III, and C. Stelmach (2009), Trans-polar observations of the morphological properties of Arctic sea ice, *J. Geophys. Res.*, 114, C00A04, doi:10.1029/2008JC004892.

Steffen, K. (1986), Atlas of the sea ice types. Deformation processes and openings in the ice. North Water project, *Zürcher Geographische Schriften, 20,* Geographisches Institut, Eidgenössische Technische Hochschule, Zürich.

Toyota, T., S. Takatsuji, and M. Nakayama (2006), Characteristics of sea ice floe size distribution in the seasonal ice zone, *Geophys. Res. Lett., 33*, 1–4.

Tucker, W. B. III, A. J. Gow, D. A. Meese, H. W. Bosworth, and E. Reimnitz (1999), Physical characteristics of summer sea ice across the Arctic Ocean, *J. Geophys. Res., 104*, 1489–1504.

Tunik, A. L. (1994), Route-specific ice thickness distribution in the Arctic Ocean during a North Pole crossing in August 1990, *Cold Reg. Sci. Technol., 22,* 205–217.

van Franeker, J. A. (1992), Top predators as indicators for ecosystem events in the confluence zone and marginal ice zone of the Weddell and Scotia seas, Antarctica, November 1988 to January 1989 (EPOS Leg 2), *Polar Biol., 12*, 93–102.

Vinje, T. (2001), Anomalies and trends of sea ice extent and atmospheric circulation in the Nordic Seas during the period 1864–1998, *J. Climate, 14*, 255–267.

Wadhams. P., M. A. Lange, and S. F. Ackley (1987), The ice thickness distribution across the Atlantic sector of the Antarctic Ocean in midwinter, *J. Geophys. Res., 92,* 14,535–14,552.

Weissling, B., S. Ackley, P. Wagner, and H. Xie (2009), EISCAM—Digital image acquisition and processing for sea ice parameters from ships, *Cold Reg. Sci. Technol., 57*, 49–60.

Worby, A. P. (1999), *Observing Antarctic Sea Ice: A practical guide for conducting sea ice observations from vessels operating in the Antarctic pack ice (CD-ROM Antarctic Sea Ice Processes and Climate (ASPeCt) program),* Australian Antarctic program, Hobart, Tasmania.

Worby, A. P., and I. Allison (1991), Ocean-atmosphere energy exchange over thin, variable concentration Antarctic pack-ice, *Ann. Glaciol., 15*, 184–190.

Worby, A. P., C. Geiger, M. J. Paget, M. van Woert, S. F Ackley, and T. DeLiberty (2008), Thickness distribution of Antarctic sea ice, *J. Geophys. Res., 113*, C05S92, doi:10.1029/2007JC004254.

Worby, A. P., M. O. Jeffries, W. F. Weeks, K. Morris, and R. Jaña (1996), The thickness distribution of sea ice and snow cover during late winter in the Bellingshausen and Amundsen Seas, Antarctica, *J. Geophys. Res., 101*(C12), 28,441–28,455.

Worby, A. P., R. A. Massom, I. Allison, V. I. Lytle, and P. Heil (1998), East Antarctic sea ice: A review of its structure, properties, and drift, in Antarctic sea ice physical processes, interactions and variability, *Antarct. Res. Ser., 74,* edited by M. O. Jeffries, pp. 41–67, American Geophysical Union, Washington, DC.

World Meteorological Organization (1970), *Sea Ice Nomenclature: Terminology, Codes, and Illustrated Glossary*, WMO/OMM/BMO 259, TP 145, World Meteorological Organization, Geneva, Switzerland.

# Chapter 3.13

# Automatic Measurement Stations

*Don Perovich*

## 3.13.1 INTRODUCTION

In situ observations are critical to understand sea ice properties and behavior and to employ sea ice as a platform for activities. There are numerous methods and platforms for making in situ measurements. There are land-based stations around the periphery of the ice cover at several locations, including Barrow, Prudhoe Bay, Resolute Bay, Alert, Eureka, Ny Alesund, Longyearbyen, and Tiksi. Long, detailed time series can be obtained at these stations. Ice camps can be deployed within the ice cover by aircraft or ship. Icebreakers can make survey cruises or be frozen in the ice as a drifting ice station. These platforms provide excellent opportunities for examining, in detail, the properties and processes of the ice-ocean-atmosphere system. However, for all of their merits, these platforms have limitations. Land stations only allow access to the periphery of the ice cover. Ice camps and icebreakers are expensive and are typically used in a campaign mode, rather than a long-term observation mode. Logistical and personnel demands also constrain the number of field camps that can operate at a given time. Additional capabilities are needed for long-term, large-scale studies.

An automatic measurement station (AMS) can provide this capability. There are many different automatic measurement stations currently operating on sea ice. They are designed to operate autonomously for years at a time and are basic building blocks for a sea ice observing network. Stations range in complexity from a simple buoy measuring air temperature to an ensemble of instruments determining the properties of the atmosphere, ice, and ocean. This is an area of ongoing research and development, where technology is advancing rapidly.

There are a few ways of implementing an AMS. Often it is possible to purchase an existing AMS that will satisfy the requirements of the project. The AMSs described in Section 3.13.3 are all available for purchase. It is also possible to customize an existing AMS by adding sensors to it, while in some cases a new AMS must be designed and developed. Contacting the developers and users of arctic

AMS is useful, since they are a valuable resource for advice and insight. The International Arctic Buoy Program (http://iabp.apl.washington.edu) provides an overarching community and coordination for arctic AMS activities.

## 3.13.2 APPROACHES AND TECHNIQUES

The overall design and development of an AMS should be governed by science questions. The science questions dictate the parameters that need to be measured and the duration of the measurements. Many factors come into play when designing and developing an AMS, including power, data transmission, weight, and survivability. Automatic measurement stations have different sensor packages and can take many different forms, with varying complexity and mission capability. However, all AMSs have some commonality. The stations have to function in an austere environment characterized by cold temperatures, occasional high winds, long periods of darkness and daylight, and limited access for repairs and maintenance. The lower atmosphere is usually saturated and instruments become covered with snow, frost, rime, or rain. It takes power to clear the instruments, and power is limited. There is also irksome wildlife, such as hungry foxes and curious polar bears. A well-designed AMS must overcome all of these difficulties.

As with many things in sea ice research, simpler is better. The smaller the above-ice profile of an AMS, the less likely it is to be damaged by wildlife. Foxes will try to chew anything, but it is possible to fox-proof an AMS using metal conduit to armor all cables. Bear proofing is problematic. At best an AMS can be made inconspicuous and less likely to attract a bear's attention. More separate components of an AMS increase the likelihood of damage from animals and ice deformation.

At its core an AMS must replace a person. The particular components may vary and packaging may be different, but there are basic elements that an AMS needs (Figure 3.13.1). It must provide control, communication, and power. Control comes from a data logger or a rugged, field-ready computer. Communication is usually through satellite links. The ARGOS satellite network provides one-way communication from an AMS to a home base. Two-way communication can be obtained using the Iridium satellite system. Using Iridium, data can be retrieved and measurement program parameters can be adjusted. Power is a key concern for a long-term automatic station. Batteries are the simplest power source. Lead acid and lithium batteries are commonly used. Lithium batteries are more expensive than lead acid, but they have greater power density and perform better in the cold. If power demands are large, batteries are augmented by wind generators and solar cells. Typically, to improve survivability, all of these components are placed in a secure housing that is structurally sound and weatherproof.

An AMS needs software to control the instruments, communicate results, and manage power. The type of software depends primarily on the controller used and the complexity of the AMS. Some controllers have a propriety operating system

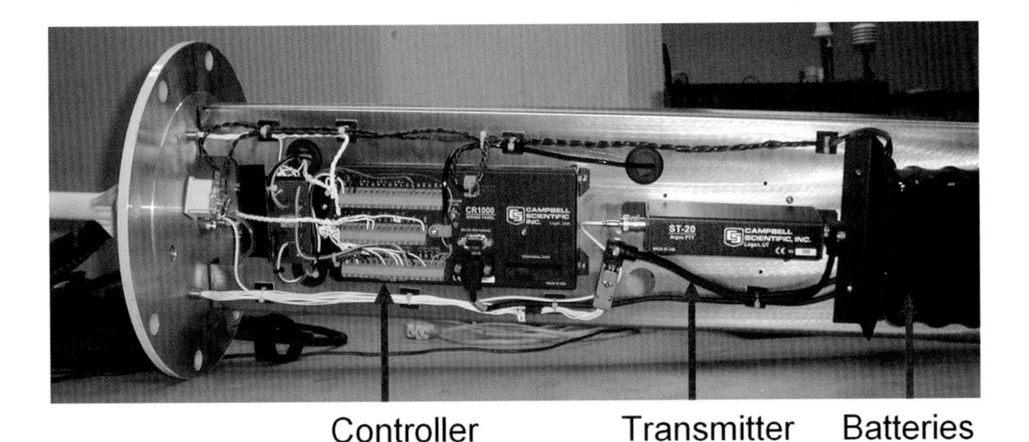


**Figure 3.13.1.** Fundamental elements of an automatic measuring system showing the controller, satellite transmitter, and battery pack

and are programmed in an assembly-like language. Others run flavors of UNIX and can be programmed in C or other languages. If the AMS has two-way communications, programs can be remotely modified or even replaced. Sampling strategies can be adjusted in light of the observations.

Webcams are an exciting addition to an AMS. They contribute scientifically by showing the state of the ice surface, sky conditions, and the AMS. They are also a marvelous tool for outreach to students and the public. Figure 3.13.2 shows images from a webcam placed at the North Pole Environmental Laboratory.

**Figure 3.13.2.** Webcam photographs from the North Pole Environmental Observatory, courtesy of Pacific Marine Environmental Laboratory. The photographs were taken on 24 April 2004 (top) and 1 July 2008 (bottom). Melt is underway in the second image, with melt ponds forming.

## 3.13.3 METHODS

Five specific examples are now presented to explore the variety and capabilities of automatic measuring stations. An AMS can also be a combination of individual stations. For example, the North Pole Environmental Observatory includes an automatic weather station, an ice mass balance buoy, and ice-tethered profilers as part of its AMS. Of the five examples given below, four are Lagrangian drifters; they are frozen into the ice and drift with the pack. The last example, the ice-profiling sonar (IPS), is different. It is an Eulerian platform that rests on the seafloor and monitors the ice that passes overhead.

### 3.13.3.1 Basic Unit

The most basic and widely used AMS is a single buoy that measures just three parameters: position, air temperature, and barometric pressure. These buoys can be deployed from the surface or airdropped and are the backbone of the International Arctic Buoy Program (http://iabp.apl.washington.edu/index.html) (Figure 3.13.3). The simplicity of these buoys belies the importance of their data. Over the past few decades hundreds of these buoys have been deployed, greatly advancing our understanding of sea ice drift and of Arctic Ocean pressure and temperature fields. For example, results from these buoys have demonstrated that changes in the basic drift pattern and in the age of arctic sea ice of have occurred. Data from these buoys are also assimilated in real time into the Global Telecommunication System of the World Weather Watch to improve forecasts of weather and sea ice conditions.

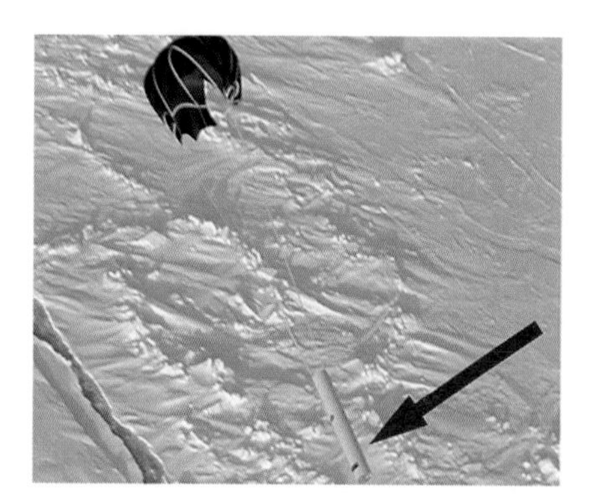

**Figure 3.13.3.** Photographs showing two types of a basic, air-deployed buoy that measures position, barometric pressure, and air temperature. Such buoys have been deployed for the past few decades as part of the International Arctic Buoy Program.

### 3.13.3.2 Automatic Weather Station

Automatic weather stations are used around the world in both populated and remote areas. In the polar regions, there are land-based networks in Greenland and the Antarctic (http://uwamrc.ssec.wisc.edu/aws.html). Stations have also been deployed on arctic sea ice. Figure 3.13.4a shows a weather station installed as part of the North Pole Environmental Observatory. Instrumentation includes temperature and wind sensors at a height of 2 m, a barometer, and longwave and shortwave radiometers. The radiometers are cleverly packaged to avoid the buildup of frost or water on the domes (Figure 3.13.4b). Most of the time the radiometers are in a sealed container. To take a reading they rotate upwards, equilibrate, measure the incident radiation, and then rotate back into the container.

**Figure 3.13.4.** Photograph of an automatic weather station deployed as part of the North Pole Environmental Observatory. Close-up photograph of longwave and shortwave radiometers plus solar cells. Photographs courtesy of Jim Overland.

### 3.13.3.3 Ice Mass Balance Buoy

Sea ice is an integrator of both the surface heat budget and the ocean heat flux, and thus the growth and melt of sea ice is a key climate change indicator. If there is net warming over time, then there will be thinning of the ice. Conversely, a net cooling leads to thicker ice. Consequently, there is interest in measuring the amount the ice grows in winter and how much it melts in summer, both on the surface and on the bottom. This information provides insight into the nature of changes in the sea ice

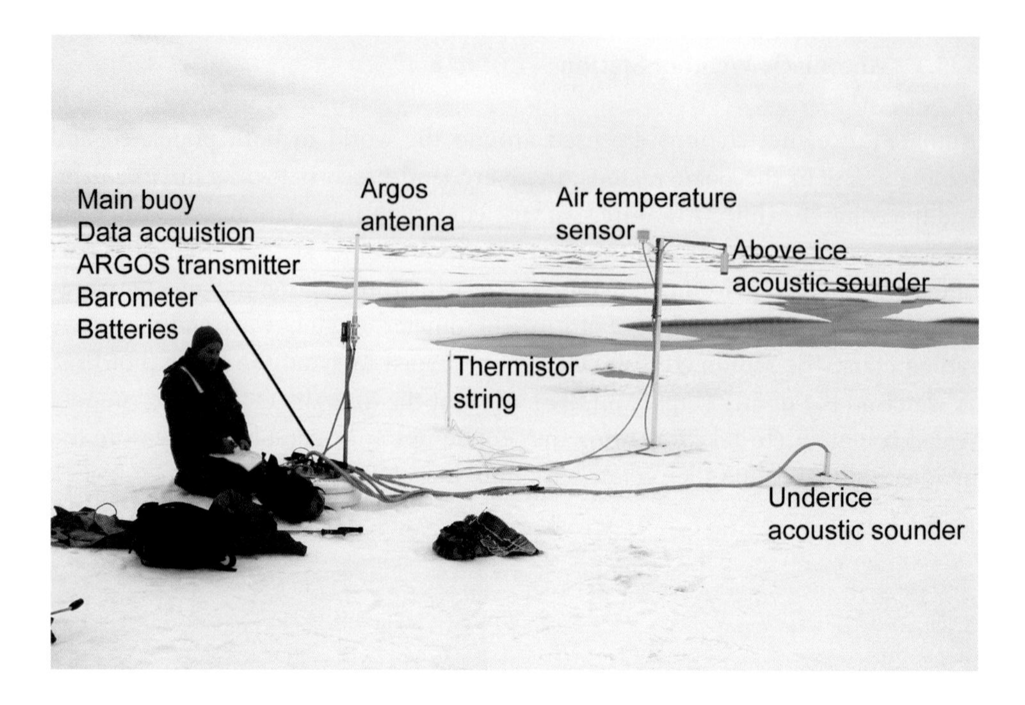


**Figure 3.13.5.** Photograph of an ice mass balance buoy

cover and can help determine relative impacts of dynamics and thermodynamics; of atmosphere versus ocean forcing; and winter changes relative to summer.

Figure 3.13.5 shows an annotated photograph of an ice mass balance buoy (IMB). IMBs typically consist of a Campbell scientific data logger, an Argos transmitter, a thermistor string, and above-ice and below-ice acoustic sounders measuring the positions of the ice surface and bottom to within 5 mm. Thermistor strings are made of PVC rod, with thermistors spaced every 10 cm. They extend from the air through the snow and ice into the upper ocean. The thermistor accuracy is better than 0.1°C. The buoys may also have a GPS, a barometer, and an air temperature sensor. Lithium battery packs provide power for up to three years of operation. It is straightforward to add other sensors to an IMB and measure quantities such as the spectral distribution of sunlight in and under the ice; changes in the hydrostatic displacement of the floe; internal ice stress; and the temperature and salinity of the upper ocean. More information is available at http://imb.crrel.usace.army.mil and http://www.metocean.com/buoy_polar_imb.jsp.

### 3.13.3.4 Ice-Tethered Profiler

AMS have been developed and deployed to monitor the properties of the ocean under the ice cover. One such station is the ice-tethered profiler (ITP) (Figure 3.13.6). An ITP consists of a buoy placed on the surface of the ice; an 800 m long cable suspended below the buoy; a ballast weight at the end of the cable; and a

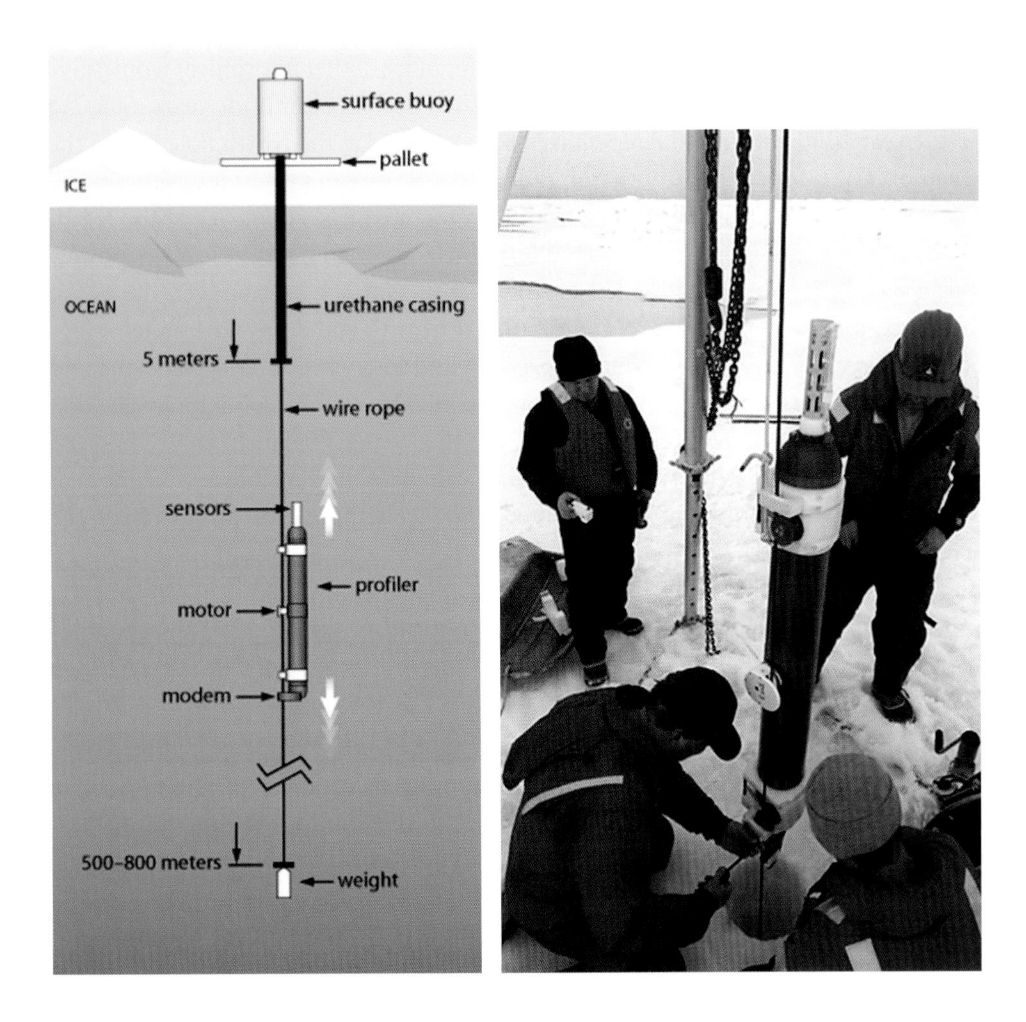


**Figure 3.13.6.** Schematic illustrating an ice-tethered ocean profiling system and a photograph showing the system being installed in the ice. Figure courtesy of Woods Hole Oceanographic Institute.

profiler that moves up and down the cable driven by a motor. The profiler contains the instruments, the controller, the drive system, and the batteries. The ITP can measure vertical profiles of conductivity, temperature, pressure, dissolved oxygen, photosynthetically available radiation, and chlorophyll fluorescence. The battery pack provides enough power for 1.5 million meters of vertical profiling or three years of a daily down and up profile pair. Data is transmitted using the Iridium satellite system (http://www.whoi.edu/page.do?pid=20756).

### 3.13.3.5 An Integrated System

One of the strengths of an AMS is the ability to easily integrate individual components into a single station. Such integration greatly enhances the value of the observations. For example, combining the three systems discussed above (an automatic weather station, an ice mass balance buoy, and an ice-tethered profiler) would give an AMS that could monitor the atmosphere-ice-ocean column for years. Starting in the year 2000, integrated AMSs have been deployed at the North Pole and in the Beaufort Gyre, forming the beginnings of an Arctic Ocean Observing Network.

Figure 3.13.7 illustrates results from an integrated AMS deployed at the North Pole Environmental Observatory (NPEO). The combined data set is quite powerful. It has the information needed to calculate the surface heat budget, the ocean heat flux, snow accumulation and melt, and ice growth and decay. The meteorological instruments measure incident shortwave and longwave fluxes. Webcams and surface temperatures are used to estimate the albedo. The outgoing longwave flux is computed using surface temperatures from the ice mass balance buoy. Observed air temperature and wind speed are used to calculate the sensible and latent heat fluxes. Conduction in the ice is computed using ice temperature profiles from the ice mass balance buoy.

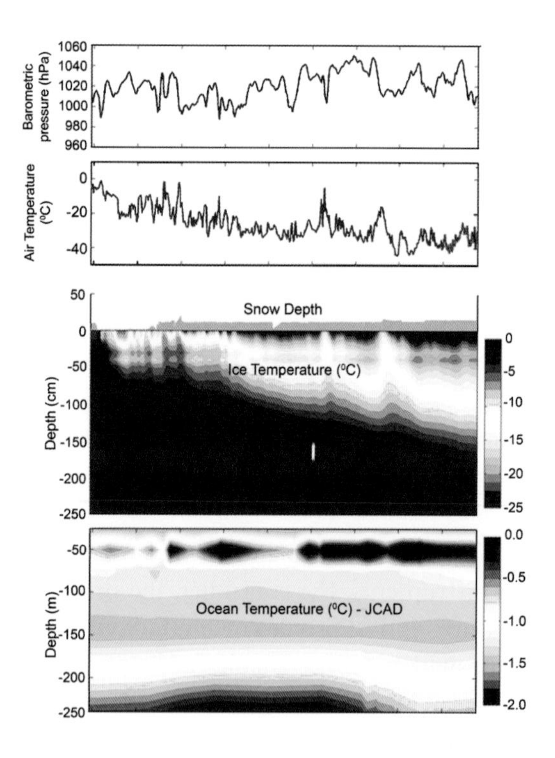


**Figure 3.13.7.** Data from an integrated AMS that includes a meteorological station, an ice mass balance buoy, and an ice-tethered ocean profiler. Data are from 2003 to 2004 and are courtesy of the North Pole Environmental Observatory.

The heat flux from the ocean to the ice is measured by an autonomous ocean flux buoy. Sensors on an ice-tethered profiler monitor vertical profiles of the salinity, temperature, and heat content of the ocean to a depth of 800 m. This data delineates the contributions to the heat flux at the bottom of the ice from solar heating of the upper ocean and heat from deeper in the ocean. It also monitors the depth of the ocean mixed layer and the freshwater content of the ocean.

An integrated AMS yields an extraordinary data set. Air temperature and barometric pressure are immediately assimilated into forecast models. The combined data set can be used for evaluation of reanalysis products, and as input and validation data for large-scale sea ice models and general circulation models. A network of integrated AMSs would provide a basinwide data set to explore the spatial variability and temporal evolution of sea ice conditions.

### 3.13.3.6 Ice Profiling Sonar

The AMSs discussed above are all Lagrangian drifters mounted on the ice and drifting with the ice. There is also a bottom-moored ice profiling system (IPS) that provides an Eulerian record of ice draft and velocity (Figure 3.13.8). The primary instrument is a narrow-beam distance-ranging sonar that measures ice draft. The ice draft data are converted to ice thickness by assuming that the ice is in isostatic equilibrium. An IPS typically has enough battery power for two years of operation. One limitation of the IPS is that there is no communication from deployment to retrieval precluding real-time data access. The instrument must be retrieved to obtain the data. Instrument retrieval is relatively straightforward if no ice is present and can be done using divers if ice is present. These devices are commercially available (http://www.aslenv.com/iceprofiler.htm).

**Figure 3.13.8.** A schematic of an upward looking ice profiling system

## 3.13.4 AN APPLICATION

One of the goals of the 2007–2009 International Polar Year has been to develop an Arctic Observing Network (AON). Automatic measuring systems are a major component of this network. Figure 3.13.9 shows components of the Arctic Ocean portion of this network, including all of the instruments discussed in the previous section. In October 2008 there were Lagrangian drifters installed on ice floes at sixty-four locations (Figure 3.13.9a) measuring position, air temperature, and barometric pressure. At sixteen of these sites there were ice-tethered ocean profilers measuring currents and temperature and salinity of the upper ocean. Eight of the sites also had ice mass balance buoys recording snow accumulation and melt, ice growth and decay, and vertical profiles of ice temperature. These stations use sea ice as a platform both literally as a place to put instruments and figuratively as the basis of a conceptual framework for arctic observations. Taken together these data provide a time series of the basic properties of the atmosphere, ice, and ocean at several locations in the Arctic Ocean.

Figure 3.13.9b is a schematic illustrating the ice-tethered ocean profilers and how they can serve as navigation beacons for underwater gliders. The gliders move between the stations, providing a means to observe the spatial variability of the salinity, temperature, and density of the upper ocean. In addition to these ice-based drifting platforms, there are also moorings placed through the Arctic and adjacent seas measuring profiles of ice thickness, ocean temperature, salinity, and current. Most important, all of these individual components are integrated into an observing network. Much of the data is being archived at the Cooperative Arctic Data and Information Service (CADIS) (http://www.eol.ucar.edu/projects/aon-cadis).

This nascent network is already serving a large range of stakeholders. Meteorological observations are being incorporated into the global telecommunications system, resulting in improved weather forecasts for the Arctic. Observing network data are assimilated into sea ice diagnostic and forecast models. Ice motion and ice mass balance results are improving the interpretation of satellite observations of the ice cover. In situ measurements from the observing network are being incorporated with satellite observations and model results to improve short-term forecasts of sea ice extent and conditions. These forecasts are used in decisions regarding Arctic Ocean shipping. Ice thickness observations are relevant to resource-extraction efforts in the arctic continental shelf. Finally, by providing observations of the ongoing changes in the Arctic Ocean, the network is enhancing our understanding of the nature and causes of the changes.

**Figure 3.13.9 (facing page).** Top: Map of drifting automatic measuring stations operating in October 2008 as part of the International Polar Year. Red dots are basic buoys, green dots are ocean profilers, and black dots are ice mass balance buoys. Bottom: schematic illustrating and autonomous Arctic Ocean observing network deployed during IPY as part of the Developing Arctic Modeling and Observing Capabilities for Long-term Environmental Studies (DAMOCLES) program.

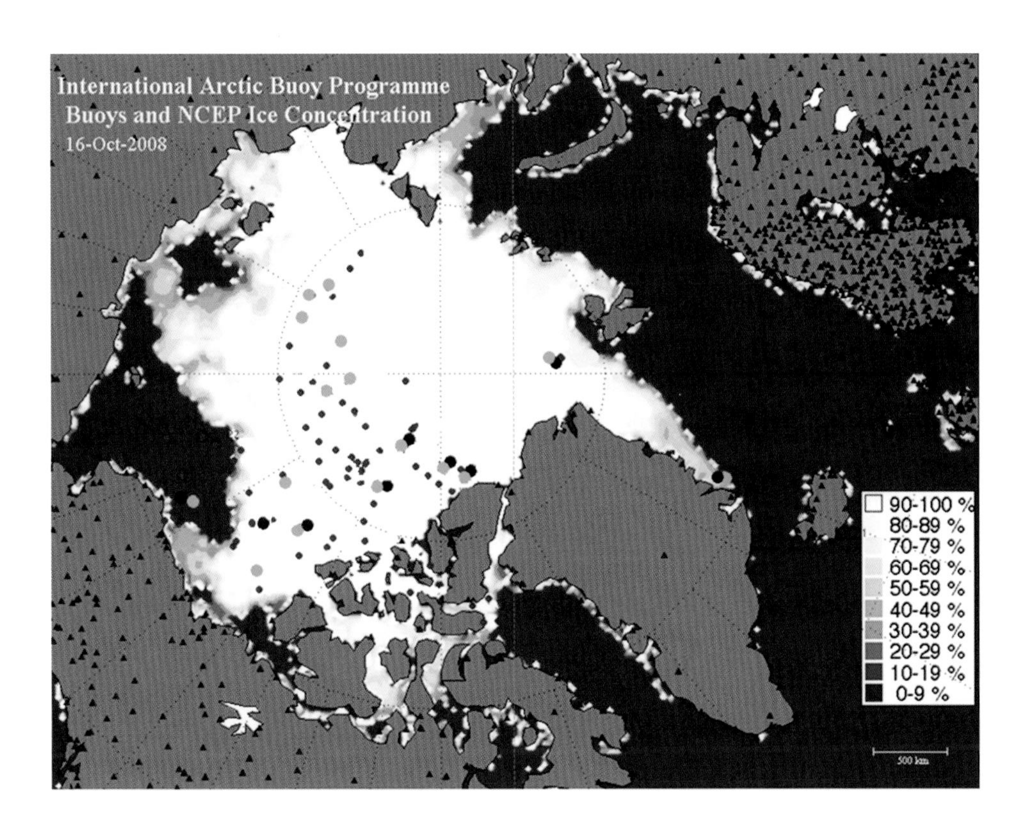
International Arctic Buoy Programme
Buoys and NCEP Ice Concentration
16-Oct-2008
90-100 %
80-89 %
70-79 %
60-69 %
50-59 %
40-49 %
30-39 %
20-29 %
10-19 %
0-9 %
500 km

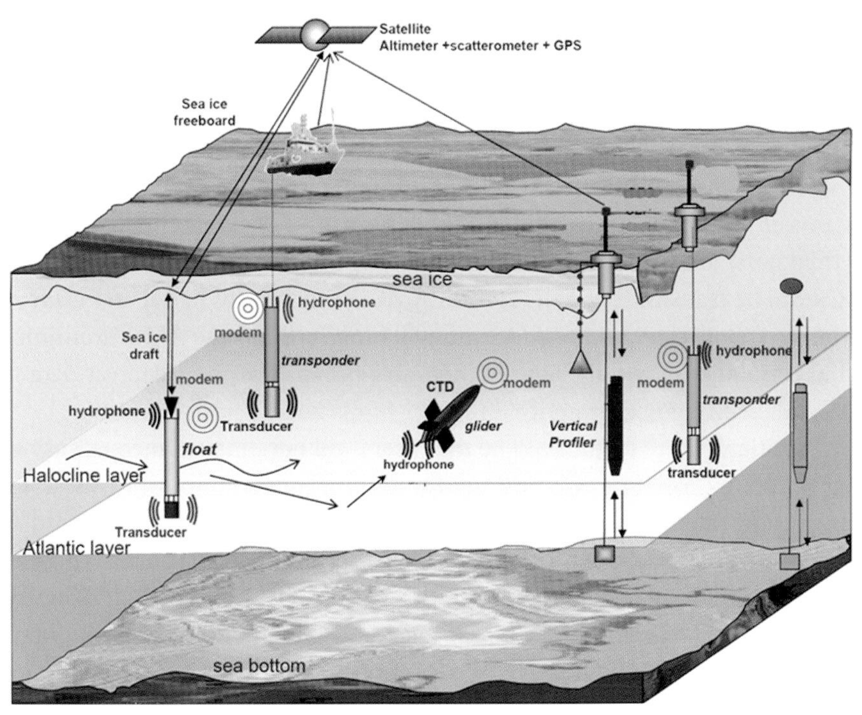
Satellite
Altimeter +scatterometer + GPS
Sea ice
freeboard
sea ice
hydrophone
modem
Sea ice
draft
transponder
modem
CTD
modem
hydrophone
modem
hydrophone
Transducer
float
glider
Vertical
Profiler
transponder
Halocline layer
hydrophone
transducer
Transducer
Atlantic layer
sea bottom

## 3.13.5 FUTURE DIRECTIONS

During the IPY in 1957, sea ice automatic measurement stations were extremely limited. Typically, it was just an instrument that could run by itself for a few days. In 1957 AMSs were routinely visited to replace batteries and change the paper in the strip chart recorder. By the 2007 IPY, the role of sea ice AMSs increased greatly. Figure 3.13.9 shows the array of arctic sea ice automatic measuring stations deployed during IPY 2007. They were widely used and were the basic components of the nascent Arctic Observing Network.

In IPY 2057 some of the sensors will be similar to those used in 2007, just as there was little change in the all-wave radiometers between 1957 and 2007. However, there will be many advances in AMS technology and their contributions will be much greater in IPY 2057. Advances in computing power, communications, and miniaturization will herald a vast expanse of new opportunities. Improvements in batteries will provide power to allow the deployment of high-energy-demand instruments and to keep sensors free of ice and water. Bandwidth will not be an issue, allowing large data files to be exchanged easily and high-resolution color video to be streamed back from the field.

A hypothetical sea ice AMS ready for deployment in IPY 2057 will be designed addressing a central aspect of the 2057 ice cover: a minimal amount of perennial ice and a summer Arctic that is nearly ice-free. The stations will function well in both ice and open water. The core of the AMS is a suite of instruments to monitor the atmosphere-ice-ocean column. The instruments will include the sensor suites discussed in Section 3.13.3 plus a cloud radar, a cloud lidar, an acoustic Doppler current profiler, and seismic-acoustic sensors. There will also be biological sensors measuring parameters including photosynthetically available radiation, oxygen content of the water, primary productivity, and marine mammal activity. Smart motes, tiny low-power sensor packages, will allow time-series measurements of snow depth, ice thickness, and ice temperature at many different sites. Data from the smart motes will be transmitted to the central AMS. Autonomous aerial and underwater platforms will be based at the AMS and will range around the AMS, examining the spatial variability of the atmosphere, ice, and ocean. They will deposit data to the AMS and receive power from it.

A critical future feature will be that users will have instantaneous, easy access to all of this information. Computing trends will improve not only our ability to make measurements but also our ability to access those measurements and seamlessly incorporate them into activities using sea ice as a platform. The greatest challenge in creating the autonomous observing network of IPY 2057 lies not in technology but in cooperation. It will take a sustained cooperative international effort to develop a comprehensive Arctic Observing Network.

**Chapter 3.14**

# Data Management Best Practices for Sea Ice Observations

*Florence Fetterer*

## 3.14.1 DATA STEWARDSHIP

The goals of scientific data management are to preserve data for the long term and to facilitate use of the data by those other than the data collector. Both goals (preservation and sharing or access) require that the data be described well and thoroughly with metadata and documentation. Metadata is a structured summary of information about the data, the brief "who, what, where, when, and why" of the collection of data. Submitting data to a national data center or other national archive when possible and submitting metadata to a catalog like the NASA Global Change Master Directory are good steps to take, and indeed are required by some funding agency sponsors. Data sharing is easier when data are in a standard format of some kind, and machine-to-machine data sharing within a sophisticated data portal for visualizing, subsetting, and analysis demands a standard, ideally self-describing, format for data. (The NOAA Live Access Server, http://ferret.pmel.noaa.gov, is one such portal.)

Whose responsibility is it to steward data in this way? The National Science Board (NSB 2005, pp. 25–29) lays out the responsibilities of data authors, managers, and others. Those of the authors (scientists) include "conform to community standards for recording data and metadata that adequately describe the context and quality of the data and help others find and use the data." Those of data managers include "be a reliable and competent partner in data archiving and preservation, while maintaining open and effective communication with the served community," and "participate in the development of community standards" including those for format, content, and metadata. Data stewardship is a shared responsibility that includes funding agencies and data users as well.

Documenting data so that it can be used appropriately by a researcher fifty years from now, who may not know your discipline, can be laborious. General guidelines for this exist (and will be paraphrased or referenced in this chapter), but it is helpful to be aware of the ways in which professional Earth science data management is evolving to assist individual investigators and research groups in preserving and sharing data.

The National Science Board categorizes digital data as either reference data, resource or community data, or research data. Here we are speaking of research data:

> Research data collections are the products of one or more focused research projects and typically contain data that are subject to limited processing or curation. They may or may not conform to community standards, such as standards for file formats, metadata structure, and content access policies. Quite often, applicable standards may be nonexistent or rudimentary because the data types are novel and the size of the user community small. Research collections may vary greatly in size but are intended to serve a specific group, often limited to immediate participants. There may be no intention to preserve the collection beyond the end of a project. One reason for this is funding. These collections are supported by relatively small budgets, often through research grants funding a specific project. (NSB 2005; p. 20)

In contrast, the budgets for resource or community data collections and for reference collections generally are provided through direct funding from agencies and are often large. Resource data collections (for example, synthetic aperture radar data from the Alaska Satellite Facility) serve a specific community, while reference data collections serve a broad range of users with far-reaching impact (such as the NCAR/ECMWF Reanalysis Project).

Research data may have been collected for a small group and limited purpose, but science agencies recognize the potential importance of smaller data sets to a wider group of researchers and insist that those data be preserved and shared. National archives like the NOAA National Data Centers and the National Snow and Ice Data Center (NSIDC), or international archives like members of the World Data System, are not usually supported to receive small data sets from numerous individual investigators, put them in a standard format, provide them with metadata and documentation, and make them available through online catalogs. And yet if any of these steps are not taken, data may remain difficult to find and use.

Data managers are working to surmount this difficulty with tools that make it easier for project scientists to author metadata and submit data. Professional data managers speak at scientific meetings about the importance of open source standards and other aspects of the cyber infrastructure that holds digital data, moves data, and allows researchers to visualize and download data. "Interoperability" is touted as a goal. For a scientist, interoperability means that the data they need are easy to find and use with their analysis tool of choice, be it Microsoft Excel, IDL, Ferret, or something else.

On an international level, the World Data Centers (WDCs), founded to house and exchange data collected as part of the International Geophysical Year in 1957–1958, are challenged to adapt to the changing ways in which data are collected and managed. The fifty extant WDCs are part of a restructuring for a more networked and interoperable World Data System, with improved online access to data and supporting services that cut across disciplinary boundaries (ICSU 2008). Other efforts, including those of the Group on Earth Observations and the Global Information Commons for Science Initiative, are working for international standards and best practices that will facilitate data preservation and access.

### 3.14.2 MANAGING DATA FROM FIELD PROGRAMS

A few simple considerations for good data management are listed below. These follow from the collective experience of data managers at the National Snow and Ice Data Center and at other data archives. See Parsons et al. (2004), Parsons and Duerr (2005), and the references therein for more information on the reasoning behind these recommendations. Following these recommendations can be considered following "best practices"[2] in that the recommendations are steps needed to ensure the desired outcome of securely archived and easily shared data.

1. Coordinate data collection plans with partners. Talk about your plans with data managers at the institution that is most likely to house your data permanently. Involve data managers early in the experiment design process and where possible involve data managers directly in the data collection efforts.
2. Arrange to use a common, well-defined, or standard format for data files, and plan to use community standards or conventions, where those exist, when defining the content of your data files.
3. Think about how uncertainty will be described. Develop quality-control procedures.
4. Back up collected data with at least two copies in separate locations as soon as possible.
5. Describe the data collection with standard formatted metadata. Metadata authoring tools can make this easier. Tools may be included in some GIS and other data-analysis packages or exist as stand-alone tools. NASA's Global Change Master Directory (GCMD) docBUILDER is one good choice. It is available both online and as an off-line tool that can be used in the field. If your data are collected as part of an agency-funded program of observations, assistance with metadata may be available through a dedicated data management component of the program. For example, the Cooperative Arctic Data and Information Service (CADIS) offers a tool for writing metadata, and personal assistance, to investigators currently within

the NSF-funded component of the Arctic Observing Network (http://aon-cadis.org). The metadata should contain an explicit reference to where the data are archived, so that anyone who finds the metadata will have a link to the data as well.

6. Submit (i.e., publish) the metadata to a clearinghouse or metadata catalog. The GCMD and Geospatial One Stop (GOS) are two possibilities. (GOS specializes in geographical information system data and maps.) Clearing-houses often "harvest" each other's metadata, so that, for instance, GOS metadata catalog entries also show up in searches for data on the GCMD. Data centers like the National Snow and Ice Data Center upload metadata records to these larger clearinghouses as well, so that if you are working with a data management project or data center, this step may be taken for you. Many polar field projects are part of the International Polar Year and can use the IPY Data and Information Service (IPYDIS; http://ipydis.org).
7. Document the data thoroughly so that those outside your community could use it. Reference other documents when doing so improves under-standing of the data (e.g., relevant chapters in this handbook for how sea ice measurements were taken).
8. Place the data along with documentation in a secure archive. If you are not part of a project that provides a pathway for this step, opportunities to archive your data may be available with a relevant World Data Center (list at http://www.ngdc.noaa.gov/wdc) or with a national data center (e.g., one of the NOAA data centers in the United States).
9. Update the documentation and inform the archive when you make any changes or updates to the data.

It is worth discussing some of these steps in more detail.

### 3.14.2.1 A Common or Standard Format for Files and Standard Data File Content

"File format" can mean the same as "file type" (e.g., binary data, ASCII text, PDF, Microsoft Excel, netCDF). It is better to archive easily described files in a non-proprietary format than to archive, say, a Microsoft Word file that may become unreadable in the future (Duerr et al. 2004). A simple comma-separated value ASCII file, with a header record describing the fields or columns, is more useful for more people than the same data in a proprietary Excel file

"File format" can also refer to the specific implementation of the format: the structure of data within a data file, or some other aspect defining the data, such as the map projection and gridding scheme for gridded data. NSIDC uses a format for some gridded data called Equal Area Scaleable Earth Grid or EASE-Grid (Brodzik and Knowles 2002). The file type for EASE-Grid data is usually flat binary, 1 byte unsigned byte format, but the EASE-Grid format refers as well to a grid interpolation and projection combination particular to EASE-Grid files.

Putting data in a standard format where format refers to the structure of data within the file, or in a self-describing format like netCDF, greatly improves data interoperability. Two examples are the ArcticRIMS hydrology data explorer (http://RIMS.unh.edu), which uses a very precisely defined gridded ASCII format, and the NOAA Earth Systems Research Laboratory's data visualization and analysis tool (http://www.cdc.noaa.gov/PublicData), which uses data in netCDF with Climate and Forecast (CF) extensions.

NetCDF is worth special consideration because it has emerged as the dominant self-describing format for much of the physical science data that polar researchers use. There are many tools for subsetting and mathematically manipulating data sets in netCDF, especially when the files are constructed with a standard method of describing file content called Climate and Forecasts extensions. Unidata maintains the netCDF standard, facilitates communication among a large group of users and developers, and collects best practices for writing netCDF files (http://www.unidata.ucar.edu/software/netcdf/docs/BestPractices.html). The format was developed for array data, but to facilitate interoperability, there is a need to render station or point data in netCDF. The Sea Ice Experiment–Dynamic Nature of the Arctic in the Beaufort Sea (SEDNA) project has been a pioneer in this area, and has proactively asked that all data be archived in netCDF (http://research.iarc.uaf.edu/SEDNA/netcdf.php). The DAMOCLES (Developing Arctic Modeling and Observing Capabilities for Long-term Environmental Studies) project requires investigators to submit data in netCDF.

In addition to a standard format, data files for the same types of data should have standard content. This content can be determined by a task force from the community that authors and uses the data files. "Contents Recommended for Data Files From Moored Ice-Profiling Sonar" (at http://nsidc.org/noaa/moored_uls/), developed at a WCRP Climate and the Cryosphere (CliC) workshop, is an example that illustrates the role that CliC and similar coordination groups can play in developing these standards. The partnership between the National Geophysical Data Center and the marine geologists collecting seafloor data is another example. Here, data managers and oceanographers founded a curators' consortium in 1977, and began by designing an eighty-column fixed-field data format. The format has evolved over the years to take advantage of geospatial database and GIS capabilities, but the close partnership between marine geologists and data managers has remained the crucial element for ongoing data stewardship (Moore and Habermann 2006).

#### 3.14.2.2 Metadata

Those unfamiliar with the concept of metadata may want to review the Introduction to Metadata by the Marine Metadata Interoperability Project (Neiswender

2009). For some quick answers to the question "what is metadata good for," see the brief overview (http://147.66.8.18/Articles/Stickscarrots.html) from the Joint Committee on Antarctic Data Management (Belbin 1998). The points in the overview are still valid, but the essay underplays the critical role metadata now has in building interoperable data management systems.

Standards for metadata are evolving and sometimes merging. Earth science archivists are tending toward using ISO19115 (http://www.iso.org/iso/iso_catalogue/catalogue_tc/catalogue_detail.htm?csnumber=26020) for basic collection metadata and PREMIS, the PREservation Metadata framework (http://www.oclc.org/research/projects/pmwg/default.htm), for additional preservation information. These standards have implications for cyberinfrastructure and for digital data preservation, but should not be of great concern to those collecting data. Metadata can often be mapped from one standard to another, and using a good authoring tool like the GCMD docBUILDER, along with writing complete documentation, will help data managers who may handle your data in the future to adhere to the best standard of the day.

Metadata authoring tools often compel the metadata author to choose from a fixed selection, or controlled vocabulary, of descriptors. The GCMD calls these science keywords and associated keywords (Olsen et al. 2007). They are lists of disciplines, parameters, platforms, regions, instruments, and other characteristics that can be associated with data. Ideally, terms in the vocabulary are referenced to precise definitions (one example would be Schaepman-Strub et al. [2006] for reflectance quantities). This handbook may be a source of definitions for keywords associated with ice observations. Using a common and well-defined vocabulary helps users discover and understand data and how different collections are related to each other. As controlled vocabularies and other semantic mechanisms become more sophisticated, tools can be developed that use computer reasoning to help users discover and use unfamiliar data. This is an active and growing area of informatics research.

Data sets can and should be cited like reference papers. This credits those responsible for the intellectual effort required to collect and organize the data, and properly associates a research result with the data that were used. Metadata should include enough information to create a citation for the data. See "How to Cite a Data Set" (http://ipydis.org/data/citations.html) for more information.

For examples of metadata, go to AWI Moored ULS Data, Greenland Sea and Fram Strait, 1991–2002 (http://nsidc.org/cgi-bin/get_metadata.pl?id=g02139) and Mooring and Drifting Buoy Data: Beaufort Gyre Freshwater Experiment, 2003 (http://nsidc.org/cgi-bin/get_metadata.pl?id=arcss165). Both metadata records are in the NSIDC online catalog, but the first describes data stored at NSIDC, while the second directs users to data archived elsewhere.

### 3.14.2.3 Document the Data Thoroughly

Throughly documenting data with metadata and, often, supplementary material referenced by and accessible through the metadata (usually with a linked URL) is necessary for preserving data. It is relatively easy to write good documentation with the user community for the data in mind. But many data sets end up having unanticipated uses. Passive microwave data from Defense Department satellites used for sea ice studies years after collection are one example cited by Parsons and Duerr (2005), who go on to recommend preparing data and documentation in a way that facilitates "broad but appropriate use."

*Metadata* can describe a collection of data files rather than individual data files. For example, an array of five moorings in the Beaufort Sea could be archived as a single collection, with the metadata "location" field giving a latitude/longitude bounding box. Or, each mooring could have its own metadata record, and the "location" field would have the latitude and longitude at which the mooring was deployed or retrieved. Generally, it is best to have complete metadata for every mooring or observation site, so that the context for the observation is less likely to be separated from the observation in a data archive. In either case, information on the location of the moorings should be in the documentation as well.

Sometimes, for very simple observations, the metadata serves as documentation and nothing more is needed. Usually, however, additional documentation is required in order to preserve the usability of the data along with the data by giving the data useful context. For example, the strengths and limitations of a data set like the National Ice Center Arctic Sea Ice Charts and Climatologies in Gridded Format (http://nsidc.org/data/g02172.html) are difficult to convey without supplemental documentation. The European Sea Search program for marine data management offers guidelines for writing documentation (http://www.sea-search.net/guidelines-practices/guidel05.htm), as does the IPY Data and Information Service (http://ipydis.org/data/data_documentation_template.html).

### 3.14.2.4 Local and Traditional Knowledge and Community-Based Observations

Local and indigenous peoples make ice observations and are involved in collecting, preserving, and exchanging knowledge that sometimes demands special data management considerations. The Exchange for Local Observations and Knowledge of the Arctic (ELOKA) is addressing this need while fostering collaboration between local and international researchers. The ELOKA Web site (http://eloka-arctic.org/index.html) includes a metadata submission form for project contributors. See Chapter 3.11 and Mahoney and Gearhard (2008) for more information.

### 3.14.2.5 Oceanographic Data Management

Polar research tends to be cross-disciplinary by nature, and this characteristic has to some extent slowed the development of efficient procedures for managing data collected in the field. Much of scientific data management has to do with description, and sea ice measurements often do not fit in neat categories (for instance, is a measurement of snow on sea ice grouped with "ocean" measurements or with "land" measurements where all the other snow measurements are?) Relatively few observations are made on sea ice each year, so there has not been much impetus for developing data management procedures to steward these research data. In contrast, data management systems for reference and community data are highly efficient and evolved. Satellite observations of sea ice from Earth Observing System sensors through the NASA DAAC (http://nsidc.org/daac) are one example.

Research and observation programs like the Study of Environmental Arctic Change (SEARCH), DAMOCLES, and the Arctic Observing Network (AON) are pushing for change in the way sea ice research data are managed. The oceanographic community is developing advanced data systems for observation networks that offer excellent data access and interoperability; see the Integrated Ocean Observing System (IOOS) testbed (http://www.openioos.org) and the related Alaska Ocean Observing System (AOOS; http://www.aoos.org), both part of the Global Ocean Observing System). The Marine Metadata Interoperability Project (http://marinemetadata.org) assists in this endeavor by offering specific guidance on how to use the standard metadata and data formats that enable data systems like these to function.

IOOS and AOOS are focused mainly on delivering real-time physical data from remote-sensing and automated in situ instruments. Biological data are more difficult to collect and share, but the challenge is being met within the new field of ocean biodiversity informatics, driven largely by the requirements of international efforts to assess the biodiversity of world's oceans (Costello and Vanden Berge 2006). As with physical oceanographic data and sea ice observations, the keys to data access and preservation are standard metadata, formats, and protocols for data exchange. Authoritative lists of species names are a needed component of developing controlled vocabulary for biodiversity metadata.

The Ocean Biogeographic Information System (http://www.iobis.org) and the Global Biodiversity Information Facility (http://www.iobis.org) demonstrate how interoperability through standards makes biological data available. This in turn is helping promote a new culture of data sharing in marine biology.

## 3.14.3 CONCLUSION

Sea ice observations from field experiments are research data with relatively high overhead costs when it comes to managing them as data sets: They vary considerably and do not offer the economy of scale of remote-sensing data, for instance. But they are critical to polar research and must be broadly accessible and usable to realize full value. Researchers and data managers need to work closely together to establish protocols, controlled vocabularies, and standard file formats and contents for sea ice observations.

## REFERENCES

Belbin, L. (1998), Sticks and carrots. The Joint Committee on Antarctic Data Management, electronical document, http://147.66.8.18/Articles/Stickscarrots.html, accessed 14 March 2009.

Brodzik, M. J., and K. W. Knowles (2002), EASE-Grid: A versatile set of equal-area projections and grids, in *Discrete Global Grids,* edited by M. Goodchild, National Center for Geographic Information and Analysis, Santa Barbara, CA. Available online at http://www.ncgia.ucsb.edu/globalgrids-book/ease_grid/

Costello, M. J., and E. Vanden Berghe (2006), "Ocean biodiversity informatics": A new era in marine biology research and management, *MEPS, 316,* 203–214.

Duerr, R., M. A. Parsons, M. Marquis, R. Dichtl, and T. Mullins (2004), Challenges in long-term data stewardship, *Proceedings of the 21st IEEE Conference on Mass Storage Systems and Technologies,* IEEE, pp. 47–67.

International Council for Science (ICSU) (2008), *Ad hoc* Strategic Committee on Information and Data. Final Report to the ICSU Committee on Scientific Planning and Review.

Mahoney, A., and S. Gearhard (2008), Handbook for community-based sea ice monitoring. *NSIDC Special Report 14* (PDF, 23.3 MB).

Moore, C. J. and R. E. Habermann (2006), Core data stewardship: A long-term perspective, *Geological Society, London, Special Publications*; *267,* 241–251, doi: 10.1144/GSL.SP.2006.267.01.18. Available at http://nsidc.org/pubs/special/nsidc_special_report_14.pdf

Neiswender, C. (2009), Introduction to metadata, electronic document, http://marinemetadata.org/guides/mdataintro, accessed 14 March 2009.

National Science Board (2005), Long-lived digital data collections enabling research and education in the 21st century, NSB-05-40, National Science Foundation, Washington, DC.

Olsen, L. M., G. Major, K. Shein, J. Scialdone, R. Vogel, S. Leicester, H. Weir, S. Ritz, T. Stevens, M. Meaux, C. Solomon, R. Bilodeau, M. Holland, T. Northcutt, and R. A. Restrepo (2007), NASA/Global Change Master Directory (GCMD) Earth Science Keywords, version 6.0.0.0.0.

Parsons, M. A., M. J. Brodzik, and N. J. Rutter (2004), Data management for the cold land processes experiment: Improving hydrological science, *Hydrol. Process., 18,* 3637–3653.

Parsons, M. A, and R. Duerr (2005), Designating user communities for scientific data: Challenges and solutions, *Data Science Journal, 4,* 31–38.

Schaepman-Strub, G., M. E. Schaepman, T. H. Painter, S. Dangel, and J. V. Martonchik (2006), Reflectance quantities in optical remote sensing—definitions and case studies, *Rmt. Sens. Env., 103,* 27–42, doi:10:1016/j.rse.2006.03.002.

## Endnotes

1. The Electronic Geophysical Year *Declaration for an Earth and Space Science Information Commons* lists eight issues of concern for data stewardship (http://www.egy.org/declaration.php), and gives an overview of scientific data stewardship, with emphasis on geographically distributed archives, potential technological solutions, and formats and standards.
2. The Group on Earth Observations promotes best practices for in situ data as well as remotely sensed observations because doing so will contribute to the success of the Global Earth Observation System of Systems (GEOSS). To that end, a best practices wiki (http://wiki.ieee-earth.org) has been established. Practices are reviewed and refined, so "best" can be made better.

# Chapter 3.15

# Principal Uses of Remote Sensing in Sea Ice Field Research

*Robert A. Massom*

## 3.15.1 INTRODUCTION

In this chapter, we examine the key role that satellite remote sensing plays in modern sea ice field research in both hemispheres, where in situ measurement and remote sensing are intimately intertwined. Over the past thirty to forty years, extraordinary developments in both remote sensing technologies and the ground segment (data acquisition, archiving, and dissemination) have revolutionized our ability to observe and monitor sea ice (Lubin and Massom 2006). This in turn has greatly advanced our understanding of its distribution, properties, and variability; its complex interaction with the ocean and atmosphere; and its central role in the global climate system and in structuring high-latitude ecosystems. Satellites alone can measure and monitor remote and vast areas in a sustained, consistent, systematic, repetitive, and cost-effective fashion, and on a variety of spatiotemporal scales. These attributes enable sustained monitoring of sea ice for research, operational, and other purposes. In fact, remote sensing can provide information that is difficult, if not impossible, to obtain from the surface. Within the overall sea ice system services theme, remote sensing plays an important role in integrating between the local and regional scales. For sea ice field research, satellites are now routinely used in both the planning and the execution of the field operations, as discussed towards the end of this chapter.

Satellite remote sensing is a powerful research tool, complementing rather than replacing in situ measurements in the sea ice zone. Field data remain crucial for validating models, for understanding geophysical and biological processes, and for validating or calibrating satellite-derived sea ice parameters. The latter is a key endeavor and has been the focus of a number of recent field programs both in Antarctica (e.g., Massom et al. 2006) and the Arctic (e.g., Cavalieri et al. 2006). Moreover, not all important sea ice properties can be measured from space. The vertical structure of sea ice, and its associated biological communities and biogeochemical properties, for example, can only be measured in situ. The same is true for snow cover properties (see Chapter 3.1), although a satellite-derived snow depth product has recently become available (see Section 3.15.6). Knowledge of these snow and ice properties is a key to understanding the variable remote-sensing signatures of the ice and to interpreting these signatures in terms of sea ice geophysical quantities.

Sea ice and its snow cover are challenging materials to measure and monitor. Their distribution and properties vary both spatially and temporally due to high variability in the environmental conditions and processes that form, redistribute, modify, and melt the ice. This leads to considerable small-scale horizontal heterogeneity. Crucial from a sea ice field research and broader-scale system services perspective (as summarized in Chapter 2, Table 2.1) is accurate and consistent knowledge of key variables that include: ice extent and the location and nature of the ice-edge zone (for both moving pack and stationary landfast ice); ice concentration and the location and orientation of ice openings (including leads and recurrent flaw leads and polynyas); thickness distribution and associated stage of development and degree of deformation (including amount of pressure ridging); ice kinematics and dynamics; floe-size distribution; ice-season duration; stage of decay during the summer melt season; and snow-cover thickness and properties. In addition, satellite remote sensing provides critical information on sea ice as a habitat for a wide range of organisms, and in support of human operations and field research focusing on interaction processes within the sea ice zone. Sea ice responds rapidly to changes in atmospheric and oceanic forcing, and as such is a sensitive indicator of climate variability and change, as well as playing a crucial role in the global climate system. As stated elsewhere in this book, the impact of sea ice change can be far-reaching and profound, and better understanding of the complexity of the system necessitates use of coordinated field research in tandem with carefully planned satellite remote sensing. This field research ideally includes other forms of remote sensing such as aerial photography to ensure data collection that is intermediate between the detailed but spatially limited surface measurements and the broader-scale satellite observations (Massom et al. 2006).

The overall aim of this chapter is to provide practical information on the main sources of satellite data available for sea ice field research, and how to interpret and use them. The chapter begins with a brief overview of hemispheric differences in sea ice as a remote sensing target. This is followed by an overview of fundamental concepts and principles of sea ice remote sensing, and the electromagnetic properties of snow and sea ice, the aim being to provide the reader with a basic grounding in the different remote-sensing systems and techniques. This is followed by tips on how to choose the optimal satellite data set(s) for a given application, which leads into an evaluation of the major sensor classes and their application to field research, including their strengths and limitations. The subsequent section entails a brief review of the different methods and/or standard protocols used to derive key sea ice variables and parameters from satellite data. The chapter continues with a case study that illustrates an application of the methodology in the context of conducting field research and sea ice system services. It ends with a brief discussion on where potential progress is required, what the challenges are, and how the methodology is likely to evolve over the coming decades. This is by no means an exhaustive overview, and the reader interested in gaining more information on the wider sea ice application of polar remote sensing is encouraged to consult the following: Bamber and Kwok (2004); Carsey (1992); Haykin et al. (1994); Comiso (2003); Jackson and Apel (2004); Lubin and Massom (2006); Rees (2005); and Tsatsoulis and Kwok (1998). The provision of detailed information on the characteristics of the different satellite and sensor systems is beyond the scope of this book, and the reader should consult Kramer (2002) for this information.

### 3.15.2 HEMISPHERIC DIFFERENCES IN SEA ICE AS A REMOTE SENSING TARGET

At the outset, we wish to highlight the fact that although the sea ice covers of both polar regions are superficially similar, there are in fact marked differences in their overall physical characteristics due to their contrasting geographical settings and the environmental conditions therein. These physical differences, which translate to fundamental differences in the remote-sensing signatures (particularly in the microwave region), include the following:

- Melt ponds are a ubiquitous feature of the Arctic in summer, where they can cover up to 60 percent of the ice surface (Eicken et al. 2004), but are largely absent from Antarctic sea ice—due (it is thought) to the combined effects of a colder, drier, and windier atmosphere over the Southern Ocean (Andreas and Ackley 1981) and a deeper snow cover. Antarctic sea ice largely melts laterally and from the bottom up, due in part to relatively large ocean heat fluxes (Martinson and Iannuzzi 1998), and those areas that survive summer melt tend to retain a snow cover and maintain a relatively high albedo (Brandt et al. 2005). The more extensive surface melt and weathering processes in the North lead to the "rounding off" of blocky surface features on perennial sea ice, whereas Antarctic sea ice tends to retain a more blocky (rougher) topographic appearance—although substantial melt occurs within the Antarctic sea ice cover in summer to create coarse-grained icy snow layers and extensive superimposed ice (Haas et al. 2001). The latter is formed by the refreezing on the ice surface of downward-percolating snow meltwater.
- Due to more turbulent ice-growth conditions in the Southern Ocean, the Antarctic sea ice cover comprises a higher proportion of frazil ice i.e., 50 to 60 percent compared to 5 to 20 percent in the Arctic Ocean (Tucker et al. 1992).
- Ice surface flooding is widespread in the Southern Ocean, as a result of snow loading combined with ice deformation and a relatively warm and

porous ice cover (Massom et al. 2001 and references therein). Such flooding is near absent in the Arctic, although this situation may change as climate warming progresses.

- By virtue of being predominantly seasonal, the Antarctic sea ice zone comprises a high proportion (~80%) of first-year (FY) ice that melts back each summer, compared to <50 percent for the Arctic (Comiso 2003) up until a few years ago. Moreover, the residence time of sea ice has been up to five to seven years in the Arctic Ocean, compared to one to two years in the Southern Ocean. However, both the relative proportions of FY and multiyear (MY) ice, and the ice residence time, are changing dramatically in the central Arctic (Maslanik et al. 2007) to create an ice cover that is increasingly more like its southern counterpart in terms of its annual growth-decay cycle and age.
- In the Arctic, MY ice is significantly less saline than FY ice due to the strong evolution of its salinity and microstructural properties. This largely results from downward percolation of meltwater from melt ponds, which occurs via interconnecting brine drainage channels and tubes, which coalesce as the ice warms (Golden et al. 1998). As a result, brine is flushed out of the ice, and brine pockets are gradually replaced over time by (sub) millimeter-scale air bubbles (Eicken 2003). These processes are much less pronounced in the Antarctic, resulting in less contrast between the physical and hence electromagnetic properties of FY and perennial ice there (Comiso et al. 1992).
- In the Arctic, sea ice plays a major role in the entrainment and transport of particulates and potential pollutants (Lange 2002), with sediments entrained over the shallow, broad Arctic shelves (Stierle and Eicken 2002) and aeolian deposition of soot (Grenfell et al. 2002). This is not the case in the Antarctic, where the sea ice and snow cover are relatively pristine. In addition to their relevance to ice services, the presence of sediments and other material within the Arctic sea ice and snow cover has a major impact on their optical properties (Light et al. 1998) and hence remote sensing signature.
- The input of river discharge also contributes to the overall freshwater budget and sea ice properties of the Arctic (Macdonald et al. 1999), a factor that is again lacking around Antarctica.
- In certain regions of the northern hemisphere e.g., the Baltic Sea, the sea ice cover tends to be more brackish (Granskog et al., 2006).

## 3.15.3 BASIC CONCEPTS AND PRINCIPLES OF SATELLITE SEA ICE REMOTE SENSING

In this section, we provide a brief overview of fundamental concepts, principles, and nomenclature of satellite remote sensing relevant to the measurement of various sea ice variables (for more information, see Rees 2005). Such information is necessary to be able to understand and evaluate the different data sets and techniques available for a given application. The basic premise of remote sensing is to acquire information about the surface without being in contact with it—by measuring (sensing) and recording reflected or emitted *electromagnetic radiation* (EMR), then processing, analyzing, and applying that information

### 3.15.3.1 The Electromagnetic Spectrum and Atmospheric Transmission Windows

Satellite remote-sensing systems suitable for observing sea ice operate at wavelengths/frequencies corresponding to *atmospheric transmission windows* (transmission bands) in the electromagnetic spectrum, within which atmospheric contamination effects are relatively small—in other words, the cloud-free atmosphere is largely transparent to EMR. The main windows in the visible, infrared (IR), and microwave parts of the electromagnetic spectrum are illustrated in Figure 3.15.1, which also shows the major wavelength ranges exploited by Earth-observing remote sensing. Sensors operating outside the windows provide key information on the vertical structure/characteristics of the atmosphere (Lubin and Massom 2006). Within the atmospheric windows, instruments are designed to measure radiation within discrete wavelength ranges or spectral bands, depending on their

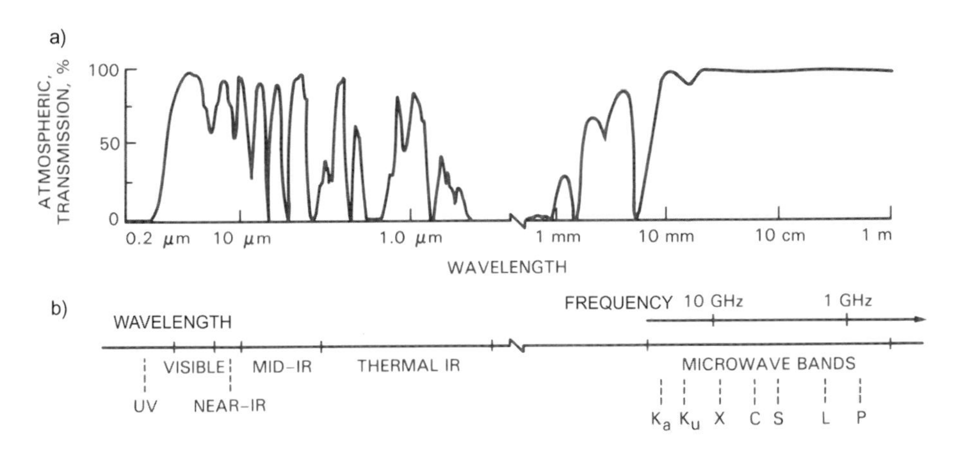


**Figure 3.15.1.** The electromagnetic spectrum, showing (a) atmospheric transmission windows and (b) spectral regions exploited by the different sensor classes (after Carver 1989). ©NASA 1989. Modified and reproduced with permission from NASA.

application. Spectral ranges exploited by the main remote sensing classes are given in Table 3.15.1.

**Table 3.15.1.** The spectral range of divisions of the electromagnetic spectrum used by the main remote sensing classes, with the quantities measured

| Spectral range | Wavelength range | Quantity measured |
|---|---|---|
| **Visible** | | |
| Blue | 0.4–0.5 μm | Reflectance |
| Green | 0.5–0.6 μm | Reflectance |
| Red | 0.6–0.73 μm | Reflectance |
| **Infrared (IR)** | | |
| Near IR | 0.7–1.3 μm | Reflectance |
| Middle IR | 1.3–3.0 μm | Reflectance + thermal emission |
| Thermal IR | 3.0–5.0 and 8.0–14.0 μm | Thermal emission |
| **Microwave** | 1 mm–1 m | |
| Active | | Reflectance + Backscatter |
| Passive | | Thermal emission |

While *clear-sky atmospheric effects* are minimal within the atmospheric windows, they are not entirely absent. In fact, variable atmospheric effects contribute to satellite-derived measurements of the surface and often require correction (Lubin and Massom 2006). As a broad rule of thumb, the "transparency" of atmospheric windows decreases with an increase in atmospheric water vapor content. Although the atmosphere at high latitudes is often very cold and therefore dry, this is not always the case. Warm moist air incursions regularly traverse the Antarctic sea ice zone even in winter with the passage of cyclones (Massom et al. 2001), for example. Moreover, the Antarctic sea ice zone extends as far north as ~55–60°S in winter, and as such comes into contact with considerable cyclonic activity (Simmonds et al. 2003). Low latitudes are also attained by sea ice in the northern hemisphere, for example, in the Sea of Okhotsk, Baltic Sea, and Labrador Sea.

Here, we concentrate on polar-orbiting rather than geostationary satellites, as they are most useful for high-latitude applications. As geostationary satellites sit ~36,000 km above given points on the Equator, their polar coverage is geometrically distorted and limited, although they provide key information on cloud/weather patterns over peripheral seas. *Polar-orbiting satellites* in Low Earth Orbit (LEO) mostly operate in a sun-synchronous orbit, whereby the satellite always crosses the Equator at the same local solar time. Orbital convergence at high latitudes means that it is often possible (cloud permitting for visible to thermal IR systems) to view a given high-latitude location several times per day with the same sensor, depending among other things on the satellite inclination and the sensor swath width and look angle. This is one of the attributes that make wide-swath satellite remote sensing such a powerful tool for high-latitude work, while narrow-swath sensing offers much lower frequency of coverage over smaller areas but is powerful in other ways, such as generally offering higher spatial resolution.

### 3.15.3.2 Sensor Classes

Broadly speaking, Earth-observing satellite systems fall into two main categories, namely *passive* and *active*. A brief introduction to these different classes is necessary to introduce the important concept of how EMR measured by them interacts with the surface—which in turn forms the basis of inference of surface physical properties from the electromagnetic signatures measured at the satellite. Active systems transmit pulses of EMR towards the surface and record the intensity, time interval, and characteristics of the return signal scattered or reflected back to the sensor from the target surface. These include *imaging radar* instruments and *ranging instruments*. Imaging radars operate in the microwave part of the electromagnetic spectrum and illuminate the surface to provide maps of the backscatter characteristics. *Backscatter* is the portion of the transmitted radiation that is redirected (scattered/reflected) back towards the sensor. Ranging instruments, on the other hand, are non-imaging sensors designed to measure surface topography/elevation along a narrow profile by using the time interval information to accurately determine the distance from sensor to surface. Key spaceborne examples are *radar* altimeters and *lidar* sensors, the former operating in the microwave and the latter in the visible and near-infrared.

In contrast to active instruments, passive systems measure naturally occurring radiation. These include sensors operating at (1) *visible* to *near-infrared* wavelengths that measure reflected solar radiation, and (2) *thermal infrared* and *passive microwave* instruments that detect EMR that is thermally emitted from the surface. There is some overlap, in that systems operating at middle IR wavelengths measure both reflected solar and emitted thermal radiation. As we shall see, a crucially important attribute of satellite Earth-observing microwave instruments (both passive and active) is their ability to "see through" both polar darkness and cloud cover, depending on frequency.

The primary quantities measured by the broad sensor classes outlined above are given in Table 3.15.1. *Reflectance* is defined as the ratio of the intensity of radiation reflected by a target (surface) to that of the radiation incident upon it. Surface reflectance is the primary variable measured by optical systems. This can be computed for cloud-free areas if the amount of radiation incident upon that surface is known. This also requires correction of variable clear-sky atmospheric effects. Even though satellite sensing of the Earth exploits atmospheric windows, three top-of-the-atmosphere (TOA) components are received at the sensor under clear-sky conditions. These are (1) direct (unscattered and surface reflected), with TOA radiation modified by atmospheric transmittance along the solar path; (2) skylight where the radiation is down scattered and surface reflected; and (3) path radiance (path-scattered).

*Emission* is the process by which a body or material naturally radiates electromagnetic energy, and is determined by its *kinetic temperature* and emissivity. Emissivity is a dimensionless number that expresses the ratio of radiation naturally emitted by a natural surface to that emitted by a *black body* (perfect radiator) at the same kinetic temperature. As shown in Table 3.15.1, the detection of EMR emitted by a surface is the realm of thermal-IR and passive-microwave systems, and underlies techniques extracting ice concentration and ice type information from passive microwave data. Generally speaking, passive systems designed to detect reflected solar and thermally emitted radiation measure the *radiance*. This is the amount of energy radiated by an object and reaching the sensor in a particular waveband (in general, radiance is a function not only of spectral wavelength but also viewing angle, expressed as energy per solid angle). In the case of sensors that detect thermal radiation, the radiance is generally expressed as a *brightness temperature*, defined as the temperature of a black body that would emit the same amount of radiation. Active microwave systems, on the other hand, measure surface reflectance in terms of its *backscatter coefficient* ($\sigma^0$). The latter is a normalized measure of the strength of the radar return from an illuminated area, and is usually expressed in decibels (dB). Importantly from the perspective of sea ice remote sensing, $\sigma^0$ is both dependent on the properties of the surface and varies with radar wavelength, *polarization*, and imaging geometry (sensor incidence angle) (Ulaby et al. 1982) to form the basis of sea ice detection and classification using imaging radar data. Polarization describes the direction of orientation in which the electrical field vector of EMR vibrates, and is a key consideration in active- and passive-microwave sea ice remote sensing.

#### 3.15.3.3 Interactions of Electromagnetic Radiation with Sea Ice and Its Snow Cover

Inference of accurate sea ice information from EMR received at a satellite sensor depends upon knowledge of the interaction of the radiation with the snow and ice layers, and its dependency on surface and internal ice properties and processes and instrument parameters such as wavelength, polarization, and incidence angle. The nature of the interaction of EMR with sea ice and its snow cover, and therefore the remote-sensing signatures of the latter, depend in turn on their electromagnetic properties, which will be covered in the next section. Detailed treatment of this field is beyond the scope of this chapter (see Hallikainen and Winebrenner 1992; Lewis et al. 1994; Lubin and Massom 2006; Rees 2005; Tucker et al. 1992 for in-depth information on sea ice and snow-cover physical properties relevant to remote sensing). However, certain generalizations can be made, and we provide a brief overview of fundamental concepts to aid understanding of how and why the different sensor classes marked in Figure 3.15.1 provide fundamentally different yet complementary information.

As a starting point, let us consider the mechanisms by which EMR from an external energy source, for example, the Sun or a spaceborne radar, that is incident on an ice/snow surface interacts with that surface. These are threefold, namely reflection, transmission, and absorption. *Reflection* is generally referred to as being *diffuse* or *specular*. Diffuse reflection refers to the scattering in all directions of incident radiation by a surface that is rough relative to the wavelength of the radiation. An ideally rough surface in this respect is termed a *Lambertian surface*, and the reflection from it *Lambertian reflection*. Conversely, specular reflection describes scatter of the incident radiation in a mirrorlike fashion away from the source by a surface that is smooth relative to wavelength. These two cases represent the extremes, and natural surfaces tend to exhibit intermediate reflection characteristics.

Of the other important terms describing the interaction of incoming radiation with a given surface, *transmission* or *transmittance* is the ratio of energy transmitted by a body to that incident upon it. *Absorption* refers to the loss of energy from EMR as it propagates through an absorbing medium, and/or a measure of the ability of a surface to absorb incident radiation (primarily determined by its dielectric properties). Moreover, *scattering* occurs not only from the surface but also within its volume in the form of multiple scattering, in which case it is termed volume scattering. Scattering refers to the multiple reflection of EMR off surfaces or particles/scattering centres.

### 3.15.3.4 Electromagnetic Properties of Sea Ice and Snow

Major determinants of the relative contributions of the various EMR-surface interactions outlined above as they affect the remote-sensing signature of sea ice are the dielectric (material) and scattering (geometrical) properties of the ice and its snow cover. These vary considerably with ice type, stage of development, and—if present—the state of the snow cover, as well as environmental factors. The dielectric properties of sea ice and snow, which are covered in detail in Chapter 3.4 and in Hallikainen and Winebrenner (1992), determine the propagation and attenuation of electromagnetic waves. These in turn govern both the optical properties of the ice (see Chapter 3.6) and sea ice signatures in remote-sensing datasets.

*Optical Properties*

Sea ice reflectance (and albedo) at optical wavelengths is determined by the sum of scattering and specular reflection at the surface, together with absorption and scattering in the ice volume (Chapter 3.6; Perovich 2001; Eicken 2003). As a result, it is strongly dependent on the small-scale structural properties of the ice, that is, the number and size distribution of brine and gas inclusions. In general, the number of inclusions acting as volume scatterers is correlated with ice thickness, and a pronounced increase in ice albedo typically accompanies an increase in ice thickness and age. As pointed out by Eicken (2003), this relationship is implicit in the standard nomenclature of thin (new) sea ice types, namely dark and light nilas, gray, gray-white, and white ice (see Chapter 3.12). Albedo is defined here as the integral of the ratio of reflected to incoming radiation over all angles as a function of solar zenith angle, wavelength, viewing zenith angle, and viewing azimuth angle.

Another major determinant of sea ice spectral reflectance is the presence or absence of a snow cover. Snow reflectance is highly sensitive to its depth (when shallow) and the presence of absorbing impurities (e.g., soot) at visible wavelengths and grain size and liquid-water content—the latter via its impact on grain size—at near-infrared wavelengths (see Chapters 3.1 and 3.6). It also has angular and surface-roughness dependence. It follows that satellite-derived maps of sea ice reflectance and/or albedo are useful for assessing the characteristics of the surface and the condition of the snow cover. Fresh snow typically has a high albedo, while old snow, which is more granular and has undergone significant metamorphism, has a lower albedo. Information on physical properties of snow on sea ice that is relevant to remote sensing is given by Massom et al. (2001) for the Antarctic and Sturm and Massom (in press) for the Arctic; Warren (1982) and Dozier (1989) give information on optical properties.

During the Arctic summer melt season, the initial appearance of small amounts of liquid water in the snow cover, followed by slush formation and then melt-pond formation, significantly reduces the ice albedo. Arctic melt ponds typically have an albedo of 0.15–0.30 and melting snow 0.77, compared to values of about 0.52 for bare Arctic FY ice and 0.87 for fresh snow (Perovich 2001).

### *Radar-Backscatter Properties*

In microwave remote sensing, the scattering of EMR from sea ice is determined by the combined effect of four parameters as a function of instrument wavelength (frequency), polarization, and incidence angle (Onstott and Shuchman 2004). These are:

1. the dielectric constant of the sea ice and its snow cover (if present);
2. the presence of dielectric discontinuities or discrete scatterers within the ice volume, such as gas bubbles;
3. the surface roughness; and
4. the geometry of surface features relative to the sensor look angle.

Hallikainen and Winebrenner (1992) provide a review of quantitative relationships between sea ice physical and dielectric properties, and their frequency and incidence angular dependence.

An important concept in the interpretation of both radar and passive microwave data, and one that is intimately linked to dielectric constant and scattering, is *penetration depth*, which is defined as the depth below the surface at which the incident EMR has been attenuated to $1/e$ (37 percent) of its original field strength. Importantly, there is a significant contrast between the dielectric loss factors of seawater and sea ice in the microwave range of 1–100 GHz (Chapter 3.4; Eicken 2003). Moreover, the dielectric constant has been empirically parameterized as a function of sea ice fractional brine volume (Chapter 3.4; Hallikainen and Winebrenner 1992), which is governed by the temperature and salinity of the ice (Chapter 3.3; Eicken 2003)—with the latter varying with ice age/type, that is, stage of development. It follows that penetration depth is strongly dependent on ice salinity and temperature. This is illustrated in Figure 3.15.2, which shows that while radar penetration depths in saline Arctic FY ice are up to only a few centimeters at a temperature of –5°C, radar waves can penetrate more than a meter into low-salinity MY ice. Under these conditions, and assuming that the snow cover is cold, dry, and homogeneous (i.e., transparent to EMR), a number of wavelength-, polarization- and incidence angle-dependent relationships hold that form a basis of SAR image interpretation:

- For saline young and FY ice, ice surface scattering is predominant and radar backscatter coefficients are largely determined by the ice surface roughness.
- Radar backscatter signatures of Arctic second-year and MY ice are dominated by strong volume scattering from gas bubbles and brine inclusions (see Eicken 2003), which approximate the wavelength of the radar waves (depending on frequency). This leads to a relatively high radar backscatter coefficient, resulting in a substantially brighter target in radar imagery compared to FY ice (and at C-band in particular).

This fundamental contrast enables discrimination of predominantly smooth FY ice and MY ice in radar data (both SAR and scatterometer imagery) under cold, dry conditions and in the Arctic. This tends to be more difficult in the Southern Ocean, where the properties of MY and FY ice are generally similar due to the lack of brine "flushing" by downward-percolating meltwater as the ice ages; that is, there is relatively little backscatter contrast, although there may be roughness and snow-cover differences that enable FY-MY ice distinction in radar data.

Interpretation of microwave data is complicated by the fact that snow makes a variable contribution to the observed remote sensing signature, depending on its age, structural characteristics and state (Barber et al., 1998). The dielectric constant of dry snow is largely dependent on its density and grain size. However, the appearance of even a small amount of free water (2–3 percent by volume) in the snow cover leads to a high dielectric loss (Winebrenner et al. 1998), thereby increasing absorption (attenuation) of the radar signal. Radar penetration depth decreases considerably and surface scattering becomes dominant, the most typical outcome being a reduction in $\sigma^0$ and darkening in a SAR image (Onstott and Shuchman 2004).

Periodic increases, and diurnal and semi-diurnal fluctuations, in $\sigma^0$ (and microwave emission) can also occur in the Arctic at the onset of seasonal melt in

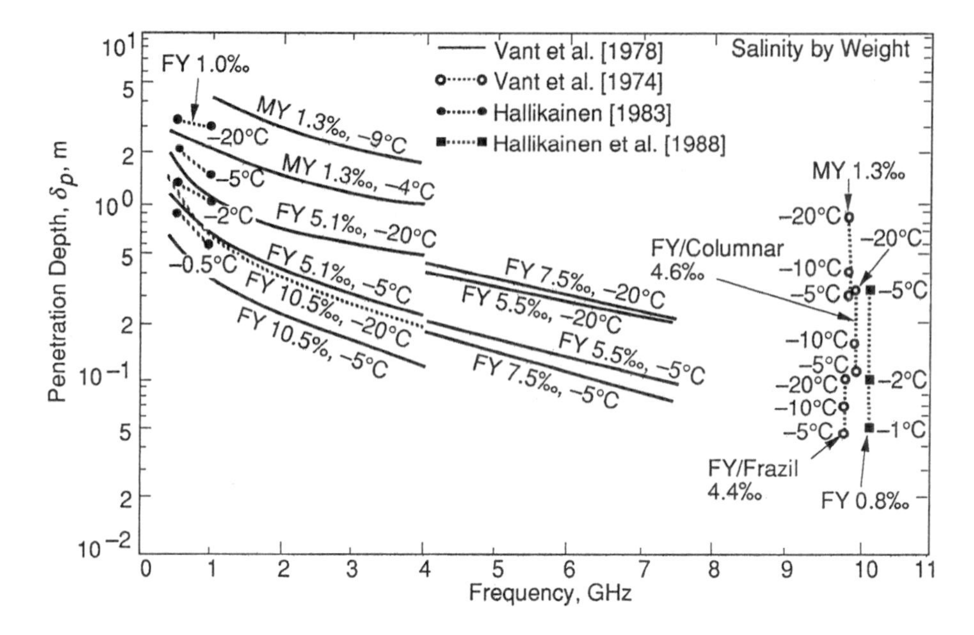


**Figure 3.15.2.** Penetration depth at microwave frequencies for arctic FY and MY ice as a function of frequency and ice temperature and salinity. The frequency- (and polarization-) dependence of sea ice dielectric properties and associated penetration depth opens the door for enhanced differentiation between ice types using multi-frequency and multipolarization satellite sensors (from Hallikainen and Winebrenner 1992). ©American Geophysical Union 1992. Reproduced by permission of American Geophysical Union.

late spring, related to thaw-freeze cycling and the formation of a rough layer of *superimposed ice* formed by snow meltwater refreezing on contact with the ice surface (Gogineni et al. 1992). A subsequent decrease in $\sigma^0$ coincides with the appearance of melt ponds, which can cover up to 60 percent of the ice surface area (Eicken et al. 2004). The subsequent drainage of melt ponds through the ice in late summer leaves behind a granular and highly porous ice surface, resulting in an increase in radar backscatter (Holt and Digby 1985). At these times, access to additional information is essential to accurately interpret satellite microwave data. With a return to freezing dry conditions in autumn, the dominant backscatter mechanism changes back from surface to volume scattering, leading to a fairly rapid return of $\sigma^0$ to winter levels (Winebrenner et al. 1998).

Regarding surface roughness, a rule of thumb for radar remote sensing is that rough sea ice surfaces typically act as diffuse reflectors, reflecting a significant proportion of incident radiation back in the direction of illumination, resulting in high backscatter and a bright radar target. Smooth surfaces such as calm open water and newly formed sea ice, on the other hand, act as specular reflectors to reflect a significant proportion of radiation away from direction of illumination. This results in a relatively low backscatter and a darker target. The strong frequency- (wavelength-)

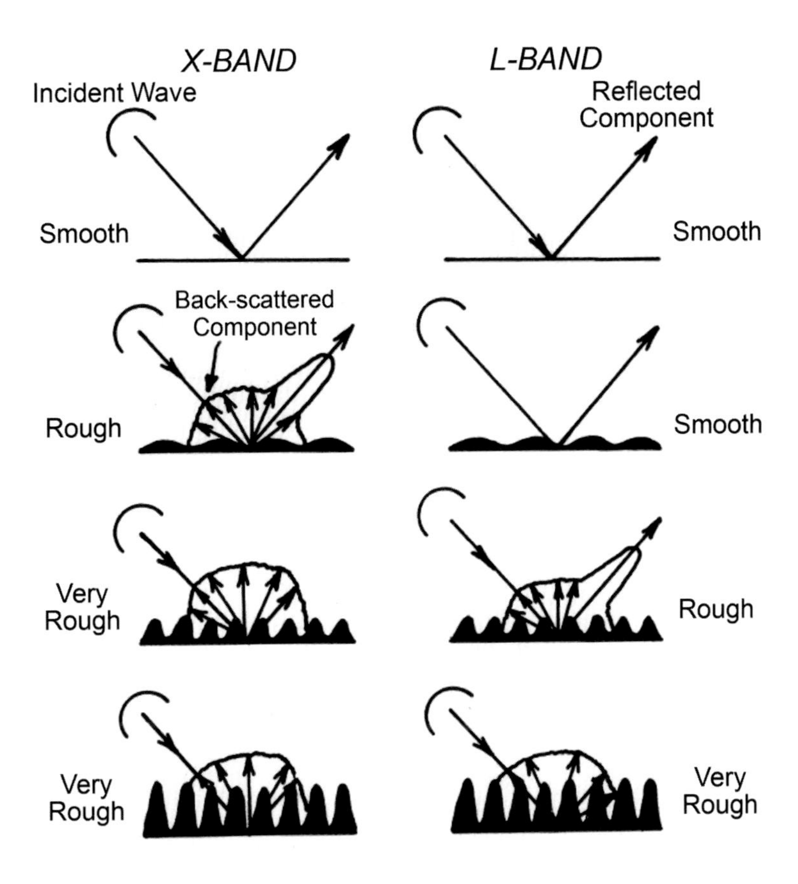


**Fig 3.15.3.** Schematic of the scattering behavior of electromagnetic radiation at X- and L-band frequencies (λ = ~3 and ~23 cm respectively) incident on smooth, rough and very rough snow/ice surfaces. These two frequencies form the end members of, recent, current ,and near-future spaceborne SARs. Scattering behavior at intermediate frequencies, e.g., C-band (λ = 6 cm), will be intermediate. Adapted from Onstott and Shuchman (2004). ©NOAA 2004. Modified and reproduced with permission from NOAA/NESDIS Center for Satellite Applications and Research.

dependence of reflection characteristics of smooth to very rough surfaces for X- and L-band radars is shown in Figure 3.15.3. The frequency and wavelength ranges of designated radar bands are given in Table 3.15.2.

**Table 3.15.2.** Frequency and wavelength of IEEE (Institute of Electrical and Electronics Engineers) radar band designation

| Band | Frequency (GHz) | Wavelength (cm) |
|---|---|---|
| P-band | 0.3–1 | 30–100 |
| L-band | 1–2 | 15–30 |
| S-band | 2–4 | 7.5–15.0 |
| C-band | 4–8 | 3.75–7.5 |
| X-band | 8–12 | 2.5–3.75 |
| Ku-band | 12–18 | 1.67–2.5 |
| K-band | 18–27 | 1.11–1.67 |
| Ka-band | 27–40 | 0.75–1.11 |

### *Thermal Infrared Emission Properties*

Thermal IR and passive microwave instruments detect and measure snow and sea ice based on their thermal emission characteristics. Thermal radiation dominates the wavelength interval from 8–15 μm. According to *Wien's law*, the *maximum radiant exitance* from the Earth's surface (at a temperature of ~300K) occurs at ~10 μm.

At these wavelengths and according to *Stefan's law*, the radiant exitance $M$ from a black body at absolute temperature $T$ is:

$$M = \sigma T^4 \quad \text{(Equation 3.15.1)},$$

where $\sigma$ is the *Stefan-Boltzmann constant* ($5.67051 \times 10^{-8}$ W m$^{-2}$ K$^{-4}$) and $M$ is defined as the total energy radiated in all directions from a body of unit area per unit time. For natural surfaces such as snow and sea ice, the radiant exitance will always be less than that from a black body, with the departure of the material from perfect emission being determined by its *emissivity* $\varepsilon$. This is a dimensionless number that expresses the ratio of radiation emitted by a natural surface to that emitted by a black body at the same kinetic temperature. As a result, the relationship between radiant exitance and temperature for sea ice and snow surfaces becomes $M = \varepsilon\sigma T^4$, where $\varepsilon$ is wavelength dependent. It follows that the physical (skin) temperature of the surface can be calculated with knowledge of its emissivity and measurements of its brightness temperature; that is, using satellite thermal IR radiometers (Key et al. 1997). Values of $\varepsilon$ for dry snow are typically 0.965 to 0.995, with an angular dependence (Dozier and Warren 1982), whereas that of sea ice is typically 0.98 (Rees 2005). This is in contrast to snow and sea ice microwave emissivities in the microwave region, which exhibit a wide range to greatly impact interpretation of satellite data (see below). At thermal IR wavelengths, $\varepsilon$ is largely insensitive to snow density, grain size, liquid-water content, impurities, depth, and temperature over the range generally encountered (Dozier and Warren 1982).

Thermal-IR remote-sensing systems operate in the 3–15 μm region, and typically exploit atmospheric-transmission windows at 3–5 μm and 8–14 μm. Measurements in the latter band provide information on the surface physical temperature (see Section 3.15.6), while those in the 3–5 μm region contain both solar-reflected and thermally emitted contributions, and are most commonly used to distinguish clouds from a snow/ice background during daylight hours. The radiation measured by a sensor in space is again modified by propagation upwards through the intervening atmosphere. An additional component is thermal radiation emitted downwards by the atmosphere and reflected by surface objects (Rees 2001).

*Passive-Microwave Emission Properties*

Passive microwave radiometers also measure the intensity of naturally emitted thermal radiation from the surface (brightness temperature of the incident radiation), in a similar fashion to thermal infrared radiometers but at much longer (i.e., centimeter) wavelengths. At these longer wavelengths, the *Rayleigh-Jeans approximation to the Planck distribution* is valid (see Rees 2001), whereby the brightness temperature

measured by the sensor $T_B$ is a linear function of surface temperature $T_S$:

$$T_B = \varepsilon T_S \quad \text{(Equation 3.15.2)},$$

where ε is the microwave emissivity of the surface material and assuming that atmospheric contributions are negligible. Importantly from a sea ice remote sensing perspective, microwave emissivity exhibits considerable variability depending on ice type, and is a function of ice microstructure, thickness/age, temperature and salinity, and the thickness, age, wetness, structure, and grain size of its snow cover. Moreover, a strong frequency- and polarization-dependent contrast exists between the microwave emissivities of sea ice and open ocean (Figure 3.15.4), to provide the means of routinely deriving ice concentration from satellite passive-microwave data under cold, dry conditions.

Unfortunately, Arctic sea ice in particular exhibits considerable seasonal variability that greatly affects its microwave signature and therefore the accuracy of geophysical parameter retrieval from microwave data. The impact of seasonal changes in Arctic sea ice properties on passive microwave emissivities at 18 and 37 GHz over an annual cycle is illustrated schematically in Figure 3.15.5. Although a dry snow cover is largely transparent at frequencies of <37 GHz (Eppler et al. 1992), this is not the case when it becomes wet. When snow contains ~3 percent liquid water by volume, it becomes opaque and its emissivity approaches unity; that is, it acts radiometrically like a black body, leading to a dramatic increase in observed at all frequencies (Barber et al. 1998). In contrast, the subsequent formation of melt ponds in the Arctic significantly lowers the sea ice microwave emissivity (Comiso and Kwok 1996).

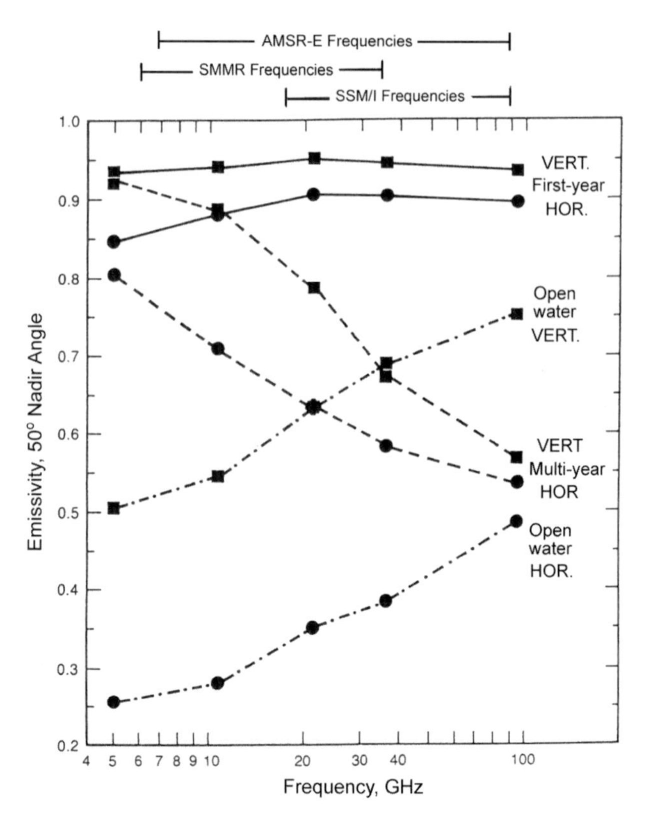


**Figure 3.15.4.** Frequency-dependence of the microwave emissivity of arctic FY and MY ice and calm open ocean for both vertical (V) and horizontal (H) polarizations. The range of frequencies exploited by spaceborne passive-microwave radiometers is marked. After Svendsen et al. (1983). ©American Geophysical Union 1983. Reproduced by permission of American Geophysical Union.

Subsequent melt pond drainage and refreezing in autumn enhance the contribution of volume scattering from ice that survived the summer melt. As a result, Arctic second-year (SY) and MY ice are characterized by lower microwave emissivities than FY ice. These relationships do not hold for the Antarctic sea ice zone, where environmental conditions differ. Regarding the passive microwave signature of Arctic SY and MY ice, multiple scattering both lowers the $T_B$ for a given physical temperature and decreases the difference between vertically and horizontally polarized $T_B$s (Hallikainen and Winebrenner 1992). The overall effect of these physical changes, of reducing the stability of the frequency- and polarization-dependent microwave emissivity of Arctic FY ice under a range of summertime ice conditions, is apparent in Table 3.15.3. Comiso et al. (1992) and Drinkwater (1995) provide information on Antarctic seasonal considerations, which are complicated by the year-round occurrence of flooding and the relative lack of wholesale snowmelt on those regions of ice that survive the summer melt.

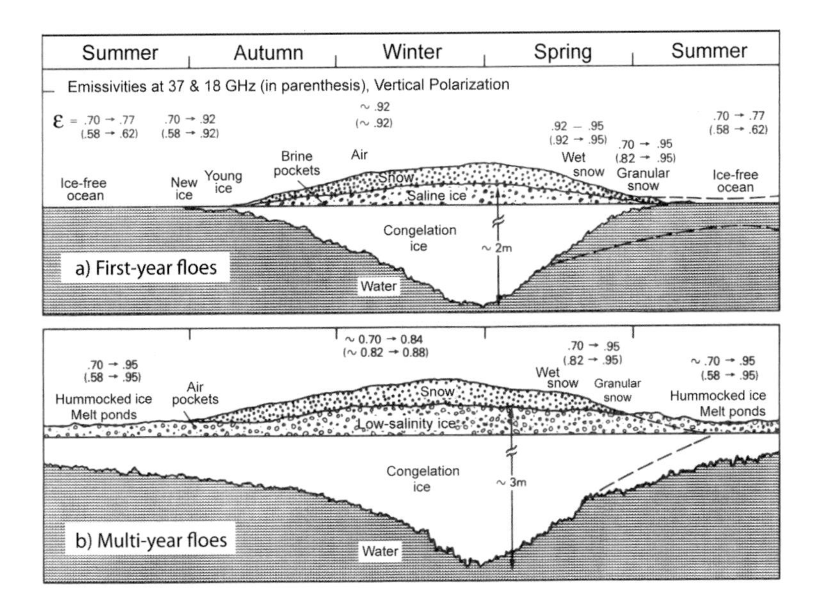


**Figure 3.15.5.** Schematic diagram of the seasonal dependence of physical and passive microwave (emissive) characteristics of arctic FY and MY ice at 18 and 37 GHz (V-polarization) over an annual cycle. After Comiso (1985). ©A. Deepak Publishing Ltd. 1985. Modified and reproduced by permission of A. Deepak Publishing Ltd., Hampton, Virginia, USA.

**Table 3.15.3.** Microwave emissivities measured in situ for various arctic sea ice conditions in summer (compared to winter), as a function of microwave frequency and polarisation for channels equivalent to the SSM/I and AMSR-E. V and H refer to vertical and horizontal polarization respectively. From Markus and Dokken (2002).

| Surface type | 19V | 19H | 37V | 37H | 85V | 85H |
|---|---|---|---|---|---|---|
| Winter FY ice | 0.936 | 0.905 | 0.932 | 0.903 | 0.928 | 0.900 |
| Late spring[a] | 0.950 | 0.905 | 0.920 | 0.845 | 0.890 | 0.760 |
| Early summer[b] | 0.950 | 0.905 | 0.950 | 0.920 | 0.945 | 0.935 |
| Mid-summer[c] | 0.960 | 0.895 | 0.960 | 0.920 | 0.955 | 0.935 |
| Late summer[d] | 0.960 | 0.910 | 0.965 | 0.930 | 0.975 | 0.950 |
| Rainy weather[e] | 0.950 | 0.870 | 0.950 | 0.895 | 0.950 | 0.935 |
| Moist snow[f] | 0.955 | 0.895 | 0.960 | 0.910 | 0.955 | 0.940 |
| Frozen crust in snow | 0.960 | 0.950 | 0.925 | 0.910 | 0.740 | 0.705 |
| Frozen melt pond | 0.969 | 0.877 | 0.970 | 0.970 | 0.876 | 0.818 |

*a. The temperature of the ice and snow is ~0°C, the snow is moist, and there is a superimposed ice layer (i.e., refrozen meltwater) at the snow/ice interface.*

*b. The air temperature approximates 0°C, and the snow cover is wet (~4% liquid water by volume).*

*c. Considerable snowmelt has occurred, creating a slush layer.*

*d. Extensive melt ponds and an icy crust on elevated snow surfaces.*

*e. Saturation in the upper part of the snowpack.*

*f. ~5% liquid water by volume.*

## 3.15.4 CHOICE OF THE OPTIMAL SATELLITE DATA SETS: FUNDAMENTAL CONCEPTS, CONSIDERATIONS, AND TRADE-OFFS

No all-purpose, ideal satellite sensing system exists to meet the needs of the sea ice system services community or to help plan and carry out field research. Rather, each satellite sensor data set or derived set of geophysical products has its own set of inherent attributes, strengths, and limitations. Choice of the optimal data set(s) necessarily involves an evaluation of the nature and temporal and spatial scales of the ice parameters being investigated versus the characteristics of the satellite sensor and data set under consideration. Fundamental trade-offs exist regarding the latter, knowledge of which is essential when choosing the appropriate data set for a given application. An important consideration is *spatial resolution*. This is a measure of the smallest object that can be seen (resolved) by a sensor, generally expressed as the most closely spaced line-pairs per unit distance that can be distinguished in the image. For imagery, spatial resolution is typically given as the smallest unit area discrete picture elements (*resolution cells*), represented by pixels. The area on the Earth's surface corresponding to an image pixel is referred to as a *resolution element* or *t*. Actual spatial resolution may, however, differ from the nominal resolution of an imaging system, based on the interplay of spatial and spectral resolution (defined below). For example, linear features narrower than the nominal spatial resolution of a given imaging sensor, and with radiometric characteristics that contrast strongly with the surrounding area, can often be resolved. An important example is leads: Stone and Key (1993) reported that leads as narrow as 0.4 km could be resolved by 1 km-resolution NOAA Advanced Very High Resolution Radiometer (AVHRR) if atmospheric conditions are conducive.

Fortunately, a number of general rules of thumb can be used to help choose the best sensor(s)/data set(s) for a given application (Lubin and Massom 2006):

- Due to sensor scanning configurations, telemetry bandwidth (as it affects data transmission from the satellite), and other technological limitations, high-resolution sensors typically offer limited spatial coverage over a narrow swath (generally ≤100 km), with ultra-high-resolution (≤5 m) sensor coverage typically over a ≤50 km swath. *Swath*, or *range*, refers to the width of the strip on the ground over which data are collected during a satellite overpass.
- The swath width of medium- to coarse-resolution (≥250 m) instruments, on the other hand, is significantly larger, at >1,500 km. Such systems are therefore more suited to large-scale studies.

This trade-off between spatial resolution and data coverage has a number of implications:

- True complete polar coverage can normally only be achieved for fixed look-angle sensors if their swath width is wide, as most geosynchronous satellites orbit to a latitudinal limit of 81.6°. This is particularly crucial for coverage of the central arctic sea ice cover.
- Generally speaking, high-resolution narrow-swath sensors are best suited to localized studies where a high level of detailed information is essential.
- Swath width and orbital configuration (inclination and altitude) also determine the time interval between repeat/successive opportunities to observe a given location or *temporal resolution* (frequency of coverage). In general, the wider the swath, the shorter the *revisit interval* of a given location and the finer the temporal resolution. This is a major consideration in process studies or ice-monitoring activities requiring fine temporal resolution such as sea ice dynamics. High temporal resolution can be specified as being less than twenty-four hours to three days, medium as four to sixteen days and less than sixteen days. In practical terms, revisit interval also depends on latitude, cloud cover (when measuring surface characteristics using optical instruments), and whether the sensor has a fixed or flexible viewing geometry. An increasing number of modern sensors possess a more flexible coverage capability, whereby a steerable antenna enables coverage of a given point more frequently than is possible with the nominal revisit interval in fixed look angle mode. Such coverage must, however, be programmed ahead of time through consultation with the satellite operating agency. Examples are the multiple modes of operation of the advanced SAR sensors onboard Radarsat-1 and -2, Envisat, and the Advanced Land Observing Satellite (ALOS).

Other important satellite-sensor parameters include the sensor wavelength/frequency range and the number of discrete bands available, the related spectral resolution, polarization of the bands (channels), and the radiometric resolution of the data. *Spectral resolution* is the sensitivity of a sensor's response to EMR in a specific wavelength/frequency range, referred to as a *band. Radiometric resolution* is the intensity or amplitude of the detected signal; the finer the radiometric resolution, the more sensitive a sensor is to detecting small differences in emitted or reflected energy; that is, the better the quality of the image. For radar remote sensing, radiometric resolution can be improved by averaging, but at the expense of spatial resolution—yet another trade-off. In practical terms, radiometric resolution refers to the number of possible brightness values in each band of data and is determined by the number of bits into which the recorded energy is divided (digitized). In 8-bit data, for example, the brightness values can range from 0 to 255 (or 28) (i.e., 256 total possible gray-level values), whereas 14-bit data correspond to up to 16,384 (or 214) intensity values.

Additional key factors affecting the practical value of satellite data are availability, timeliness, and cost, and whether the derived geophysical products of interest are standardized and routinely produced or research and development only. Timeliness is a crucial determinant of data applicability, particularly if the data are required in real time or near real time, as is the case with sea ice field operations. Real-time acquisition requires access to an appropriate receiving station antenna to automatically downlink data from a satellite as it overpasses. In fact, many modern icebreakers are equipped with satellite data-reception facilities, enabling the downloading of key information on sea ice and meteorological conditions in real or near real time. This has resulted in safer navigation, immense savings on fuel bills, and, as we shall see, much-improved information with which to carry out field operations. Other data are available in near real time, that is, within a few hours after acquisition, via archives or directly from the space agencies or consortia such as PolarView (http://polarview.org). Finally, data only available from archives weeks to months after acquisition are largely suitable for research purposes only, although they also provide crucial information on "climatological" ice conditions. In other words, analysis of past sea ice conditions is a crucial initial stage when planning field experiments and operations. Many important satellite data are freely available, such as passive microwave-derived ice concentration data and Terra and Aqua MODerate-resolution Imaging Spectroradiometer (MODIS) imagery. Other data are not, however, and the cost of acquiring certain satellite data at times needs to be factored into field research and sea ice systems services campaigns, and can be prohibitive.

As a general overriding comment, remote sensing is at its most powerful when complementary data from different sources are synergistically combined. This invariably yields greater levels of information and insight than can be derived from individual sensor data sources alone. Good examples are the use of satellite visible-thermal IR imagery in combination with SAR imagery (Steffen and Heinrichs 1994; Kwok et al. 2007) and passive microwave with radar scatterometer data (Walker et al. 2006). Of particular benefit is analysis of satellite data combined with environmental information such as wind speed and direction. Moreover, information on change and variability can obviously be more readily obtained from time series of images rather than isolated snapshots.

## 3.15.5 THE MAJOR SENSOR CLASSES: APPLICATIONS, ATTRIBUTES, AND LIMITATIONS

In this section, we provide a more detailed evaluation of the different satellite sensors and data sets used in sea ice field research and applicable to studies of sea ice system services, highlighting general considerations and relative advantages and disadvantages. These are medium-resolution visible-thermal IR sensors, passive microwave radiometers, and synthetic aperture radars (SARs). Pointers are also given to aid image data product interpretation. More detailed information is available in Lubin and Massom (2006).

### 3.15.5.1 Passive Remote Sensing

*Optical to Thermal Infrared*

*Moderate-resolution visible-thermal IR radiometers*, with a spatial resolution of 0.25–4 km and swath width of ~1500–2800 km, have been extensively used in sea ice field research, because of their easy accessibility, low cost, frequent coverage, and broad applicability. Primary sensors are the NOAA Advanced Very High Resolution Radiometer (*AVHRR*, 1978–present); the Optical Linescan System (*OLS*) onboard Defense Meteorological Satellite Program (DMSP) satellites (1979–present); *MODIS* sensors onboard the NASA satellites Terra (1999–present) and Aqua (2002–present); the Medium Resolution Imaging Spectrometer (*MERIS*) and Advanced Along-Track Scanning Radiometer (*AATSR*) onboard ESA's Envisat (2002–present); and Visible/Infrared Imager Radiometer Suite (*VIIRS*) sensors onboard National Polar Orbiting Environmental Satellite System (NPOESS) satellites (starting with the NPOESS Preparatory Project or NPP satellite in 2010). Detailed sensor characteristics for these as well other sensors discussed below are given in Lubin and Massom (2006). A key feature of all is that they are multispectral sensors and, apart from MERIS, simultaneously measure in the visible-near IR and thermal IR.

Advantages include ease of accessibility (in real to near real time) and frequent coverage (multiple satellites and short revisit intervals), low cost, relative ease of interpretation (compared to radar data, for example), and multiple applications (Steffen and Heinrichs 1994). The latter include mapping of pack and fast ice distribution (Massom et al. 2009); open water and thin ice regions (including leads and polynyas) and ice production rates; albedo and its relationship to ice stage of development/type (for thin sea ice types); ice motion; characterization of the marginal ice zone; and ice skin-surface temperature (from thermal IR data). Disadvantages, in addition to their relatively coarse spatial resolution, are an inability to "see" ice through thick cloud cover and fog (a major limitation in polar regions in general, although automated procedures have been developed to detect sea ice under optically thin cloud using multichannel data from the AVHRR [Williams et al. 2002]); and for the visible and near-IR channels, an inability to operate during periods of darkness. This effectively rules out the wintertime use of visible/near-IR imagery, although thermal-IR channel data are not reliant on reflected solar radiation and can be used to map sea ice at this time, clouds permitting. A key prerequisite to using the latter is accurate cloud masking and the ability to accurately composite cloud-free image sections. Although persistent cloud cover is a major limiting factor, the availability of many overlapping scenes at high latitudes increases the probability of obtaining cloud-free coverage over a field site. Moreover, wide-swath visible-thermal IR sensors provide useful real-time meteorological information in support of field operations on the approach of storms and/or clear conditions (revealed in cloud patterns) and can yield important information on cloud fraction/type—a crucial factor affecting the surface energy balance and the optical properties of the sea ice cover.

Other limitations include the lack of onboard sensor calibration (for the AVHRR in particular) combined with satellite-specific sensor performance degradation over time and the relative inaccuracy of image navigation achievable at high latitudes. This factor, which is exacerbated by the lack of recognizable ground control points in the sea ice zone, is noticeable when accurately geolocated coastal masks are superimposed on the imagery, in the form of a significant offset (of several kilometers) between the actual coastline and that in the image. This can to some extent be rectified by incorporating spacecraft orbital ephemeris data that are available from NOAA (see Emery et al. 1989 for details) and by manually "nudging" the image relative to the coastline.

Operating in six spectral bands (covering 0.58–12.5 μm) at a 1.1 km spatial resolution (at nadir for direct broadcast data) and at a relatively high radiometric resolution (i.e., 10-bit quantization), the AVHRR has been a workhorse sensor in sea ice research for about thirty years in spite of these limitations. In contrast, the DMSP OLS provides visible and thermal data from only two broad spectral channels but at improved 0.55 km spatial resolution that is consistent across the width of the swath. In terms of applicability to sea ice field research, both sensors have to a large extent been superceded by the MODIS sensors onboard Terra and Aqua. This instrument provides significantly higher radiometric resolution (12 bit) and spectral resolution, with thirty-six spectral bands ranging in wavelength from 0.405 to 14.385 μm. This provides the potential for more detailed and accurate information on snow and ice characteristics. The MODIS sensors are also equipped with sophisticated onboard calibration. Two bands are imaged at a nominal resolution of 250 m at nadir, 5 at 500 m, and the remaining twenty-nine bands at 1 km. The 2,330 km swath provides almost complete global coverage on a daily basis at high latitudes. A disadvantage compared to the AVHRR (and OLS) is that MODIS generates substantially larger data sets and requires a more sophisticated X-band satellite data reception facility for direct data download (whereas a standard L-band antenna is sufficient for AVHRR data reception). Unfortunately, X-band facilities cannot currently operate onboard ships, meaning that MODIS (and radar) imagery cannot be received in real time on ships during sea ice field research operations. Standard MODIS snow and sea ice data products (Hall et al. 2006) are routinely available from the U.S. National Snow and Ice Data Center (NSIDC) at http://nsidc.org/daac/modis/index.html. These products include daily global sea ice extent maps at a 1 km spatial resolution and maps of ice surface temperature (IST). These data also have a valuable role to play as a means of interpreting coincident/near-coincident microwave data, particularly from the new generation of SARs with multipolarization and polarimetric capabilities (Envisat ASAR, ALOS PALSAR, and Radarsat-2 SAR), as discussed later.

One of the most useful contributions of visible-thermal IR remote sensing to sea ice field research is in the provision of important information on the distribution of open water and new/thin sea ice, based on both their optical and thermal signatures (Yu and Rothrock 1996; Kwok et al. 2007). Particularly useful are maps of surface albedo, which can be retrieved from satellite radiance data after accurately masking clouds, correcting for clear-sky atmospheric effects, and factoring in the bi-directional reflectance effect caused by viewing geometry and surface angular anisotropy (Jin and Simpson 2000; Nolin and Frei 2001). As noted earlier, sea ice spectral reflectance depends upon its age and thickness, and particularly the presence/absence of a snow cover. This provides a potential means of classifying ice types in satellite cloud-free optical imagery. For example, Brandt et al. (2005) integrated in situ spectral albedo measurements of the East Antarctic sea ice zone over narrow bands to obtain representative albedos of various sea ice "types" for AVHRR channels 1 and 2 (visible and near-IR), and for the spring and summer periods. These values, computed using the satellite spectral response functions (for NOAA 11), are given in Table 3.15.4. Values are generally lowest for thin or melting ice and highest for thick snow-covered MY ice. Note the relative lack of seasonal change in albedo, in stark contrast to the Arctic in spring/summer (Perovich 2001). The strong sensitivity of near-IR radiation to snow grain-size growth with melting further enables the potential detection/monitoring of seasonal melt-refreeze progression (although microwave methods may be more effective overall in the complex sea ice environment, particularly given that they can "penetrate" cloud cover—see below).

**Table 3.15.4.** Representative albedos of surface types in the east antarctic sea ice zone in spring (September–November, SON) and summer (December–February, DJF), for NOAA AVHRR channels 1 (0.58–0.68 μm) and 2 (0.725–1.10 μm) under clear-sky conditions. Values in bold were derived from actual measurements, while all others were interpolated. After Brandt et al. 2005.

| | | | Thin snow (<3 cm) | | | | Thick snow (>3cm) | | | |
|---|---|---|---|---|---|---|---|---|---|---|
| | No snow | | Channel 1 | | Channel 2 | | Channel 1 | | Channel 2 | |
| **Ice type** | **Channel 1** | **Channel 2** | **SON** | **DJF** | **SON** | **DJF** | **SON** | **DJF** | **SON** | **DJF** |
| Open water | **0.07** | **0.07** | – | – | – | – | – | – | – | – |
| Grease | **0.10** | **0.08** | – | – | – | – | – | – | – | – |
| Nilas | **0.17** | **0.14** | **0.49** | **0.46** | **0.44** | **0.39** | – | – | – | – |
| Young gray | **0.30** | **0.24** | 0.64 | 0.60 | 0.59 | 0.52 | 0.79 | 0.74 | 0.78 | 0.68 |
| Young gray-white | 0.38 | 0.28 | 0.73 | 0.68 | 0.68 | 0.59 | 0.86 | 0.81 | 0.83 | 0.73 |
| FY <0.7 m | 0.47 | 0.32 | **0.85** | **0.80** | **0.80** | **0.70** | 0.96 | 0.90 | 0.88 | 0.77 |
| FY >0.7 m | **0.54** | **0.33** | **0.93** | **0.87** | **0.88** | **0.77** | **0.96** | **0.90** | **0.88** | **0.77** |

Regarding thermal IR information, detection of ice versus open water and classification of thin ice types is based on the thermal contrast between open water within leads, etc. (~271.2 K or −1.8°C) and thick snow-covered sea ice (the *IST* of which approximates the surface air temperature), while the *IST* of thin newly formed ice without a snow cover falls somewhere in between. Additionally, due to the fact that thermal contrasts between water and ice are not as large as reflectance (albedo) differences, infrared imagery generally requires image enhancement. Unfortunately, detailed ice-type classification is not possible once the sea ice acquires a snow cover, as the latter largely masks the thermal contrast of the underlying ice. *IST* is then very similar across all thicker ice types; that is, there is little or no thermal contrast at TIR wavelengths. As we shall see, such classification is, however, possible using passive and active microwave remote sensing. Moreover, difficulties in interpretation of TIR data can occur during periods of surface melt when the *IST* can approximate the SST open water regions.

Effective cloud detection and masking in optical-thermal IR imagery is a difficult problem, as clouds can be indistinguishable from a snow and ice background at optical to near-IR wavelengths, while the temperature contrast can also be low in thermal IR imagery. Moreover, the thermal contrast can diminish in the polar winter due to the formation of strong near-surface temperature inversions. Also, thin cirrus cloud remains unresolved, meaning that the observed radiance represents a mix of the surface and cloud radiances. A number of different cloud detection and masking methods are available, as outlined in Chapter 5 of Lubin and Massom (2006). These generally involve either multispectral analysis or pattern-recognition techniques. Another technique exploits the relatively rapid motion of clouds relative to sea ice in order to detect and mask the former in time-series image sequences such as AVHRR data (Comiso et al. 2003). Compositing procedures are then used to create nominally cloud-free images over periods of several days during which the surface conditions may be considered static. Unfortunately, however, no one method is universally applicable, and unresolved (i.e., sub-pixel scale) cloud remains one of the largest error sources in the retrieval of geophysical parameters from satellite visible to thermal-IR data, particularly during the polar night.

Removing atmospheric effects on the retrieved surface temperature data, including those of water vapor and aerosol, is also a challenge. Clear-sky atmospheric correction is most important for optical and near-IR data as the atmospheric contribution to the surface-leaving radiation is relatively high at these wavelengths. This typically involves correction for water vapor absorption, aerosol (Mie) scattering, and Rayleigh scattering (Kaufman 1989). *Mie scattering* occurs where the atmospheric particle dimensions are equivalent to the wavelength of the radiation, whereas in *Rayleigh scattering* the particles are small compared to the wavelength (Rees 2001). The most straightforward means of atmospheric correction is to difference data from spectrally adjacent channels with different absorption responses to water vapor, for example, 4 (10.3–11.3 μm) and 5 (11.5–12.5 μm) of the AVHRR/3. This technique, which is termed the *split-window* method, is typically used in determination of IST (Key et al. 1997). Two other methods are used for atmospheric correction of TIR imagery: *dual-look* viewing and physical modeling (Cracknell and Hayes 1991). With the former, the satellite instrument (e.g., the Envisat AATSR) views the Earth's surface in two directions to calculate and subtract the atmospheric contribution. The modeling approach generally uses an atmospheric transmission model such as MODTRAN or LOWTRAN, but requires accurate local data on atmospheric water vapor profiles, which can be derived from satellite atmospheric sounding data and/or radiosondes launched from the surface. The accuracy of IST retrievals again also depends on effective cloud detection and masking—a major challenge given the similarity in reflective and thermal properties of snow/ice and clouds.

*High-resolution visible to near-IR imagers* acquire images with a high geometric precision over a narrow swath (<200 km) with a nominal repeat cycle of sixteen to twenty-six days, and a spatial resolution of ~5 to <100 m. Current commercial examples include the Enhanced Thematic Mapper Plus (ETM+) on Landsat 7 (1999–present), and the High-Resolution Geometric (HRG) and High-Resolution Stereoscopic (HRS) sensors onboard SPOT-5 (2002–present, but lacking TIR channels). Non-commercial examples include the EOS Terra Advanced Spaceborne Thermal Emission and Reflection Radiometer (ASTER) and the Advanced Visible and Near-Infrared Radiometer 2 (AVNIR-2) on the Japanese ALOS mission. This sensor class is potentially useful in sea ice field research in that it offers localized but fine-scale coverage. In general, the imagery needs to be requisitioned ahead of time over the region of interest, which necessitates that subsequent fieldwork then occurs within this region at the time of image acquisition. While cloud cover and darkness are once again severe limitations, these are the data of choice for work on floe-size distribution and melt pond coverage determination (Markus and Dokken 2002), for example, as well as fine-scale characteristics of the marginal ice zone (Comiso 2003), landfast ice, and polynyas. These data have made a significant contribution to the validation of coarser-resolution satellite data products. They also have an important role to play in regional process studies requiring very high-resolution coverage and to support interpretation of SAR data (with a similar spatial resolution). They are, however, limited in their applicability to studies of lead opening and closing, given their typically long repeat interval. Similarly, commercial ultra-high-resolution sensors (e.g., Ikonos-2 and EROS-A1) are of limited applicability, given their narrow swath coverage of < 20 km, although SPOT-5 has a 2.5 m resolution mode with 60 km coverage. Cost is also an overriding factor with these data.

### *Passive Microwave Radiometers*

With excellent wide-swath and multifrequency coverage uninterrupted by clouds and darkness, satellite passive microwave (PMW) radiometers are the primary source of large-scale information on sea ice concentration, extent, and type at a 6.25–25.0 km resolution (type classification in the Arctic only, for reasons outlined below). Important derived products are ice-season length and large-scale maps of ice motion. The continued routine monitoring of the cryosphere by PMW sensors has generated a unique climatological data set that dates back to 1978 for sea ice concentration retrievals based on multifrequency data. Important current sensors include the DMSP Special Sensor Microwave/Imager (SSM/I series, 1987–present) and the Aqua Advanced Microwave Scanning Radiometer-EOS (AMSR-E, 2002–present). Additional products derived from the latter are snow depth on sea ice and ice temperature (Comiso et al. 2003). This sequence will continue into the future with the launch of the AMSR-2 instrument onboard the Japanese Global Change Observation Mission-Water (GCOM-W) in 2011 and Microwave Imager Sounders (MIS) onboard the NPOESS series in 2016 and launches planned to NPOESS C-4 in ~2022 (http://www.ipo.noaa.gov).

A key characteristic of modern PMW sensors is that they are multichannel systems, simultaneously measuring the same target in several different spectral bands. This is highly beneficial for sea ice applications, as it enables extraction of considerably more information than is possible from single-channel sensors. In addition, PMW radiometers and SARs are designed to operate at different polarizations or combinations of polarizations. Importantly, horizontally-polarized (H) waves interact in a different manner with sea ice and snow compared to vertically polarized (V) waves. This again enables retrieval of additional and complementary information from multipolarization satellite data to aid discrimination of various ice types. As we shall see, this also forms the basis of a number of algorithms designed to extract important sea ice information (e.g., ice concentration) from satellite microwave data.

Passive microwave systems measure thermal emission of EM radiation in the millimeter to centimeter range from the surface. At these wavelengths, a strong contrast exists between the microwave emissivities of ice-free ocean and different ice types (Figure 3.15.4), for example, ~0.4 for open ocean, ~0.90 for FY ice, and ~0.70 for MY ice at 37 GHz horizontal (H) polarization and in the Arctic under freezing dry conditions (Eppler et al. 1992). This enables retrieval of ice concentration and extent, and also of information on the fraction of MY ice in the Arctic. Note that in the Antarctic, the different sea ice growth processes noted earlier lead to a much smaller contrast in the emissivities of FY and MY ice (Comiso et al. 1992), enabling the extraction of total ice concentration only. Methods for deriving ice concentrations are evaluated in 3.15.6. As a general comment, PMW data tend to be unsuitable for fast ice studies, due to their coarse resolution (effectively set by antenna size [the diffraction limit]), although they are very useful as a means of monitoring wider coastal polynyas.

### 3.15.5.2 Active Remote Sensing

Of the three types of radar systems currently used to measure sea ice—imaging radars, radar scatterometers, and radar altimeters—the former (and specifically synthetic aperture radar or SAR) are most important from a sea ice field research perspective. Although radar and laser altimeters have the potential to provide critical new information on sea ice thickness (see Lubin and Massom 2006 for details), they have limited applicability to field research, given their footprint size of 0.25–6.0 km and long revisit intervals (although in situ campaigns are playing a key role in calibrating and validating altimeter data acquired over sea ice). Operational application of advanced sensor technology and near-real-time access to altimeter data may, however, add another tool for ice-system services in the future. Satellite radar scatterometers, on the other hand, are increasingly used in sea ice research and operational analyses (Long et al. 2001), in addition to providing wind velocity measurements over ice-free oceans. These low-resolution radars complement SAR with their broader swath and more frequent coverage. Moreover, they provide information at a similar spatial scale as passive microwave radiometers, but can provide better information on the extent of the Arctic perennial ice cover (Walker et al. 2006) as well as complementary information on other properties due to different sensitivities to differences in the ice surface (e.g., due to melt and variable snow characteristics). Their strength lies in their excellent wide-swath coverage; low data rate; near-real-time availability; all-weather, day-and-night operation; near-daily complete polar coverage; and a continuous time series of stable and accurately calibrated data dating back to 1991 (the ERS series). Although their nominal spatial resolution is poor (25–50 km), it has been improved to 8–10 km by recent processing advances (Long 2003), enabling post-processing of data back in time. As such, scatterometers are best suited to large-scale systematic observations of sea ice extent, ice type regimes, and ice motion. Important current missions include the SeaWinds on QuikSCAT (Ku-band, 1999–present) and the Advanced Scatterometer (ASCAT) on MetOp (2006–present). Radar scatterometers are a primary source of information on large-scale melt onset and progression, plus subsequent seasonal refreeze (Drinkwater and Liu 2000).

*Synthetic Aperture Radar*

First launched on Seasat in 1978, the Earth-observing spaceborne synthetic aperture radar (SAR) is a powerful research tool that provides all-weather and day-night high-resolution imagery of radar backscatter with which to map and interpret surface and near-surface characteristics. SAR differs from spaceborne real-aperture radar (e.g., onboard Russian Kosmos Okean satellites, which are limited to a ground resolution of ~1–2 km), by exploiting the satellite's along-track motion and using the range and Doppler shift of multiple radar-return pulses to synthesize a large antenna in space and increase the resolution in azimuth to meters or tens of meters (Rees 2001). Coverage is typically over a relatively narrow swath of 75–100 km in conventional mode and 400–500 km in reduced-resolution ScanSAR mode, but much narrower for new systems operating in high-resolution and/or polarimetric or cross-polarization modes. This again highlights the important trade-off for SAR between areal coverage (swath width) and spatial (azimuth) resolution. Key current missions include:

- Envisat Advanced SAR or ASAR (C-band, 5.33 GHz, choice of five polarization modes [VV, HH VV/HH, HV/HH, or VH/VV], nominal resolution 8–100 m over a swath width range of 45–500 km respectively, 2002–present);
- Radarsat-1 (C-band, 5.3 GHz, HH polarization, nominal resolution 8–100 m over a swath width range of 45–500 km respectively, 1997–present);
- ALOS PALSAR (L-band, 1.27 GHz, nominal resolution 7–100 m over a swath width range of 30–360 km respectively, 2005–present);
- TerraSAR-X (X-band, 9.65 GHz, λ 3.1 cm, nominal resolution 1–16 m over a swath width range of 5–100 km respectively, choice of single, dual, or experimental quad polarization, operational January 2008–present, http://www.dlr.de/en/desktopdefault.aspx/); and
- Radarsat-2 (C-band, 5.405 GHz, operational January 2008–present, HH, HV, VH, and VV polarisation with a choice of selective copolarized, cross-polarized, and fully polarimetric data [see Lubin and Massom 2006 for an explanation of these terms], nominal resolution 3–100 m over a swath width range of 20–500 km respectively, http://www.radarsat2.info).

Whereas TerraSAR-X represents the first spaceborne X-band SAR, the Envisat ASAR, ALOS PALSAR, and Radarsat-2 SAR are based upon the heritage of the SARs onboard ERS-1/-2, JERS-1 and Radarsat-1 (see Lubin and Massom 2006), but with additional capabilities, modes of operation, and flexibility of coverage. With this comes an apparently bewildering array of choices for the user regarding data products.

The steerable beam capability of modern SARs enables more frequent image acquisition of regions of interest than is possible with fixed look-angle spaceborne SARs. SAR images can also be accurately geo-registered compared to medium-resolution optical-TIR instruments (particularly AVHRR). This enables accurate location of major sea ice morphological features when in the field, to aid choice of suitable field sites and overlay of ship tracks and/or field stations if SAR imagery is available in near real time.

The main sea ice field research applications of SAR are ice type classification and open water–thin ice discrimination (leads and polynyas), lead opening and closing, ice motion/deformation based on automated floe tracking in image time series, ice-surface roughness (including pressure ridge information), landfast ice extent, wave-ice interaction, seasonal melt/freeze-up detection, detailed ice-edge characteristics and wind speed and direction, and iceberg and ice island detection and tracking. This represents a powerful combination, making wide-swath SAR the most suitable and preferred sensor for regional sea ice mapping and monitoring.

Limitations include the relative complexity of interpretation and the lack of routine SAR data availability (particularly in the Antarctic), due in large part to the lack of receiving capacity/stations. Although SARs are generally unsuited to routine large-scale coverage, reduced-resolution data are routinely available in near real time from the Envisat ASAR operating in a background Global Monitoring Mode (1 km resolution over a 400 km swath). Regarding data availability, it should be noted that Radarsat-2 and TerraSAR-X are commercial satellites. Data are obtainable for scientific purposes, however, and the product Web sites provide additional guidance.

Given their high horizontal spatial (azimuth) resolution (of the order of tens of meters in general and up to 1 m for TerraSAR-X), SARs can resolve leads and detect pressure-ridge systems, which also have different backscatter signatures than the surrounding ice. Composed of piled-up blocks of ice, ridge sails comprise a high backscatter target, with a further contribution from increased volume scattering due to a general increase in ridge porosity with age. SAR detection of pressure ridges is also facilitated by their linear shape. Smaller ridges also contribute to the observed backscatter in certain SAR pixels (Haas et al. 1999). These factors enable retrieval of the amount of ridging from SAR scenes, based upon analysis of the backscatter distribution in certain areas. Knowledge of the number and density of ridges and the degree of deformation are relative measures of sea ice thickness (see Chapter 3.2; Haas 2003).

Leads are in general readily distinguishable from surrounding older ice by virtue of their distinctive open water signature when newly formed (dark); low to intermediate backscatter when refreezing; and their elongated linear shape. These factors also enable monitoring of the opening and closing of leads in time series of SAR images (Kwok et al. 1999). Wavelength- and polarization-dependent variations occur, however, in the form of wind roughening of the open water and the growth of extensive mats of highly saline frost flowers on the surface of nilas under cold, calm, and clear-sky conditions. Both phenomena lead to a major increase in backscatter, resulting in a brighter target that can be confused with older and thicker ice. Knowledge of wind speed can greatly aid interpretation of this ambiguity.

A calibrated SAR image of an ice-covered ocean represents a "snapshot" map of the spatially varying intensity of the radar reflectance, or backscatter coefficient, of the surface. As discussed above, the measured backscatter is determined by a complex range of factors, and ice-type mapping with SAR depends on the discrimination of surface versus volume scattering, surface roughness characteristics, and large-scale deformation and ice structures, and their change as the ice ages (Drinkwater 1995). Fortunately, a number of general relationships exist to aid interpretation of SAR images of ice-covered oceans under freezing conditions (following Shuchman et al. 2004 and Onstott and Shuchman 2004 and others):

- Strong backscatter contrasts generally exist between calm open water and nilas (i.e., low backscatter plus strong specular reflection and therefore a dark appearance), smooth new ice (i.e., moderate backscatter, appearing medium-bright), and thick deformed ice (i.e., a strong backscatter, resulting in a relatively bright target).
- For young to FY ice, the relatively high salinity largely prevents radar penetration at shorter wavelengths e.g., C- and X-band, and backscatter is dominated by small-scale surface roughness.
- For Arctic SY and MY ice, on the other hand, the low salinity in the ice freeboard results in greater microwave penetration into the ice under dry-snow conditions, and the observed backscatter is dominated by volume scattering (from dielectric discontinuities or discrete scatterers such as gas bubbles). In winter, Arctic MY ice has a stable signature that contrasts (by 2–4 dB at C-band) with that of FY ice, and represents a significantly brighter target.
- This contrast is not as apparent in Antarctica, where winter backscatter signatures for MY and thick FY ice are similar, and the largest contrast in Antarctic backscatter signatures is between new/young ice and older, thicker ice types.
- In the Antarctic summer, the backscatter response is largely characterized by backscatter fluctuations that are correlated with melt and refreeze cycles associated with the passage of storms.
- Surface scattering associated with surface roughness dominates the backscatter contribution from a wet snow cover and/or flooded/melt-ponded ice.
- Ice-type discrimination under freezing and dry conditions is best accomplished at C-band rather than L-band, as the wavelength is equivalent to scatterer dimensions within the ice, but at the expense of diminished surface-roughness information. Also, X-band produces similar results to C-band. The optimal polarization combinations for ice-type discrimination is HH or HV.
- Of current spaceborne copolarized SARs, X-band data are the optimal choice for detecting thin FY sea ice such as nilas and gray ice.
- Radars operating at L-band are less sensitive to small-scale surface roughness and penetrate much deeper into the sea ice. As a result, the volume scattering largely emanates from internal ice inhomogeneities to yield information on the morphological and structural features of the ice, with ridges and fractures appearing as bright streaks. Ice-type and floe-boundary discrimination at this frequency is best achieved at VV polarization, while optimal surface topographic information comes from HV polarization.

- In their comparison of C-band ERS (VV polarization) and Radarsat-1 (HH polarization) data, Dokken et al. (2000) established that VV polarization is preferred for thin ice discrimination, while HH polarization is most useful for ice/open water discrimination, that is, ice-concentration calculations; the ability to discriminate drops off at higher frequencies in particular under melt conditions.
- A cold, dry snow cover is largely transparent to the radar waves at L-band and even C- and X-band, although the presence of icy layers and crystal enlargement by metamorphism can affect the backscatter contribution at the higher frequencies.

Optimal incidence-angle considerations for given sea ice parameter retrievals as a function of frequency and polarization are summarized in Lubin and Massom (2006).

In spite of these known relationships, the interpretation of sea ice signatures in SAR data remains a challenging issue, given the complex, highly variable, and often unpredictable nature of backscatter returns from different ice types and wind-roughened open water (Shuchman et al. 2004). In particular, large and abrupt variations occur in Arctic sea ice and snow-cover backscatter in response to changing environmental conditions to greatly limit the retrieval of accurate geophysical parameters during the annual melt season (Figure 3.15.6). The wet snow contribution at this time can dominate to effectively mask the backscatter contrasts between underlying ice types, such as Arctic MY and FY ice (Comiso and Kwok 1996). This effect increases with frequency (Hallikainen and Winebrenner 1992), with Drinkwater et al. (1991) suggesting that L-band may be the optimal frequency for discriminating thin and thick ice over the seasonal melt period due to its lower absorption loss compared to higher frequencies. In spite of these difficulties, the abrupt sea ice backscatter response to seasonal melt/refreeze can be used advantageously to detect and track their timing across the Arctic. Together with the fairly rapid return of $\sigma^o$ to winter levels in autumn, the abrupt onset of melt signature can be used advantageously to map the onset of seasonal melt, its spatiotemporal progression, and subsequent autumnal refreezing in time series of radar images.

Moreover, SAR-based classification of Antarctic sea ice is even more challenging, due to the lack of backscatter contrast between different ice types even under freezing conditions (Drinkwater 1998). In summer, snowmelt and refreeze can actually lead to an increase in backscatter by creating highly reflective icy layers within the snowpack (Haas 2001). Also in contrast to the central Arctic, episodic incursions of warm maritime air over the Antarctic ice pack can cause almost complete snowmelt even in winter (Massom et al. 1997). Such events can have a significant impact on the microwave backscattering characteristics of sea ice, but this effect tends to be ephemeral.

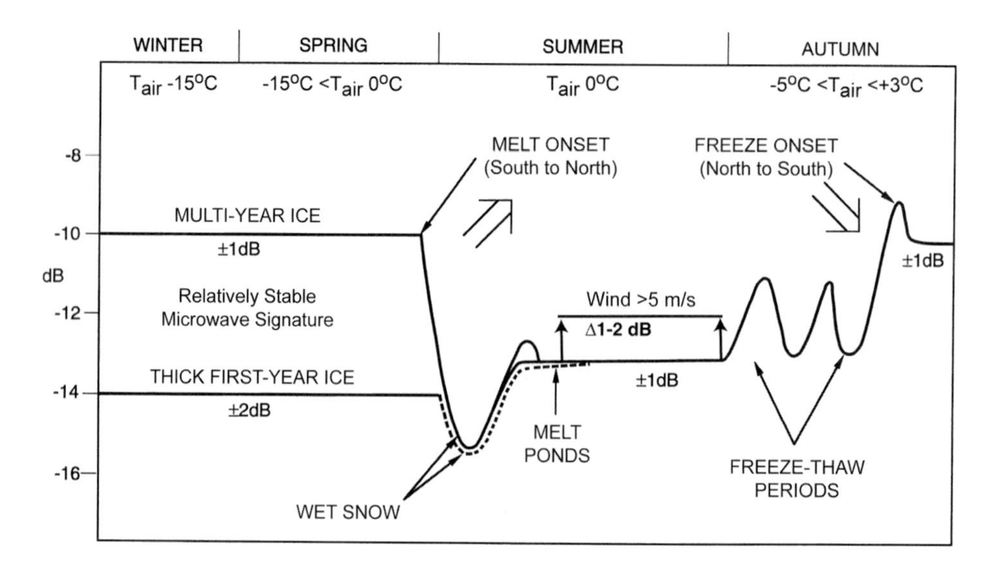


**Figure 3.15.6.** Schematic representation of the seasonal response of radar backscatter measured by the ERS C-band VV-polarization SAR to changes in physical characteristics of arctic FY and MY ice over an annual cycle. Approximate ranges of backscatter and seasonal air temperature are given. Note the effect of an increase in wind speed on the backscatter of the surface during summer. Courtesy of Ben Holt (NASA Jet Propulsion Laboratory), modified from Soh et al. (1998). ©Springer-Verlag 1998. Modified and reproduced by permission of Springer-Verlag.

Other factors complicating the unambiguous application of copolarized SAR data include:

- the similarity in the backscatter signatures of MY and pancake ice (Drinkwater 1998);
- the sensitivity at all times of year of open-water returns (i.e., within leads and polynyas) to wind speed. Such regions exhibit a wide range of returns, from approximately –20 to –5 dB, which often overlap with signatures from all ice types (Soh et al. 1998);
- variations in snow/ice temperature and surface salinity resulting in large dielectric fluctuations (Nghiem et al. 1998); and
- even when dry, the snowcover can make a variable and at times significant backscatter contribution as it evolves, particularly at frequencies >5 GHz (Barber et al. 1998).

Extensive coverage of frost flowers on snow-free ice in its early stages of growth can represent a rough surface at shorter wavelengths such as C-band (Onstott and Shuchman 2004). Frost flowers are strongly backscattering at C-band and higher wavelengths due to both surface roughness and their high salinity (to 110 psu), which causes them to be highly reflective to radar energy. This can diminish the backscatter contrast between thin ice and thicker FY ice, and can even create a backscatter similar to that of MY ice in the Arctic, an effect that is diminished using longer-wavelength SARs such as L-band (Drinkwater et al. 1991).

Due to the overlapping and often unpredictable backscatter returns from different ice types and wind-roughened open water, and poor contrast between new ice and open water, classification using simple thresholding techniques that assign unique backscatter ranges to different ice classes is generally not successful when applied to conventional copolarized SAR imagery. This is particularly true for thinner (younger) ice types, although improvements in classification accuracy are expected by using better textural measures from very high spatial resolution systems such as TerraSAR-X and Radarsat-2 as additional information in the classification procedure. Unfortunately, this high resolution comes at the expense of narrow swath coverage.

Given the frequency and polarization dependence of sea ice dielectric properties, these ambiguities can potentially be minimized and different ice types distinguished more accurately by using more sophisticated multipolarization radars (see Lubin and Massom 2006). These same principles apply to interpretation of thermal emission at microwave frequencies, and form the foundation of algorithms that discriminate between Arctic FY and MY sea ice in satellite passive microwave data (Comiso 2003). As outlined above, either multipolarization and/or fully polarimetric data are now available from the advanced new SARs onboard Envisat-1, Radarsat-2, ALOS, and TerraSAR-X. These data show great promise as a means of deriving improved ice edge detection, ice type discrimination, and ice topography and structure information (Dierking et al. 2004). In theory, the acquisition of data in quad-pol mode will facilitate the computation of a range of key variables that relate to the polarization, strength, and/or phase of the backscatter. The following summary of potential benefits at C-band and for freezing Arctic conditions is compiled from van der Sanden (2004) and Lubin and Massom (2006), and references therein:

- HV improves ice edge detection and ice concentration mapping, and also iceberg detection.
- HV also gives enhanced information on sea ice structure and topography, and this may lead to improved discrimination of smooth (relatively undeformed) and deformed ice.
- The HH/VV copolarization ratio yields the best differentiation between thin newly forming ice types and open water. For both HV and HH/VV, the enhanced contrast is based upon the fact that the cross-polarization backscatter response of open ocean is relatively independent of wind-induced surface roughening.
- The HH/HV ratio could improve FY-MY ice discrimination, as MY ice provides a significantly stronger depolarized response.

Given the complexity of SAR data analysis, using simultaneous information from other sensors (preferably on the same satellites) can greatly aid SAR interpretation and improve ice mapping capabilities for field research, for example, Steffen and Heinrichs (1994). In another example, Drucker et al. (2003) combined a calibrated Radarsat-1 ScanSAR image with a near-coincident AVHRR ice skin surface temperature (IST) image to circumvent the problem of distinguishing them from wind-roughened open water in an important Arctic polynya (St. Lawrence Island) based upon the copolarized backscatter signature alone. In an Antarctic study, Kwok et al. (2007) assessed thin ice distribution and thickness using a combination of Envisat ASAR, AMSR-E, and MODIS satellite imagery with atmospheric data.

## 3.15.6 KEY GEOPHYSICAL SEA ICE VARIABLES EXTRACTED FROM SATELLITE DATA

In this section, we present a short summary of primary sea ice related variables that can be derived from satellite data and are relevant to field research (much more detailed information is given in Lubin and Massom 2006). Some are routinely and operationally produced, while others are currently research and developmental only. Information is given on techniques (algorithms) used and data sources, as well as on error sources and relative strengths and weaknesses. Archived data are also discussed as these are an essential element of the planning of field experiments (see Section 3.15.8). A more complete summary of sea ice parameters and satellite sensor classes used to measure them is provided in Table 3.15.5. In this section, we look in more detail at methods used to retrieve sea ice concentration from passive microwave data. Much of this material is derived from Chapter 5 of Lubin and Massom (2006) and the citations therein. In all cases, the applicability and suitability of a satellite sensor/data set to the measurement of a given sea ice related parameter or process depends upon the nature and scale of the parameter and the attributes of the satellite sensor data. Much depends on the factors and trade-offs outlined in 3.15.4.

**Table 3.15.5.** Summary of sea ice related applications of the major satellite sensor classes, with current primary satellites and sensors given

| Sensor types | Passive micro-wave | SAR (hi-res.) | Med.-low res. vis-TIR# | Hi-res. vis-TIR# | Ultra-high res. vis (≥5 m) | Radar scatt. (low res.) | Radar alt | Laser alt# |
|---|---|---|---|---|---|---|---|---|
| Current and near-future primary satellites and sensors variables | DMSP SSM/I, EOS Aqua AMSR-E, NPOESS MIS, SMOS | Envisat ASAR, Radar-sat-1 and -2, ALOS PALSAR, Terra SAR-X and -L, GMES Sentinel-1 | NOAA AVHRR, Terra and Aqua MODIS, Envisat AATSR, DMSP OLS, SeaWiFS, NPP and NPOESS VIIRS | Landsat, SPOT, Terra ASTER, ALOS AVNIR -2 | Quick-Bird, For-mosat-2, Geo-Eye-1, SPOT-5, World-View -1 and -2 | QuikScat Sea Winds, MetOp ASCAT | Envisat Cryo Sat-2, GMES Senti-nel-3 | IceSat GLAS |
| Sea ice concentration | P | S | S | S | S | S | – | – |
| Sea (pack) ice extent | P | S | S | S | – | S | S | R |
| Sea ice season length | P | – | – | – | – | – | – | – |
| Fast ice extent | – | P | P | P | S | – | – | – |
| Polynya size | P | P | P | P | S | S | – | – |
| Lead detection & orientation | – | P | P | S | S | – | – | – |
| Iceberg size/ tracking | – | P | P | P | S | S | S | S |
| Ocean color (ocean Chl-*a* content & primary production)# | – | – | P | – | – | – | – | – |
| Sea ice motion | P | P | P | S | – | P | – | – |
| Sea ice dynamics & kinematic parameters | S | P | S | – | – | S | – | – |
| Sea ice thickness | R | R | R | – | – | – | R | R |

| | | | | | | | | |
|---|---|---|---|---|---|---|---|---|
| Sea ice classification – thicker ice types | P | P | S | S | – | S | – | – |
| Thin/new sea ice distribution | R | P | P | S | S | S | – | – |
| Snowcover thickness | P/R* | – | – | – | – | – | – | – |
| Surface roughness/ topography | – | R | – | R | – | S | – | R |
| Floe-size statistics | – | P | S | P | S | – | – | – |
| Surface "skin" temp. | – | – | P | S | – | – | – | – |
| Snow-ice interface temp. | P/R* | – | – | – | – | – | – | – |
| Broadband albedo | – | R | P | S | – | – | – | – |
| Snow grain size | – | – | P | – | – | – | – | – |
| Snow impurity content | – | – | P | – | – | – | – | – |
| Annual melt onset & freezeup, & their progression | P | S | S | S | – | R | – | – |
| Melt pond coverage | – | R | – | R | P | – | – | – |
| Wave-ice interaction | – | R | – | – | – | R | – | R |
| Ice-edge characteristics (banding, eddies, etc.) | – | P | P | P | S | R | – | – |

Vis is visible to near infrared; TIR is thermal infrared; Scatt is scatterometer; Alt is altimeter; SMOS is Soil Moisture and Ocean Salinity (launch in 2009); SeaWiFS is Sea-viewing Wide Field-of-view Sensor; NPP is NPOESS Preparatory Project (launch ~2010); GMES is Global Monitoring for Environment and Security; GLAS is Geoscience Laser Altimeter System; # = cloud-affected (visible to near-infrared sensors are also darkness-affected); P = primary data source; S = secondary data source; - = no application; and R = research and development (at this stage).

*Although operationally produced (with data from the EOS AMSR-E), these are new products that require intensive validation. Icebergs are included as they have a significant effect on sea ice distribution, and primary production is included because it is closely linked to sea ice distribution and melt.

# Only MODIS, SeaWiFS, and VIIRS can be applied to ocean color measurement.

### *Sea Ice Concentration and Extent*

Concentration and extent are fundamental parameters describing the two-dimensional distribution of the sea ice cover, and are routinely derived from satellite passive microwave data. Sea ice concentration algorithms generally combine $T_b$s measured at different frequencies (e.g., 19, 37, and 85/89 GHz) and both horizontal (H) and vertical (V) polarization (detailed information on the different algorithms and their comparative strengths and weaknesses is given in Lubin and Massom 2006). This results in the retrieval of total ice concentration at ground resolutions of 6.25 to 25 km. An important general trade-off is this: the higher the frequency (shorter the wavelength), the better the spatial resolution but the greater the atmospheric contribution (particularly above 37 GHz).

Passive microwave ice concentration maps provide a key backdrop for sea ice field operations. More detailed information can be gained from analysis of high-medium resolution visible-thermal IR data (from MODIS in particular) and SAR data (see Lubin and Massom 2006 for information on techniques).

For the multichannel passive microwave sensors, the NSIDC archives gridded ice concentration products developed at the NASA Goddard Space Flight Center (GSFC) and at NSIDC using two different algorithms, the *Bootstrap* and *NASA Team* (see http://nsidc.org/data/seaice/pm.html for links to the data). For looking at past conditions, it is recommended that the GSFC combined SMMR-SSM/I data set is used, as this entails the longest continuous data record and is subject to the highest quality control. The latter includes removal of pixels containing coastal contamination, additional weather filtering to remove spurious ice-over-water pixels, and adjustment of algorithm tie points to ensure inter-satellite consistency. The standalone SSM/I time series, on the other hand, runs from 1987 to the near present, being updated in the NSIDC archive every three to six months. Both are available at a spatial resolution of 25 km. As the combined SMMR-SSM/I product, which extends back to 1979, is updated only every one or two years, the time series needs to be supplemented with more recent data from either the SSM/I or AMSR-E if near-present time periods are required. For studies requiring recent data rather than long time series, daily polar-gridded SSM/I ice concentration images are also available in near real time (one to two days after data acquisition), derived using the NASA Team algorithm but with lesser quality control (http://nsidc.org/data/nsidc-0081.html). The AMSR-E data set (available since 2002) is also available in near-real time from the University of Bremen (http://www.iup.uni-bremen.de:8084/amsr/amsre.html). It is generally preferred to the SSM/I, as it is available as a higher-resolution (i.e., 12.5 km versus 25 km; see Figure 3.15.7) product derived using improved algorithms (Comiso et al. 2003), and 6.25 km resolution using the University of Bremen algorithm (Spreen et al. 2008).

Regarding the choice of algorithm for SSM/I ice concentration retrieval (Bootstrap versus NASA Team), the following guidelines are given by the NSIDC

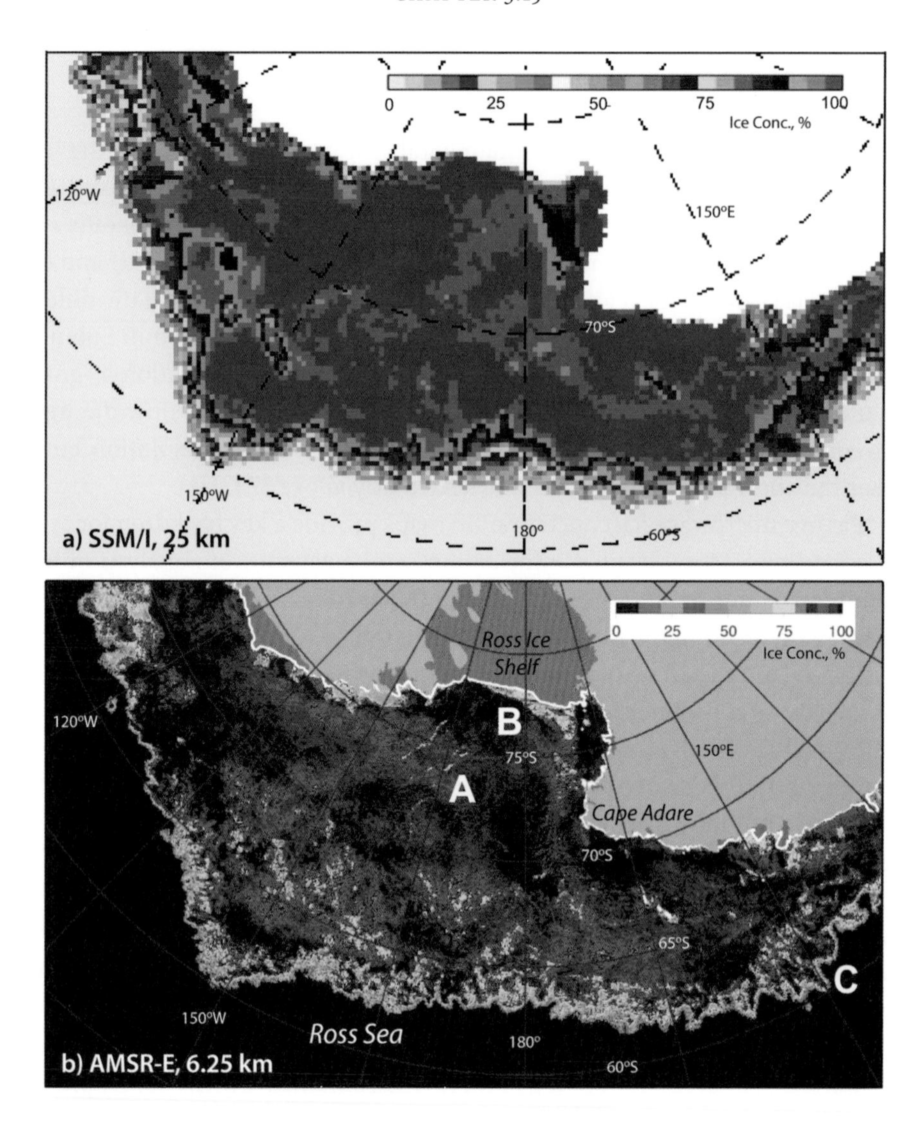


**Figure 3.15.7.** A comparison of daily maps of sea ice concentration from (a) the DMSP SSM/I at a spatial resolution of 25 km (derived from the NASA Bootstrap algorithm dataset and obtained from http://nsidc.org), and (b) EOS Aqua AMSR-E at a resolution of 6.25 km. Note the significant additional detail in the higher-resolution AMSR-E image e.g., large leads (marked A), the Ross Sea Polynya (marked B), and ice edge/marginal ice zone characteristics e.g., bands and eddy-like features in the region marked C and around the pack ice perimeter in general. Imagery courtesy of NASA Earth Observing System Distributed Active Archive Center (DAAC) at the U.S. National Snow and Ice Data Center (University of Colorado) and the University of Bremen.

(http://nsidc.org/data/seaice/faq.html#3). Either algorithm/product is suitable for general qualitative views of the ice. The Bootstrap data set is however believed to be more useful for sea ice related processes as it is generally free of residual errors that could not be removed by conventional techniques, and a temporally more consistent time series of sea ice concentrations is provided. More specifically, the NASA

Team algorithm (which is based on ratios of brightness temperatures) is thought to be sensitive to both layering within the ice and snow cover (a common phenomenon over Antarctic sea ice; Massom et al. 2001) and thin ice. Highest errors in ice concentration estimates occur in regions of predominantly thin ice and where melt or refreeze occurs. Its major strength is that it is insensitive to changes in surface physical temperature. The Bootstrap algorithm, on the other hand, is sensitive to changes in temperature and uses tie points that are multi-seasonal (switch by date). Highest errors for it occur in regions with very cold surface temperatures (e.g., parts of the Antarctic coastal zone in winter). A strength of the Bootstrap algorithm is, however, that it is less sensitive to thin ice and layering within the ice and snow. For in-depth scientific analyses or input into model fields, it is therefore recommended that users first study the differences between the algorithms (for information, see http://nsidc.org/data/docs/daac/nasateam/index.html and http://nsidc.org/data/docs/daac/boot- strap/index.html). Biases exist between algorithms and between products, so they should not be used interchangeably.

Error sources common to the two algorithms include:

- unresolved ice types within the large sensor footprint, with thin/new ice resulting in a low-concentration bias (Grenfell et al. 1992). In an Arctic comparison with Radarsat SAR data, Kwok (2002) showed that overestimates of open-water fraction by about three to five times can occur in winter in the Arctic;
- surface melt effects during late spring to summer in the Arctic, which cause significant overestimates of open-water fraction (Comiso and Kwok 1996). Ephemeral synoptic-scale warming events can also occur over the Antarctic pack as storms pass over during the autumn and winter, when much of the outer pack attains relatively low latitudes (Massom et al. 1997);
- surface wetness throughout the year, including flooding throughout the antarctic pack due to the effect of snow on the ice isostasy. Particularly large concentration underestimates of up to tens of percent can occur in the MIZ, where wave overwashing and floe buffeting and deformation create an ice cover composed predominantly of brash ice with a wet surface (Massom et al. 1999). This can in turn affect the accuracy of ice-extent determinations using ice concentration thresholds (e.g., 15%);
- the contamination of coastal pixels with terrestrial/ice-sheet signals; and
- frequency- and polarization-dependent temporal and spatial variations in microwave emissivity $\varepsilon$, as a function of thickness, age, salinity, temperature, structure, snow cover, and surface conditions overall (Eppler et al. 1992). This represents the largest source of error in sea ice concentration retrievals outside the melt season (Comiso and Steffen 2001).

Overall ice concentration errors of up to 10–20 percent have been reported for SMMR and SSM/I data in the summer, dropping to an average of 5–15 percent in winter (Comiso and Steffen 2001).

For AMSR-E data (2002–present), the NASA Antarctic 12.5 km sea ice concentration product is generated using the improved *Bootstrap Basic Algorithm*, while the 25 km product uses the *AMSR Boostrap Algorithm*, or ABA (Comiso et al. 2003). As noted above, a higher-resolution (6.25 km) product is also available in near real time from the University of Bremen, created by incorporating 87 GHz channel data. While these higher-resolution data are generally preferable for field operations, the use of the higher-frequency data means that they are more prone to atmospheric contamination effects, that is, spurious ice distributions at times, although these become apparent through analysis of time series. Work carried out during the ARISE remote-sensing validation experiment in the East Antarctic pack in September–October 2003 (see Section 3.15.7) shows the AMSR-E ice concentration retrieval to be accurate to within 5 percent under cold dry conditions in the inner pack (Massom et al. 2006). Further research is underway to establish the accuracy over the wider sea ice domain (both hemispheres) and in different seasons.

*Landfast Ice Extent*

Medium-resolution visible- and thermal-IR data are useful in terms of detecting and monitoring larger-scale landfast ice extent (e.g., Enomoto et al. 2002; Massom et al. 2009), with narrow-swath high-resolution imagery (e.g., from Landsat, ASTER, or SPOT) being suitable for more detailed localized studies. Cloud cover is a severe limitation, however, and an important prerequisite is accurate cloud masking and the ability to accurately composite cloud-free image sections for work requiring large-scale coverage (Fraser et al. in press). Recent work has exploited the all-weather, day-night, and high-resolution capability of spaceborne SARs operating in ScanSAR (wide-swath) mode to detect and monitor fast ice. Unfortunately, fast ice is generally indistinguishable from surrounding pack ice based upon its normalized backscatter signature alone. A way around this is to apply feature-tracking methods to pairs of carefully coregistered ScanSAR image pairs (e.g., maximum cross-correlation), in order to determine sea ice motion; that is, fast ice is then mapped as being the stationary component of the sea ice (Giles et al. 2008). An alternative technique, used in the Beaufort Sea, has been to measure coastal landfast ice width, and its interannual variability, along series of transects set across high-resolution Radarsat-1 ScanSAR data (Mahoney et al. 2007a). Another approach, again applied in Alaska, has been to map landfast ice behavior and characteristics using land-based marine radar (Mahoney et al. 2007b).

*Sea Ice Motion and Dynamics*

A range of large-scale ice-motion products are available from satellite data, both high-resolution (SAR) and low-resolution (visible-thermal IR radiometers, passive microwave radiometers, and radar scatterometers) (see Lubin and Massom 2006). Coordinated drifting ice buoy launches remain important by providing the only high-resolution information available on sub-daily (i.e., tidal and inertial) ice motions, which have a major impact on sea ice thickness (Heil et al. 2001a). They also provide a means of validating satellite-derived ice motion fields (Heil et al. 2001b), and supply important information that cannot be measured by satellite sensors over sea ice, for example, surface atmospheric pressure and wind speed and direction. Unfortunately, buoy coverage is spatially and temporally sparse, being limited by their expense, the logistical difficulty of deployment, and their relatively short operating lifetimes (within the divergent Antarctic sea ice zone in particular).

Unfortunately, many of the satellite-derived ice motion products are research and development only at this stage. Although synoptic-scale ice drift patterns consistent with winds can typically be observed in satellite-derived ice motions sampled on intervals as fine as twenty-four hours, the noise level of daily displacements is high. This noise can be effectively reduced through time averaging, which produces useful mean fields and climatologies that portray both hemispheric patterns and regional variability. In their comparison of SSM/I- and buoy-derived ice drift data in the Weddell Sea, Drinkwater et al. (1999) determined the root-mean-square (rms) error of the satellite product to be ~8 cm $s^{-1}$ for one day, reducing to 1.45 cm $s^{-1}$, or ~18 percent, for monthly mean velocities. In a comparison of the mean distributions of daily velocity derived from drifting buoy versus SMMR and SSM/I data from the East Antarctic sea ice zone, Heil et al. (2001b) further found that while the broad-scale velocity patterns agree, the magnitude of the satellite-derived sea ice velocities is significantly lower (typically by 40 percent or less) than the velocities derived from buoy measurements. Improvements (reduced uncertainties) are likely using improved-resolution AMSR-E data (2002–present) (Kwok et al. 2007).

In the Arctic, the Radarsat Geophysical Processor System (RGPS) provides high-resolution repeat coverage of the entire Arctic Ocean every three days using calibrated and geolocated ScanSAR Wide B backscatter images with a 500 km swath and 100 m resolution. This has enabled the routine processing of the image data into fields of geophysical variables, including ice motion, sea ice age, sea ice thickness of young ice, sea ice deformation, open water fraction, and (twice/year) dates of melt onset and freeze-up (Kwok 1998). A data users' handbook for the RGPS is given by Kwok and Cunningham (2000), with Kwok et al. (2000) providing RGPS product specifications. Similar coverage of the Southern Ocean by an Antarctic RGPS-like system does not exist, and remains a difficult proposition due to logistical constraints and the fact that Antarctic sea ice represents a more homogeneous radar target (see Section 3.15.3) and is therefore more difficult to track in sequential images (Drinkwater 1998).

*Sea Ice Thickness*

No one technique currently exists for directly measuring this key parameter accurately and systematically over large-scales from space. To date, our somewhat-fragmented though improving knowledge of seasonal and regional variability is based on compilations of data from a number of different and diverse in situ and non-satellite remote-sensing sources (see the review in Lubin and Massom 2006). In essence, the only hope of gaining systematic information on sea ice thickness distribution is by satellite remote sensing using radar and laser altimeters. Recent measurements of ice or ice plus snow freeboard, using spaceborne radar and laser altimeters respectively, have demonstrated the potential to provide estimates of sea ice thickness in both the Arctic (Laxon et al. 2003) and Antarctic (Zwally et al. 2008). This is based on the hydrostatic relationship between the above and below sea-level proportions of the ice cover, requiring accurate knowledge of snow thickness and ice and snow density. In the Antarctic, however, additional uncertainties are present. One obstacle is the complex relationship between ice elevation and thickness in the Southern Ocean. The presence of significant areas of flooded ice with seawater at the ice surface–snow interface affects the density used to compute ice thickness from snow elevation.

Uncertainties in snow depth and density introduce uncertainties into the sea ice thickness estimates, and better snow depth and density data are needed to improve sea ice thickness calculated from satellite-derived sea ice freeboard and elevation data. Moreover, additional in situ measurements coincident with satellite altimeter and AMSR-E measurements are required to further calibrate/validate the snow depth, freeboard, and thickness results. Validation is difficult with point measurements (particularly in the presence of sea ice drift) but is underway using data from coordinated field experiments with both airborne and ground-based observations, for example, the Australian 2003 ARISE cruise in east Antarctica (Massom et al. 2006), as discussed in the subsequent section.

Of major importance is the upcoming CryoSat-2 mission (launch 2009), which is dedicated to monitoring precise changes in the thickness of sea ice (using new synthetic aperture technology). Once again, validation is essential to quantify and understand errors associated with the conversion of sea ice freeboard to ice thickness. A remaining issue with both spaceborne laser and radar altimeters is that their narrow surface coverage combined with their orbital parameters means that complete coverage of the entire Antarctic sea ice cover takes many weeks at best.

*Snow Depth on Sea Ice*

In response to the need for improved information on the large-scale distribution of snow depth on sea ice, and its variability, an algorithm was developed to derive snow depth from AMSR-E passive microwave brightness temperature ($T_B$) data. Snow depth $h_s$ (in centimeters) is determined from the spectral gradient ratio (GRV) of $T_B$ data from the AMSR-E 18.7 and 36.5 GHz channels, vertical (V) polarization (Comiso et al. 2003), as:

$$h_S = a_1 + a_2\mathrm{GRV(ice)} \quad \text{(Equation 3.15.3)},$$

where $a_1$ (= −2.34) and $a_2$ (= −771) are coefficients derived from the linear regression analysis, and GRV(ice) is the gradient ratio corrected for ice concentration $C$:

$$\mathrm{GRV(ice)} = \frac{\left[T_B(36.5\mathrm{V}) - T_B(18.7\mathrm{V}) - k_1(1-C)\right]}{\left[T_B(36.5\mathrm{V}) + T_B(18.7\mathrm{V}) - k_2(1-C\right]} \quad \text{(Equation 3.15.4)},$$

where $k_1 = \left[T_{BO}(36.5\mathrm{V}) - T_{BO}(18.7\mathrm{V})\right]$ and $k_2 = \left[T_{BO}(36.5\mathrm{V}) + T_{BO}(18.7\mathrm{V})\right]$ and $C$ is derived from the same data set by applying the NT2 algorithm (see Section 7.1). The open-water brightness temperatures are used as algorithm constraints and represent mean values from ice-free regions. It can be seen from Equation 3.15.4 that the performance of the snow depth algorithm depends on accurate knowledge of ice concentration. Moreover, snow thicknesses greater than 0.5 m cannot be detected using the current algorithm, due to the limited penetration depth at the frequencies used, and snow depth retrievals are not carried out over Arctic MY ice, as the microwave signature of the latter strongly resembles that of snow on FY ice (Comiso et al. 2003). Marginal ice zones are also excluded.

These measurements date back to June 2002 and are routinely available on a daily basis at a 12.5 km resolution, mapped polar stereographically for direct comparison with the coincident AMSR-E sea ice concentration and temperature products. Unfortunately, recent validation campaigns suggest that this product underestimates actual snow depth by as much as a factor of two to three (Massom et al. 2006; Worby et al. 2008), due to ice surface roughness effects (Stroeve et al. 2006). Work is underway to improve this accuracy, possibly by incorporating ice surface roughness information acquired on a similar scale by satellite radar scatterometers. While the current AMSR-E snow depth product may be inaccurate in terms of providing absolute values, it does provide considerable insight into large-scale spatial patterns of snow depth variability that correspond to similar patterns in ice type and age.

*Polynya Distribution and Extent*

Polynyas have been measured and monitored a number of ways (see Lubin and Massom 2006 for details). Applying sea ice concentration thresholds to satellite passive microwave data, Massom et al. (1998) detected and monitored twenty-eight polynyas along the East Antarctic coastline over the period 1987–1994, and also determined the factors affecting their recurrence and maintenance. Problems in applying passive microwave data to polynya studies relate to their coarse resolution (12.5–25 km) and coastal pixel contamination effects, given that many polynyas are small relative to pixel size. Other ambiguities stem from the difficulty in deriving accurate estimates of ice concentration in regions of predominantly new/thin ice, raising an important issue concerning the definition of polynya extent.

An alternative method, the Polynya Signature Simulation Method or PSSM, has been applied to detection of subpixel-scale polynyas in SSM/I data using 85 and 37 GHz $T_B$ data, creating an enhanced resolution of ~6.25 km (Markus and Burns 1995). The PSSM algorithm in fact yields the area not only of open water but also that of new ice (up to an estimated thickness of ~0.06 m). If atmospheric effects can be properly accounted for (a problem using 85 GHz data), this likely produces a more realistic estimate of polynya "core" area than the simple thresholding techniques. Arrigo and Van Dijken (2003) developed a variation on this technique, in that they used PSSM-derived daily images of sea ice distribution to construct a circumpolar Antarctic map of the percentage of days that a given pixel remained ice-free over the June to October period from 1997 through 2001. Coastal regions with consistently low sea ice concentrations were then identified as polynyas, based on the qualifying threshold of 50 percent ice-free days in winter. Another approach, by Tamura et al. (2007), combines NOAA AVHRR data with polarization ratio information from the 37 and 85 GHz channels of the SSM/I. Further improvements in both cases are likely using improved-resolution AMSR-E data.

Higher-resolution visible- to thermal-IR data (e.g., from the AVHRR) have been used to better resolve polynya areal extent (e.g., Adolphs and Wendler 1995). Such techniques are based upon the strong contrast in the surface temperature and albedo of open water/thin ice versus thicker snow-covered ice. The major limitation is again cloud cover, and even under apparently cloud-free conditions, high thin clouds and ice fog can obscure polynya regions and bias polynya measurements using these data (Martin et al. 2004). In spite of these limitations, the ready availability of AVHRR and MODIS data makes them attractive to polynya studies, especially when combined with microwave satellite data.

The use of satellite SAR circumvents the limitations of visible-thermal IR methods, with its all-weather, day-night capabilities. Moreover, a high spatial resolution is desirable to determine the area of open water and spatial distribution of sea ice thickness classes present, and the shape of the polynya and the nature of its boundaries. Polynya detection is based here on the backscatter contrast between open water/frazil ice and older ice types (see Lubin and Massom 2006 for details). Unfortunately, this approach is limited to some extent by the impact of wind roughening on the backscatter characteristics of open-water surfaces (see Section 3.15.5). This limitation also applies to using SAR to delineate and monitor lead distributions. However, it can be overcome by combining the SAR data with near-coincident thermal-IR satellite imagery and derived thin sea ice thickness maps (Drucker et al. 2003; Kwok et al. 2007). Considerable improvements are also potentially possible using new polarimetric/multipolarization data from advanced new-generation SARs (see Section 3.15.5). However, a major disadvantage regarding these data is that they are acquired over very narrow swaths.

*Sea Ice (Snow) Physical Temperature*

The ABA sea ice concentration algorithm described above also allows for the calculation of ice physical temperature (Comiso et al. 2003). This gridded product is also available on a daily basis at 25 km resolution from the NSIDC (http://nsidc.org/data/ae_si25.html), dating back to June 2002. Inherent sources of error in the retrieval of sea ice temperature include errors in ice concentration and spatial and temporal variations in the ice emissivity at 6.9 GHz from different ice types and surface characteristics (Cavalieri and Comiso 2004). Broad-scale patterns are observable in comparison with the coincident AMSR-E sea ice concentration and snow depth maps. The accuracy of the AMSR-E ice temperature product is currently unestablished, however, although the product agrees to within 0.5 K with a somewhat limited in situ data set acquired from the East Antarctic pack during the ARISE remote-sensing validation experiment (Massom et al. 2006). Further validation under a broader range of conditions and seasonally is essential before this new standard product can be confidently used.

This product is not to be confused with sea ice/snow surface temperature (IST), which is derived from satellite thermal infrared data, for example, from the AVHRR and the MODIS. Daily MODIS IST products are available from NSIDC at a spatial resolution of 4 km for the period July 2002–present (see http://nsidc.org/data/myd29e1d.html). The accuracy of this IST product is estimated to be 0.3 to 2.1 K by Key et al. (1997) and ±1.5 K by Scambos et al. (2006). The AMSR-E product is assumed to represent the physical temperature of the ice surface (i.e., snow/ice interface) for Antarctic FY ice, whereas the IST product is a measure of the skin-surface temperature of the snow/ice surface. Detailed information on both products, and algorithms used, is available in Lubin and Massom (2006).

### 3.15.7 CASE STUDIES OF SATELLITE DATA USE IN SEA ICE FIELD RESEARCH

In this section, examples are provided of the crucial and central role played by satellite remote sensing in two recent multinational and multidisciplinary fieldwork campaigns in the East Antarctic sea ice zone. These were the Antarctic Remote Ice Sensing Experiment (ARISE) of September–October 2003 (Massom et al. 2006) and the Sea Ice Physics and Ecosystem eXperiment (SIPEX) cruise of September–October 2007 (Worby 2008). Both were conducted from the Australian icebreaker *R/V Aurora Australis* and involved not only detailed in situ measurement of sea ice and snow-cover physical (and biogeochemical) properties at multiple ice-floe stations but also the deployment of drifting ice-buoy arrays to determine ice motion and dynamics/deformation, together with considerable remote-sensing work from helicopters. The latter involved sub-meter resolution digital aerial photography and thermal infrared radiometry on both ARISE and SIPEX, airborne ice-thickness measurement using an electromagnetic induction "bird" on ARISE, and detailed measurement of sea ice/snow-cover surface topography using a scanning laser altimeter on SIPEX. A primary aim of ARISE was to acquire data with which to validate AMSR-E sea ice and snow-cover geophysical products (ice concentration, temperature, and snow depth), with a secondary aim of validating both ice freeboard and derived thickness estimates from the NASA IceSat Geoscience Laser Altimeter System (GLAS) and ice skin-surface temperature retrievals from the AVHRR and MODIS thermal-IR channels. By the same token, a main objective of SIPEX was validation of the IceSat ice freeboard and thickness estimates. Satellite remote sensing also provided key large-scale sea ice information in support of the other components of the multidisciplinary fieldwork.

As in most modern sea ice field research campaigns, satellite data played a critical role in the three important stages, namely initial planning, actual fieldwork operations (including aerial survey), and subsequent data analysis. During initial planning, analysis of time series of archived satellite images from previous years and also the lead-up period prior to the experiments, combined with knowledge of the regional conditions, enabled us to home in on potentially suitable areas to conduct the fieldwork. The satellite data used for this purpose were high-resolution SAR (Radarsat-1 ScanSAR and Envisat ASAR wide swath), coarse-resolution DMSP SSM/I and Aqua AMSR-E ice concentration imagery, and visible-thermal IR imagery from Terra and Aqua MODIS (0.25–1.0 km resolution) and NOAA AVHRR (1.0 km resolution)—all mapped to a common grid. Combining different yet complementary satellite data sets yields substantially more information on ice distribution and conditions (and their variability) than is attainable from a single source. Moreover, the use of time series rather than single "snapshot" images is critical to gaining insight into the dynamic nature of sea ice conditions and the developmental history of the study area. The information so derived was then provided to the ship's captain, to facilitate logistical and operational planning of the overall cruise track and the expected field stations.

During the actual fieldwork operations, a range of complementary and georeferenced satellite images (mapped to a common grid) of the fieldwork region were acquired from the PolarView consortium (http://www.polarview.org) by prior agreement and e-mailed to the ship in near real time during the 2007 experiment, while the ARISE cruise obtained similar near-real-time images from other sources (including AMSR-E images from NASA). These data included high-resolution Envisat ASAR swath imagery and 1 km resolution Global Monitoring Mode (GMM) image composites, MODIS and Envisat MERIS imagery, and sea ice concentration maps derived from DMSP SSM/I data (resolution 25 km) and AMSR-E (6.25 km resolution), the latter acquired from the University of Bremen. They were supplemented by real-time acquisition of SSM/I ice concentration and AVHRR multispectral imagery by the SeaSpace TeraScan system onboard the ship.

During ARISE, we also gained access in near-real time to the AMSR snow-depth product imagery, which enabled us to identify and sample "hot spots" in the imagery. Having access to this suite of satellite products in a timely fashion allowed us to effectively fine-tune and adapt our sampling strategy and procedures "on the fly," depending upon the local and regional ice conditions. This approach also enables flexibility in acquiring satellite data that cover the experimental region, in a manner that is not possible when relying on the requisition of narrow-swath satellite data ahead of time based on best estimates of future ship/experimental location (unless this is fixed in space). This is a particular issue when ordering SAR and/or high-resolution visible-thermal IR imagery ahead of time, because such narrow-swath data may fall outside of the study region which has to adapt to ever-shifting ice conditions. Nevertheless, such data remain important in the final analysis when the exact location of the measurements has been established and ice-drift trajectories (derived from satellites or buoys) enhances the ability to match up sampled ice with available data masks. In addition, the satellite-derived information on regional ice conditions and prevailing ice motion was critical in determining the location of the buoy array deployments and, for ARISE, the 50 × 100 km box (demarcated by the initial deployment of nine buoys) that formed the focal point of the field measurement and helicopter remote-sensing campaign. In both cases, it was crucial that the buoy deployments took place to the south of the Antarctic Divergence within the drifting coastal current (to ensure that the buoys drifted westwards rather than being swept northwards into open ocean), but not too far south (to avoid being crushed in the highly dynamic shear zone between moving pack and stationary fast ice). This was achieved with the aid of the near-real-time SAR, MODIS, and MERIS imagery.

Access to real-time buoy drift location information enabled ship- and helicopter-based sampling to continue within the ARISE buoy array as it drifted westwards and deformed (Figure 3.15.8a), and also regular helicopter remote-sensing runs around the array to measure changes in ice conditions such as floe-size distribution and thin ice fraction. One such transect is shown in Figure 3.15.8b, superimposed on a largely cloud-free portion of a near-coincident MODIS channel 2 image. While this imagery is restricted to 250 m resolution, large floe conglomerations are apparent, as are substantial regions of open water/thin ice. Combination of the helicopter/aircraft and satellite data in this manner greatly enhances interpretation of the former in the context of the wider spatiotemporal domain, while floe size and sea ice type distribution, ice concentration, surface temperature, and ice thickness information derived from the helicopter remote-sensing campaign are compared to similar though larger-scale information in the satellite imagery to place the field observations in the larger wider regional ice type domain. The combined use of SAR and passive microwave ice concentration and SAR imagery proved to be invaluable under cloudy conditions throughout.

Important satellite- and model-derived meteorological data also provided critical information on weather patterns over and around the experimental regions, enabling safe helicopter operations to take place and adaptation of the fieldwork to the prevailing weather conditions. In a GIS environment, meteorological fields such as forecast wind speed and direction vectors can be superimposed on high-resolution SAR imagery and/or ice concentration maps, for example to provide invaluable insight on why the ice is drifting where it is, and what local and regional ice conditions are likely to be in the subsequent few days in response to the forecast meteorological conditions. Moreover, ocean depth contour information overlain on high-resolution SAR imagery combined with information from the latter on iceberg distributions enables determination of the location of grounded iceberg fields, which in turn form anchor points for landfast ice buildup. This satellite-based information was crucial during SIPEX for the successful planning and execution of a number of field stations within the landfast ice zone.

Taken as a whole, this satellite-derived information proved to be of immense importance to not only logistical but also scientific operations, the two being intimately linked. Satellites also played a key role in transmitting positional and environmental data from the drifting buoy array via the global positioning system, which also provided information on helicopter positions during remote-sensing transect runs. This information is crucial to the subsequent analysis of the data sets, and enables transect (and ship track) information to be superimposed on contemporary satellite imagery to aid direct comparison. This approach greatly facilitates interpretation of the spatially limited helicopter data in terms of the wider regional ice conditions. The analysis stage of the fieldwork involves coalescing the various data sets and acquiring additional important satellite data that were not supplied in near real time, that is, from archives. At this stage as during the experiment, additional environmental information acquired during the cruise is critical to the interpretation of the satellite data, which in turn spatiotemporally extend the field observations.

In both experiments, the sea ice sampling strategy was based on a hierarchical approach, whereby detailed though spatially limited in situ measurements were used to validate helicopter remote-sensing data, which were then used to spatially extrapolate the in situ observations (Figure 3.15.9). This information was then applied to validation of the relevant satellite data sets, which in turn enabled spatial and temporal extrapolation of sea ice and snow cover information derived from the combined in situ and helicopter-derived data. This entails one approach to addressing the considerable challenge of stepping up from the surface to space and validating coarse-resolution and large-scale satellite data products with detailed but spatially limited surface measurements. As an example, localized in situ snow depth

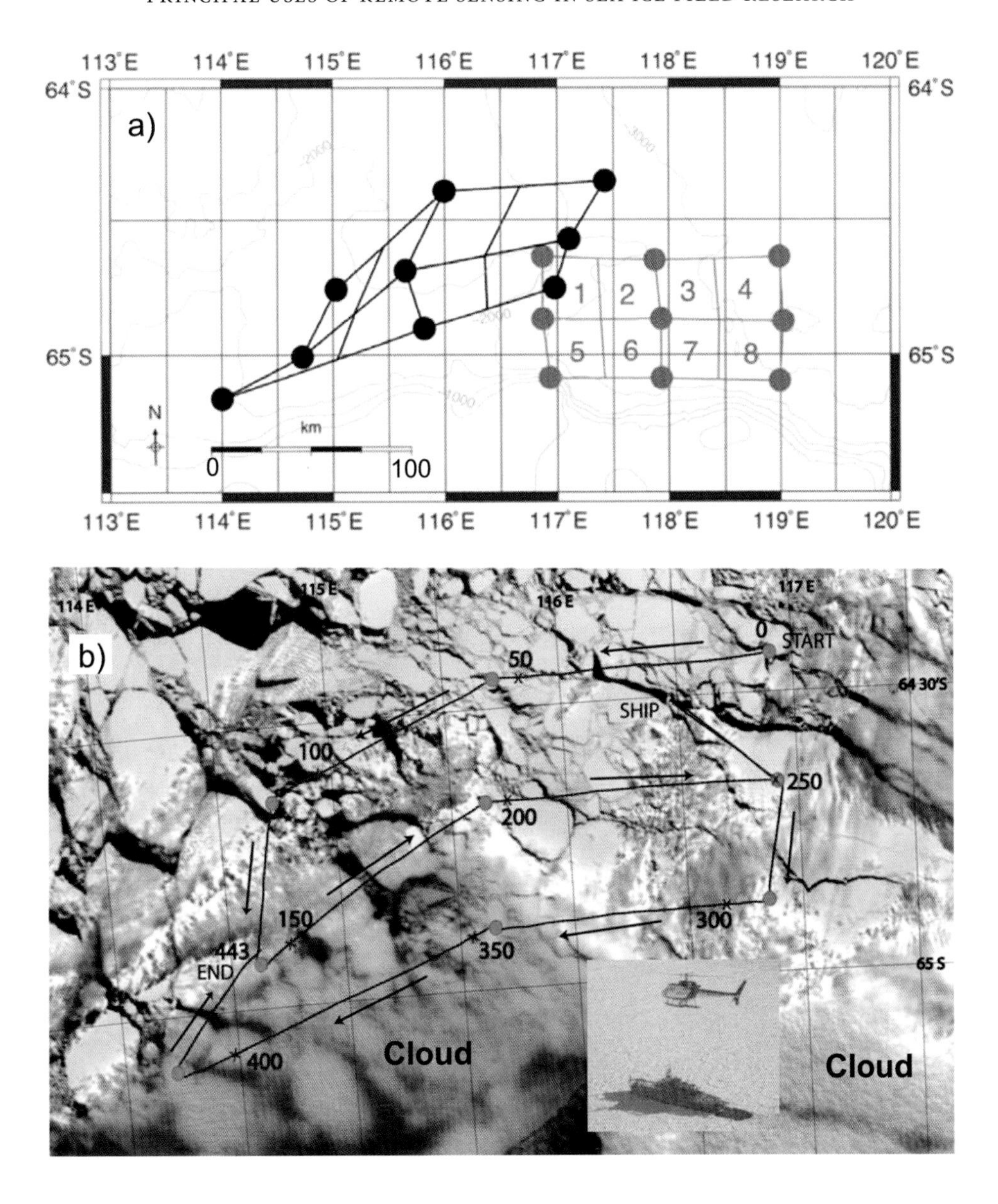


**Figure 3.15.8.** (a) The ARISE study region, east antarctic sea ice zone, showing the original deployment locations (gray circles) of nine drifting buoys deployed on ice floes on 26 September 2003, and their subsequent locations (black circles) on 16 October 2003. The location of the original 100 x 50 km grid was based on analysis of regional ice conditions using satellite data sent to the ship in near-real time. (b) MODIS image over the study region at 0110 GMT on 8 October, 2003. The locations of the nine drifting buoys are shown by green dots, and the helicopter aerial photography flight lines over the buoy array is marked in red, with arrows indicating the flight direction. The location of the ship and the start and end of the aerial photography flight line are also shown. The red numbers correspond to the approximate location of the sequential digital aerial images taken during the flight, marked with a red cross. After Worby et al. (2008). ©American Geophysical Union 2008. Reproduced by permission of American Geophysical Union.

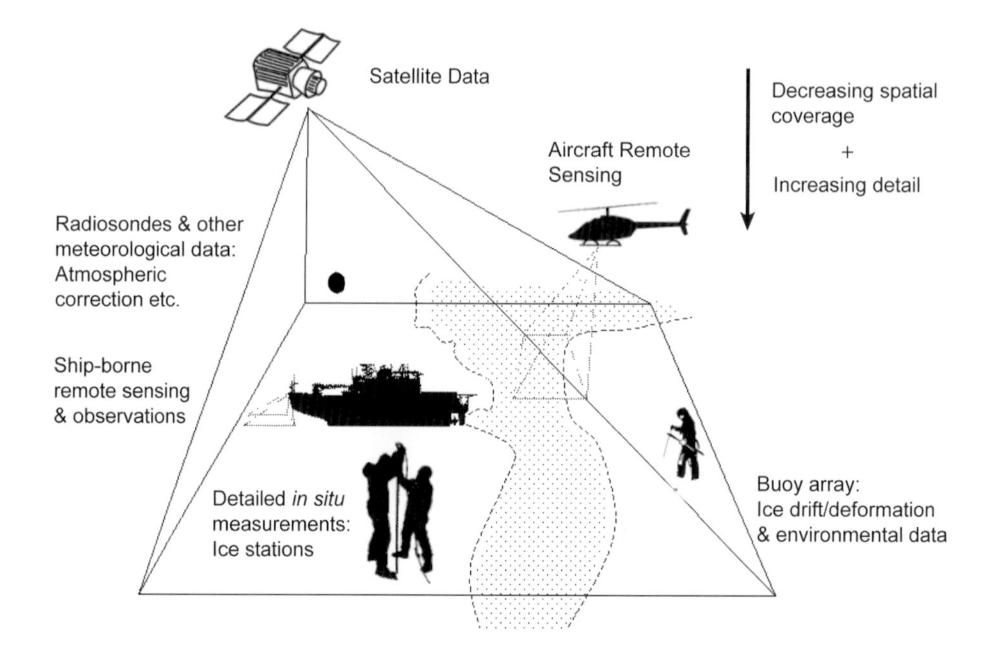


**Figure 3.15.9.** Schematic of the hierarchical approach to sea ice sampling adopted during the ARISE 2003 field experiment. Adapted from original drawing courtesy of Alice Giles.

data have been combined with large-scale sea ice roughness and type information derived from the aerial photography to validate the AMSR-E snow depth product. This study has revealed an apparent discrepancy between actual and AMSR-E derived snow depths, likely due to sea ice surface roughness effects (Massom et al. 2006; Worby et al. 2008).

## 3.15.8 FUTURE

While sea ice system services have benefited greatly from the immense advances that have occurred in satellite remote sensing over the past few decades, several key shortcomings remain. These have been summarized in the recent Integrated Global Observing Strategy (IGOS) Cryosphere Theme Report (Key et al. 2007). An overarching issue is the critical need to maintain key elements of the satellite observing system, without breaks. This is in jeopardy, given the delay of the transition of the U.S. civilian and military operational polar-orbiting satellite system (comprising NOAA and DMSP satellites) to the National Polar-Orbiting Operational Environmental Satellite System (NPOESS). Current issues include technical problems with the primary sensor, namely the Visible Infrared Imaging Radiometer Suite (VIIRS). A strong case can also be put for the launch, as necessary, of dedicated small satellite missions. In contrast to large-scale missions, small satellites have a relatively short development time to enable highly focused investigations using the most modern sensor technical innovations (Baker and Worden 2008).

No all-encompassing single sensor exists for sea ice field research, nor will it be developed; rather, the combination and synthesis of data from different yet complementary sensors remains essential, and underlines the critical importance of maintaining continuity and the continued provision of satellite data in a timely fashion. It is likely that sea ice field research in future will benefit from greater use of routine fusion/integration of satellite data, for example, microwave plus optical/thermal or new/thin ice detection (Kwok et al. 2007), and passive microwave radiometer plus radar scatterometer (particularly useful for improved ice-type classification such as FY versus MY ice) and ice-edge delineation (Tonboe and Haarpaintner 2003). Methods need to address temporal and spatial (resolution and coverage) differences between data types.

Significant trade-offs will remain to govern the appropriate choice of SAR data for a given task. The continued availability of wide-swath SAR (ScanSAR) coverage over Arctic sea ice is a critical factor, and the improved coverage of Antarctic sea ice is a high priority (ideally at repeat intervals of one to three days to enable RGPS-like ice tracking)—with a current lack of receiving capacity seriously limiting the amount of data available. There are, however, significant ambiguities in the single-polarization backscatter signature, making interpretation difficult at times, for example, during periods of melt and windy conditions. Multipolarization and polarimetric SAR shows immense promise in terms of improved ice type and ice-ocean discrimination, although new methodologies need to be developed to take full advantage of their capabilities (backed up by extensive airborne validation). These new operating modes are unlikely to supplant standard ScanSAR imaging, however, due to the swath-width (coverage) limitations (typical swath width ~20–70 km) and data interpretation complexity. Their application is therefore largely limited to localized regions where more precise information is required (Dierking et al. 2004). Combination of these data with coincident satellite high-resolution visible-thermal IR imagery could create a powerful sea ice system services tool.

Effective cloud detection and masking remains a challenging problem for visible-thermal infrared imagery acquired over snow- and ice-covered regions, particularly where the cloud is optically thin (e.g., cirrus, or sub-pixel scale) and at nighttime (Lubin and Massom, 2006). For daylight conditions, the inclusion of a 1.6 μm channel is highly beneficial, given the high reflectance of cloud compared to snow in this spectral region (Scorer 1989). The probability of cloud-free scenes remains lower for narrow-swath visible-thermal IR sensors with their poor temporal resolution (long revisit interval) than wide-swath systems such as MODIS.

Use of medium-coarse-resolution satellite visible-thermal IR imagery in sea ice field research and operations has been hampered in the past by difficulties in georeferencing the imagery in the absence of identifiable and stable reference points and accurate satellite orbit ephemeris information, which is typically the case. As a result, geolocation errors of a few kilometers can occur, for example, in AVHRR imagery. Accurate geolocation becomes a critical factor for studies requiring accurate georeferencing and overlay of different data sets, as is typically the case in a number of key applications. Fortunately, the situation is improving as more sophisticated tracking devices are included in satellite payloads.

Regarding sea ice concentration and extent, current passive microwave (and radar scatterometer) products are too coarse to obtain fine-scale details of the sea ice cover, although the situation has improved with the launch of AMSR(-E). The coarse resolution results in uncertainties due to mixed surface types within a sensor footprint, particularly at the ice edge and near the coast where coastal polynyas dominate. Future acquisitions at a high resolution (optimally 1–3 km) are highly desirable, not least to resolve lead and polynya systems and narrow bands of coastal fast ice as well as better characterize the ice edge. Improved spatial resolutions may be achievable in the future through the developing technology of synthetic aperture passive microwave systems (Ruf et al. 1998). In addition, rigorous evaluation and consolidation of products are needed, together with formal estimates of algorithm uncertainties/errors (Key et al. 2007).

The accurate synoptic-scale measurement of sea ice thickness from space remains a challenge and a high priority. Although promising results have recently emerged from satellite radar and laser altimeter data (as outlined above), these are compromised by a lack of accurate information on snow depth and density and their variability/evolution. This introduces considerable uncertainties in total ice thickness estimates derived from freeboard observations by satellite altimeters, and underlines the need for dedicated validation experiments such as those described in Section 3.15.7. More accurate freeboard measurements are promised by the launch of a technologically advanced radar altimeter on the dedicated CryoSat-2 (launch 2009), but the issues remain to be resolved. Critical to this is improvement in the performance of the AMSR-E snow depth algorithm, which currently underestimates snow thickness largely as a result of ice-roughness effects (Section 3.15.6). Satellite-based snow depth products should be extended to perennial sea ice. Dual-frequency radar altimeter sensors, and combined information from radar and laser altimeter sensors, may offer new and independent estimates to complement passive microwave techniques.

Another deficiency in the current sea ice observing system is the lack of a standard ice motion product that is available in near-real time (particularly in the Southern Ocean). It is likely that improvements in large-scale ice motion maps will come from combining radar scatterometer with improved-resolution passive microwave brightness temperature data, assimilating drifting buoy information where possible. For finer-scale measurements over more localized areas, the potential may exist to accurately measure sea ice motion through interferometric processing of data from closely spaced along-track configurations of SARs, or the innovative split-antenna mode of TerraSAR-X, for example.

Technological changes in the ground sector are about to occur to affect real-time data availability. While routine real-time acquisition of satellite data such as

AVHRR imagery has been possible via access to fairly simple L-/S-band receiving antennae (stations), data download from the more sophisticated and higher data-rate sensors onboard the upcoming (replacement) NPOESS suite of satellites will require upgrading to larger X-band systems (the latter are also required for accessing current MODIS imagery in real time). Another important issue affecting data availability of SAR data in particular is the recent trend towards commercialization, for example, of the follow-on to Radarsat-1 (Radarsat-2) (Staples et al. 2004). A further key factor determining near real-time data provision to ships and field camps is continuation of important systems such as PolarView that provide mapped imagery from various sensor sources. As noted earlier, availability of such data has revolutionized sea ice field research in recent years.

As a final comment, it should be stressed again that the sea ice field research observations themselves are crucial for satellite validation purposes in general (and to provide information on key sea ice parameters that cannot be measured from space). A standardized globally applicable methodology needs to be developed to enable consistent and efficient acquisition of data in forms that are optimal for the calibration and validation of satellite data algorithms, specific to the sea ice geophysical quantities in question. Many such observations have tended not to be well integrated in the past, and considerable effort is being put into ensuring coordination of validation efforts in the present-future time frame. In this sense, developments will be mutually beneficial to both fields, to maintain satellite remote sensing as an extraordinarily powerful and useful tool in sea ice field research for decades to come (assuming that developing challenges are met).

## Acknowledgments

This work was supported by the Australian Government's Cooperative Research Centres Programme through the Antarctic Climate and Ecosystems Cooperative Research Center (ACE CRC), and contributes to Australian Antarctic Science Projects 3024 and 2298. I am indebted to Hajo Eicken (University of Alaska Fairbanks, USA) for his superb help and encouragement in putting this chapter together. Sincere thanks are also due to numerous colleagues in the field of sea ice remote sensing—most notably Joey Comiso (NASA Goddard Space Flight Center, USA), Dan Lubin (Scripps Institute of Oceanography, USA), and Gareth Rees (Scott Polar Research Institute, UK). Thanks to Phil Reid (Australian Bureau of Meteorology and ACE CRC) and Glenn Hyland (AAD and ACE CRC) for their helpful comments on the manuscript, and to my colleagues in the sea ice group at the ACE CRC (notably Ian Allison, Petra Heil, Tony Worby, Kelvin Michael, and Jan Lieser). This chapter is dedicated to my wife, Yuko, and children Adelle and Kaiki.

## REFERENCES


Adolphs, U., and C. Wendler (1995), A pilot study on the interactions between katabatic winds and polynyas at the Adélie Coast, eastern Antarctica, *Ant. Sc., 7*, 307–314.

Andreas, E. L., and S. F. Ackley (1981), On the differences in ablation seasons of the Arctic and Antarctic sea ice, J. *Atmos. Sci., 39*, 440–447.

Arrigo, K. R., and G. L. van Dijken (2003), Phytoplankton dynamics within 37 Antarctic coastal polynyas, *J. Geophys. Res., 108*(C8), 3271, 10.1029/2002JC001739.

Baker, D. N., and S. P. Worden (2008), The large benefits of small-satellite missions, *EOS, Trans. Am. Geophys. Union*, 89(33), 301–302.

Bamber, J., and R. Kwok (2004), Measurement Techniques, in *Mass Balance of the Cryosphere*, edited by J. Bamber and A. Payne, pp. 59–116, Cambridge University Press, Cambridge.

Barber, D. G., A. K. Fung, T. C. Grenfell, S. V. Nghiem, R. G. Onstott, V. Lytle, D. K. Perovich, and A. J. Gow (1998), The role of snow on microwave emission and scattering over first-year sea ice, *IEEE Trans. Geosc. Rem. Sens., 36*(5), 1750–1763.

Brandt, R. E., S. G. Warren, and A. P. Worby, and T. C. Grenfell (2005), Surface albedo of the Antarctic sea ice zone, *J. Clim., 18*, 3606–3622.

Carsey, F. D. (ed.) (1992), Microwave Remote Sensing of Sea Ice, in *Geophysical Monograph, 68*, pp. 462, American Geophysical Union, Washington, DC.

Carver, K. (ed.) (1989), SAR. Synthetic Aperture Radar. *Earth Observing System Instrument Panel Report, Volume IIf*, NASA, Washington, DC.

Cavalieri, D., and J. Comiso (2004), *AMSR-E/Aqua Daily L3 12.5 km Tb, Sea Ice Concentration, and Snow Depth Polar Grids V001, March 2004*. National Snow and Ice Data Center, Boulder, CO. Digital media.

Cavalieri, D. J., T. Markus, J. A. Mastanik, M. Sturm, and E. Lobl (2006), March 2003 EOS Aqua AMSR-E Arctic Sea ice field campaign, *IEEE Trans. Geosc. Rem. Sens., 44*(11), 3003–3008.

Comiso, J. C. (1985), Remote sensing of sea ice using multispectral microwave satellite data, in *Advances in Remote Sensing Retrieval Methods*, edited by A. Deepak, H. E. Fleming, and M. T. Chahine, pp. 349–369, A. Deepak Publishing, Hampton, VA.

Comiso, J. C. (2003), Large-scale characteristics and variability of the global sea ice cover, in *Sea Ice: An Introduction to Its Physics, Chemistry, Biology, and Geology*, edited by D. Thomas and G. Dieckmann, pp. 112–142, Blackwell Science, Oxford.

Comiso, J. C., and R. Kwok (1996), Surface and radiative characteristics of the summer Arctic sea ice cover from multi-sensor satellite observations, *J. Geophys. Res., 101*(C12), 28,397–28,416.

Comiso, J. C., and K. Steffen (2001), Studies of Antarctic sea ice concentrations from satellite observations and their applications, *J. Geophys. Res., 106*(C12), 31,361–31,385.

Comiso, J. C., T. C. Grenfell, M. Lange, A. W. Lohanick, R. K. Moore, and P. Wadhams (1992), Microwave Remote Sensing of the Southern Ocean Ice Cover, in *Microwave Remote Sensing of Sea Ice*, edited by F. D. Carsey, pp. 243–259, Geophysical Monograph 28, American Geophysical Union, Washington, DC.

Comiso, J. C., D. J. Cavalieri, and T. Markus (2003), Sea ice concentration, ice temperature, and snow depth, using AMSR-E data, *IEEE Trans. Rem. Sens., 41*(2), 243–252.

Cracknell, A. P., and L. W. B. Hayes (1991), *Introduction to Remote Sensing*, CRC Press, Boca Raton, FL.

Dierking, W., H. Skriver, and P. Gudmandsen (2004), On the improvement of sea ice classification by means of radar polarimetry, in *Remote Sensing in Transition*, edited by R. Goossens, pp. 203–209, Millpress, Rotterdam.

Dokken, S. T., B. Hakansson, and J. Askne (2000), Intercomparison of Arctic sea ice concentration using Radarsat, ERS, SSM/I and in-situ data, *Can. J. Rem. Sens., 26*(6), 521–536.

Dozier, J. (1989), Remote sensing of snow in visible and near-infrared wavelengths, in *Theory and Applications of Optical Remote Sensing*, edited by G. Asrar, pp. 527–547, John Wiley and Sons, New York.

Dozier, J., and S. G. Warren (1982), Effect of viewing angle on the infrared brightness temperature of snow, *Water Resources Research, 18*(5), 1424–1434.

Drinkwater, M. R. (1995), Airborne and satellite SAR investigations of sea ice surface characteristics, in *Oceanographic Applications of Remote Sensing*, edited by M. Ikeda and F. W. Dobson, pp. 339–357, CRC Press, Boca Raton, FL.

Drinkwater, M. R. (1998), Satellite microwave radar observations of Antarctic sea ice, in *Analysis of SAR Data of the Polar Oceans*, edited by C. Tsatsoulis and R. Kwok, pp. 145–187, Springer-Verlag, Berlin.

Drinkwater, M. R., and X. Liu (2000), Seasonal to interannual variability in Antarctic sea ice surface melt, *IEEE Trans. Geosc. Rem. Sens., 38*(4), 1827–1842.

Drinkwater, M. R., R. Kwok, D. P. Winebrenner, and E. Rignot (1991), Multifrequency polarimetric synthetic aperture radar observations of sea ice, *J. Geophys. Res., 96*(C11), 20,679–20,698.

Drinkwater, M. R., R. Kwok, J. A. Maslanik, C. W. Fowler, and C. A. Geiger (1999), Quantifying surface fluxes in the ice-covered polar oceans using satellite microwave remote sensing data, in Proc. OCEANOBS '99, *An International Conference on the Ocean Observing System for Climate, Volume 1*, San Raphael, France, 18–22 October 1999, Centre National D'Etudes Spatiales, Toulouse, France.

Drucker, R., S. Martin, and R. Moritz (2003), Observations of ice thickness and frazil ice in the St. Lawrence Island polynya from satellite imagery, upward-looking sonar, and salinity/temperature moorings, *J. Geophys. Res., 108*(C5), 3149, doi:10.1029/2001JC001213.

Eicken, H. (2003), From the Microscopic to the Macroscopic to the Regional Scale: Growth, Microstructure, and Properties of Sea Ice, in *Sea Ice: An Introduction to Its Physics, Biology, Chemistry, and Geology*, edited by D. N. Thomas and G. S. Dieckmann, pp. 22–81, Blackwell Science, Oxford.

Eicken, H., T. C. Grenfell, D. K. Perovich, J. A. Richter-Menge, and K. Frey (2004), Hydraulic controls of summer Arctic pack ice albedo, *J. Geophys. Res., 109*, C08007, doi:10.1029/2003JC001989.

Emery, W. J., J. Brown, and Z. P. Nowak (1989), AVHRR image navigation: Summary and review, *Photogramm. Eng. Rem. Sens., 55*(8), 1175–1183.

Enomoto, H., F. Nishio, H. Warashina, and S. Ushio (2002), Satellite observation of melting and break-up of fast ice in Lützow-Holm Bay, East Antarctica, *Polar Met. Glac., 16*, 1–14.

Eppler, D. T., L. D. Farmer, A. W. Lohanick, M. A. Anderson, D. Cavalieri, J. Comiso, P. Gloersen, C. Garrity, T. C. Grenfell, M. Hallikainen, J. A. Maslanik, C. Mätzler, R. A. Melloh, I. Rubinstein, and C. T. Swift (1992), Passive microwave signatures of sea ice, in *Microwave Remote Sensing of Sea Ice*, edited by F. D. Carsey, pp. 47–71, Geophysical Monograph, 68, American Geophysical Union, Washington, DC.

Fraser, A. D., R. A. Massom and K. J. Michael (in press), A method for compositing MODIS satellite images to remove cloud cover, *IEEE Trans. Geosc. Rem. Sens.*

Giles, A. B., R. A. Massom, and V. I. Lytle (2008), Fast-ice distribution in East Antarctica during 1997 and 1999 determined using Radarsat data, *J. Geophys. Res., 113*, C02S14, doi:10.1029/2007JC004139.

Gogineni, S. P., R. K. Moore, T. C. Grenfell, D. G. Barber, S. Digby, and M. Drinkwater (1992), The effects of freeze-up and melt processes on microwave signatures, in *Microwave Remote Sensing of Sea Ice*, edited by F. D. Carsey, pp. 329–341, Geophysical Monograph, 68, American Geophysical Union, Washington, DC.

Golden, K. M., S. F. Ackley, and V. I. Lytle (1998), The percolation phase transition in sea ice, *Science, 282*, 2238–2241.

Granskog, M., H. Kaartokallio, H. Kuosa, D.N. Thomas and J. Vainio (2006), Sea ice in the Baltic Sea—A Review, *Estuarine Coastal Shelf Sc., 70*, 145–150.

Grenfell, T. C., D. L. Cavalieri, J. C. Comiso, M. R. Drinkwater, R. G. Onstott, I. Rubinstein, K. Steffen, and D. P. Winebrenner (1992), Considerations for microwave remote sensing of thin sea ice, in *Microwave Remote Sensing of Sea Ice*, edited by F. D. Carsey, pp. 291–301, Geophysical Monograph, 68, American Geophysical Union, Washington, DC.

Grenfell, T. C., B. Light, and M. Sturm (2002), Spatial distribution and radiative effects of soot in the snow and sea ice during the SHEBA experiment, *J. Geophys. Res., 107*(C10), 8032, doi 10.1029/2002JC000414.

Haas, C. (2001), The seasonal cycle of ERS scatterometer signatures over perennial Antarctic sea ice and associated surface ice properties and processes, *Ann. Glac., 33*, 69–73.

Haas, C. (2003), Dynamics versus Thermodynamics: The Sea Ice Thickness Distribution, in *Sea Ice: An Introduction to Its Physics, Biology, Chemistry and Geology*, edited by D. N. Thomas and G. S. Dieckmann, pp. 82–111, Blackwell Science, Oxford.

Haas, C., Q. Liu, and T. Martin (1999), Retrieval of Antarctic sea ice pressure ridge frequencies from ERS SAR imagery by means of in situ laser profiling and usage of a neural network, *Int. J. Rem. Sens., 20*(15–16), 3111–3123.

Haas, C., D. N. Thomas, and J. Bareis (2001), Surface properties and processes of perennial Antarctic sea ice in summer, *J. Glac., 47*, 623–625.

Hall, D. K., G. A. Riggs, and V. V. Salomonson (2006), MODIS Snow and Sea Ice Products, in *Earth Science Satellite Remote Sensing, Volume I: Science and Instruments*, edited by J. J. Qu, W. Gao, M. Kafatos, R. E. Murphy, and V. V. Salomonson, pp. 154–181, Springer, New York.

Hallikainen, M., and D. P. Winebrenner (1992), The physical basis for sea ice remote sensing, in *Microwave Remote Sensing of Sea Ice*, edited by F. D. Carsey, pp. 29–46, Geophysical Monograph, 68, American Geophysical Union, Washington, DC.

Haykin, S., E. O. Lewis, R. K. Raney, and J. R. Rossiter (1994), *Remote Sensing of Sea Ice and Icebergs*, John Wiley and Sons, New York.

Heil, P., I. Allison, and V. I. Lytle (2001a), Effect of high-frequency deformation on the sea ice thickness, in *IUTAM Symposium on Scaling Laws in Ice Mechanics and Ice Dynamics*, edited by J. P. Dempsey and H. H. Shen, pp. 417–426, Solid Mechanics and Its Applications 94, Kluwer Academic Publishers.

Heil, P., C. W. Fowler, J. A. Maslanik, W. J. Emery, and I. Allison (2001b), A comparison of East Antarctic sea ice motion derived using drifting buoys and remote sensing, *Ann. Glac., 33*, 139–144.

Holt, B., and S. A. Digby (1985), Processes and imagery of first-year fast sea ice during the melt season, *J. Geophys. Res., 90*(C3), 5045–5062.

Jackson, C. R., and J. R. Apel (eds.) (2004), Synthetic Aperture Radar Marine User's Manual, U.S. National Oceanic and Atmospheric Administration.

Jin, Z., and J. J. Simpson (2000), Bidirectional anisotropic reflectance of snow and sea ice in AVHRR channel 1 and 2 spectral regions. Part II. Correction applied to imagery of snow on sea ice, *IEEE Trans. Geosc. Rem. Sens., 38*, 999–1015.

Kaufman, Y. J. (1989), The atmospheric effect on remote sensing and its correction, in *Theory and Applications of Optical Remote Sensing*, edited by G. Asrar, pp. 336–429, Wiley Series in Remote Sensing, John Wiley and Sons, New York.

Key, J., J. Collins, C. Fowler, and R. Stone (1997), High-latitude surface temperature estimates from thermal satellite data, *Rem. Sens. Envir., 61*, 302–309.

Key, J., M. R. Drinkwater, and J. Ukita (2007), *Integrated Global Observing System (IGOS) Cryosphere Theme Report*, World Climate Research Programme, Geneva, Switzerland.

Kramer, H. J. (2002), *Observation of the Earth and Its Environment: Survey of Missions and Sensors*, 4th ed., Springer-Verlag, Berlin and Heidelberg.

Kwok, R. (1998), The Radarsat geophysical processor system, in *Analysis of SAR data of the Polar Oceans: Recent Advances*, edited by C. Tsatsoulis and R. Kwok, pp. 235–258, Springer-Verlag, Berlin and Heidelberg.

Kwok, R. (2002), Sea ice concentration estimates from satellite passive microwave radiometry and openings from SAR ice motion, *Geophys. Res. Lett., 29*(9), 1311, doi:10.1029/202GL014787.

Kwok, R., and G. F. Cunningham (2000), *Radarsat Geophysical Processor System Data User's Handbook (Version 1.0)*, JPL D-19149, NASA Jet Propulsion Lab., Pasadena, CA.

Kwok, R., G. F. Cunningham, and S. Yueh (1999), Area balance of the Arctic Ocean perennial ice zone: October 1996–April 1997, *J. Geophys. Res., 104*(C11), 25,747–25,759.

Kwok, R., G. F. Cunningham, and D. Nguyen (2000), *Alaska SAR Facility Radarsat Geophysical Processor System: Product Specification (Version 2.0)*, JPL D-13448, NASA Jet Propulsion Lab., Pasadena, CA.

Kwok, R., J. C. Comiso, S. Martin, and R. Drucker (2007), Ross Sea polynyas: Response of ice concentration retrievals to large areas of thin ice, *J. Geophys. Res., 112*, C12012, doi:10.1029/2006JC003967.

Lange, M. A. (2002), Sea ice contamination: A Review, in *Ice in the Environment, Proceedings of the 16th International Symposium on Ice*, edited by V. Squire and P. Langhorne, Dunedin, 2–6 December 2002, vol. 3, pp. 152–162, IAHR and University of Otago, Dunedin, New Zealand.

Laxon, S., N. Peacock, and D. Smith (2003), High interannual variability of sea ice thickness in the Arctic region, *Nature, 425*, 947–950, doi10.1038/nature02050.

Lewis, E. O., C. E. Livingstone, C. Garrity, and J. R. Rossiter (1994), Properties of Snow and Ice, in *Remote Sensing of Sea Ice and Icebergs*, edited by S. Haykin, E. O. Lewis, R. K. Raney, and J. R. Rossiter, pp. 21–96, John Wiley and Sons, New York.

Light, B., H. Eicken, G. A. Maykut, and T. C. Grenfell (1998), The effect of included particulates on the optical properties of sea ice, *J. Geophys. Res., 103*(C12), 27,739–27,752.

Long, D. G. (2003), Reconstruction and resolution enhancement techniques for microwave sensors, in *Frontiers of Remote Sensing Information Processing*, edited by C. H. Chen, pp. 255–281, World Scientific Publishing Company, New Jersey.

Long, D. G., M. R. Drinkwater, B. Holt, S. Saatchi, and C. Bertoia (2001), Global ice and land climate studies using scatterometer image data, *Eos, Trans. Am. Geophys. Union, 82*(43), 503, 23 October 2001. Includes Eos Electronic Supplement at http://www.agu.org/eos_elec/010126e.html.

Lubin, D., and R. A. Massom (2006), *Polar Remote Sensing*, Volume 1: Atmosphere and Polar Oceans, Praxis/Springer, Chichester and Berlin.

Macdonald, R. W., E. C. Carmack, M. F. A., K. K. Falkner, and J. H. Swift (1999), Connections among ice, runoff and atmospheric forcing in the Beaufort Gyre, *Geophys. Res. Lett., 26*, 2223–2226.

Mahoney, A., H. Eicken, A. G. Gaylord, and L. Shapiro (2007a), Alaska landfast sea ice: Links with bathymetry and atmospheric circulation, *J. Geophys. Res., 112*, C02001, doi:10.1029/2006JC003559.

Mahoney, A., H. Eicken, and L. S. Shapiro (2007b), How fast is landfast sea ice? A study of the attachment and detachment of sea ice near Barrow, Alaska, *Cold Reg. Sc. Technol., 47*(3), 233–255, doi:10.1016/j.coldregions.2006.09.005.

Markus, T. and B. A. Burns (1995), Method to estimate sub-pixel scale coastal polynyas with satellite passive microwave data, *J. Geophys. Res., 100*(C3), 4473–4487.

Markus, T., and S. T. Dokken (2002), Evaluation of late summer passive microwave Arctic sea ice retrievals, *IEEE Trans. Geosc. Rem. Sens., 40*(2), 348–356.

Martin, S., R. Drucker, R. Kwok, and B. Holt (2004), Estimation of thin ice thickness and heat flux for the Chukchi Sea Alaskan coast polynya from Special Sensor Microwave/Imager data, 1990–2001, *J. Geophys. Res., 109*, C10012, doi:10.1029/2004JC002428.

Martinson, D. G., and R. A. lannuzzi (1998), Antarctic ocean ice interaction: Implications from ocean bulk property distributions in the Weddell Gyre, in *Antarctic Sea Ice Physical Processes, Interactions and Variability*, edited by M. Jeffries, pp. 243–271, Antarctic Research Series 74, American Geophysical Union, Washington, DC.

Maslanik, J. A., C. Fowler, J. Stroeve, S. Drobot, H. J. Zwally, D. Yi, and W. Emery (2007), A younger, thinner Arctic ice cover: Increased potential for rapid, extensive sea ice loss, *Geophys. Res. Lett., 34* (L24501), doi:10.1029/2007GL032043.

Massom, R. A., M. R. Drinkwater, and C. Haas (1997), Winter snow cover on sea ice in the Weddell Sea, *J. Geophys. Res., 102*(C1), 1101–1117.

Massom, R. A., P. T. Harris, K. J. Michael, and M. J. Potter (1998), The distribution and formative processes of latent heat polynyas in East Antarctica, *Ann. Glac., 27*, 420–426.

Massom, R. A., J. C. Comiso, A. P. Worby, V. I. Lytle, and L. Stock (1999), Regional classes of sea ice cover in the East Antarctic pack from satellite and in situ data during the winter time period, *Rem. Sens. Envir., 68*(C1), 61–76.

Massom, R. A., H. Eicken, C. Haas, M. O. Jeffries, M. R. Drinkwater, M. Sturm, A. P. Worby, X. Wu, V. I. Lytle, S. Ushio, K. Morris, P. A. Reid, S. Warren, and I. Allison (2001), Snow on Antarctic sea ice, *Rev. Geophys., 39*(3), 413–445.

Massom, R. A., A. P. Worby, V. I. Lytle, T. Markus, I. Allison, T. Scambos, H. Enomoto, K. Tateyama, T. Haran, J. C. Comiso, A. Pfaffling, T. Tamura, A. Muto, P. Kanagaratnam, and B. Giles (2006), ARISE (Antarctic Remote Ice Sensing Experiment) in the East 2003: Validation of satellite derived sea ice data products, *Ann. Glac., 44*, 288–296.

Massom, R. A., K. Hill, C. Barbraud, N. Adams, A. Ancel, L. Emmerson and M. Pook (2009), Fast ice distribution in Adélie Land, East Antarctica: Interannual variability and implications for Emperor penguins (*Aptenodytes forsteri*), *Mar. Ecol. Res. Prog., 374*, 243–257.

Nghiem, S. V., R. Kwok, S. H. Yueh, A. J. Gow, D. K. Perovich, C.-C. Hsu, K.-H. Ding, J. A. Kong, and T. C. Grenfell (1998), Diurnal thermal cycling effects on microwave signatures of thin sea ice, *IEEE Trans. Geosc. Rem. Sens., 36*(1), 111–124.

Nolin, A. W. and A. Frei (2001), Remote sensing of snow and snow albedo characterization for climate simulations, in *Remote Sensing and Climate Simulations: Synergies and Limitations, Advances in Global Change Research*, edited by M. Beniston and M. M. Verstraete, pp. 159–180, Kluwer Academic Publishers, Dordrecht and Boston.

Onstott, R. G., and R. A. Shuchman (2004), SAR measurements of sea ice, in *Synthetic Aperture Radar Marine Users Manual*, edited by C. R. Jackson and J. R. Apel, pp. 81–115, NOAA, Washington, DC.

Perovich, D. K. (2001), UV radiation and optical properties of sea ice and snow, in: *UV Radiation and Arctic Ecosystems*, edited by D. Hessen, pp. 73–89, Springer-Verlag, Berlin and Heidelberg.

Rees, W. G. (2001), *Physical Principles of Remote Sensing*, Cambridge University Press, Cambridge.

Rees, W. G. (2005), *Remote Sensing of Snow and Ice*, CRC Press, Boca Raton, FL.

Ruf, C. S., C. T. Swift, A. B. Tanner, and D. M. LeVine (1998), Interferometeric synthetic aperture microwave radiometry for the remote sensing of the Earth, *IEEE Trans. Geosc. Rem. Sens., 26*, 597–611.

Scambos, T., T. Haran, and R. A. Massom (2006), Validation of AVHRR and MODIS ice surface temperature products using in situ radiometers, *Ann. Glac., 44*, 345–351.

Scorer, R. S. (1989), Cloud reflectance variations in channel 3, *Int. J. Rem. Sens.*, 10, 675–686.

Shuchman, R. A., R. G. Onstott, O. M. Johannessen, S. Sandven, and J. A. Johannessen (2004), Processes at the ice edge—the Arctic, in *Synthetic Aperture Radar Marine Users Manual*, edited by C. R. Jackson and J. R. Apel, pp. 373–395, NOAA, Washington, DC.

Simmonds, I., K. Keay, and E.-P. Lim (2003), Synoptic activity in the seas around Antarctica, *Monthly Weather Review, 131*, 272–288.

Soh, L.-K., C. Tsatsoulis, and B. Holt (1998), Identifying ice floes and computing ice floe distributions in SAR images, in *Analysis of SAR Data of the Polar Oceans*, edited by C. Tsatsoulis and R. Kwok, pp. 9–34, Springer-Verlag, Berlin.

Spreen, G., L. Kaleschke, and G. Heygster (2008), Sea ice remote sensing using AMSR-E 89 GHz channels, *J. Geophys. Res.*, 113, 002803, doi:10.1029/2005JC003384.

Staples, G., J. Hornsby, W. Branson, K. O'Neill, and P. Rolland (2004), Turning the scientifically possible into the operationally practical: Radarsat-2 commercialization plan, *Can. J. Rem. Sens.*, 30(3), 408–414.

Steffen, K., and J. Heinrichs (1994), Feasibility of sea ice typing with synthetic aperture radar (SAR): Merging of Landsat Thematic Mapper and ERS-1 SAR imagery, *J. Geophys. Res., 99*(C11), 22,413–22,424.

Stierle, A. P., and H. Eicken (2002), Sedimentary inclusions in Alaskan coastal sea ice: Small-scale distribution, interannual variability and entrainment requirements, *Arct. Antarct. Alpine Res.*, 34(4), 103–114.

Stone, R., and J. Key (1993), The detectability of winter sea ice leads in thermal satellite data under varying atmospheric conditions, *J. Geophys. Res.*, 98(C7), 12,469–12,482.

Stroeve, J. C., Markus, T., Maslanik, J.A., D. J. Cavalieri, A. J. Gasiewski, J. F. Heinrichs, J. Holmgren, D. K. Perovich, and M. Sturm (2006), Impact of surface roughness on AMSR-E sea ice products, *IEEE Trans. Geosc. Rem. Sens.*, 44(11), 3103–3117.

Sturm, M., and R. A. Massom (in press), Snow on Sea Ice, in *Sea Ice*, edited by D. Thomas and G. Dieckmann, Blackwell Science, Oxford.

Svendsen, E., K. Kloster, B. Farrelly, O. M. Johannessen, J. A. Johannessen, W. J. Campbell, P. Gloersen, D. Cavalieri, and C. Mätzler (1983), Norwegian Remote Sensing Experiment: Evaluation of the Nimbus 7 Scanning Multichannel Microwave Radiometer for sea ice research, *J. Geophys. Res., 88*(C5), 2781–2991.

Tamura, T., K. I. Ohshima, T. Markus, D. J. Cavalieri, S. Nihashi, and N. Hirasawa (2007), Estimation of thin ice thickness and detection of fast ice from SSM/I data in the Antarctic Ocean, *J. Atmos. Oc. Technol.*, 24(10), 1757–1772.

Tonboe, R., and J. Haarpaintner (2003), Implementation of QuikScat SeaWinds data in the EUMETSAT Ocean and Sea Ice Ice Product, *Danish Meteorological Institute Technical Report 03-13*, Copenhagen, Denmark.

Tsatsoulis, C., and R. Kwok (eds.) (1998), *Analysis of SAR Data of the Polar Oceans: Recent Advances*. Springer-Verlag, Berlin and Heidelberg.

Tucker, W. B., D. K. Perovich, A. J. Gow, W. F. Weeks, and M. R. Drinkwater (1992), Physical properties of sea ice relevant to remote sensing, in *Microwave Remote Sensing of Sea Ice*, edited by F. D. Carsey, pp. 9–28, Geophysical Monograph, 68, American Geophysical Union, Washington, DC.

Ulaby, F. T., R. K. Moore, and A. K. Fung (1982), *Microwave Remote Sensing: Active and Passive, Volume II: Radar Remote Sensing and Surface Scattering and Emission Theory*, Addison-Wesley Publishing Company, Reading, MA.

Van der Sanden, J. J. (2004), Anticipated applications of Radarsat-2 data, *Can. J. Rem. Sens., 30*(3), 369–379.

Yu, Y., and D. A. Rothrock (1996), Thin ice thickness from satellite thermal imagery, *J. Geophys. Res.*, 101(C10), 25,753–25,766.

Walker, N. P., K. C. Partington, M. L. Van Woertz, and T. L. T. Street (2006), Arctic sea ice type and concentration mapping using passive and active microwave sensors, *IEEE Trans. Geosc. Rem. Sens., 44*(12), 3574–3584.

Warren, S. G. (1982), Optical properties of snow, *Rev. Geophys. Space Phys., 20*(1), 67–89.

Williams, R. N., K. J. Michael, S. Pendlebury, and P. Crowther (2002), An automated image analysis system for determining sea ice concentration and cloud cover from AVHRR images of the Antarctic, *Int. J. Rem. Sens., 23*(4), 611–625.

Winebrenner, D. P., D. G. Long, and B. Holt (1998), Mapping the Progression of Melt Onset and Freeze-up on Arctic Sea Ice Using SAR and Scatterometry, in *Analysis of SAR Data of the Polar Oceans: Recent Advances*, edited by C. Tsatsoulis and R. Kwok, pp. 129–144, Springer-Verlag, Berlin and Heidelberg.

Worby, A. P. (2008), Sea Ice Physics and Ecosystem eXperiment, *Austr. Ant. Mag., 14*, 14–15.

Worby, A. P., T. Markus, A. D. Steer, V. I. Lytle, and R. A. Massom (2008), Evaluation of AMSR-E snow depth product over East Antarctic sea ice using in situ measurements and aerial photography, *J. Geophys. Res., 113*, C05S94, doi:10.1029/2007JC004181.

Zwally, H. J., D. Yi, R. Kwok, and Y. Zhao (2008), ICESat measurements of sea ice freeboard and estimates of sea ice thickness in the Weddell Sea, *J. Geophys. Res., 113*, C02S15, doi:10.1029/2007JC004284.

## Chapter 3.16

# The Use of Models in the Design and Interpretation of Field Measurements

*Jennifer K. Hutchings, Chris Petrich, Ron Lindsay, Andrew Roberts, and Jinlun Zhang*

### 3.16.1 INTRODUCTION

The history of sea ice modeling is firmly based in field observations. Observations of sea ice drift during the Fram Expedition demonstrated that ice is transported at typically 2 percent of wind speed and 15 degrees to the wind direction (Nansen 1902; Sverdrup 1928). These observations led to Ekman's theory of wind-driven transport at the sea surface. This is perhaps one of the first models of sea ice drift to be developed.

Arctic ice camps during the mid-twentieth century contributed greatly to the development of the sea ice models we use today. Notable contributions are from the Arctic Research Laboratory Ice Stations (ARLIS) in the 1960s and Arctic Ice Dynamics Joint Experiment (AIDJEX) in the 1970s. Investigations of the temperature profile through sea ice at ARLIS led to the formulation of the Maykut and Untersteiner (1971) model of thermodynamic ice growth and melt. A simplified version of this model (Semtner 1976), with modifications (Parkinson and Washington 1979; Ebert and Curry 1993), became the basis of thermodynamic sea ice modeling at large scales (regional, arctic, antarctic, and global) until the advent of the full enthalpy-conserving model by Bitz and Lipscomb (1999). During AIDJEX it was found that long-term mean ice motion, in the summer when ice concentration is low, results from the balance of four forces on the ice (Hunkins 1975; McPhee 1979). This is called *free drift*, when the significant forces on the ice are wind stress, ocean stress, the Coriolis force, and acceleration down the sea surface tilt. The residual term in the measured force balance is found to be comparable to the Coriolis force, and an order of magnitude smaller than the wind stress (Hunkins 1975). This residual is attributed to interactions between ice floes. In areas of ice convergence, resistance to compression becomes important, and ice can no longer be considered to be in free drift. Although the idea of using a rheology to describe this residual force was not new (see, e.g., Reed and Campbell 1960; Nikiforov et al. 1967), aerial observations that ice floes behave in much the same way as particles in a granular fluid (Marco and Thomas 1977) led to the concept that on large (>100 km) scales internal forces within the ice pack could be modeled as a granular plastic. Hibler (1977) demonstrated that on appropriate temporal and spatial scales, this plastic property could be approximated as a viscous-plastic, leading to the Hibler (1979) model that has underpinned most subsequent basin-scale model developments.

This condensed and select history of sea ice modeling illustrates that the theory describing sea ice formation, decay, and motion is founded in field observations. Some of the models used are empirical, such as rheologies describing the plastic deformation of sea ice. Section 3.16.2 of this chapter provides a case study of a model development (Petrich 2005), demonstrating the close relationship between theoretical and field investigations. Models have been developed for many phenomena on many scales. Our understanding of specific phenomena in the system comes from detailed investigation in the field and with models. Not only are models developed from consideration of observational data, they also aid in the interpretation of that data.

Models allow us to consider processes on scales much larger than field investigations can resolve. The development of generalized models of sea ice thermodynamics and dynamics has led to regional and global sea ice models with which the role of sea ice in the climate and ecological system has been investigated. Such models necessarily require small-scale processes to be parameterized. Model-observation investigations, such as the one described in Section 3.16.2, are invaluable in developing simplified and parameterized model components. There is also a need to validate that large-scale models correctly reproduce large-scale phenomena. In Section 3.16.3 case studies on model validation efforts are given, to demonstrate the scope of sea ice modelers' observational needs.

Finally we come full circle, as a good large-scale sea ice model can be used to aid in the design of sea monitoring campaigns. Section 3.16.3 describes a case study by Lindsay and Zhang (2006), which considers cost-effective strategies for monitoring the long-term variability of sea ice thickness.

## 3.16.2 MODELING SEA ICE

Seawater freezes when a net energy loss from the ocean to the atmosphere cools the sea surface below about −1.8°C. The resultant ice growth forms a thermal barrier between the ocean and the atmosphere and increases the solar reflectivity above that of a bare ocean surface. Snow accumulating on sea ice amplifies both of these effects. Wind and currents advect sea ice along the sea surface, causing convergence of floes and resulting in ridge and raft formation, or else causes divergence that creates open water regions within the pack. Ridge, raft, and open water creation is typically heterogeneous and anisotropic, and produces a highly fractured sea ice cover. Meantime, an uneven and ever-changing distribution of surface snowdrifts and internal brine drainage channels can cause meter-scale, thermodynamically driven heterogeneity, evident, for example, in distributions of melt ponds during thaw (Perovich et al. 2003).

The superposition of dynamic and thermodynamic influences generates a set of floe sizes, ice thicknesses, and lead fractions particular to different sea ice zones on Earth's surface. One could model sea ice as a granular material at a spatial resolution of, say, one meter in order to represent sea ice conditions on basin scales and account for local heterogeneities. However, this is a task beyond the reach of the most powerful supercomputers available today. Consequently, changes in sea ice cover at distances extending to basin scales are modeled as a continuum following Nikiforov (1957). Ice velocity and mass evolution are estimated as spatial averages within grid cells with resolutions typically in the range of kilometers to hundreds of kilometers. This continuum approach requires sub-grid-scale assumptions about the distribution of ice thickness and velocity, with evolution of sea ice mass, momentum, and enthalpy approximated on a model's mesh with a standard set of differential equations. These equations include parameterizations of a variety of processes—such as the stress release in fracturing of the ice (yielding if one assumes a plastic model), evolution of the surface topography (melt pond distribution in summer, ridging in winter), and the thermal conductivity of sea ice. The interested reader can learn more from books by Pritchard (1979), Untersteiner (1986), Wadhams (2000), and Lepparanta (2005).

### 3.16.3 MODEL DEVELOPMENT

We will follow the development of a fluid dynamics model as an example for small-scale modeling that is a direct result of field observations. Throughout history, observations have left people pondering about the essence of the phenomena they observed: the movement of celestial bodies, the colors of the rainbow, the depression of the freezing point. Hypotheses were developed to explain the observed patterns. Many hypotheses were formulated as mathematical models that allowed people to make predictions that could be tested. A particularly well-tested and well-established hypothesis would have reached the status of a theory. More formally, in this section we shall be concerned with mathematical models as representations of the essential aspects of systems which present knowledge in usable form (Eykhoff 1974).

This work emerged from a project on sea ice flexural strength. Sea ice flexural strength and its load-bearing capacity are of concern to everybody standing, working, or landing planes on sea ice. In the course of investigating the flexural strength of sea ice (Squire et al. 1988; Haskell et al. 1996), the potential importance of the ubiquitous frozen cracks emerged (Langhorne and Haskell 2004; Williams and Squire 2006). There are only a few systematic investigations of frozen cracks, so

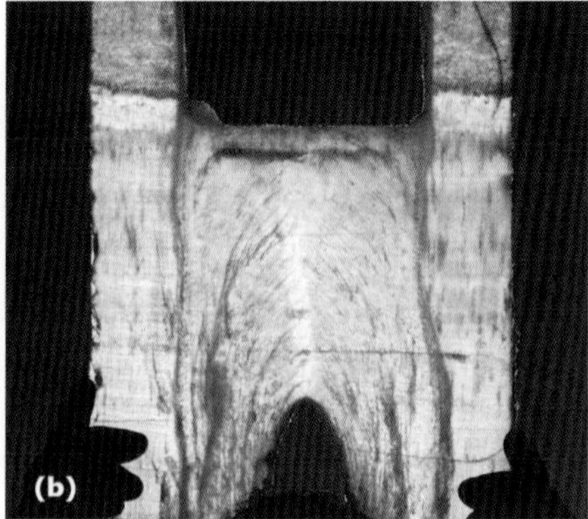


**Figure 3.16.1.** (a) Frozen crack in landfast sea ice off the coast of Barrow, Alaska, in January 2008. The glove is for scale. (b) Vertical thick section of a partially refrozen man-made crack in McMurdo Sound, Antarctica (approx. 150 mm wide); photo courtesy Pat Langhorne.

we decided to focus on a subset of these cracks, straight-sided flooded cracks that froze. These cracks can be found in landfast sea ice, for example, off the coast of Barrow, Alaska (Figure 3.16.1a). Some aspects of the formation and freezing process of these cracks in McMurdo Sound, Antarctica, were described by both Shackleton's and Scott's expeditions about one century ago (Shackleton 1909; David and Priestley 1914).

Vertical cross sections of frozen cracks show an arch-shaped alignment of brine inclusions (Figure 3.16.1b). The origin of this structure is not immediately clear as the closest features described in sea ice seem to have been slanted, linear alignments of inclusions (e.g., Niedrauer and Martin 1979; Cole et al. 2004). However, arch-shaped arrangements of inclusions are observed in alloys that cool from several directions (Flemings 1974). Possible hypotheses included that the alignment was remnant of the arch-shaped ice-ocean interface where saline pockets might have been trapped by advancing ice lamellae; that brine pocket migration took place, changing vertical brine channels into arch-shaped ones; and that the path of fluid motion through the very porous skeletal layer was actually curved, and these pockets are remnant of that path. We decided to test if the process of the third hypothesis could make a significant contribution, encouraged by another curious observation: The time series of thermistors frozen into man-made cracks was unusual. Usually, thermistors frozen into unidirectionally growing sea ice reveal a rapid decline in temperature as the sea ice interface passes (Lake and Lewis 1970). However, we detected an extended period of comparatively slow decrease of the ice temperature followed by the expected rapid decrease in temperature (Petrich et al. 2007). Further, several thermistors arranged vertically along the center of the crack registered the initially slow decrease at the same time. Could both observations mean that we were dealing with a volume of high-porosity sea ice that is substantially larger than the usual 10 to 30 mm skeletal layer found in unidirectionally growing sea ice?

We decided to approach this question with a numerical model. Based on well-defined physical processes, we would be able to identify the relevant processes if they were included. Our method of choice was to develop a porous medium-continuum model based on the volume-integrated Navier-Stokes equations (Bear and Bachmat 1991). We gave the conventional fluid dynamics approach preference over the volume-integrated Lattice-Boltzman approach to porous media flow since the development of the latter was still in its infancy at the time we started our work.

Without knowledge of the detailed microscopic geometry of the pore network, we describe fluid transport in a porous medium based on the macroscopic continuum approach (Bear and Bachmat 1991). At this scale state variables can be defined that are measurable, continuous, and, conveniently for analysis, differentiable. We assume that the porous domain consists of two microscopic phases: a stationary solid matrix of ice and a liquid that completely fills the remaining volume. The air content of growing sea ice is generally very low and will be neglected here. Each phase is interconnected; that is, we can reach any point in the liquid by traveling solely through the liquid phase. In the continuum approach we assume that the solid is distributed throughout the porous medium such that a macroscopic representative elementary volume (REV) can be found that contains both solid and liquid, no matter where we place it in the porous medium. The size of this volume is such that the geometric characteristics of the microstructure it contains do not depend on the size of the volume; specifically, the REV has to be considerably larger than the scale of microscopic inclusions, yet smaller than the scale of macroscopic inhomogeneities. Basically we treat sea ice as a homogeneous porous medium with interconnected pore space, and we look at sea ice at a scale much larger than the pore size yet smaller than the features we want to resolve.

We should emphasize that the continuum approach to sea ice is a model in the sense that it captures essential (but not all) aspects of the system and that it yields usable results. The most apparent aspects that are not captured shall be mentioned here. While it is clear that sea ice contains brine inclusions from the micrometer to the meter scale (Cole and Shapiro 1998; Light et al. 2003), there is only limited evidence that young sea ice can be treated as homogeneous material. Cole et al. (2004) demonstrate that the horizontal pore area fraction appears to be correlated with the pore volume fraction (Delesse principle), which is consistent with the assumption of a homogeneous pore space but it is not sufficient. Hence, sea ice may be better characterized as a fractured porous medium if it is penetrated by large brine channels, for example, channels that develop during spring and summer. Further, currently it is not clear to what extent the pore space of sea ice can be considered interconnected. For example, Golden at al. (2007) show X-ray tomography images that suggest that sea ice undergoes a transition between a connected and a disconnected state around 5 percent porosity. A more fundamental problem is that the continuum approach precludes macroscopically discontinuous material properties, for example, macroscopic interfaces like the sea ice–ocean transition. An accurate calculation of state variables across the macroscopic interface is not possible without further considerations.

These three points can be addressed by, first, limiting the application of the model to resolutions at the centimeter scale and assuming that prominent brine channels are not crucial features during the process considered, that is, ice growth. Second, we note that it is not clear what the implications are of a disconnecting pore space for fluid motion. While the connectivity has been linked to the apparent limit of desalination of sea ice to porosities larger than 5 percent (Golden et al. 1998; 2007), the bound on desalination has been replicated with continuum models, whether they drastically reduce the permeability below 5 percent porosity (Petrich et al. 2006b, 2007) or not (Petrich et al. 2006a). Third, we accept that the calculation of the momentum at the interface will be at error somewhat. This is to be expected anyway because of our poor knowledge of the permeability at the interface and simplifications made in the derivation of the governing equations (Ganesan and Poirier 1990).

Fluid dynamics simulations of growing sea ice have been performed before (Medjani 1996; Oertling and Watts 2004), but we decided to implement a model that includes different densities for ice and water (expected to improve the accuracy of salinity calculations) and to treat flow through ice of high porosity as porous medium fluid flow rather than flow of a suspension of particles. The governing equations were derived rigorously from first principles (cf. Ganesan and Poirier 1990; Petrich 2005) which resulted naturally in equations that contain the phase factor at the correct positions. As an example, we compare the momentum conservation equation in vertical direction

$$\rho\left[\frac{\partial(fw)}{\partial t}+\frac{\partial(fuw)}{\partial x}+\frac{\partial(fww)}{\partial z}\right]=\mu\left[\frac{\partial^2(fw)}{\partial x^2}+\frac{\partial^2(fw)}{\partial z^2}\right]-f\frac{dp}{dz}+\rho g-f\frac{\mu}{\Pi}(fw)$$

(Equation 3.16.1),

with the solute conservation equation

$$\frac{\partial(fC)}{\partial t}+\frac{\partial(fuC)}{\partial x}+\frac{\partial(fwC)}{\partial z}=D\left[\frac{\partial}{\partial x}\left(f\frac{\partial C}{\partial x}\right)+\frac{\partial}{\partial z}\left(f\frac{\partial C}{\partial z}\right)\right]$$

(Equation 3.16.2).

Here, $x$ and $z$ are the horizontal and vertical directions, respectively; $t$ is time; $f$ is the porosity, that is, the volume fraction of the liquid; $u$ and $w$ are the horizontal and vertical interstitial flow rates, respectively; $\mu$ is the kinematic viscosity; $g$ is the vertical acceleration; $p$ is the pressure; $\rho$ is the density of the liquid; $\Pi$ is the permeability (e.g., Eicken et al. 2004); $C$ is the solute concentration of the brine; and $D$ is the diffusion coefficient. Looking at the stress term in Equation 3.16.1 (first term on the right-hand side), we see that the change in momentum depends on the change of the volume flow rate *fw*, while the diffusion term in the solute conservation Equation 3.3.16 (right-hand side) depends on the change of the gradient of brine concentration *C*. Mathematically, this difference results naturally from different assumptions at the microscopic interface between ice and brine inclusions: In the derivation of Equation 3.16.1 we assumed that the fluid velocity at the microscopic ice-water interface is zero, while in Equation 3.16.2 we assumed that the solute concentration at the microscopic ice-water interface is equal to the local mean solute concentration of the brine (Petrich 2005).

We discretized the governing equations with the finite volume method (FVM) on a staggered grid (Patankar 1980) and solved them with a multigrid solver. This approach is conceptionally simple and apparently quite robust. As it turns out, we obtain results that are unaffected by artifacts reported occasionally (cf. the study of thermal convection in a mixed fluid-porous medium by Le Bars and Worster 2006).

Applied to the study of freezing cracks we find that our effective medium approach reproduces both arch-shaped alignments of inclusions and the peculiar temperature-time records of our field experiments (Petrich et al. 2007). Observation of the simulated freezing process shows that arches of elevated salinity and porosity develop inside the porous medium; that is, we need to invoke neither crystal structure nor brine pocket migration to reproduce the observations.

In spite of the discrepancies between model and pore structure of sea ice discussed above, the continuum porous medium fluid dynamics model appears to reproduce well the fluid dynamics inside growing sea ice. Arguably, a large portion of the pack ice is in the form of level ice that grows under unidirectional heat removal. Considering that intrinsic fields and their evolution are directly accessible in the fluid dynamics model, the model provides a wide range of opportunities for reference simulations. For example, relationships for the stable salinity of sea ice can be derived (Petrich et al. 2006a), and bounds on the nutrient flux for biological communities in the ice can be found (Petrich et al. 2006a). Further, conceptional insights into ice growth can be gained. Simply applying the macroscopic model to microscopic processes is clearly questionable unless done very carefully. For example, it should be questioned if the macroscopic permeability is meaningful at this scale.

### 3.16.4 MODEL VALIDATION

The previous section illustrates a study where a theoretically based model was used to test a hypothesized explanation of observed behavior at the ice freezing front across a narrow crack. It was found that the model qualitatively reproduced the observed phenomena, and this study shed light on the mechanisms controlling the shape of the freezing front. In order to reach his conclusion, Petrich had to perform some level of validation of his model against field observations. Validation, often against data collected in situ or in laboratory experiments, is an important step in the development of models that capture sea ice phenomena on all scales from the microstructure to global circulation.

Large-scale sea ice models are most often validated by direct comparison against observations of sea ice concentration, extent, velocity, and thickness. At a minimum, models used in climate study should re-create the spatial distribution and seasonal cycle of ice extent, thickness, and transport. In reality most large-scale models do not simulate the ice state equally well for each variable, and the models need to be tuned to minimize the difference between simulated and observed ice state. An example of model validation and sensitivity study used to tune such a model is SIMIP, which is outlined in Section 3.16.1. More recently, with advances in observing technology that allow more detailed monitoring of pack ice, field experiments have been designed that test particular parameterizations and empirical models employed in large-scale sea ice models. A description of a recent field campaign with these goals, SEDNA, is given in Section 3.16.2.

### 3.16.4.1 The Sea Ice Model Intercomparison Project (SIMIP)

Climate modelers have instigated a large number of model intercomparison projects (MIPs) that compare the performance of a variety of climate models in representing specific phenomena. SIMIP was a project in the mid-1990s that evaluated the dynamic sea ice models incorporated into the then-current generation of climate models. More recently, SIMIP2 focuses on intercomparison of one-dimensional thermodynamic sea ice models, using in situ data collected during the Surface Heat and Energy Budget of the Arctic (SHEBA) experiment to constrain models.

The strategy taken by SIMIP1 and 2 is to isolate the model components solving for ice velocity (momentum equation) and temperature profile (1-D thermodynamic equation), respectively. The isolated equation is solved as a stand-alone model, with prescribed surface forcing. Each participant model in the project runs a set of numerical experiments with the prescribed forcing. A standard set of observational data is provided for comparison with model output. The results from each model validation against the standard observational data are then compared to determine which of the parameterizations and numerical solutions tested perform best in simulations.

The main goal of SIMIP1 (Kreyscher 2000) was to identify which rheological model of sea ice deformation, describing the divergence of internal ice pack stress tensor, most faithfully reproduces observed ice drift patterns and sea ice cover. The rheological models tested were those commonly used in global climate models (GCMs) at the time: free drift (no internal ice stress); free drift with stoppage (ice velocity goes to zero when a maximum thickness is obtained); cavitating fluid (ice has no shear strength [Flato and Hibler 1992]); and the viscous-plastic rheology with elliptical yield curve (Hibler 1979). Perhaps not surprisingly, the

Hibler rheology, which had been developed based on in situ observations of ice failure, performed the best when validated against buoy drift and satellite passive microwave–derived maps of ice area. Results from SIMIP1 encouraged many GCMs to change to a Hibler (1979) type rheology.

Investigations were also made of the sensitivity of simulated ice to changes in key parameters in the dynamic model: the air and water drag coefficients, and the assumed magnitude of ice strength. It is possible to develop metrics describing model performance. For example, SIMIP1 used the difference between simulated and observed buoy drift trajectories, averaged across the model domain over two decades. A set of runs were made with the model varying and the minimum of the model-observations metric found in parameter space over the set of model runs. This procedure allows tuning of parameters within a model to provide the most faithful representation possible of the ice pack. It is a useful procedure when a parameter's value is not exactly known. However, the method can result in unphysical values being chosen, and it is necessary for the inverse problem posed in the sensitivity study to be well constrained by observations.

The models tested in SIMIP1 have been superseded by recent developments, though many GCMs still use the rheological model recommended by SIMIP1 today. One weakness of the SIMIP1 project was a lack of ice thickness observations against which to validate the models, especially in the Canada Basin, where small changes in the parameterization of ice strength, yield curve shape, and mechanical redistribution influence the simulated ice thickness distribution. Zhang and Rothrock (2005) used more recently available submarine upward-looking sonar ice draft data to investigate the performance of various yield criteria in reproducing observed ice thickness. They tested these yield curves: teardrop, parabolic lens, ellipse (Hibler 1979), and the Mohr-Coulomb ellipse (Hibler and Schulson 2000). They found that yield curves that incorporate biaxial tensile stress tend to result in increased model ice thickness. The fatness of a yield curve is related to the frequency of large shear stress events, with fatter curves having more frequent events, reducing ice speed and export and increasing ridging in the arctic interior. In their study Zhang and Rothrock found that the teardrop yield curve provided the lowest spatial bias between simulated and submarine-observed ice draft. The choice of yield criteria considerably affects simulated ice motion, deformation, and thickness—and hence surface energy budgets. Although the results from this single model validation study are not conclusive as to which yield criteria is most realistic, they do indicate that choice of plastic rheology can be used in model tuning.

The standard model validation practices used in SIMIP1 focus on the model's ability to re-create the large-scale sea ice state; that is, for example, the ice thickness distribution, concentration, extent, and Fram Strait outflow. This methodology for model validation does not guarantee that the model is reproducing the observed sea ice state for the correct reasons. Data that resolves specific processes within the sea ice model is useful in validation, especially of processes that are parameterized in large-scale models. Field campaigns are often designed to measure specific processes, and such investigation, model development, and model validation can be catered to with a well-planned measurement set. A good example of such a field campaign is the Surface Heat and Energy Budget of the Arctic (SHEBA). Primary goals of SHEBA were to understand the ice-albedo and cloud-radiation feedbacks, and to use that understanding to improve models. This entailed acquiring a year-long data set of a column of ocean, ice, and atmosphere with horizontal dimensions representing a GCM grid cell. SIMIP2 is currently underway and will use the SHEBA data set to validate 1-D thermodynamic sea ice models used in GCMs.

### 3.16.4.2 Field Campaign in Support of Model Validation

SIMIP1 is an example of model validation using existing observations of global variables such as ice extent, velocity, and concentration. Such validation methods give an idea of how well a particular model reproduces the observed sea ice state; however, there are limitations to the method. Chiefly, the global data sets available are limited and might not be sufficient to test specific parameterizations. With careful planning it is possible to design field campaigns that provide measurements to constrain the validation and/or tuning of specific parameterizations in large-scale sea ice models. By necessity these field campaigns are large collaborative efforts; as when isolating a single component out of a dynamic-thermodynamic sea ice model, we need observations of a large number of variables describing the ice pack to both drive and validate the isolated model. Below we present an example of such a field campaign.

*Sea Ice Experiment: Dynamic Nature of the Arctic (SEDNA)*

SEDNA commenced with a campaign in the Beaufort Sea in spring 2007, designed to improve our understanding of the contribution of sea ice dynamics to the ice mass balance of the Arctic Ocean. During winter the ice thickness distribution evolves due to ice growth and mechanical redistribution of ice thickness (ridging and rafting). Mechanical redistribution occurs abruptly, and is manifest as organized linear regions of deformation in leads and ridges. Large-scale sea ice models relate thickness redistribution to strain rate (divergence, convergence, and shear), which in turn is related to stress propagation through the ice pack and hence the constitutive relation or rheological model assumed.

The AIDJEX campaign in the 1970s led to several assumptions that are still taken today in most rheologies used in large-scale sea ice models. A recent publication (Coon et al. 2007) has found that the three key assumptions arising from AIDJEX are inadequate. It is assumed that pack ice deformation is isotropic, such that the yield curve can be described in the 2-D space of the stress principal components. The basis for this assumption was that linear deformation features (leads and ridges) are randomly distributed over spatial scales of one model grid cell. Recent advances in satellite observation have shown that linear deformation features can be hundreds of kilometers long, and tend to align with weather systems (Kwok 2001). This challenges the assumption of isotropy, and there is a need to identify if isotropic yield curves adequately simulate pack ice deformation.

There is a recent trend towards large-scale sea ice models with increasing spatial resolution, down to 10 km, for example. Laboratory results (Hibler and Schulson 2000) and aerial surveys (Schulson and Hibler 2004) suggest that sea ice deformation is scale invariant; however, this has not been quantitatively examined on geophysical scales. As sea ice modelers are making the assumption that at spatial scales from kilometers to hundreds of kilometers a single model of sea ice failure can be applied, this assumption needs to be validated.

There are a variety of redistribution models describing ridging and rafting of pack ice. These typically relate thickening of the ice due to convergence and shear rates, taking an assumed shape (often triangular) for the resultant ridges or rafts. It is often assumed that there are no voids in ridges, that some ridging occurs under pure shear, and that there is always a little open water present under closing. The last two concepts are scale dependent, and may not be relevant for high-resolution (e.g., <10 km) models. The shape of ridges and concept of volume conservation during ridging are constructs from theoretical consideration of energy release in ridging (Rothrock 1975). We identified that there is a need for data to validate ridging parameterization schemes, relating the observed strain rate of the ice pack to changes in the thickness distribution. Our main questions are: Do typical ridge shape parameterizations allow realistic representation of the ridged ice probability distribution function (pdf)? Must voids in ridges be considered in this pdf? Over what scales do we need to parameterize ridging and leads under closing and shearing? And what is the decorrelation length scale for ridged ice pdf?

SEDNA was designed to directly validate constitutive relationships between strain rate and internal ice stress and redistribution parameterizations, on scales appropriate for climate modeling. Unlike previous campaigns—such as AIDJEX, out of which modern sea ice dynamic models were born—SEDNA attempts to directly measure geophysical-scale stress strain-rate relations for pack ice. The measurements collected have to provide information about all phenomena controlling ice mass balance over a defined region of ice that we track in time. The measurements need to separate the relative effect of thermodynamics and dynamics, and determine the relative effect of dynamic processes on new ice growth, ridging, and rafting. This requires measurement of deformation (strain rate), stress propagation through the ice, and the evolution of the ice thickness distribution in time. The measurements were coordinated in a single campaign that tracks a region of ice, the size of which is representative of the grid resolution in large-scale sea ice models. As we are interested in scaling effect, we chose to take a nested set of measurements that track the ice pack over a set of cascading scales: 1 km, 10 km, 70 km, and the Beaufort Sea.

**Table 3.16.1.** The methods of measurement for pack ice mass balance variables resolved by the SEDNA field campaign. The table illustrates the scales over which each variable is estimated. The acronyms are: Ice Mass Balance buoy, IMB; electromagnetic, EM; Acoustic Doppler Current Profiler, ADCP; National Center for Atmospheric Research NCAR; National Center for Environmental Prediction, NCEP; European Centre for Medium Range Weather Forecasting, ECMWF; North American Regional Reanalysis, NARR; Synthetic Aperture RADAR, SAR; Global Positioning System, GPS; International Arctic Buoy Program, IABP; Autonomous Underwater Vehicle, AUV; Upward Looking Sonar, ULS; Light Detection and Ranging, LIDAR.

| Variable | | Point | 1 km | 10 km | 70 km | Regional |
|---|---|---|---|---|---|---|
| Growth/Melt | $f$ | IMB buoy | Calibration transects | EM-bird | EM-bird | Model |
| Surface stress | $F_O$<br>$F_A$ | Wind tower, ADCP | | | NCAR/NCEP, ECMWF, NARR reanalysis | NCAR/NCEP, ECMWF, NARR reanalysis |
| Internal ice stress | $\sigma$ | Stress sensor | | Stress senor array | Stress senor array | Stress senor array |
| Strain rate | $\dot{\varepsilon}$ | | SAR | GPS, SAR | GPS, SAR | SAR, IABP |
| Ridge parameters | $\varphi$ | On foot, diving transects | AUV ULS | LIDAR | Submarine ULS, LIDAR | Model |
| Thickness distribution | $\gamma$ | Drill holes | Calibration transects, EM | LIDAR, submarine ULS, EM-bird | LIDAR, EM-bird | ICEsat, model |

We took advantage of a U.S. Navy ice camp, which substantially reduced logistics costs. Our field campaign was centered on this ice camp, and was designed to measure the strain rate of the ice pack, stress propagation through the ice, and sea ice thickness distribution. Table 3.16.1 gives a summary of measurement methods available and the spatial scales over which they were applied. All measurements were designed to provide information about the dynamic changes to the ice pack, and are outlined in Figure 3.16.2. A single ice mass balance buoy (Richter-Menge et al. 2006) was deployed at the ice camp to track thermodynamic changes to the ice mass balance.

Measurements of sea ice thickness have specific limitations, and no specific measurement method provides an accurate estimate of the ice thickness distribution over the spatial scales we are interested in. In designing our field campaign, careful thought was taken on how best to use available technology to resolve the thickness distribution. We coordinated laser altimeter retrievals (lidar) (Forsberg et al. 2002); submarine and autonomous underwater vehicle (AUV) upward-looking sonar (ULS) swaths (Wadhams 1988); electromagnetic induction ice thickness measurements (EM), both airborne EM-bird (Pfaffling et al. 2004) and surface EM (Eicken et al. 2001); and in situ drill hole measurements. A set of 1 km calibration

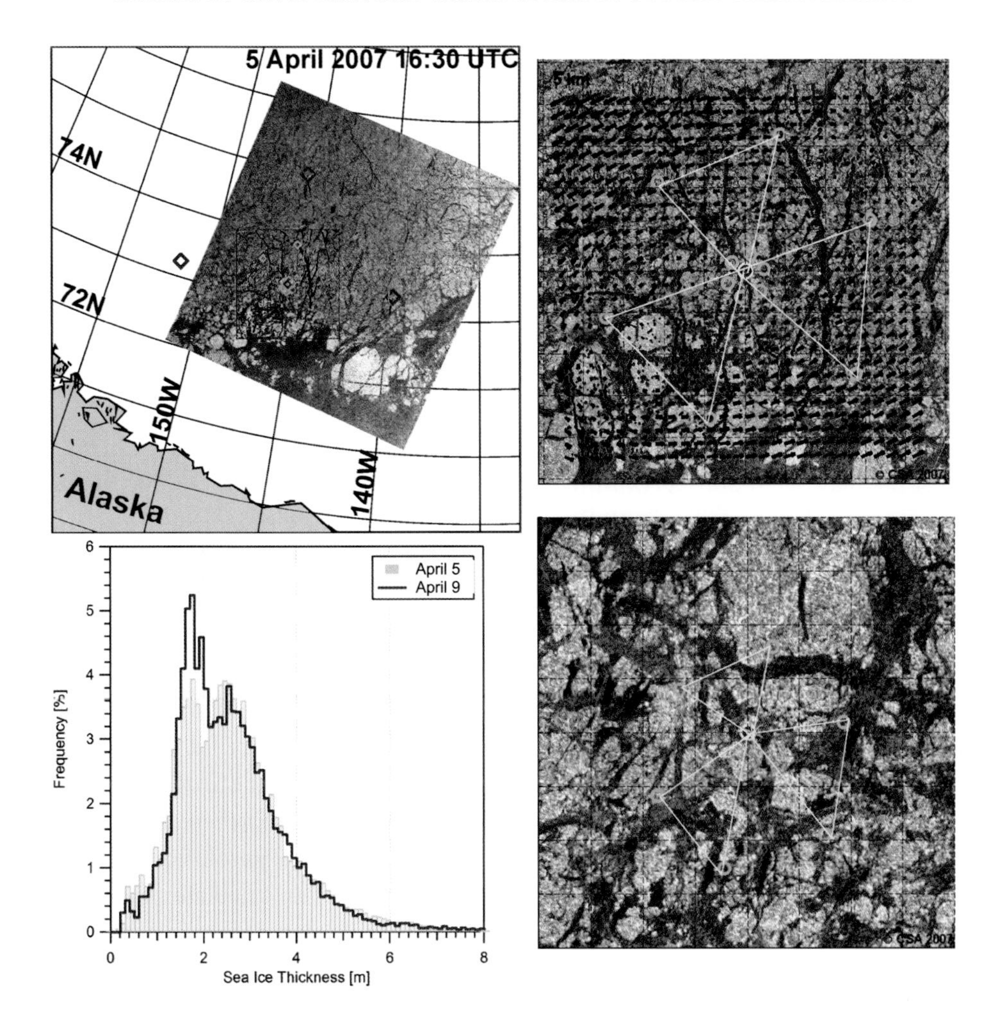


**Figure 3.16.2.** The SEDNA array on April 5, 2007. Position of buoys overlaid on a RADARSAT scanSAR-B scene. Red diamonds are International Arctic Buoy Program meteorological beacons, green diamonds are GPS drifters, yellow dots are stress sensors, the blue dot is an ice mass balance buoy located close to the ice camp. Pink dots are North West Research Institute GPS drifters, which were clustered along individual leads (personal communication M. Pruis). Discontinuities in ice motion field, calculated with the University of Delaware (UD) map of moving topography (MMT) analysis between two SAR images on April 5 and April 8 are shown as red lines. MMT ice motion vectors are plotted every 6.4 km in the top right panel. Top right and bottom right show the outer, 70 km radius, and inner, 10 km radius, buoy arrays respectively. These two arrays formed the basis of the SEDNA scaling experiment. LIDAR (not shown) and EMI (yellow lines) aerial surveys of ice thickness were performed over the two scales. There was a submarine based upward looking sonar survey of the 10 km region (not shown). A set of 1 km calibration transects (green lines) of in situ measurements and a detailed ridge study, in the vicinity of the ice camp, allowed inter-comparison of all thickness measurements. Bottom left: Ice thickness distribution was estimated by EM-bird on April 5 and 9.

transects were laid out, over which all measurement methods were intercompared. Before arriving on the ice, we knew that EM methods do not fully resolve ridges, and ULS processing requires a detailed knowledge of the ice bottom topography.

We hoped that we could combine the two methods to fully resolve the ice thickness distribution over 20 km$^2$ around the ice camp. So we included campaigns to better understand the limitations within our field plan. In the end, we were unable to line up measurements over the 20 km$^2$ scale to resolve deformation events as the AUV was not ready to run untethered under the ice pack. We did, however, measure the thickness distribution before and after a deformation event with airborne EM and laser, and performed a detailed ULS and EM study of a ridge that formed within our study area. The design of our measurement campaign allowed the first intercomparison of all airborne ice thickness measurement methods, and so will aid in design of future field campaigns.

To investigate the relationship between internal ice stress, strain rate, and failure of the pack ice required coordination in the deployment of autonomous stress buoys and ice drifting buoys with Radarsat synthetic aperture radar (SAR) swaths. The timing of lead reorganization events (new lead formation and ridging) can be identified in the record of pressure gauges monitoring internal ice stress with stress buoys (Richter-Menge et al. 2002). By knowing this timing we can identify pairs of SAR images showing the formation of new leads. Then we can identify the deformation events in the calculated deformation of an array of global positioning system (GPS) drifting buoys (Hutchings and Hibler 2008). It is useful to combine drifting buoy and SAR image analysis estimates of strain rate, as the SAR data product has low temporal resolution and excellent spatial resolution, whereas the buoy data has higher temporal resolution but only regional spatial resolution (Geiger and Drinkwater 2005). By coordinating three data sets we can calculate strain rate for times before, during, and after failure events, and we can determine the corresponding magnitude of stress and principal stress components. It will, we hope, be possible to plot out the stress states for a series of failure events and relate these to strain rate at failure. This will provide a direct estimate of the stress–strain rate relation for a 20 km$^2$ region of ice.

To demonstrate how the field experiment design allows resolution of pack ice deformation-driven thickness redistribution events, we discuss an example of one synoptic event that occurred during the field campaign. A low-pressure weather system passed over the camp between April 5 and 9, causing advection of the ice situated along a linear region extending from Point Barrow into the Central Beaufort, into the Chukchi Sea. This created a shear zone in the vicinity of the ice camp. We estimate pack ice deformation (the four components of strain rate: divergence, pure shear, normal shear, and vorticity), at two scales, with a Green's theorem analysis of buoy drift. Ridging, opening, and shearing events are discernible, and apparently occur in response to changes in large-scale wind stress on the ice pack, with a time delay of two to three days after the onset of each new weather system (Figure 3.16.3). Motion analysis of SAR imagery (Figure 3.16.2) and the deformation of the buoy arrays (Figure 3.16.3) show that shearing and closing of the ice pack occurred between April 4 and 6. These ridging events were directly observed at the ice camp,

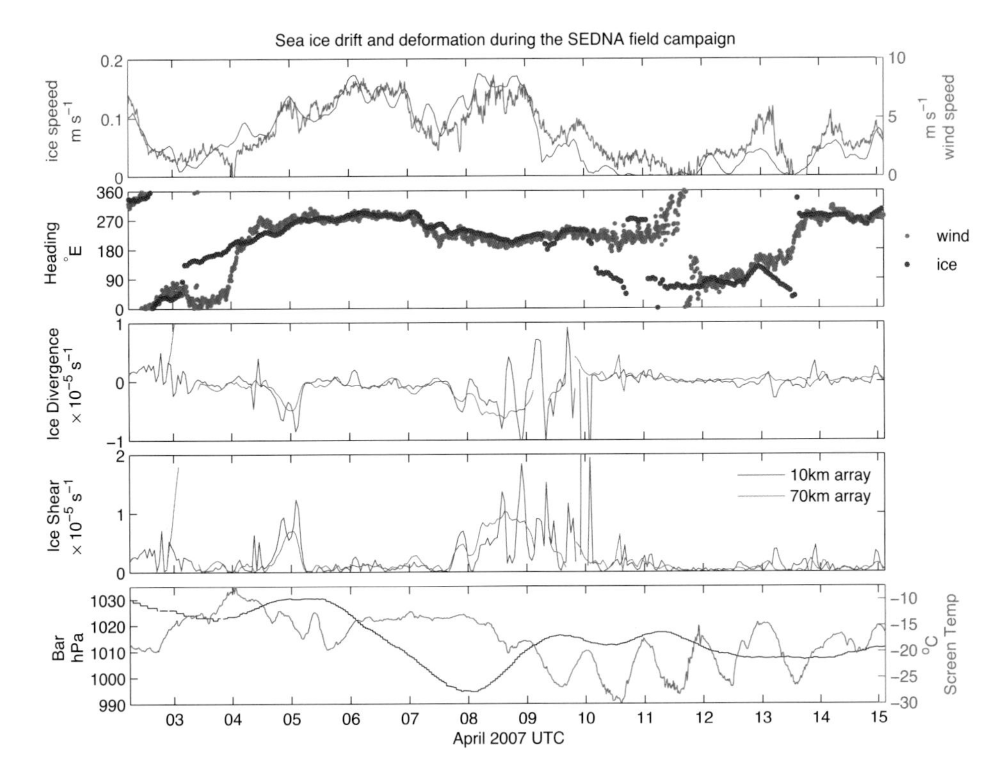


**Figure 3.16.3.** Time series of speeds (top panel) and headings (second panel) of hourly mean ice (blue), and three meter wind (green) at the ice camp. The divergence and maximum shear strain rates are shown for the 10 km and 70 km buoy arrays (panels 3 and 4). The bottom panel shows sea level pressure and 2 m air temperature observed at the camp.

as ridges formed at the edges of frozen leads that had opened two weeks prior to the camp. On April 8 another shearing event occurred, which was accompanied by several opening and closing events. Comparison of coincident EM-bird flights, within the 10 km array area between April 5 and 9, provide an estimate of mechanical ice thickness redistribution during this particular synoptic event. Figure 3.16.2 shows that thin lead ice (<1 m) was redistributed into the distribution's modal peak (at 1.7 m, level first-year ice). A closer analysis of the EM-bird data will illuminate the change of ridge sail and keel statistics, such as the redistribution of thin ice, through ridging, to the second prominent mode (2.5 m, multiyear ice), and the deformed ice tail of the distribution.

It will take several years, and the maintenance of collaborative relationships between field workers, to finish analysis of the SEDNA data set. The outcome of this analysis will be a database of measurements and derived ice thickness products for dynamic model validation. In particular, this data set will, we hope, be sufficient to constrain stand-alone solutions of constitutive relationship and mechanical redistribution with observed forcing and ice state changes. Of course, it is the nature of field experiments to reveal new knowledge that might require further investigation and a different observation set. In which case we would not have achieved our dream of the perfect sea ice dynamic model validation data set, but we will be a step closer towards realistic modeling of dynamic sea ice processes.

## 3.16.5 USING MODELS IN DESIGN OF MONITORING CAMPAIGNS

From the case studies in the previous two sections, you may see that it is very common for a field campaign to be designed to validate or develop a particular model. Models can also be useful in interpreting field measurements. This can lead to new theoretical insight when, as is often the case, our field data does not behave as we expected it would.

Once we are happy that a model performs well at simulating the ice pack, against whatever benchmark is appropriate, we may use the model for many purposes. Typically large-scale and global models are used to simulate changes and variability in the ice pack in recent history or for a paleoclimate. Some numerical weather forecast models include sea ice in their medium-range weather simulations. Global climate models are also used to project the future changes to the ice pack this century. Our confidence in these models is improved with better insights into the performance of parameterizations within the model and with advances in the observational data against which the models may be validated.

Large-scale models can also be of great use to the sea ice observational community. Recent changes to arctic sea ice have been dramatic (e.g., Rothrock et al. 1999; Wadhams and Davis 2000; Tucker et al. 2001; Stroeve et al. 2005; 2007). Hence there is a great need to maintain and implement monitoring systems for the sea ice mass balance. In a tight funding climate we would wish such a monitoring system to be efficient, as cheap as possible, and yet resolve the major modes of variability of the ice pack. The following case study outlines a model investigation that provides insight into where to place moored upward-looking sonar to monitor ice thickness. Although this study is one of a kind, we expect to see an increase in the use of models to design monitoring systems in the future.

### 3.16.5.1 Arctic Ocean Ice Thickness: Modes of Variability and the Best Locations from Which to Monitor Them

One technique to determine the major modes of variability of the ice extent or thickness is through the use of empirical orthogonal functions (EOFs). For example, analysis of passive microwave satellite imagery provides maps of ice concentration, on which EOF analysis may be performed (Partington et al. 2003). Recommended textbooks outlining the statistical methods used in EOF and principal component analysis are Wilks (1995) and Cressie (1993), for the mathematically minded.

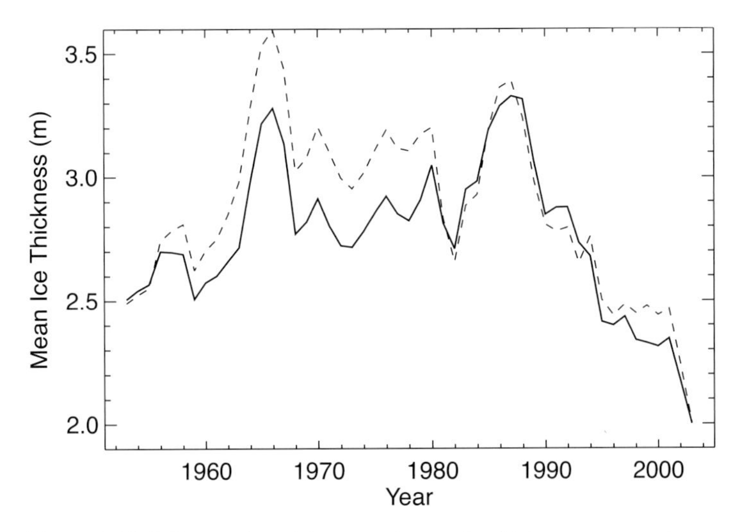


**Figure 3.16.4.** Time series of the annual mean ice thickness averaged over the Arctic Ocean. The dashed line is from a model-only simulation, and the solid line is from a simulations that includes assimilation of ice concentration and, starting in 1979, ice velocity.

Direct observations of the ice thickness are few and far between and are far too sparse to permit an analysis of the modes of variability. However, estimates of the ice thickness at all times and locations can be made with a model (Figure 3.16.4). We use a coupled ice-ocean model that is forced with observed surface pressure and temperature fields and, in addition, uses data assimilation methods for ice concentration and ice velocity (for more details, see Lindsay and Zhang [2006]). Consequently the simulated ice thickness and concentration are thought to be reasonably faithful re-creations of the actual parameters. The simulated ice thickness can then be used to determine the major modes of variability of the ice thickness and to determine how the weighting of the modes changes over time.

But models alone are not sufficient because of uncertainties in their representations of the ice thickness. Direct observations of the ice thickness distribution are essential to document the true evolution of the air-ice-ocean system in a changing global climate. A long-term observing system could help to verify the interannual variability of the total ice volume and how the regional distribution of the ice mass changes from year to year. Well-placed observations will both document the changing state of the arctic system and provide information to validate and improve coupled ice-ocean models.

Coupled ice-ocean models can aid in the design of an observing system that is optimal for addressing specific science questions. Model simulations of the ice thickness can help estimate where to place moorings with upward-looking sonars to most efficiently monitor the thickness of the ice pack. The methods can include ongoing measurement programs.

### *Model Analysis: Principal Component Analysis*

The spatial and temporal variability of the ice thickness fields may be partitioned in a manner that most efficiently represents the information content of the fields. This method, called principal component analysis, represents the time history of the ice thickness as a weighted sum of a set of spatial maps. The maps, or empirical orthogonal functions (EOF), are independent of time and are all orthogonal to one another in that the correlation of the points in any one map with those in any other is always zero. The weights given to each map in the weighted sum that re-creates the orginal data are the time-dependent principal components (PC). There is a set of PCs associated with each map and each set is orthogonal to all of the others. The variance of each set of PCs is the variance explained by the companion EOF. The EOFs are ordered by the fraction of the variance of the data set explained from highest to lowest.

The major modes of spatial variability of the simulated annual mean ice thickness are illustrated with the first three EOFs (Figure 3.16.5). The spatial domain used to determine the EOF patterns included only the Arctic Ocean. The maps show the correlations of the annual mean ice thickness for all ice-covered locations with each of the first three PC time series. The first mode, representing 30 percent of the Arctic Ocean total ice thickness variance, shows much of the basin increasing and decreasing in ice thickness in a coordinated fashion. The maximum of the pattern is near the center of the basin, in part because points near the edges have fewer neighbors with which to build high covariability. This bull's-eye pattern with diminished weighting near the edges of the domain is ubiquitous for fields in which a major mode of variability has a spatial scale commensurate with the size of the domain. This pattern likely reflects the basin-wide interannual variability in the thermal forcing. The averages of submarine-measured ice draft for entire cruises shown by Rothrock et al. (2003) differ from year to year in a way that is consistent with the large-scale pattern of interannual variability as seen in the first EOF.

The second and third EOF patterns, explaining 18 percent and 15 percent of the variance, respectively, together represent modes of lateral variability, meaning points on opposite sides of the basin have an element of anticorrelation. The second EOF shows a mode with opposite signs centered near the Beaufort Sea and the Laptev Sea. It is similar in shape and location to the East-West Arctic Anomaly Pattern (EWAAP) identified by Zhang et al. (2000). The third pattern has centers in the East Siberian Sea and off Svalbard in an orientation orthogonal to the second EOF.

The interpretation of the EOFs is made more difficult because the time series of ice thickness fields is far from stationary and in recent years the nature of the thickness field is changing rapidly. The EOFs for the most recent 16 years of the study, 1988–2003 when the ice is thinning rapidly, are shown in Figure 3.16.5. The first EOF, explaining 41 percent of the variance, is quite different from that of the fifty-one-year period. It is more broadly situated, and the largest loading is in a

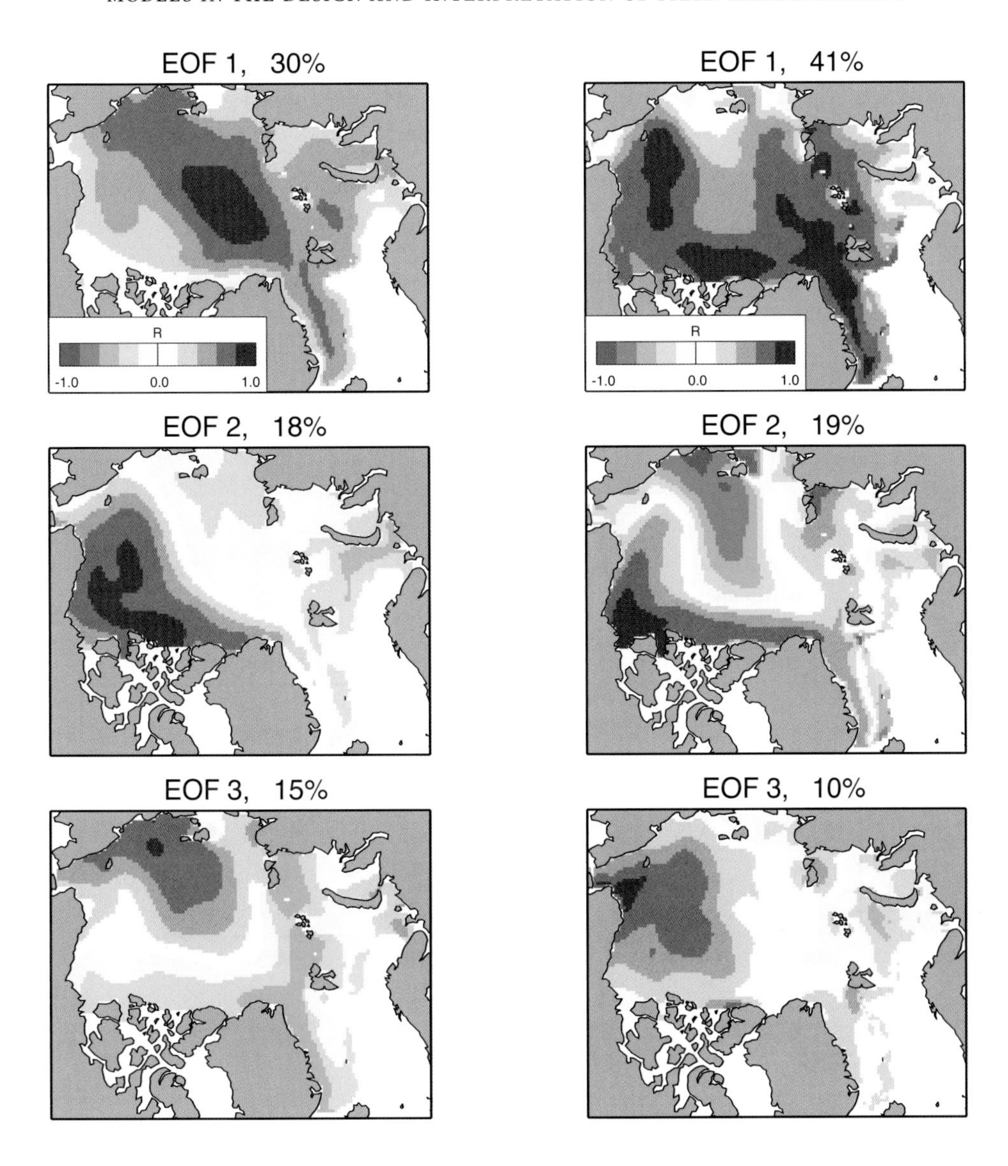


**Figure 3.16.5.** Correlation of the annual mean ice thickness at each location with the first three PCs of the annual mean ice thickness for the Arctic Ocean. Panels on the left are computed from fifty-one years of simulation, 1953–2003. Right panels show the same calculations from sixteen years of the simulation, 1988–2003. The percentages indicate the fraction of the total variance of the annual mean ice thickness explained by each of the EOFs.

band stretching from the Chukchi Sea to north of Canada to Fram Strait. This band is roughly coincident with the region of the greatest thinning rates for this same period seen in Lindsay and Zhang (2005). This is consistent with the fact that about 80 percent of the ice thickness variance during this period is accounted for by a thinning trend as opposed to just 11 percent over the fifty-one-year period. The second EOF is still similar to the EWAAP, though the pattern is more accentuated, and it accounts for a similar amount of the variance as in the fifty-one-year period. The third EOF is very different from that seen in the longer period and accounts for less of the total variance.

The time series of the first PCs from the fifty-one-year analysis are shown in Figure 3.16.6. The first component resembles the time series for the mean ice thickness (Figure 3.16.4), reflecting the fact that it accounts for a substantial portion of the variance in the ice thickness fields. The two prominent maxima in the ice thickness correspond to maxima in the first PC and moderate values of the other two. In recent years it is in an extreme negative mode; much of the recent observed thinning of the basin-wide mean ice thickness is accounted for by this mode. The second EOF was in a strong positive phase in the early 1990s, shortly after the 1987 maximum, and has since returned to near zero values while the third had a strong positive phase shortly before the 1987 maximum. The autocorrelations (not shown) indicate that the dominant time scale of the variability of the PCs is greatest for the first and least for the third.

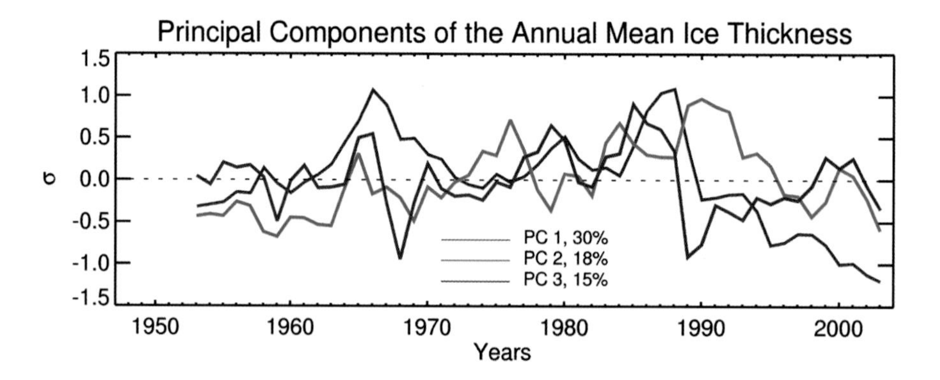


**Figure 3.16.6.** Time series of (top) the first three PCs of the annual mean ice thickness

*Monitoring the Basin-Wide Mean Ice Thickness and the Thickness Spatial Distribution*

The model output may be used as a simulated observing system to guide where to best locate fixed moorings to monitor the changing mean ice thickness of the basin as well as the changing spatial variability of the thickness. In determining the best locations to place new moorings, it is prudent to consider ongoing measurement programs. Beginning in the spring of 2001 a mooring has been deployed at the North Pole as part of the North Pole Environmental Observatory (Morison et al. 2002). A second mooring was deployed by the National Oceanic and Atmospheric Administration (NOAA) in the Chukchi Sea in the summer of 2003. How representative of the basin-wide mean thickness are these locations? Figure 3.16.7a shows the squared correlation of the annual mean ice thickness at each model node with the basin-wide mean value as seen in Figure 3.16.7. The maximum variance

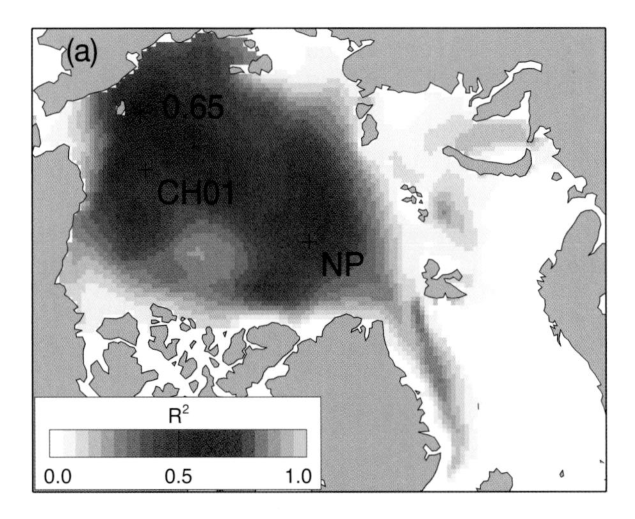


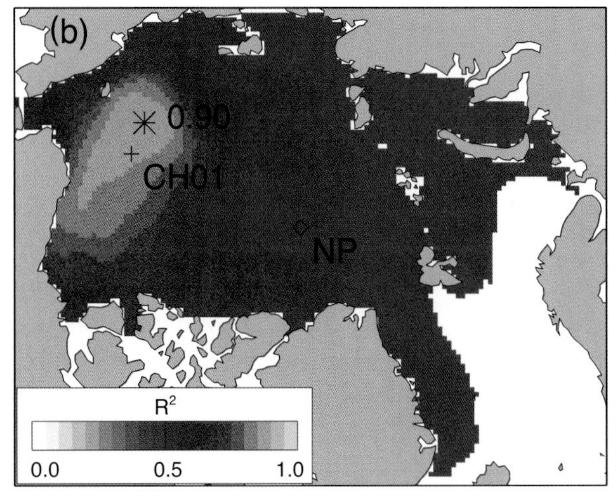


**Figure 3.16.7.** (a) Variance explained ($R^2$) by one-point correlations of the annual mean ice thickness for each location with the annual mean for the whole Arctic Ocean and (b) the variance explained by two points including one at the North Pole. The maximum fraction of the variance explained by two points is 0.90. CH01 is the location of the new mooring first deployed by NOAA in August 2003.

explained (65 percent) is at a point near the Chukchi and East Siberian seas; however, the North Pole is not a bad place for a first measurement, where the annual mean ice thickness explains 53 percent of the variance of the basin-wide mean. Assuming a measurement is made at the pole, where might a second point be placed to optimally predict the basin-wide mean thickness? Figure 3.16.7b shows the additional variance explained by a second point given an observation at the pole. The two points explain 90 percent of the interannual variance of the basin-wide mean ice thickness. The location of the second maximum is not far from the location of the new Chukchi Sea mooring where the fraction of the variance explained is 0.86 if the measurement at the North Pole is included.

This analysis may help to determine the locations to best monitor the mean ice thickness of the basin but does not determine how to best monitor the temporal changes in the spatial patterns of the ice thickness. One way to find the locations that best represent both the temporal and spatial variability is to determine the points that maximize the explained variance for the most important PCs. Each PC represents the annual weighting (i.e., the temporal variability) of the spatial pattern of the corresponding EOF (i.e., the spatial variability). The PCs and EOFs are an efficient representation of both the spatial and the temporal variability of the ice thickness estimated for all 4,400 model grid cells that represent the Arctic Ocean. Here we use only the first ten PCs, which embody 86 percent of the variance.

First, each of the PCs is normalized so that its variance $\sigma_i^2$ is the same as the fraction of the total variance explained by the corresponding EOF, thus the first PC has a variance of 0.30, the second a variance of 0.18, etc. The mean annual thickness at each grid location *h(x,y,t)* is then correlated with each of the first ten PCs, $\phi_i(t)$, $i = 1$ to 10, where *(x,y)* is the location and *t* is the year. The correlation coefficient at each location is $R_i(x,y)$. The fraction of the total variance explained by each correlation is $R_i(x,y)^2\sigma_i^2$ and the total variance explained by all ten PCs is

$$\sigma_{T1}^2(x,y) = \sum_{i=1}^{10} R_i^2(x,y)\sigma_i^2 \quad \text{(Equation 3.16.3).}$$

Figure 3.16.8 shows the spatial pattern of the total variance explained by one point. The maximum (28 percent) is in the Siberian sector of the central basin at a location that shows strong loading on the first EOF (27 percent) and lesser loading on both of the next two (4 percent each). Figure 3.16.5 shows the spatial patterns associated with these three PCs.

Rather than find the optimum location for two points, which would require testing all pairs of grid points, we recognize the fact that there is an ongoing ice draft measurement program at the North Pole and investigate where a second point might be located assuming a measurement there. A two-variable linear regression for each of the PCs with the ice thickness at the pole *h(0,0,t)* and with each of the other ice-covered grid points *h(x,y,t)* yields $R_{2,i}(x,y)^2\sigma_i^2$ as the fraction of the total variance explained by each of these two-point correlations. The total variance explained at each location is then found as in Equation 3.16.3. Figure 3.16.8b shows the variance explained by the ice thickness measured at two locations, one of which is the North Pole. The maximum total variance explained (44 percent) is located in the Beaufort Sea, but many locations in the Beaufort and Chukchi seas do well. This location primarily measures the second PC because the North Pole location is well situated to monitor the first.

Finally, we recognize that there is a new measurement site in the Chukchi Sea established by NOAA and we ask where a third location might be placed if measurements are available from both the North Pole and from the new mooring, CH01. Here a three-variable multiple regression is performed for each grid location. Figure 3.16.8c shows the variance explained by each location. The maximum is in the East Siberian Sea and, with the two established moorings, it accounts for 57 percent of the total variance. This location monitors primarily the second PC (6 percent additional variance explained) and the third PC (7 percent).

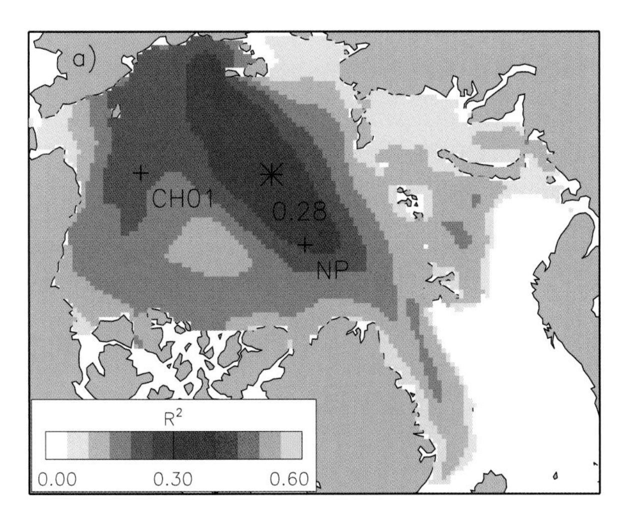


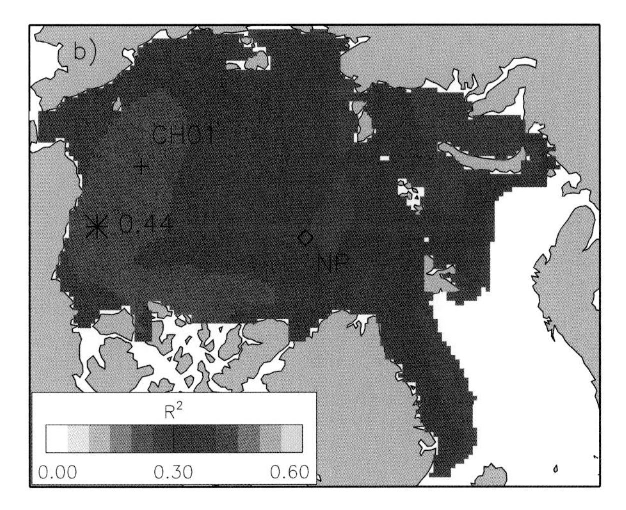


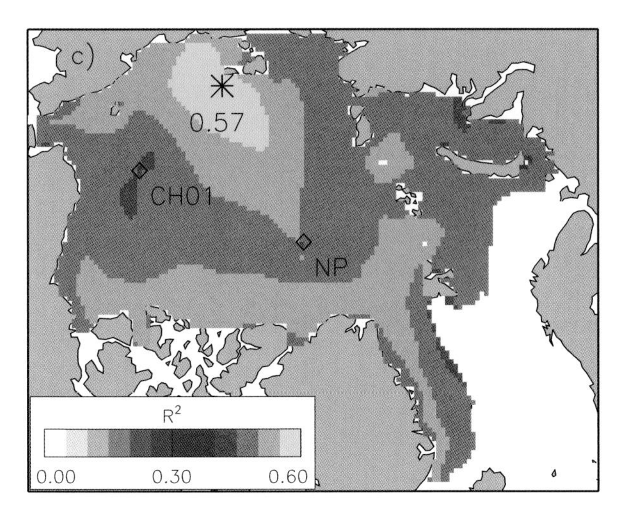


**Figure 3.16.8.** (a) Fraction of both the spatial and temporal variance of the annual mean ice thickness explained by one point, (b) the fraction explained by two points including one at the North Pole (the maximum is 0.44, and at CH01 the fraction is 0.43), and (c) the fraction explained by three points including one at the North Pole and one at CH01 (the maximum is 0.57).

### 3.16.5.2 Comments

While according to this analysis the optimal new location to monitor the basin-wide spatial variability of the annual mean thickness is in the East Siberian Sea, other factors might also be important for selecting ice thickness monitoring locations. Among these is the monitoring of locations that show large differences in the simulations from different ice models so as to provide good information about which models do well. Also, there is little information about the mean and variability of the ice thickness in the region north of the Canadian Archipelago, a region that may be particularly important if it is one of the last bastions of multiyear ice in an ice-diminished Arctic Ocean. Of course, logistics requirements will always play an important role in selecting monitoring sites.

While we have shown how EOF analysis can be used to determine where best to monitor both the temporal and the spatial variability of the ice cover, there are two important caveats. First, the method is based on model simulations, which may be flawed, and second, perhaps more important, it is based on an assumption of a stationary time series: The patterns of variability in the past will be representative of those in the future. As we saw with the EOFs, the patterns of variability are very different in the most recent 16 years compared to the whole fifty-one-year period. There is abundant evidence that the ice cover in the Arctic is changing rapidly from one of predominantly multiyear ice to one with much more extensive summer open water and winter first-year ice. This dramatic regime change makes analyses of the historic ice cover difficult to interpret when planning future monitoring efforts.

## 3.16.6 SUMMARY

There are many good examples where models can be used in the design and interpretation of field measurements. We have provided just a few in this chapter. Models are useful tools in generating hypothesis-explaining observed behavior, or may be used to provide information that is not available observationally. There is a close relationship between modeling and field measurements, where models are used to summarize and generalize observed phenomena. Theoretical investigations of sea ice also can lead to insight into phenomena that we would like to observe with field measurements. Hence model development and fieldwork often sit hand in hand.

With the advent of modern computing, sea ice modeling has evolved from process models focused on the theoretical understanding of single phenomena to simulations of sea ice in the global climate and ecological systems. Such large-scale models require validation against observations with global coverage, extending several decades. Satellite observations do not provide coverage of all variables we wish to validate in models. Some parameters can only be measured in situ. This highlights the importance of field measurements in improving large-scale model simulations. There is great need for consistency in these observations, maintaining monitoring networks and ensuring uniform methodology over many past and future decades, and for data centers that collate global field observations.

In the last two decades there has been a move towards field campaigns that are specifically designed to validate and improve parameterizations of sea ice in large-scale models and GCMs (e.g., SHEBA and SEDNA). These campaigns recognize that sub-grid-scale processes must be parameterized, and that special data sets are required to develop and validate these parameterizations.

As our understanding of the importance of sea ice in the Earth system has grown, so has our desire and need to monitor the global ice packs. Large-scale models provide insight into the physical processes driving sea ice change and can be used to forecast such change. The same models can also be used to design global observing systems for the ice pack. We could be entering an era of a rapidly diminishing arctic ice pack and increasing economic and human activity in the Arctic. The efficient development and maintenance of ice pack observing systems would be an asset to the safety and well-being of those using the Arctic, and will provide much-needed clarification of the role sea ice plays in the global climate. Models play an important role in such a monitoring system, as does improved monitoring technology and the coordination of field measurements into accessible databases.

## REFERENCES

Bear, J., and Y. Bachmat (1991), Introduction to Modeling of Transport Phenomena, in *Porous Media, Theory, and Application of Transport in Porous Media*, vol. 4, Kluwer Academic Publishers, Dordrecht, The Netherlands.

Bitz, C. M., and W. H. Lipscomb (1999), An energy-conserving thermodynamic model of sea ice, *J. Geophys. Res.., 104*, 15,669–16,677.

Cole, D. M., H. Eicken, K. Frey, and L. H. Shapiro (2004), Observations of banding in first-year Arctic sea ice, *J. Geophys. Res., 109*, C08012. doi:10.1029/2003JC001993.

Cole, D. M., and L. H. Shapiro (1998), Observations of brine drainage networks and microstructure of first-year sea ice, *J. Geophys. Res., 103*(C10), 21,739–21,750.

Coon, M., R. Kwok, G. Levy, M., Pruis, H. Schreyer, and D. Sulsky (2007), Arctic Ice Dynamics Joint Experiment (AIDJEX) assumptions revisited and found inadequate, *J. Geophys. Res., 112*, C11S90, doi:10.1029/2005JC003393.

Cressie, N. A. C. (1993), *Statistics for Spatial Data*, Wiley Series in Probability and Statistics, John Wiley, New York.

David, T. W. E., and R. E. Priestley (1914), Geology, in *British Antarctic Expedition 1907–1909, vol. 1*, pp. 9, 185–187, William Heinemann, London.

Ebert, E. E., and J. A. Curry (1993), An intermediate one-dimensional thermodynamic sea ice model for investigating ice-atmosphere interactions, *J. Geophys. Res., 98*(C6), 10,085–10,109.

Eicken, H., T. C. Grenfell, D. K. Perovich, J. A. Richter-Menge, and K. Frey (2004),

Hydraulic controls of summer Arctic pack ice albedo, *J. Geophys. Res., 109*, C08007, doi:10.1029/2003JC001989.

Eicken, H., W. B. Tucker III, and D. K. Perovich (2001), Indirect measurements of the mass balance of summer Arctic sea ice with an electromagnetic induction device, *Annals of Glaciology, 33*, 194–200.

Eykhoff, P. (1974), *System identification: Parameter and state estimation*, John Wiley & Sons, Chichester, UK.

Flato, G., and W. D. Hibler III (1992), Modeling pack ice as a cavitating fluid, *J. Phys. Oceanog., 22*, 626–651.

Flemings, M. C. (1974), *Solidification Processing*, McGraw-Hill, New York.

Forsberg, R., K. Keller, S. M. Jacobsen, F. LeDrew, and Ellsworth (chairperson) (2002), Airborne lidar measurements for Cryosat validation, in *24th Canadian ASymposium on Remote Sensing: June 24–28, Toronto, Ontario; International Geoscience and Remote Sensing Symposium*, pp. 3, 1,756–1,758.

Ganesan, S., and D. R. Poirier (1990), Conservation of mass and momentum for the flow of interdendritic liquid during solidification, *Metallurgical Transactions B, 21B*(1), 173–181.

Geiger, C. A., and M. A. Drinkwater (2005), Coincident buoy- and SAR- derived surface fluxes in the western Weddell Sea during Ice Station Weddell 1992, *J. Geophys. Res., 110*, C04002, doi:10.1029/2003JC002112.

Golden, K. M., S. F. Ackley, and V. I. Lytle (1998), The percolation phase transition in sea ice, *Science, 282*(5397), 2238–2241.

Golden, K. M., H. Eicken, A. L. Heaton, J. Miner, D. J. Pringle, and J. Zhu (2007), Thermal evolution of permeability and microstructure in sea ice, *Geophysical Research Letters, 34*, L16501, doi:10.1029/2007GL030447.

Haskell, T. G., W. H. Robinson, and P. J. Langhorne (1996), Preliminary results from fatigue tests on in situ sea ice beams, *Cold Regions Science and Technology, 24*, 167–176. doi:10.1016/0165-232X(95)00015-4.

Hibler, W. D., III (1977), A viscous sea ice law as a stochastic average of plasticity, *J. Geophys. Res., 82*(27), 3932–3938.

Hibler, W. D., III (1979), A dynamic thermodynamic sea ice model, *J. Phys. Oceanog., 9*, 815–146.

Hibler, W. D., III, and E. Schulson (2000), On modeling the anisotropic failure and flow of flawed sea ice, *J. Geophys. Res. 105*(C7), 17,105–17,120.

Hunkins, K. (1975), The ocean boundary layer and stress beneath a drifting ice floe, *J. Geophys. Res., 80*(24), 3,425–3,433.

Hutchings, J. K., and W. D. Hibler, III (2008), Small scale sea ice deformation in the Beaufort Sea seasonal ice zone, *J. Geophys. Res., 113*, C08032, doi:10.1029/2006JC003971.

Kreyscher, M., M. Harder, P. Lemke. and G. M. Flato (2000), Results of the sea ice model intercomparison project: Evaluation of sea ice rheology schemes for use in climate simulations, *J. Geophys. Res., 105*(C5), 11,299–11,320.

Kwok, R. (2001), Deformation of the Arctic Ocean ice cover between November 1996 and April 1997: A survey in I*UTAM Symposium on Scaling Laws in Ice Mechanics and Ice Dynamics*, edited by J. P. Dempsey and H. H. Shen, Kluwer Academic Publishers.

Lake, R. A. and E. L. Lewis (1970), Salt rejection by sea ice during growth, *J. Geophys. Res., 75*(3), 583–597.

Langhorne, P. J., and T. G. Haskell (2004), The flexural strength of partially refrozen cracks in sea ice in *Proceedings of the 14th International Offshore and Polar Engineering Conference*, May 23–28 2004, Toulon, France, pp. 819–824.

Le Bars, M., and M. G. Worster (2006), Solidification of a binary alloy: Finite-element, single-domain simulation, and new benchmark solutions, *Journal of Computational Physics*, *216*, 247–263.

Lepparanta, M. (2005), *The Drift of Sea Ice,* Springer-Verlag, Berlin.

Light, B., G. A. Maykut, and T. C. Grenfell (2003), Effects of temperature on the microstructure of first-year Arctic sea ice, *J. Geophys. Res.*, *108*(C2), 3051, doi:10.1029/2001JC000887.

Lindsay, R. W., and J. Zhang (2005), The thinning of arctic sea ice, 1988–2003: Have we passed a tipping point?, *J. Climate*, *18*(22), 4879–2894.

Lindsay, R., and J. Zhang (2006), Arctic Ocean ice thickness: Modes of variability and the best locations from which to monitor them, *J. Phys. Oceanog.*, *36*(3), 496–506.

Marco, J., and R. Thomson (1977), Rectilinear leads and internal motions in the ice pack of the western Arctic Ocean, *J. Geophys. Res.*, *83*, 979–987.

Maykut, G. A., and N. Untersteiner (1971), Some results from a time-dependent thermodynamic model of sea ice, *J. Geophys. Res., 76*(6), 1550–1575.

McPhee, M. G., (1979), An analysis of pack ice drift in summer, in *Proceedings of ICIS/AIDJEX Symp. On Sea Ice Process and Models*, edited by R. S. Pritchard, pp. 62–75, University of Washington, Seattle.

Medjani, K., (1996), Numerical simulation of the formation of brine pockets during the freezing of the NaCl–$H_2O$ compound from above, *International Communications in Heat and Mass Transfer, 23*(7), 917–928.

Morison, J. H., K. Aagaard, K. K. Falkner, K. Hatakeyama, R. Moritz, J. E. Overland, D. Perovich, K. Shimada, M. Steele, T. Takizawa, and R. Woodgate (2002), North Pole Environmental Observatory Delivers Early Results, *Eos Trans, AGU, 83*, 357, 360–361.

Nansen, F. (1902), *The Oceanography of the North Polar Basin, V(IX), Sci. Res.*

Niedrauer T. M. and S. Martin (1979), An experimental study of brine drainage and convection in young sea ice, *J. Geophys. Res., 84*(C3), 1176–1186.

Nikiforov, Y. G. (1957), Changes in the compactness of the ice cover in relation to the dynamics of the ice cover, *Problems of the Arctic*, 2 Morskoi Transport Press, Leningrad [in Russian].

Nikiforov, E. G., Z. M. Gudkovich, Y. N. Yefimov, and M. A. Romanov (1967), Principles of a method for calculating for ice redistribution under the influence of wind during the navigation period in arctic seas, *Tr. Arkt. Antartkt. Inst., 257*, 5–25.

Oertling, A. B. and R. G. Watts (2004), Growth of and brine drainage from NaCl–$H_2O$ freezing: A simulation of young sea ice, *J. Geophys. Res., 109*, C04013, doi:10.1029/2001JC001109.

Parkinson, C. L., and W. M. Washington (1979), A large-scale numerical model of sea ice, *J. Geophys. Res., 84*(C1), 311–337.

Partington, K., T. Flynn, D. Lamb, C. Bertoia, and K. Dedrick (2003), Late twentieth century northern hemisphere sea ice record from U.S. National Ice Center ice charts, *J. Geophys. Res., 108*(C11), 3343, doi:10.01029/2002JC001623.

Patankar, S. V. (1980), *Numerical Heat Transfer and Fluid Flow*, Hemisphere Publishing Co., New York.

Perovich, D. K., T. C. Grenfell, J. A. Richter-Menge, B. Light, W. B. Tucker, and H. Eicken (2003), Thin and thinner: Sea ice mass balance measurements during SHEBA, *J. Geophys. Res., 108*(C3), 8050, doi:10.1029/2001JC001079.

Petrich, C. (2005), Growth, structure, and desalination of refreezing cracks in sea ice, unpublished PhD thesis, University of Otago, Dunedin, New Zealand.

Petrich, C., P. J. Langhorne, and H. Eicken (2006a), Fluid Dynamics Simulations of Sea Ice Structure in the Context of Biological Activity and Meltwater Percolation, *Eos Trans. AGU, 87*(52), Fall Meet. Suppl., Abstract C41D-0364.

Petrich, C., P. J. Langhorne, and Z. F. Sun (2006b), Modelling the interrelationships between permeability, effective porosity and total porosity in sea ice, *Cold Regions Science and Technology, 44*(2), 131–144.

Petrich, C., P. J. Langhorne, and T. G. Haskell (2007), Formation and structure of refrozen cracks in land-fast first-year sea ice, *J. Geophys. Res., 112*, C04006, doi:10.1029/2006JC003466.

Pfaffling, A., C. Haas, and J. E. Reid (2004), Empirical processing of HEM data for sea ice thickness mapping, expanded in *Abstracts, 10th European Meeting of Environmental and Engineering Geophysics*, Utrecht, The Netherlands. Paper A037.

Pritchard, R. S. (ed.) (1979), *Proceedings ICIS/AIDJEX Symposium on Sea Ice Process and Models*, University of Washington, Seattle.

Reed, R. J., and W. J. Campbell (1960), Theory and observations of the ice station Alpha, *ONR final report, Technical Report NR 307-250*, University of Washington, Seattle.

Richter-Menge, J. A., S. L. McNutt, J. E. Overland, and R. Kwok (2002), Relating Arctic pack ice stress and deformation under winter conditions, *J. Geophys. Res., 107*(C10), 8040, doi:10.1029JC00477.

Richter-Menge, J. A., D. K. Perovich, B. C. Elder, I. Rigor, and M. Ortmeyer (2006), Ice mass balance buoys: A tool for measuring and attributing changes in the thickness of the Arctic sea ice cover, *Annals of Glaciology, 44*, 205–210.

Rothrock, D. A., (1975), The energetics of the plastic deformation of pack ice by ridging, *J. Geophys. Res., 80*(33), 4514–4519.

Rothrock, D. A., Y. Yu, and G. A. Maykut (1999), Thinning of the Arctic sea ice cover, *Geophys. R. Lett., 26*, 3469–3472.

Rothrock, D. A., J. Zhang, and Y. Yu (2003), The arctic ice thickness anomaly of the 1990s: A consistent view from observations and models, *J. Geophys. Res., 108*(C3), doi:10.1029/ 2001JC001208.

Shackleton, E. H. (1909), *The Heart of the Antarctic, vol. 2*, William Heinemann, London.

Schulson, E. M., and W. D. Hibler III (2004), Fracture of the winter sea ice cover on the Arctic Ocean, *Comptes Rendus-Physique*, *5*, 753–767.

Semtner, A. J. (1976), A model for the thermodynamic growth of sea ice in numerical investigations of climate, *J. Phys. Oceangr., 6*, 379–389.

Squire, V. A., W. H. Robinson., P. J. Langhorne, and T. G. Haskell (1988), Vehicles and aircraft on floating ice, *Nature*, *333*, 159–161.

Stroeve, J., M. M. Holland, W. Meier, T. Scambos, and M. Serreze (2007), Arctic sea ice decline: Faster than forecast, *Geophys. Res. Lett., 34*, L09501, doi:10.1029/2007GL029703.

Stroeve, J. C., M. C. Serreze, F. Fetterer, T. Arbetter, W. Meier, J. Maslanik, and K. Knowles (2005), Tracking the Arctic's shrinking ice cover: Another extreme September minimum in 2004, *Geophys. Res. Lett., 32*, L04501, doi:10.1029/2004GL021810.

Sverdrup, H. U. (1928), *The Norwegian North Polar Expedition with Maud 1918–1925, Scientific Results, The wind-drift of the ice on the North Siberian Shelf.*

Tucker, W. B., III, J. W. Weatherly, D. T. Eppler, D. Farmer, and D. L Bentley (2001), Evidence for the rapid thinning of sea ice in the western Arctic Ocean at the end of the 1980s, *Geophys. Res. Lett., 28*(9), 2851–2854.

Untersteiner, N. (1986), *The Geophysics of Sea Ice,* Plenum, New York.

Wadhams, P. (1988), The underside of Arctic sea ice imaged by sonar, *Nature, 333*(6169), 161–164.

Wadhams, P. (2000), *Ice in the Ocean*, Taylor and Francis Ltd.

Wadhams, P., and N. R. Davis (2000), Further evidence of ice thinning in the Arctic Ocean, *Geophys. Res. Lett., 27*, 3973–3975.

Williams, T. D., and V. A. Squire (2006), Scattering of flexural-gravity waves at the boundaries between three floating sheets with applications, *J. Fluid Mech., 569*, 113–140.

Wilks, D. S. (1995), *Statistical Methods in the Atmospheric Sciences, Vol 59, First Edition: An Introduction (International Geophysics)*, Academic Press, .

Zhang, J., and D. A. Rothrock (2005), The effect of sea ice rheology in numerical investigations of climate, *J. Geophys. Res., 110,* C08014, doi:10.1029/2004JC002599.

Zhang, J., D. A. Rothrock, and M. Steele (2000), Recent changes in Arctic sea ice: The interplay between ice dynamics and thermodynamics, *J. Climate, 13,* 3099–3114.

Chapter 3.17

# Integrated Sea Ice Observation Programs

*Matthew Druckenmiller and Christian Haas*

## 3.17.1 INTRODUCTION

This chapter highlights two integrated and interdisciplinary observation programs from opposite poles: first, a "permanent" coastal sea ice observatory along Alaska's arctic coast at Barrow, and second, the Ice Station Polarstern's drift in the Antarctic's Southern Ocean. Here the term *integrated* refers to efforts to bring together data from different instruments, methods, and disciplines, but importantly also efforts to interface disparate types of information and knowledge, as is especially the case in the first example where differences in epistemology are fundamentally important. In contrast to the Antarctic, research in the Arctic is increasingly acknowledging that the systems under investigation are often not remote at the edge of the earth, but rather are part of a social-ecological system which in most cases includes indigenous peoples. In both examples, however, this objective of integrated observations is to arrive at an improved understanding of the complexity of sea ice's role in supporting ecosystem services. Often, a scientific approach of addressing issues of complexity involves targeting spatial and temporal transitional boundaries. In the first example in the Arctic, the observatory targets the coastal environment, which is an area of dramatic transition as far as ice use, ecology, and sea ice dynamics are concerned. In the second example, it will be seen that placing the drifting camp at the transition between first- and second-year ice was of scientific as well as logistical importance. Periods when observations are scientifically most important and observation intensity is increased, such as during spring melt and breakup, present challenges in terms of resources and logistics, safety, and sampling strategies, as will be seen in this chapter.

The efforts presented here are the result of many people representing a diverse assortment of skills, experience, and knowledge. There are those who brainstormed the initial ideas, planned the programs, provided expert consultation, orchestrated logistics, labored in the field, interpreted results, and, perhaps most important, mentored the programs along the way. While it is impossible to give credit to each and every person involved in these programs, this chapter presents lists of contributors and primary funding sources at the end of each section, as well as a short compilation of relevant publications to serve as a resource for more information. Also, this chapter will refer to the techniques and methods discussed in the preceding chapters of this text, but will not go into much detail. In most cases, a cross-chapter reference is provided for accessing further information.

Other examples of integrated observation programs include the Surface Heat Budget of the Arctic Ocean (SHEBA) Project in 1997–1998 and the Ice Station Weddell (ISW) in the Weddell Sea off Antarctica in 1992. Both were interdisciplinary and international efforts to better understand the complex role that sea ice plays in the climate system.

## 3.17.2 COASTAL SEA ICE OBSERVATORY AT BARROW, ALASKA

In spring, the village of Barrow, Alaska, demonstrates the most significant use of sea ice by a local community as perhaps anywhere in the world. As bowhead whales migrate up through the Bering Strait to their summer feeding waters in the eastern Beaufort Sea, dozens of Iñupiat whaling crews stage a hunt from the shorefast ice of the Chukchi Sea. The development of a predictable flaw lead (see Figure 3.17.1) through which the whales migrate allows Barrow and a handful of other coastal Alaskan communities to partake in this traditional activity that utilizes the communities' rich body of local sea ice knowledge. Given that Barrow also has a long history of working with scientists to understand and monitor the arctic environment (Brewster 1997), it represents a microcosm for understanding how science and local knowledge in the Arctic can collectively examine questions related to environmental change and the associated impacts to ecosystem services or, in this case, sea ice system services.

Since the late 1990s, a sea ice observing effort has been maintained in Barrow, such that certain components of the observatory are now providing long-term data sets revealing trends in the timing of key events in the annual evolution of the seasonal coastal ice zone. The observatory location (see Figure 3.17.2) is ideal for a number of reasons. First, the local sea ice cover undergoes the full spectrum of sea ice's transitional processes to include most major ice types at different times of the year. Shorefast ice, which is typically present from November through July, is predominantly composed of first-year ice. Traces of multiyear ice are often found incorporated into the shorefast ice; however, this occurrence has become less frequent in recent years (Huntington et al. 2001; Drobot and Maslanik 2003)—a point of concern for local ice users who typically associate the multiyear ice with a stably grounded shorefast ice cover (George et al. 2004). While the coastline is exposed to open water from July to September, it is common to encounter drifting ice during

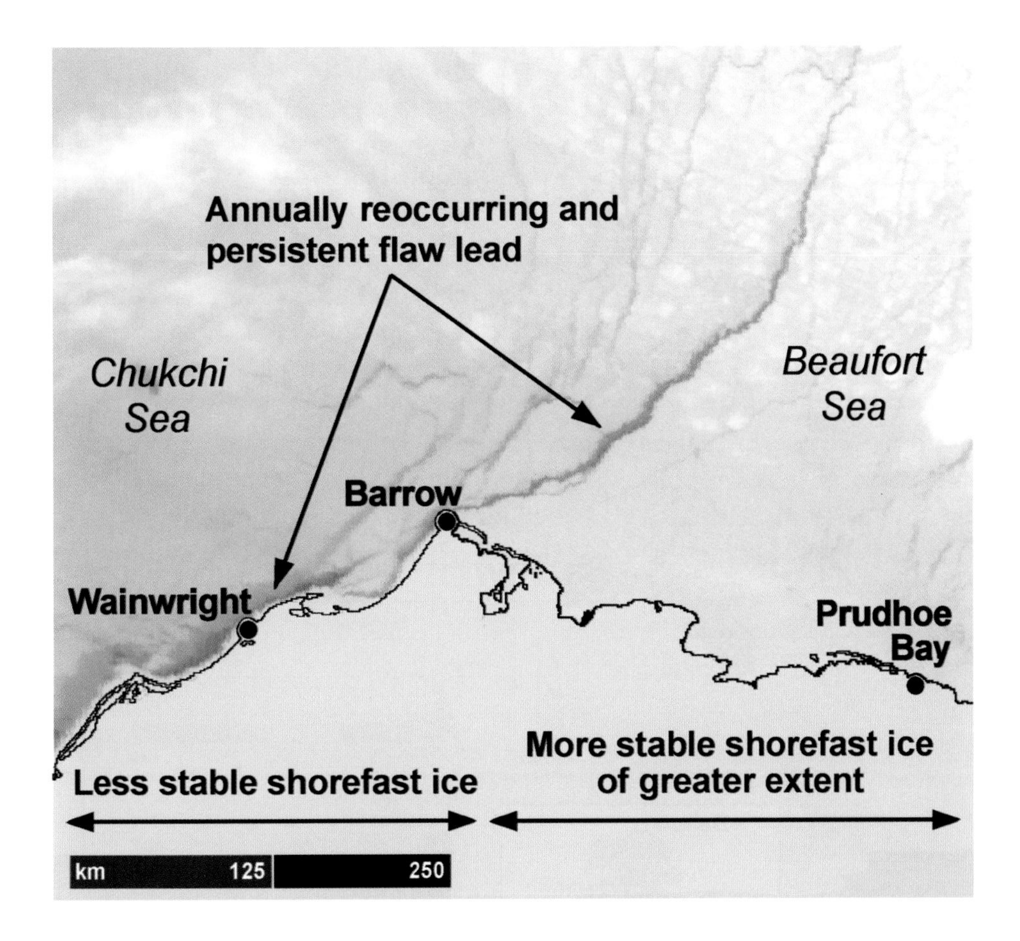


**Figure 3.17.1.** Alaska's northernmost arctic coastline overlaid on an AVHRR satellite scene from 19 February 2004. An annually recurring and seasonally persistent flaw lead exists along Alaska's northwestern coast with the Chukchi Sea and extends into the Beaufort Sea. This lead provides local hunters consistent access to bowhead whales, which make use of this open water during their eastward springtime migration. Sea ice dynamics in this region lead to relatively less stable and narrow shorefast ice along the Chukchi, while more stable shorefast ice of greater extent persists along the Beaufort coast (Eicken et al. 2006; Mahoney et al. 2007b). Barrow and Wainwright (populations of approx. 4500 and 500 respectively) are Iñupiat communities that practice spring whaling in the flaw lead. Prudhoe Bay is the origin of the Trans-Alaska Pipeline and a regional hub for near-shore and onshore oil and gas development, which typically conducts associated industrial activities from shorefast ice.

this time depending on proximity to pack ice and to areas of relatively late summer breakup (e.g., due to large grounded ridge remnants from the previous ice year). Second, the extensive list of both past and present studies in the area, ranging from coastal erosion to sea ice microorganisms to marine mammal habitat, provides broad opportunities to address questions related to the role of sea ice in supporting the current state of the ecosystem.

Following the 2000 Barrow Symposium on Sea Ice, which included bringing together a diverse assortment of sea ice experts from within and beyond the

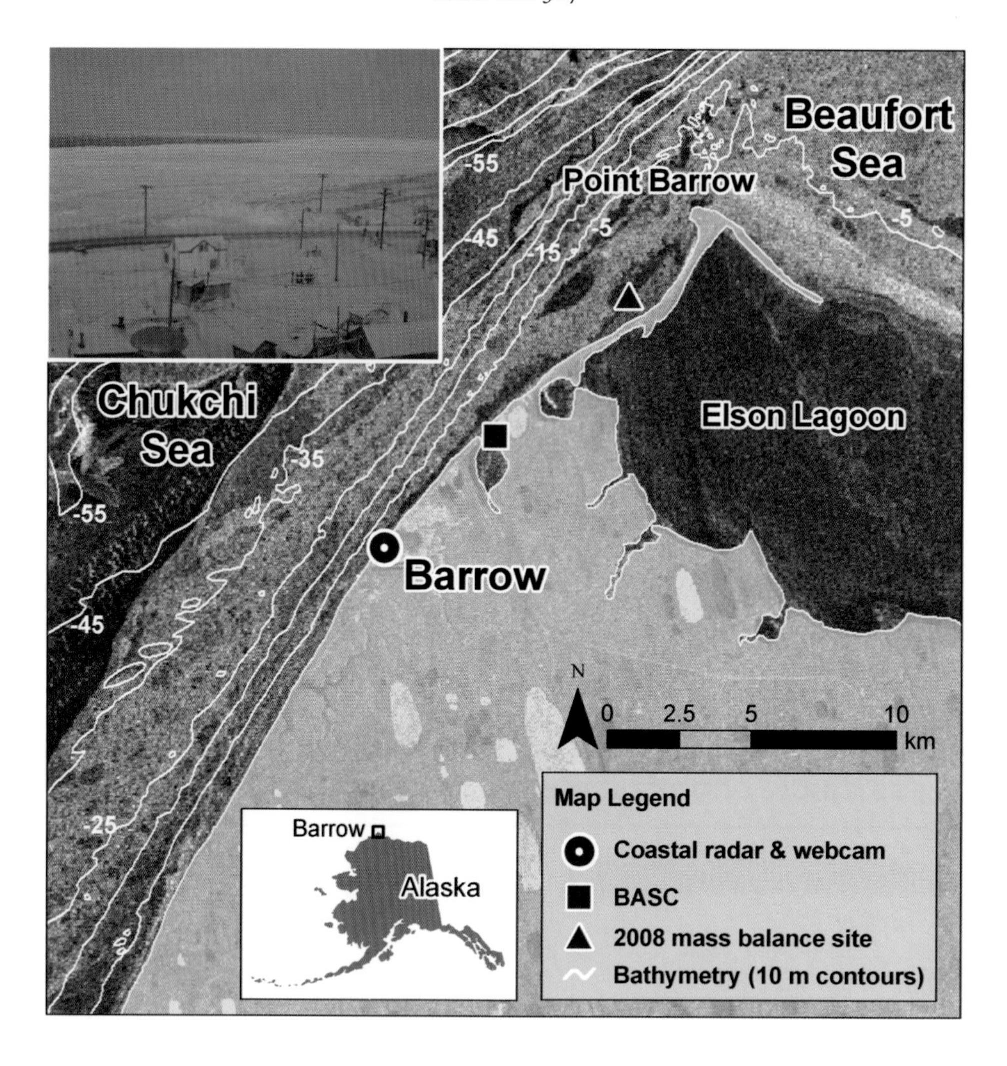


**Figure 3.17.2.** Map of the Barrow coastal sea ice observatory overlaid on an ERS-2 SAR satellite scene from 22 March 2008. A winter image from the coastal webcam, which is mounted atop a four-story building in downtown Barrow, is shown in the upper right corner. Field campaigns are staged from the Barrow Arctic Science Consortium (BASC), a few miles north of the village. This image shows the site of the mass balance site and sea-level gauge in 2008 located in level first year ice within the shorefast ice zone. The slight curvature of the Chukchi coastline immediately south of Point Barrow provides a semi-protected embayment of water less than 5 m in depth.

community (Huntington et al. 2001), formal collaborations with the North Slope Borough's Department of Wildlife Management and the Alaska Ocean Observing System (AOOS) and targeted exchanges with the scientific community (Hutchings and Bitz 2005; SEARCH 2005), the main focus areas for the observatory emerged:

1. Examine key events in the annual evolution of the ice cover, including such observations as the onset of freezeup, formation of stable ice, significant snowfalls, onset of bottom-ice melt, formation of melt ponds, and breakup.

2. Examine dynamic events within the shorefast ice zone, including ice shoves, shorefast ice breakout (or calving) events, and the building of ridges through interaction with adjacent pack ice.
3. Relate the above key and dynamic events to sea ice system services, with a focus on ice travel and subsistence activities.
4. Provide near-real-time data and information to the community both as a service and as a means to communicate observations in a collaborative manner such that related local observations and knowledge can be shared.
5. Collect a broad range of temporally and spatially resolved local-scale data sets to investigate cross-scale correlations with similar data sets at regional and pan-arctic scales.

In 1998 an ice core-sampling program began on the shorefast ice off Barrow to investigate ice growth history (Eicken et al. 2004) and was followed one year later by an annually deployed mass balance site to monitor in situ ice growth in level first-year shorefast ice (Grenfell et al. 2007). Shortly thereafter, a full-time coastal radar (see Figure 3.17.3) and webcam site (see Figure 3.17.2) was established to build upon earlier radar work by Shapiro (1976) to observe near-shore ice movement, deformation, and ridging. Throughout the years, ice thickness-profiling campaigns using electromagnetic induction sounding from the ice surface were performed to investigate shorefast ice morphology, including the anchoring strength of grounded ridges (Mahoney et al. 2007a; Druckenmiller et al. 2009). The analysis of satellite imagery, such as thermal-IR and visible-range MODIS and AVHRR, synthetic aperture radar (SAR), QuikSCAT scatterometer, and passive microwave, has provided aerial perspective for ground-based measurements and a greater understanding for how these observations fit within the regional setting in terms of shorefast ice dynamics and morphology, pack ice interaction, and timing of key events in the annual ice cycle. Lastly, a local observer program was initiated in 2006 to provide near-daily sea ice observations from the perspective of the Iñupiat hunters and to assist in relating instrument measurements to sea ice use by the local community.

The different components of the observatory are summarized in Table 3.17.1, which lists the observed parameters, spatial and temporal coverage, as well as where further information may be found in this text on the various techniques. Much of the observatory's instrument data (radar, webcam, mass balance, and sea level) are transferred to the University of Alaska Fairbanks in near real time and made available on the Internet as the primary means of data dissemination to the public. Reviewing near-real-time data also allows for other components of the observatory to be responsive to seasonally evolving ice conditions. For example, ice coring and thickness measurements are scheduled to coincide with the onset of stable shorefast ice in winter, maximum ice thickness in midspring, and the onset of melt in

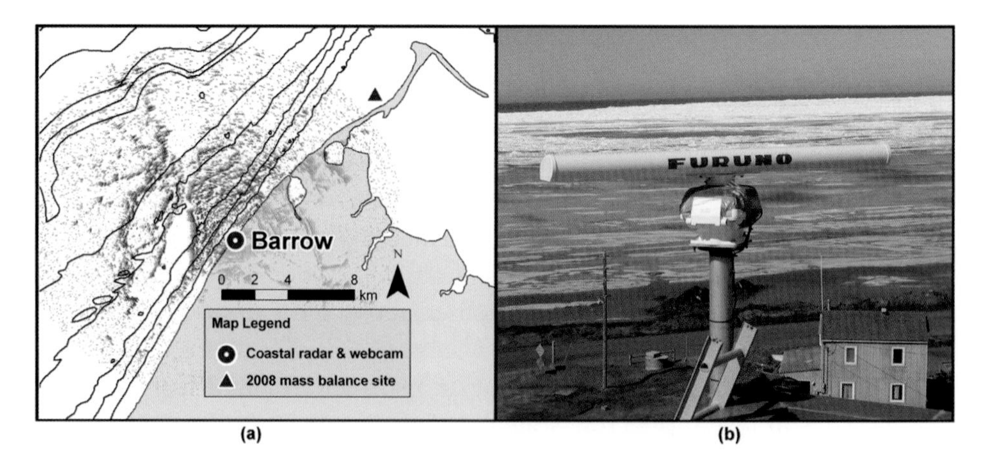


**Figure 3.17.3.** (a) Coastal radar image generated from backscatter from ridges, floe edges, and other rough features in the near-shore during spring 2007. Areas of flat sea ice or calm open water do not generate sufficient backscatter to appear in these images. Radar images are archived every three minutes to be used in daily animations of ice movement to improve upon temporal and spatial resolution relative to satellite imagery. *View the multimedia CD for animations of springtime shorefast ice detachments.* (b) The Furuno FR-7112 10 kW, X-band (3 cm, 10 GHz) marine radar with a 1.65m open array positioned 22.5 m above sea level on a building in downtown Barrow (71°17'33"N, 156°47'17"W).

late spring. Depending on ongoing projects associated with the observatory, other measurements are typically performed (e.g., snow thickness, albedo, melt pond characterization, etc.). These efforts are further supported by satellite remote-sensing products made available online within hours of acquisition through various geographic information services.

Partnerships with local ice experts and the Barrow Whaling Captains Association (BWCA) have led to unique opportunities to understand how geophysically derived data sets compare with local observations and may be of value to community activities. Combining geophysical-based interpretations with the perspective of local ice users reveals the complexity of the coastal sea ice environment and the associated epistemological difficulties that a single observer may face. The knowledge of indigenous arctic residents is often described as originating from a close connection with the natural environment (Krupnik and Jolly 2002), and therefore acknowledges a greater level of interconnectedness within natural systems than does the reductionist approach of traditional scientific research. In the case of Barrow on the coast of the Chukchi Sea, local ice experts are keenly aware of the intricate coupling of the atmosphere, ocean, and ice, and how these three components annually interact to provide access to the marine life on which they subsist. Local observers in Barrow, such as Arnold Brower Sr. (2006–2007) and Joe Leavitt (2006–2009), report on near-shore ice conditions along the Barrow coastline and the associated impacts to travel, hunting, and marine mammal behavior, as well as implications for the state of the ice later in the year. Their observations also help

**Table 3.17.1.** Components of the Barrow coastal sea ice observatory

| Component | Observed parameters and processes* | Spatial coverage | Temporal coverage | Chapter reference |
|---|---|---|---|---|
| Satellite imagery | LF ice stabilization & extent, lead occurrence, ridging, MY ice concentration | Regional scale; dependent on sensor (resolutions typically >10 m) | Dependent on sensor (data acquired daily to monthly) | 3.15 |
| Coastal radar | Ice drift, LF ice stabilization & breakout events, ridging | Within 6 km of coast | Updated online every 5 minutes | 3.17, Figure 3.17.3 |
| Coastal webcam | Presence of first ice, melt pond formation, snow cover, breakout events, open water | Within 2 km of coast, depending on visibility | Updated online every 5 minutes | 3.17, Figure 3.17.2 |
| Mass balance site | Ice thickness, snow thickness, water-ice-snow-air temperature profile, ice salinity | Point-based measurements at site on level first-year LF ice (see Figures 3.17.2 and 3.17.3) | Data updated online every 5 minutes | 3.2 and 3.13 |
| Sea level measurements | Tidal, storm surges, & wind-driven sea level fluctuations | | | N/A |
| Ice thickness & topography surveys | Ice thickness & surface elevation | Entire extent of LF ice off Barrow along approx. 20 km of coastline | 2–4 campaigns throughout spring | 3.2 |
| Ice core sampling | Salinity and temperature profiles, sediment entrainment, stable isotope analysis, permeability, etc. | Point-based measurements, typically coincident with mass balance site | 3–4 campaigns throughout spring | 3.3 |
| Local observations (journal entries & interviews) | Key events in the annual evolution of the ice cover, dynamic events, etc. | Typically within 20 km of Barrow, dependent on time of year & travel conditions | Near-daily journal entries; periodic interviews | 3.11 |

* *LF = landfast (or shorefast), MY = multiyear*

to interpret how instrument-derived data sets of ice thickness relate to community assessments of ice stability. These perspectives often shed light on aspects of the observatory's strategy that either need strengthening or are missing. For example, discussions throughout past years have identified an existing data gap for local currents, which are routinely credited for destabilizing shorefast ice through thermal ablation of grounded ridge keels in late spring.

Most of the accumulated local sea ice knowledge that resides with the Iñupiat people of Barrow has some origin or relevance to the springtime hunt of the bowhead whale—a labor-intensive activity that successfully culminates in the catching of whales in open leads using traditional skin boats and hauling the catches onto the ice edge for butchering (see Figure 3.17.4). Accessing this subsistence resource from the ice edge in a safe and responsible manner requires constant evaluation of ice conditions. The proximity and drift direction of the pack ice and the strength and direction of winds and currents are critical. It is the responsibility of a whaling captain to ensure that all people involved in the hunt make it home safely from activities on the ice, which often extend as far as 10 km offshore. During this time of year, local observations made alongside the observatory's coincident measurements, especially those of the shorefast ice thickness distribution and sea level fluctuations, provide unique opportunities for understanding perspectives on shorefast ice stability as well as for cross-cultural data and knowledge exchange. In this context, these coordinated observations may be considered a strategic means for observing the unexpected. In his book *Hunters of the Northern Ice*, an account of two

**Figure 3.17.4.** Iñupiat whalers from Barrow, Alaska, butchering a bowhead whale on the shorefast ice following a successful spring hunt. Photo by C. George.

years living with the Iñupiat hunters of Wainwright, Alaska, Richard Nelson (1969) expressed his opinion on the breadth of their sea ice knowledge when he stated that "even those statements which seem utterly incredible at first almost always turn out to be correct." Partnering with qualified local observers enhances the observatory's ability to capture the increased uncertainty and variability often associated with environmental change in the Arctic. Arctic residents not only observe variability but maintain resilient lifestyles that continually cope with change (Krupnik and Jolly 2002; Duerden 2004).

It is not well understood how the satellite-observed retreat of pan-arctic ice extent since the late 1970s (Stroeve et al. 2008) correlates with observations of the seasonal ice zone at the local scale. For example, how reductions in multiyear ice near Point Barrow in the fall impact freezeup along the coast is an important question for local residents. In a study by Mahoney et al. (2007b) of shorefast ice along Alaska's northern coastline, it was discovered that climatologically the onset of thawing is occurring earlier in spring and freeze-up is taking place later in fall. However, in this study, the increasingly late fall formation of shorefast ice was shown to best correlate with the incursion of pack ice in near-shore waters. Fitting extensive local observations in context with broader regional and pan-arctic change, while considering potential future outlooks for local-scale sea ice system services, is one of the greatest contributions that may arise from continued measurements by the Barrow sea ice observatory.

For more information related to the type of measurements and observations performed at this sea ice observatory, explore the accompanying DVD, since much of its content has been produced near Barrow, Alaska.

***Contributors***: Arnold Brower Sr., Lewis Brower, Patrick Cotter, Matthew Druckenmiller, Hyunjin Druckenmiller, Guy Dubuis, Hajo Eicken (Observatory Principal Investigator), Karoline Frey, Allison Gaylord, Craig George, Richard Glenn, Tom Grenfell, Jeremy Harbeck, Mark Johnson, Jonas Karlsson, Mette Kaufman, Joe Leavitt, Andy Mahoney, Don Perovich, Chris Petrich, Daniel Pringle, Lew Shapiro, Glenn Sheehan, Matthew Sturm, Christina Williams, Barrow Arctic Science Consortium, Barrow Whaling Captains Association, North Slope Borough Department of Wildlife Management, and many others

***Primary sources of financial and logistical support***: National Science Foundation, University of Alaska, Barrow Arctic Science Consortium, and the Alaska Ocean Observing System

***Relevant publications***:

"Ice motion and driving forces during a spring ice shove on the Alaskan Chukchi coast" (Mahoney et al. 2004)

"Hydraulic controls of summer Arctic pack ice albedo" (Eicken et al. 2004)

"Observations on shorefast ice dynamics in Arctic Alaska and the responses of the Iñupiat hunting community" (George et al. 2004)

"Sediment transport by sea ice in the Chukchi and Beaufort Seas: Increasing importance due to changing ice conditions?" (Eicken et al. 2005)
"How fast is landfast ice? A study of the attachment and detachment of nearshore ice at Barrow, Alaska" (Mahoney et al. 2007a)
"Thermal conductivity of landfast Antarctic and Arctic sea ice" (Pringle et al. 2007)
"Towards an integrated coastal sea ice observatory: System components and a case study at Barrow, Alaska" (Druckenmiller et al. 2009)

## 3.17.3 ICE STATION POLARSTERN (ISPOL): STUDY OF PHYSICAL AND BIOLOGICAL SEA ICE PROCESSES AND INTERACTIONS IN THE WEDDELL SEA, ANTARCTICA

Although only a few humans live in Antarctica, services provided by sea ice (Chapter 2) are just as important as in the Arctic. Many arctic sea ice service aspects apply as well in the Antarctic, and some could be even more important due to the close interrelation of sea ice and ice shelves, and due to the persistence of at least landfast ice throughout the summer along the coasts of the continent proper. Sea ice is used as a loading platform for resupply of research stations and a landing platform for airplanes, for example, at the U.S. station McMurdo. However, it is also an impediment for resupply ships and for the increasing number of cruise ships visiting the continent every summer.

The climate-related regulating services of antarctic sea ice might even be larger than in the Arctic, as the seasonal variability of ice extent is much larger, with most ice disappearing during the summer. The Southern Ocean and particularly the coastal polynyas and ice shelf regions are some of the most important sources of cold, deep ocean waters worldwide and drivers of the global thermohaline circulation. However, climate models fail to realistically simulate summer ice extent in particular. This may be related to insufficient representation of some of the most striking differences between arctic and antarctic sea ice, which are the occurrence of widespread flooding due to large snow/ice thickness ratios, the absence of melt ponds, contrasting seasonal cycles of microwave properties seen by satellites, and the slow increase of antarctic sea ice extent over the past three decades. These differences are the result of a relatively thick snow cover compared to the Arctic, and of generally different climatic conditions dominated by the proximity of the cold and dry antarctic ice sheet and the remoteness from other continents.

In addition to these regulating services, antarctic sea ice is an important habitat and supports high standing stocks of bacteria, algae, and zooplankton, as well as large abundances of krill, birds and penguins, seals, and whales. One of the most striking habitat-related features is the development of porous gap layers close to the ice surface during summer (Figure 3.17.5) (Haas et al. 2001; Ackley et al. 2008), which are well supplied by nutrients and light and where record concentrations

**Figure 3.17.5.** Photograph of a broken ice floe tilted by an icebreaker, showing the layered and porous structure of antarctic sea ice in summer. Green and brownish discolorations are the result of high-standing stocks of algae and other organisms.

of chlorophyll and particulate and dissolved organic matter have been measured (Thomas et al. 1998).

In order to improve the understanding of those physical and biological properties and processes at the beginning of summer and their relation to meteorological and oceanographic boundary conditions (see Figure 3.17.6), an international, interdisciplinary team of scientists conducted the Ice Station Polarstern (ISPOL) drift station project in the western Weddell Sea (Figure 3.17.7) (Hellmer et al. 2006).

**Figure 3.17.6.** Logo of the Ice Station Polarstern (ISPOL) drift station project, which studied physical and biological properties and processes on one ice floe at the beginning of summer in dependence of meteorological and oceanic boundary conditions (red and blue arrows).

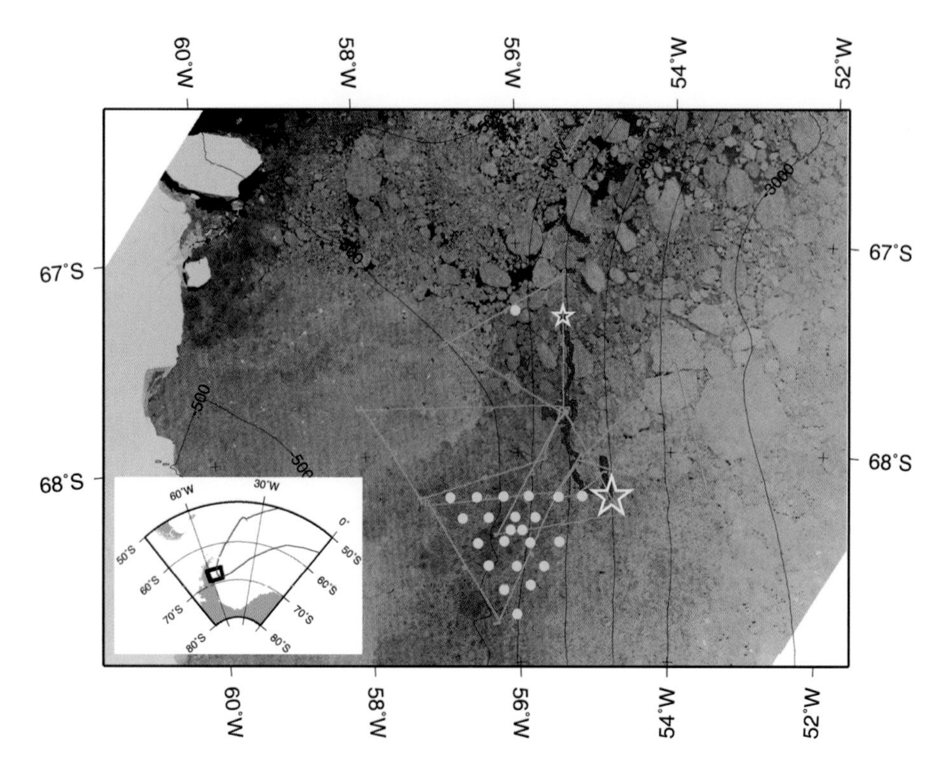


**Figure 3.17.7.** Envisat synthetic aperture radar (SAR) image acquired on 30 November 2005, showing the Ice Station Polarstern (ISPOL) study area in the western Weddell Sea (inset, plus cruise track) and the start (27 November 2004, large yellow star) and end points (2 January 2005, small yellow star) of the drift (Hellmer et al. 2006). The western border of the image shows the Larsen-C ice front; the northern boundary is close to the sea ice edge. Black contours show water depth in meters. Blue symbols indicate locations of buoy deployments, and orange lines show the tracks of HEM thickness surveys. Note the north-south extending, dark appearing band of first-year sea ice at about 56°W.

Most methods described in this handbook were applied to obtain a most complete data set of all ocean-ice-atmosphere processes and their interactions. Numerous results were published in the scientific literature, and in particular in a special ISPOL issue of *Deep Sea Research*, Part II, *Vol. 55*, No. 8–9 (Hellmer et al. 2008).

During December 2004, the German research icebreaker *Polarstern* was anchored to an ice floe and served as accommodation, laboratory, and platform for field and water column studies. The satellite radar image in Figure 3.17.7 shows the location of the station and its 98 km long (net distance) south-north drift track. Compressed files of satellite radar imagery, particularly from the QuikSCAT scatterometer and Envisat SAR (Chapter 3.15), were transmitted to the ship in near real time and played a critical role in supporting ice navigation, site selection, and interpretation of observed ice properties and seasonal changes. Based on these, the ISPOL floe location was chosen to be at the transition between two prominent bands of second- and first-year ice, allowing sampling of more varied ice types. It was also chosen to be far enough south to have not yet been affected by any melting before the onset of observations, and to prevent too early disintegration of the floe while drifting north towards the ice edge.

Long-range helicopter flights were performed to survey numbers and types of seals (Chapter 3.10), as well as the ice thickness and floe size distributions of the various ice regimes by means of EM sounding (Figure 3.17.7; Chapter 3.2) and aerial photography (Chapter 3.15). The satellite images also served the design of a triangular array of ice-deployed GPS buoys (Chapter 3.13), which were operated to study the motion and deformation of the ice and to improve the representation of these processes in sea ice models (Chapter 3.16).

Figure 3.17.8 shows a map of measurement stations on the ISPOL floe. These were selected to cover the main ice types, and to perform joint, interdisciplinary investigations of the same ice. The main activities of ice coring (for both physical

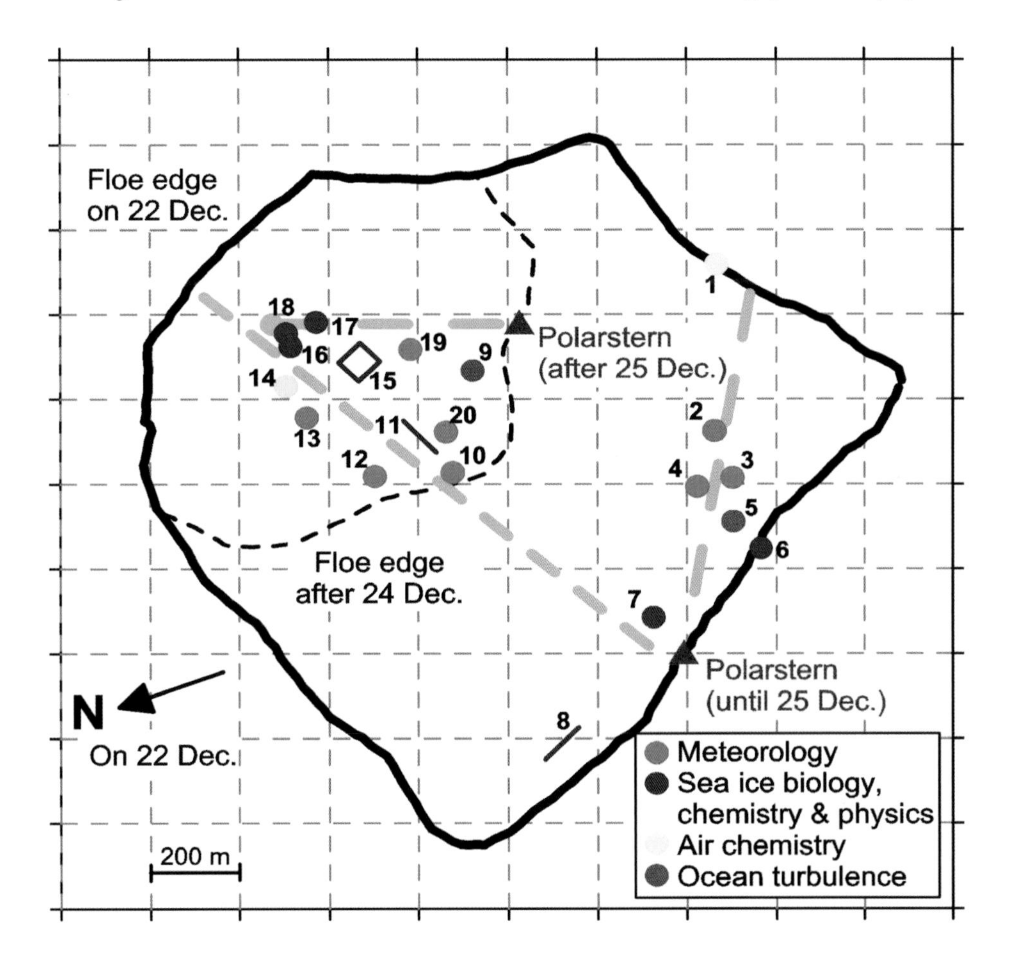


**Figure 3.17.8.** Sketch of the ISPOL floe with locations of long-term sampling sites on different ice types (thick and thin first-year ice, and second-year ice) for studies of physical and biological ocean-ice-atmosphere interactions (Hellmer et al. 2008). An aerial photograph of part of the floe detailing regions of different ice types, ice thickness, and sampling sites is shown in Figure 3.2.31.

and biological/biogeochemical parameters; Chapters 3.3, 3.8, and 3.9) and measurements of snow properties (Chapter 3.1) and ice thickness (Chapter 3.2; Figure 3.2.31) were repeated throughout the study period. Seven dedicated coring days were chosen every five days to allow for the best possible cross-comparison between sites. In addition, Conductivity-Temperature-Depth (CTD) and ocean turbulence measurements (Chapter 3.7) were performed throughout the study period to quantify ocean heat flux and to relate it to observations of ice warming and thinning. Similarly, atmospheric heat flux was measured at various locations, as well as the exchange of gases like $CO_2$ and DMS. Nets and water samplers were operated from the ship to study biology and biogeochemistry in the water column and their exchanges with the ice (Chapter 3.9). This work was partially supported by divers sampling more dedicated regions of the ice underside. Sediment traps were suspended under the ice to collect biological matter released from the ice—a potential food source for higher trophic levels.

All results obtained during ISPOL contributed unique information about ocean-ice-atmosphere properties and processes in this rarely studied sea ice region of the Southern Ocean. Although it was indeed expected that no melt ponds would form and that the snow would remain intact, arguably the most surprising result was the slowness of changes throughout the early summer observational period, in contrast to extreme conditions in the Arctic, where strong melting typically commences in early June (comparable to early December in the Southern Hemisphere) and at much higher latitudes. However, ocean and atmospheric heat fluxes averaged only a few watts per square meter, merely leading to snow and ice thinning of more than one or two decimeters. Changes in the snow cover were too small to trigger an albedo feedback that would enhance melt (Chapter 3.6). Similarly, despite significant drift and deformation of the ice field, ice concentration remained high throughout and prevented the absorption of heat in the surface water and the initiation of lateral and bottom melt.

However, the available heat led to warming of the ice and snow by up to 2 K to temperatures close to melting, which has fundamental consequences for important ice and brine properties like porosity, salinity, and stratification (Figure 3.17.9). These strongly impacted the biological productivity and chemistry within the ice, and their role in seeding growth under the ice and modifying ocean-atmosphere gas fluxes. Figure 3.17.10 shows how other parameters varied as a consequence of these changes. The higher porosity also allowed larger organisms like various zooplankton species to enter the ice proper and even the surface.

The results shown in Figure 3.17.10 also point to methodological problems of repeated, destructive ice core and snow sampling to measure small changes (Chapters 3.1, 3.3, 3.4, 3.8, and 3.9). Due to the general small-scale variability and patchiness, temporal changes were not easily separated from lateral changes, despite a sophisticated sampling strategy developed to include the sampling of several cores within a very small region and the prevention of disturbances to the original snow

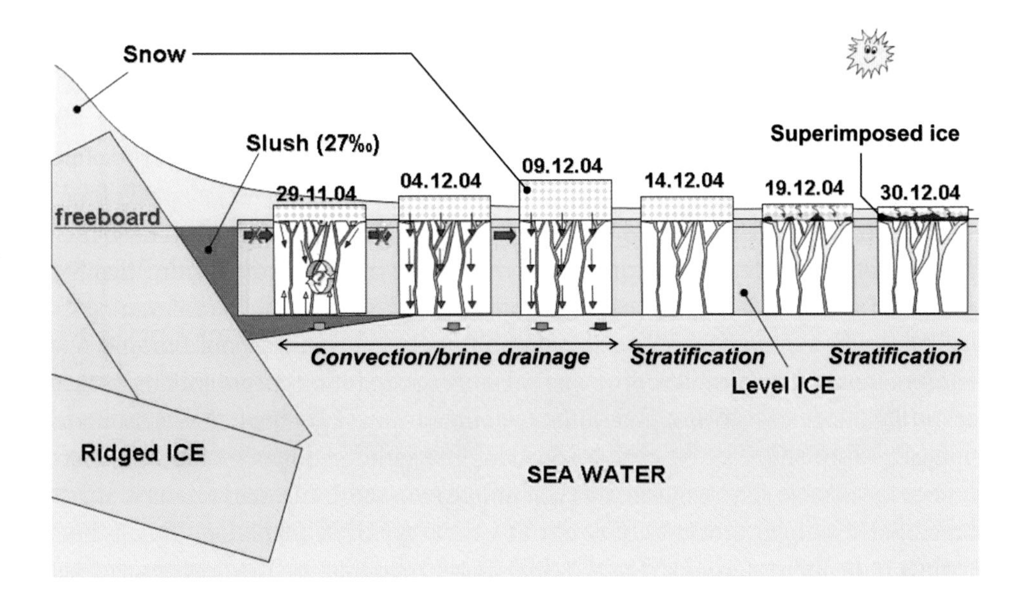


**Figure 3.17.9.** Illustration of property changes of thin first-year ice during early summer, as concluded from extensive ice coring on six sampling days throughout the ISPOL study period (Tison et al. 2008). Main changes occurred with respect to temperature-dependent ice porosity and consequent horizontal and vertical brine motion and gas exchange.

and ice surface. In addition, the thick second-year ice presented challenges for the processing of large numbers of samples within the short time available between sampling days, and led to a focus on thinner first-year ice for most biological and biogeochemical studies. The small changes occurring over the study period also affected most other applied methods, as their accuracies were hardly good enough to detect these changes. In addition, close-to-melting conditions and high irradiance affected many instruments installed on the ice for more continuous, automatic measurements.

Long-term observations were hampered by two breakup events, which made revisiting of sites difficult and partially required the relocation of some instruments. Here, the unavailability of more automatic systems for the measurement of biological and biogeochemical ice properties was a clear disadvantage. This general problem of long-term sea ice studies was, however, overcome by the use of helicopters and inflatable boats allowing continued revisiting of more remote sites and crossing of leads after the breakups. ISPOL's observations had to be concluded before any really strong melting occurred, leading to further disintegration of the ice. However, an extension of the study period was impossible because of higher-level logistical and organizational requirements, as the ship could only operate for a certain period without resupplies and also had to serve other antarctic and arctic research programs during the short summer seasons. More dedicated and extensive ship time will be required for future studies of summer melt processes in the

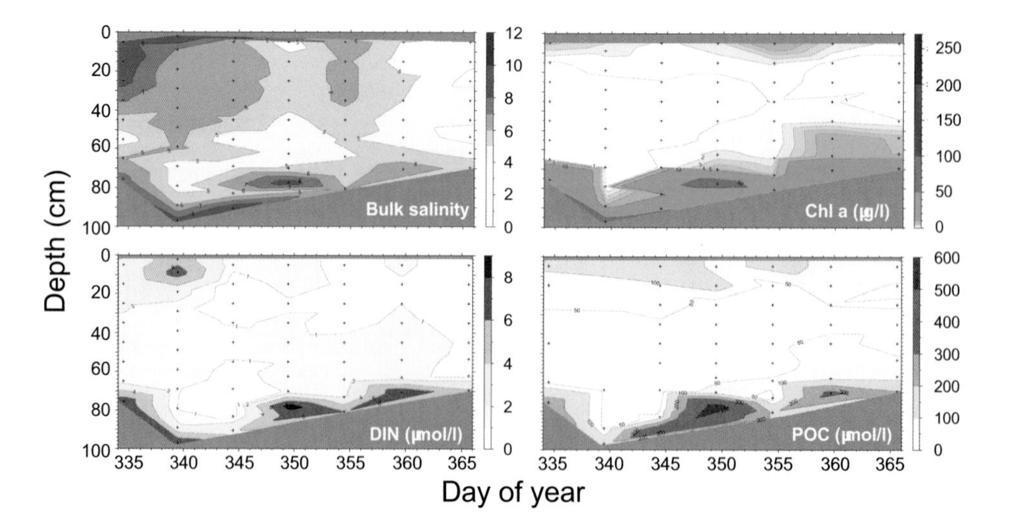


**Figure 3.17.10.** Temporal evolution of ice thickness, salinity, Chlorophyll a, dissolved inorganic nutrients (DIN) and particulate organic carbon (POC) in thin first-year ice of the ISPOL floe (Courtesy of Matthias Steffens and Gerhard Dieckmann).

Southern Ocean, as was available for the Ice Station Weddell (ISW) and Surface Heat Budget of the Arctic (SHEBA) drift stations.

***Contributors:*** ISPOL was proposed, planned, and coordinated by Gerhard Dieckmann, Christian Haas, Hartmut Hellmer, and Michael Schroeder. The cruise was led by Michael Spindler. However, ISPOL's program and success was only possible through the engagement and participation of fifty-seven scientists from eleven countries, and the support of a ship's crew of forty-three. In addition, numerous people worked on land to prepare and assist the ship before and during the cruise.

***Primary sources of support:*** ISPOL has primarily been supported by the Alfred Wegener Institute for Polar and Marine Research in Germany. However, participation and scientific data exploitation was facilitated by national funding agencies and the home institutions of the international participants.

***Relevant publications:***

There are too many publications to be listed here. However, two key publications are:

Ice Station POLarstern (ISPOL): Results of interdisciplinary studies on a drifting ice floe in the western Weddell Sea, a special ISPOL issue of *Deep Sea Research*, Part II, *Volume 55*, No. 8–9 (Hellmer et al. 2008)

The expeditions ANTARKTIS-XXII/1 and XXII/2 of the research vessel '*Polarstern*' in 2004/2005 (El Naggar et al. 2007)

## REFERENCES

Ackley, S. F., M. J. Lewis, C. H. Fritsen, and H. Xie (2008), Internal melting in Antarctic sea ice: Development of "gap layers," *Geophys. Res. Lett., 35*, L11503, doi:10.1029/2008GL033644.

Brewster, K. (1997), Native contributions to Arctic sciences at Barrow, Alaska, *Arctic 50*(3), 277–288.

Drobot, S. D., and J. A. Maslanik (2003), Interannual variability in summer Beaufort Sea ice conditions: Relationship to winter and summer surface and atmospheric variability, *J. Geophys. Res., 108*(C7), 3233, doi:10.1029/2002JC001537.

Druckenmiller, M. L., H. Eicken, M. A. Johnson, D. J. Pringle, and C. C. Williams (2009), Towards an integrated coastal sea ice observatory: System components and a case study at Barrow, Alaska, *Cold Reg. Sci. Technol.* 56(2–3), 61–72.

Duerden, F. (2004), Translating climate change impacts at the community level, *Arctic 57*(2), 204–212.

Eicken, H., R. Gradinger, A. Graves, A. Mahoney, and I. Rigor (2005), Sediment transport by sea ice in the Chukchi and Beaufort Seas: Increasing importance due to changing ice conditions?, *Deep-Sea Res. II, 52*, 3281–3302.

Eicken, H., T. C. Grenfell, D. K. Perovich, J. A. Richter-Menge, and K. Frey (2004), Hydraulic controls of summer Arctic pack ice albedo, *J. Geophys. Res., 109*, doi:10.1029/2003JC001989.

Eicken, H., L. H. Shapiro, A. G. Graves, A. Mahoney, and P. W. Cotter (2006), Mapping and characterization of recurring spring leads and landfast ice in the Beaufort and Chukchi Sea, OCS Study MMS 2005-068.

El Naggar, S., G. Dieckmann, C. Haas, M. Schroeder, and M. Spindler (eds.) (2007), The Expeditions ANTARKTIS-XXII/1 and XXII/2 of the research Vessel "Polarstern" in 2004/2005, Reports on Polar and Marine Research, 551.

George, J. C., H. P. Huntington, K. Brewster, H. Eicken, D. W. Norton, and R. Glenn (2004), Observations on shorefast ice dynamics in Arctic Alaska and the responses of the Iñupiat hunting community, *Arctic 57*(4), 363–374.

Grenfell, T. C., D. K. Perovich, H. Eicken, B. Light, J. Harbeck, T. G. George, and A. Mahoney (2007), Energy and mass balance observations of the land-ice-ocean-atmosphere system near Barrow, Alaska, November 1999–July 2002, *Ann. Glaciol., 44*, 193–199.

Haas, C., D. N. Thomas, and J. Bareiss (2001), Surface properties and processes of perennial Antarctic sea ice in summer, *J. Glaciol., 47*(159), 613–625.

Hellmer, H. H., G. S. Dieckmann, C. Haas, and M. Schröder (2006), Sea ice feedbacks observed in western Weddell Sea, *Eos Trans. AGU, 87*(18), 173–179.

Hellmer, H. H., C. Haas, M. Schröder, G. S. Dieckmann, and M. Spindler (2008), The ISPOL drift experiment, *Deep Sea Research II, 55*(8–9), 913–917, doi:10.1016/j.dsr2.2008.01.001.

Huntington H. P., H. Brower, and D. W. Norton (2001), The Barrow Symposium on Sea Ice, 2000: Evaluation of one means of exchanging information between subsistence whalers and scientists, *Arctic 54*(2), 201–204.

Hutchings, J., and C. Bitz (2005), Sea ice mass budget of the Arctic (SIMBA) workshop: Bridging regional to global scales, *Report from NSF sponsored workshop held at Applied Physics Laboratory*, University of Washington, Seattle, 28 February–2 March 2005.

Krupnik, I., and D. Jolly (eds.) (2002), *The Earth is Faster Now: Indigenous Observations for Arctic Environmental Change*, Arctic Research Consortium of the United States, Fairbanks, Alaska.

Mahoney, A., H. Eicken, L. Shapiro, and T. C. Grenfell (2004), Ice motion and driving forces during a spring ice shove on the Alaskan Chukchi coast, *J. Glaciol., 50*(169), 195–207.

Mahoney, A., H. Eicken, and L. Shapiro (2007a), How fast is landfast ice? A study of the attachment and detachment of nearshore ice at Barrow, Alaska, *Cold Reg. Sci. Technol., 47,* 233–255.

Mahoney, A., H. Eicken, A. G. Gaylord, and L. Shapiro (2007b), Alaska landfast sea ice: Links with bathymetry and atmospheric circulation, *J. Geophys. Res., 112,* C02001, doi:10.1029/2006JC003559.

Nelson, R. (1969), *Hunters of the Northern Ice*, University of Chicago Press, Chicago.

Pringle, D. J., H. Eicken, H. J. Trodahl, and L. G. E. Backstrom (2007), Thermal conductivity of landfast Antarctic and Arctic sea ice, *J. Geophys. Res., 112,* C04017, doi:10.1029/2006JC003641.

SEARCH (2005), Study of environmental Arctic change: Plans for implementation during the International Polar Year and beyond, *Report of the SEARCH Implementation Workshop*, Lansdowne, Virginia, 23–25 May 2005.

Shapiro, L. H. (1976), A preliminary study of ridging in landfast ice at Barrow, Alaska, using radar data, *3rd International Conference on Port and Ocean Engineering under Arctic Conditions, Vol. 1,* pp. 417–425, University of Alaska Fairbanks.

Stroeve, J., M. Serreze, S. Drobot, S. Gearheard, M. Holland, J. Maslanik, W. Meier, and T. Scambos (2008), Arctic sea ice extent plummets in 2007, *Eos Trans. AGU, 89*(2), 13–20.

Thomas, D. N., R. J. Lara, C. Haas, S. B. Schnack-Schiel, G. S. Dieckmann, G. Kattner, E.-M. Nöthig, and E. Mizdalski (1998), Biological soup within decaying summer sea ice in the Amundsen Sea, Antarctica, *Antarctic Research Series AGU* 73, pp. 161–171, Washington DC.

Tison, J.-L., A. Worby, B. Delille, F. Brabant, S. Papadimitriou, D. Thomas, J. de Jong, D. Lannuzel, and C. Haas (2008), Temporal evolution of decaying summer first-year sea ice in the Western Weddell Sea, Antarctica, *Deep Sea Research II, 55*(8–9), 975–987, doi:10.1016/j.dsr2.2007.12.021.

# Chapter 3.18

# Personal Field Logistics

*Alice Orlich*

## 3.18.1 INTRODUCTION

When working on sea ice, researchers find themselves in a place where humans are naturally disadvantaged by the physical environment. Many groups of indigenous peoples have lived near sea ice and have, over generations, adapted very well to traveling using the ice as a platform for hunting and gathering. Researchers, however, access the sea ice as visitors with goals specific to gathering scientific data, and therefore their skills and knowledge of the environment need to be learned and put into practice in a relatively abbreviated time scale. In addition, there are psychological issues to contend with, such as trusting the stability of the sea ice and adequately assessing the risks in the polar climate. Thus, fieldwork participants are faced with many unique concerns that dictate the complex management of field logistics in the polar regions.

These issues shape how to arrive at the site; what accommodations are made; who are members of the field team; how much time and money is budgeted to the field experience; and which safety measures are necessary to practice. As a member of a field party, you may be involved in the coordination of some or all of the planning and management stages that carry an initial proposal for study from its inception all the way through to a field campaign collecting in situ data. You may, in this situation, feel adequately informed and prepared for the intense field experience; however, pre-season negotiations and preparations do not guarantee field readiness. If you are not a member of the organizing team, then it is even more difficult to know all of the relevant information that will affect your time on ice.

This chapter addresses the awareness and responsibilities that are owned by each and every individual heading off to participate in a sea ice research program, whether you are a student, instructor, technician, principal investigator, or observer.

This chapter is designed to provide you with some of the knowledge, skills, and equipment you will need while working in the field. Specifically, this chapter addresses how an individual can endeavor to work on sea ice safely with confidence and good judgment, and how to prevent doubt or insecurities due to little or no previous experience from limiting a successful field campaign. Because conducting sea ice research is never a solo venture, the text will stress that personal field logistics requires being aware of oneself, but also assuming responsibility for others in the group. In remote fieldwork, all team members are dependent on each another. By advising on areas of preparedness, personal gear and etiquette, field equipment, and research platforms, this chapter will serve as an introduction for students, new technicians, and academics, and also as a refresher for returning experienced researchers, instructors, or PIs. Providing the same information to all participants is meant to strengthen the group dynamic so that each individual is encouraged to pursue active dialogue with fellow field party members, regardless of their position or background.

## 3.18.2 PREPAREDNESS

The concept of preparedness can be interpreted in many ways. One can become "prepared" in a number of ways, including through targeted training or academic coursework, or from previous experience. A particularly important component of preparing oneself for polar fieldwork is to recognize the value of local knowledge transmitted by indigenous or long-term residents of the Arctic. Whichever approach you relate to, keep in mind that you are likely working with and amongst others who represent experience gained in any combination of the above routes to preparedness.

In this section, each type of preparedness will be discussed in order to demonstrate the numerous opportunities available to those readying themselves for polar field exercises. The accompanying lists provide suggested references and include additional space for you to expand the list of your personal experience and that of others. Remember that any field team will be composed of individuals with diverse levels of experience and skill sets. The strength in any team is accepting that each individual has something to contribute.

*Training programs* often become available through educational institutions, government agencies, community organizations, public facilities, or private companies. The training might be offered as part of your job or as a community service, or you might seek it out to increase your aptitude in a skill area useful in field research. Your role as a member of a field team will expose you to remote, cold environments where situations considered a minor inconvenience in an urban or familiar setting could escalate into a serious emergency when on the ice. Consider examples like inappropriate clothing, unexpected weather changes, compromised communication technologies, encounters with unpredictable wildlife, medical complications, psychological stresses, and equipment or transportation failures. Understanding that these are just some of the potential obstacles of work in remote sites should motivate you to round out your spectrum of knowledge and skills to be able to react to and best manage a variety of unfortunate situations. Box 3.18.1 provides a list of suggested training courses that are available to equip you with practical skills helpful in polar research.

*Academic and vocational courses* are often written into a curriculum specific to your research discipline. However, keep an eye out for offerings from other departments, even recreation courses, special sessions, summer school, or special certifications. Exposure to other disciplines and the methods applied to fieldwork will enhance your own sea ice research. This manual and the course from which it was inspired are evidence of the benefits of a multidisciplinary approach. The accompanying chapters illustrate how field studies such as chemistry, physics, and biology are interrelated and yet often seemingly focus on specific research questions. Other academic disciplines from the social sciences—such as political science or geography—could lend a more comprehensive understanding of sea ice issues. For example, learning more about international policies, current social movements in the Arctic, or the intricacies of the unique agreement that is the International Antarctic Treaty can round out a visiting researcher's knowledge and sensitivity for a study area. Box 3.18.2 suggests a number of optional courses available on university and tech school campuses.

*Previous field experience* is the best guidance for safe, efficient, and industrious fieldwork. While training and coursework will guide you towards doing good science and fieldwork, it is only when this knowledge is put into real-life practice that true experience is gained. In essence, the familiar system of trial and error factors in greatly in any field experience. Any experienced field researcher will humbly admit that there is no way to know everything about safety, accuracy, and productivity before first stepping out on the ice. This mind-set should be celebrated, as modesty and humility are necessary traits when assessing the risks inherent in sea ice travel and work. Confidence and wisdom are gained by consciously respecting the more experienced people around you and also being honest enough with yourself to learn from any misfortunes that may befall your time on the ice. If you are working with team members that are new to the ice, watch them closely and gently and humbly share your experience as a mentor. On the other hand, if you are learning a new method, working with new people, or have little experience, decidedly seek out a mentor and let them know your concerns. Either scenario will pay dividends for both parties.

There is always a learning curve when working with both gear and people, especially in remote and inclement settings. Subtle challenges such as how best to prepare a daypack or tie down a sled may at first seem like insignificant skills. However, through repetition and innovation you can transform seemingly small tasks into reliable techniques that streamline your time and ensure your success. While these examples represent some of the smaller technical tasks involved, the same philosophy also applies to interpersonal relationships. Taking the time to communicate with fellow team members and others at your research site (locals, technicians, crew, or other science groups) will help establish a working relationship with all present. An effort like this can effectively inventory the human resources in the immediate community and create beneficial alliances.

*Local knowledge* stands on its own as the most personal and reliably tested experience. Generations of indigenous groups define their culture by time spent in the proximity of, and on, sea ice. As mentioned in Chapter 3.11, the Iñupiat of Alaska and the Inuit of Canada and Greenland have sustained their populations as hunters of sea mammals, harvesters of fish, and harvesters of the ice itself. The aptitude required to perform these subsistence activities is passed from elders sharing traditional knowledge gained from their own elders and through their own lifetime. Similarly, many other residents of the Arctic have developed and adapted technologies, practices, and understanding useful for living along the coast or in the sea ice environment. As a visiting scientist, you may not have the time or opportunity to learn the types of valuable skills in the same way or to the same degree that these groups have, but you should still view the members of the local communities as mentors with unparalleled experience. Successful early polar explorers like Amundsen and Nansen are examples of visitors who learned critical lessons about the appropriate clothing, food, and modes of transport from indigenous communities. The wisdom retained in the people of the Arctic today is a source for survival skills, observational strategies, and cooperative research. If research is to be conducted near an established arctic community, the field party members should consider how to optimize the exchange with the local residents.

## 3.18.3 PERSONAL GEAR AND ETIQUETTE

Selecting proper personal gear is dependent on the season, the mode of access to the sea ice, type or severity of environmental exposure, and the type of work planned. These factors influence which pieces and styles are most appropriate when considering gear such as footwear; head, eye, face, and hand protection; and general layering. For instance, short hiking or snowmachine day trips onto landfast ice during the spring melt season under calm sunny skies would entail considerably different clothing and daypack gear than if you were being dropped off for the day by a helicopter late in the freezeup season. Similarly, you would prepare differently for a GPS traverse than you would for coring ice or sampling water. Although these examples may be deliberately extreme, they are intended to illustrate the variables that affect the many approaches to sea ice research. The point is not to assume that all field campaigns offer the same logistical concerns.

Being able to adjust your personal gear according to weather and your field mission will guarantee a more comfortable and productive research experience in even the most taxing field conditions. Unfortunately, proper outfitting and preparation for all occasions will require that your wardrobe be quite extensive, and therefore expensive. Before investing in technical garments that you might use for only brief bouts of fieldwork, check to see if your institution or the logistics contractor (if one is involved) could supply the recommended clothing and gear. This is frequently the case and will save you time and money. The issue of how to choose the right personal gear is much like the discussion in the last section regarding preparedness—researching gear is not the same as using it in the field. Working with the gear yourself and learning what styles are preferred by seasoned sea ice users are the most direct means for you to gain familiarity and confidence about clothing and equipment. A brief discussion of some major factors influencing gear selection, along with descriptions of specific gear, is found below.

#### 3.18.3.1 Seasons, Climate, and Weather

Sea ice research is conducted during every season, in a variety of conditions. What may be referred to as the "melt season" might be interpreted as a warmer time of year. However, this could mean harsher conditions if the work is done further poleward or in areas prone to storms. Although "summer" in the polar regions can be underestimated, you can certainly count on freezeup and "winter" to pose the most severe weather. Always consider that the weather conditions can change, sometimes without warning, and therefore you should pack an assortment of gear in your daypack. In the event that the weather turns into white-out conditions, travel on ice may not be safe. Knowing how to pack for or construct a temporary shelter is suggested if you plan to be out of range of your base camp. It is best to pack as if the temperatures will be colder and the winds will be stronger; the water is always something to avoid coming into contact with.

#### 3.18.3.2 Access to Site

How you approach the site will also affect your choice of personal gear. Are you traveling by foot, snowmachine, boat, or helicopter? Consider that walking vs. riding may dictate how heavy and bulky the boot type should be. If walking, you are likely to be hauling at least your daypack, if not a sled with equipment in tow. No matter what the weather, that level of exertion will raise your body temperature and likely promote sweating. Remember to go slow, and to wear layers that can be removed or that will wick moisture and reduce chances of hypothermia. In contrast, if traveling by vehicle, the level of inaction on your part may lower your body temperature. Aside from physical discomfort, being cold can directly affect your morale and productivity. Watch for hypothermia in active as well as more sedentary activities, in yourself and in your team members. Be sure to review the safety policies with whoever is responsible for passengers traveling on a snowmachine or in a boat or helicopter. Some contractors or government agencies require helmets or specific floatation suits when aboard which are not the ideal to wear and manage once on the ice. These are some examples of how the clothing and gear you use in transit are sometimes different than the combination that is appropriate for the work once on site. Take this into consideration when packing and transporting all gear for the trip.

Another consideration is whether you will be conducting your research near established infrastructure or if you will be in a "deep field" situation, like an ice camp. If working on landfast ice, there are likely land-based accommodations in the area with elements of modernity that will improve your comfort level. The same could be said if traveling on an icebreaker, where after your day on the sea ice, the ship awaits with nearly the same conveniences found in a small town. Being able to return to a warm building or accommodations with plumbing and perhaps even laundry facilities can make life considerably easy. Conversely, if stationed at a remote ice camp, field attire tends to be more complex and sturdy, laundry facilities are rare, and unwashed garments are apt to become socially acceptable.

### 3.18.3.3 Type and Length of Exposure

Sea ice research inherently exposes you to an extreme climate. Some of the fieldwork requires staying in the same place on the ice for extended periods, often in contact with the ice, snow, and water, all the while being exposed to low temperatures and chilling wind. Preparing the site can involve repetitive chores like unpacking and adjusting equipment, shoveling, drilling, coring, and sampling snow. While these activities may warm you up temporarily, the work is not physical enough nor continuous to keep your body temperature up, and eventually you will again feel the effects of the polar environment.

As you become more comfortable with working on sea ice, it is easy to forget how vulnerable humans are in this unique environment. Respecting the elements as the hazards that they are will keep you mindful of the risks and prepared to react to a change in weather conditions or in the event of an emergency. Temperature, wind, sun, snow, water, and the presence of wildlife all have the power to influence your ability to safely accomplish fieldwork.

Low temperatures are to be expected and are the primary reason that your personal gear should be specifically designed with layers and insulation. When planning to be out for a long period of time, pack additional layers into a daypack so you are able to add layers or switch out damp socks or other compromised gear.

Wind will exacerbate the effect of the cold, requiring your outermost layer to serve as a windbreak and for you to reduce the exposure of any flesh. Be reasonable when planning your goals and expectations while working on days with an increased windchill, as work will become more difficult to perform, equipment may malfunction, and the overall time on ice may be reduced.

Falling or blowing snow will reduce visibility and the light conditions before and after a precipitation event will likely be flat, making the sea ice surface difficult to read during travel. Fresh snowfall can also camouflage cracks or other deformation features, such as ridges and rubble fields, any of which can be hazards for punching through into water or tripping over hard blocks of older ice. Using a probe or long ice tool to test the ice stability and thickness as you walk can help identify a safe route. Your instrument of choice should be of considerable weight, so that it can apply enough force when striking the ice in front of you as you test the structural integrity of the ice.

Water in melt ponds or a slushy surface should be avoided if possible, as saturation of footwear or clothing can cause major discomfort and potentially lead to the greater risks of frostbite or hypothermia. More serious is the risk of becoming partially or completely exposed to the water. In the case that your field research involves sampling near or in a pond, dress accordingly by wearing either a drysuit or neoprene waders. If the ice at the bottom of the melt pond is potentially rotten, or if you are working near a lead, a full floatation suit is recommended. Before working near the ice edge, from a boat, with a dive project, or on a dynamic ice pack, review a safety plan with the other members of your team. Quick response is the best recourse for cold water emergencies. A popular self-rescue device is a knife or ice pick that is small enough to wear but easy enough to access and use as leverage to climb out of the water. If uncertain about self-rescue procedures or other emergency practices, be sure to ask your team members, team leader, or expedition leader. Sometimes people assume you have had experience or proper briefing on protocols when you may not have. Be sure you know what skills or information you may need on any endeavor.

### 3.18.3.4 Polar Bears

Sea ice is polar bear habitat, so before venturing out, inquire about the local population, recent sightings, and the contingency plan if any were to enter your study area. Bear awareness and gun training specialty courses are available to increase your understanding of the animals as well as the methods to deter or dispatch a bear. Most often, an experienced bear guard is designated for each field party, but it pays to be prepared for and comfortable with the idea of working where bears may roam. Even if you do not plan to be the primary responder in the event of a bear encounter, strive to learn what defenses are rational and what role you will play within your group's solution to the threat.

The use of a firearm is the last resort. Avoidance, air horns, flares, and warning shots are preferred means to discourage bears from advancing. Typically, the sound and smell of engines, generators, or snowmachines will offend a nearby bear.

If a bear is closing in on your field site and you are not able to vacate, use an air horn to alert it to your presence; it is possible that it is unaware of your location. If it is determined to come closer, flares or warning shots from a gun can be aimed at the ice or sky between you and the bear. Consider that landing a shot behind a bear that is facing you may cause it to run away from the flare or bullet's point of impact and head towards you. As a safety precaution, do not travel with the flare gun loaded, as they can easily discharge in a pocket or bag. Discuss the firearm-handling policy for your group in advance; know who will be responsible for carrying the gun, when it is loaded, and who is trained to discharge it.

A good habit to develop is to routinely look up from what you are doing and scan the horizon. By taking a brief break from your work to assess the environment, you can observe changes in the weather, visibility, or presence of wildlife. This minimal effort will create a greater peace of mind and a more productive field day.

### 3.18.3.5 Type of Work

Plan ahead as to what tasks you will need to perform on the ice. The biggest concern here is to understand if your work will require you to be still, as when recording observations and measurements, or if you will be in motion, like when running a transect, or digging, coring, and drilling. When participating in passive duties, heavier insulation is preferable, but if performing physically demanding work, the clothing should be designed to offer greater mobility and breathability. Specifics on how to decide on footwear, headgear, protection for the eyes, face, and hands, as well as layering in regards to different types of work are discussed in greater detail below.

Proper footwear can vary with the condition of the snow and ice. The best features include a thick sole, waterproof exterior, and multilayered insulation. The benefits with these elements of construction are the dense barriers from the frozen surface, standing water, and relentlessly cold temperatures. Some designs come with a steel toe, which is a good choice, and sometimes required, if performing deckwork on an icebreaker or if you are working with heavy equipment on the ice. During the winter months (post-freezeup) when the snow is dry and the ice is thick, mukluks or other soft-formed boots are a reasonable choice. Remember to air out your boots and liners daily; the moisture that collects after a day's work or exposure to water can cause trenchfoot, a condition that essentially rots the flesh of the foot after being exposed to cold, damp conditions for too long. Aside from discomfort and inconvenience, the condition can lead to tissue loss.

Keeping your head covered while working in the cold is one of the best ways to maintain your body heat. Hats, hoods, and balaclavas are all viable choices, each with its own specific advantages. Your working environment may be windy, with blowing snow and ice crystals, or still and cold. The proper head covering should block the elements, be adaptable to changing environments, and not allow you to overheat. Like most essential gear, choose a variety of options and store them on your person or in an easily accessible pack. Also consider if you will be using a snowmachine or helicopter helmet or any type of communications system like a headset while working. This additional gear can complicate your clothing system, so take the time to find the best combination and have backups.

Sunglasses or goggles are a must during any daytime activity on snow or ice. Ball caps or other styles of brimmed hats are also good gear choices to block the sun. The sun's UV rays are intensified by the reflection off the bright surface. Choose a style that best fits your facial structure and that you are comfortable wearing for long periods of time. Remember to order corrective lenses for goggles or sunglasses if you normally wear glasses. At times, trying to work with the dimming effect of the glasses or goggles may seem counterproductive to detailed work, but try to be patient and perhaps adjust your expectations and pace of the work to accommodate the safety concern of protecting your eyes while still collecting necessary data. It is strongly suggested that you pack more than one pair as replacements, in your daypack, as trying to work without them in the event of loss or breakage can cause severe damage to your eyes.

Skin, particularly your face, is going to be at risk as you work in the field. The best practice is to use lip balm, sunscreen, and moisturizer all over. Keeping the skin hydrated and protected is a matter of comfort and overall health maintenance. Much like the layers of clothing that you choose, your skin is another barrier and regulator of heat. The extreme cold and dry conditions can cause frostnip, frostbite, blistering, and excessive drying and cracking of the flesh. Facial skin can be damaged easily from the cold and wind. Be careful to consider the neck, underside of the chin, and even the nostrils. These areas receive reflected rays that bounce upward off the surface, and are often forgotten when applying sunscreen. Remember that sun damage can occur even during overcast or spring seasons. Lips are particularly sensitive and can blister within a short time. Hands and feet are prone to drying and cracking, which can be irritating and debilitating, as cracks on the fingers and heels entail a prolonged healing process. Avoid these discomforts and medical emergencies by applying products as you prepare to go out and continue to apply them throughout your ice time.

The preference for mittens or gloves is influenced by whether you will be using your hands or just trying to keep them warm. There are different options depending on if you require dexterity for taking notes, operating electronics, and adjusting small pieces of equipment or if you are coming in contact with snow, ice, or water and require gear made with more insulation and waterproof materials. Hand-warming packets are small chemically charged sacks that are designed to be placed in the glove or mitten and can help maintain or rewarm the hands. These are a popular gear item for everyday use as well as in survival bags. It is a good practice to pack multiple sets of handwear to accommodate various tasks, but also as additional layers or for replacement.

The final key is to dress in layers, as the benefit is twofold: Layers provide multiple defenses and create trapped airspaces that provide warmth, and second, layers can be removed to adjust for level of activity, allowing heat to escape to avoid sweating and later getting hypothermic. With many layers to consider, try to create a routine for how you put them on and take them off. A reliable test is to take a winter trip involving hiking, skiing, or other type and level of activity similar to what you will experience in your field research. Once familiar with your favorite combinations, it becomes second nature to adjust accordingly. It is important to keep all of your layers dry. This may require you to have multiple sets so that after a day's work you can air and dry out one set and still find comfort and warmth in another. A serious concern is to select or modify gear so that it does not catch on moving parts when working with field equipment. Toggles, cords, and loose materials can cause quick and dramatic injuries if caught in a drill or corer. Another way to avoid these accidents is to always make a quick buddy check before anyone begins to operate or inspect equipment.

### 3.18.3.6 Gear from Head to Toe

There are a number of basics which should be included in any sea ice researcher's personal gear collection on ice. Some pieces will be optional based on each individual's comfort and preference. What follows are brief comments on some of the items that are pertinent to anyone who plans to work on sea ice. Always remember to pack separate clothing for camp and off-hours. A quick reference list for suggested personal gear (Box 3.18.3) is found at the end of the chapter.

***Helmet, snowmachine***: Used not only as safety equipment but also to protect the head from intense cold and chill while riding.

***Hat***: Fleece, wool, fur, wind-resistant materials are all good. Multiples recommended.

***Balaclava***: A combination of hat and neck gaiter typically made of fleece.

***Neck gaiter***: Full ring of material, often fleece, which fits over the head and protects the neck.

***Face mask***: Additional face coverage, protects nose and lower face from exposure.

***Sunglasses***: UV-blocking, avoid metal rims, options include nose guard or slide panels. Consider prescription lenses, if needed. Multiples recommended.

***Ski goggles***: Sometimes preferred over sunglasses, more complete coverage.

***Sunscreen***: Choose a high SPF rating, apply often.

***Lip balm***: Choose a high SPF rating, apply often.

***Lotion***: Personal choice, but extremely useful for protection and maintenance.

***Base layer***: Insulating shirt and pants in contact with skin, often thin yet warm. Avoid cotton.

***Mid-layer***: Insulating shirt and pants warn over base layer, bulky and warm. Avoid cotton.

***Jacket/Parka***: Choose according to conditions. Down fill only if no threat of wet weather or exposure to water. Invest in multiple designs and materials. Should be thick material resistant to wind and water.

***Windbreaker***: Additional outer layer to wear over main insulator, especially helpful in high-wind environments or prolonged exposure.

***Hand warmers***: Chemical packets that produce heat when exposed to air and shaken. Foot warmers (larger) are also available. Good, quick solution for poor circulation.

***Glove liners***: Thin, insulating layer worn under gloves or mittens.

***Gloves***: Water-resistant with insulating material. Styles designed for skiing, ice climbing, or mountaineering have a technical fit. Leather or other industrial designs can be just as reliable. Rubber insulated style great for sampling. Multiples recommended.

***Mittens***: Similar to glove options, but bulkier designs are ideal for snowmachining, or when dexterity not required. Multiples recommended.

***Snow pants***: Outer layer worn over insulating layers. Waterproof and reinforced on knees and seat helpful for sitting or kneeling. Bib style reduces cold air seepage between top and bottom layers.

***Overalls***: Styles designed for industrial use offer thick cotton duct and heavy insulation. Very warm and can take a lot of abuse, but risk absorbing water. Available as bibs or full suit with hood optional.

***Boots***: Midcalf, insulated, waterproof. Steel toe is optional.

#### 3.18.3.7 Etiquette

In addition to a strong proficiency in your research plan and competency with personal and field gear, another set of skills is required for a successful field campaign: the ability to demonstrate a healthy professional and social disposition. Proper field etiquette includes scenarios like instigating cooperation or committing to greater patience when working with individuals you may not have known before the project began or perhaps would not have chosen if you had selected your team members. Even the most familiar colleagues can become affected by the inherent stress of the field, testing a well-established, functional relationship.

Working with multiple personalities compels you to stretch your level of comfort in interpersonal relationships to avoid misunderstandings and unnecessary tension. Proceed with respect for the organization and the formal responsibilities of others, but be willing to contribute or accept suggestions and constructive criticism. When coordinating with military logistics support, strive to understand their chain of command and the expectations of how the science community is to interact when communicating issues such as requests, field plans, and safety briefings. Private logistics contractors may also have an organization that, if you take the time to understand it, may lead to a more efficiently run field campaign. What may seem like a reasonable expectation in your daily life can be impossible to create once in the field environment. If you are conducting fieldwork in proximity of an established community, invest your time in meeting curious residents and offer to present your research at their convenience. Often local guides and logistic personnel are hired based on their area knowledge and experience and they are an extension of your field team.

Choose to have the right effect on morale by maintaining an open and positive attitude. There are many reasons why the field experience can gradually or suddenly become dysfunctional. Sea ice research requires that everyone function in a cold, typically remote, complicated work environment. In order to reach the site, participants have made significant investments in planning. A sincere commitment of time and money have been made to solidify a research agenda, manifest people and gear to the site, and adjust personal lives to accommodate what is likely to be a very intense experience. If any one of these logistical or personal factors is threatened, the team or the work may inherit undue stress. Knowing how to reroute a professional or emotional trainwreck can save the participants pain and embarrassment, and ensure quality results from the field time.

Preparing your mind-set before embarking on the trip can help you focus on the strengths that you can share with the team. Because much of the work in the field is physically taxing, plan to adjust the pace and choose to make safe, efficient moves. Practice open communication, and take a moment before reacting to stressful or testy exchanges. Ask questions; do not always assume that you know how to fix a problem. Just because you may have experience with a similar scenario in the past does not entitle you to take the lead. Let the other person explain how he or she sees the issue. Sometimes spirits are low or personalities clash due to something as innocent and avoidable as dehydration, hunger, or exhaustion. To engender trust and mutual respect, make a habit of extending assistance and compassion to the others around you. You may decide that it is a good personal policy to manage your interpersonal skills with more sensitive attention to detail than you normally would in your daily routine back home.

Before heading into the field, make an attempt to become familiar with your team members. Even if you have worked with them in the past, open a dialogue about expectations, responsibilities, and contingency plans. Involve everyone in cross-communication. If you are not at the same institution, perhaps begin via e-mail or some easily accessed mode of communication. Once gathered together, sit and discuss the stages leading up to the present and then plan the remaining time you will have together. This direct approach will nurture stronger teamwork and help identify if any training or planning sessions are still needed.

If you are in a position of authority or just possess more relative knowledge or skill than others on site, extend yourself and serve as a mentor. Given the opportunity, engage volunteers to assist you and to teach others new skills. Even if you do not need others to perform your work, remember that it is always in the group's interest for everyone involved to understand the goals and methods of a project. This mode of sharing tasks can promote interdisciplinary research and other types of cooperation. When introducing students, assistants, or colleagues to new equipment or field activities, be sure to address each person's curiosities and concerns patiently. Consider that your audience is potentially insecure or uncertain of their abilities. Give them a chance to use the tool or reiterate the plan before you delegate responsibility to them.

The final comment on proper field etiquette is the most personal, as it addresses various hygiene issues. Due to the reality that fieldwork on sea ice will, if only temporarily, put you and your team out on the ice, considerations must be made for how to safely and appropriately grant privacy when members need to tend to personal hygiene. The most unconventional scenarios will occur when working on the ice, with no means to return to the comfort of a building with plumbing. Less inconvenient are the options when living and working out of an ice camp. Ship and dormitory sites can also pose complications, even though those environments characteristically possess the most modern amenities.

Inherent in the research team culture is that you will be working and living in close proximity. The best expectation is that the members will be mature and respectful enough to know how to openly communicate amongst themselves. Assuming that is the case, it is best to address the standards of practice for the area. That said, be certain to discuss the topic tactfully, yet completely, as different people have different levels of experience and comfort. Ice camps, field sites, or research ships may each have their own policies about what can be left behind on the ice. Concerning the outdoor toilet experience, the most lenient practice is to safely find someplace to relieve yourself; however, in some situations waste is collected to be transported later. If you are free to roam, learn about your team members' comfort and level of privacy needed. Agree on a distance or direction for a designated area, and be sure to identify the route, to ensure safe travel and return. Although it may be tempting to hide behind big ice blocks, realize that wildlife and cracks in the ice can present hazards as well hidden as a person seeking privacy. If the local policy imposes restrictions, respectfully follow the rules. Sometimes toilet paper or feminine hygiene products are to be packed out. In these cases, the trash is usually collected into a biohazard bag so that it is properly disposed of later. Citing the U.S. Antarctic Program waste-management system as an example, other technical equipment includes using wide-mouthed water bottles for urine or Ziploc bags for feces; these are commonly referred to as "poo bags" and are occasionally the source of unfortunate juvenile pranks and accidents. Double bagging is recommended if carried in your daypack.

If working and living in an ice camp, there will likely be temporary structures for outhouses. Depending on the ability to drill into or through the ice to provide depth, some houses are intended only for women, whereas men are expected to use others and/or urinate outside. This is meant to reduce the time it takes to fill the ice toilet. Something to remember about ice camps is that the floe upon which they are situated may experience dynamic processes that could create sudden cracks or leads between buildings. Also, polar bears could enter camp and not be easily spotted around the buildings. These are specific hazards to discuss with the camp logistic support crew while touring the camp.

Icebreakers or coastal communities are often used as bases to access sea ice. Expect to live and work in relatively small quarters, which could be dormitory-style accommodations. Depending on the size and space available, teams may share coed living, restroom, shower, or sauna facilities. Take time to fully understand the schedule or assigned rooms.

Facilities for washing hands and taking showers are sometimes limited. When freshwater is scarce, packaged moist towelettes are a fine substitute for bathing. These products are available in small pocket-sized soft packs or in larger hard-cased tubs. Be certain to keep the packaging shut tight to avoid loss of moisture. Beyond cleansing, hand sanitizers have become popular and are sometimes the only disinfectant available while on the ice. As for immediate sources of water, melt pond water is apt to be freshwater later in the melt season, and if the temperature can be tolerated, is quiet refreshing. Alternatively, the melting of snow and ice is common practice at field camps. Considering ice types, multiyear ice is preferred due to its low salinity. Since it requires energy to convert snow and ice to water, the supply is valuable. Of course, it is universally understood that the priority is to keep your body hydrated, so for the purposes of drinking, water use is unlimited. After that, washing hands and preparing food are the next essential uses (perhaps another reason to volunteer in the kitchen more often). Finally, the more formal bathing experiences, via makeshift gravity-fed shower systems, pressurized water can, or sponge bath, are precious opportunities. Ships and facilities on land are often more accommodating, but may still have resource limitations. Learn about the limits on consumption, if any, and recognize that desalinating and heating water is an expensive venture.

## 3.18.4 FIELD EQUIPMENT

Whereas the previous section focused on how to be conscious of the environment, proper clothing, and the members of your field party, this section examines how to best prepare and care for the technical and scientific equipment used in field campaigns.

Sea ice research relies heavily on specialized equipment to provide transportation, communication, sampling methods, and comfort. Although later it will be necessary to delegate specific responsibilities to each member of the field party, general field equipment issues are pertinent to consider and can instigate a dialogue for more complete planning. Avoid accidents or the loss of valuable research time or equipment by helping every member of a field party to comprehend the processes involved in organizing, maintaining, and operating the gear. Remember that adding new equipment to a study or employing inexperienced crew members will require thorough, patient, two-way instruction. Some gear will only be used by properly trained researchers. One crucial piece of equipment that each member of a field party should carry is a daypack that includes everything they will need to work and survive independently.

Details on specific equipment and sampling methods for various disciplines of sea ice research are found in the respective chapters of the handbook.

### 3.18.4.1 Supplies

To ensure that all of the necessary materials are present and ready for use when working on the ice requires that the work is well designed and mindfully planned. This process begins long before the field campaign. In working through the goals and methods with future field partners and crew you should include a discussion of preferred equipment and sampling styles. At that time, field planners should compile gear and equipment lists and consider items that may require repair kits, multiples, or alternatives in the event of failure or loss. To keep inventory and ensure access, keep a catalog of supplies throughout the stages of use: as they are selected, packed, shipped, prepared for each field day, used, consumed, maintained, replaced, retired, and returned to storage. Recognizing that each research party may organize this information differently, the simplest way to ensure cross-communication is to produce both paper and electronic documentation to be shared and updated.

### 3.18.4.2 Shipping

The complexity of packing and shipping field gear and science equipment is often underestimated. Many researchers do not realize that military and commercial transport companies often restrict certain equipment from traveling without the proper clearance and paperwork. Another consideration is whether the cargo may be transported across international borders. In this case, customs policies need to be addressed and the routing closely followed as the materials are in transport. Nothing is more stifling to a field campaign than for valuable equipment to be held at a customs office instead of reaching the destination on time. Delays and complete losses can occur even if all of the proper paperwork and communication has been aptly executed. The only insurance that a complicated shipment will join the field team is to commit to tracking it along its journey and knowing how to contact the proper handlers at each stage of transport and delivery. Given that researchers may not have the best communications systems as they venture closer to the field site, it is paramount to coordinate with someone at the home institution who is capable of monitoring the shipping and making executive decisions, should any complications arise.

### 3.18.4.3 Freezer Space

Depending on your research plan, you may require freezer space for samples during your field campaign. If working out of a field camp, samples might be stored outside and shaded from the sun. This, of course, would only be a temporary solution. More likely, you will be working on a ship or out of a small coastal community. In these cases, it is best to inquire and confirm space both before arriving and once onsite. Take the time to coordinate with whomever is managing the space and learn about the policies regarding access and maintenance of the freezer. Remember to label all samples completely and indelibly, especially if space is being shared.

### 3.18.4.4 Lab Space

Be prepared for the lab work that you do during field excursions to be influenced by the limits of space, time, and lab resources. Here, flexibility in your schedule and your sense of space is going to make life much easier. Due to the likelihood that any labs and equipment available to you on a ship or in a small camp will be shared by multiple users, allow time to meet with other users and agree on lab order, chores, and communication systems. Preemptive negotiations can nurture mutual respect and understanding among fellow researchers, and can guarantee more efficient results.

### 3.18.4.5 Storage

General storage will also require early planning and communication, and like many logistical details, it is not to be taken for granted. While maintaining order of all stored items is of obvious importance, coordination with the camp's or ship's logistics staff should be the responsibility of a field team representative. This relationship between the science crew and the support staff will assist both parties in delegating proper space and access for the needs of both the hosts and the visiting scientists. Remember that some materials are not meant to be exposed to the cold for long periods, if it all. Although awareness of the sensitivity of these materials would be expected by most people working in research or support, proper markings on packaging is a simple reminder. Proper labeling of all cargo pieces is equally recommended.

### 3.18.4.6 Fuel/Generators

Some equipment—like drills, corers, heaters, and snowmachine or boat engines—requires fuel or generators. As mentioned above, the complications of legally shipping the fuel-dependent equipment require formal paperwork and preparation of the equipment. Even though the engines are not filled with oil and fuel when in transport, they need to be documented for the residue. To properly prepare the paperwork, training and certification in transporting hazardous goods is necessary, as the preparer whose name appears on the document is held liable and can be fined thousands of dollars should there be improper transport or an unfortunate incident due to poor preparation. The seriousness of this example demonstrates how learning about all levels of logistics is relevant to a safe and productive field experience.

#### 3.18.4.7 Communication Technology

You will need to adjust from the excess of modern communication technologies that you enjoy at the office and home to the limitations of power sources and connectivity in field locations. Some variations may be available for higher costs, but mostly, the technology is limited and upgrades in products and services are not an option. Satellite communications may be spotty or unavailable beyond certain higher latitudes. Phone and Internet access may simply not be available, or might have restrictions based on time or cost of operation. Some field parties will employ HF or VHF radios to maintain communications with other groups or the base station located at the camp or ship.

Consider the power necessary to run your equipment and how it will be generated. Are you operating out of a facility, ship, or camp that can recharge your batteries or provide constant electricity? Portable solar panels are an option to charge small devices, or are sometimes available to charge batteries or automated weather stations at temporary field camps. Generators of all sizes are also an option, depending on potential concerns such as the cost and complications of transporting fuel or pollution to the site (if doing chemical sampling).

Internet access is now a daily necessity for most researchers. Going into the field requires you to sacrifice that constant connection to current and updated news, databases, satellite images, and communication with colleagues, family, and friends. Before you go, learn what the limits will be, including the windows of time available. Remember that interrupted Internet service may impair a research program that relies on satellite images for ship route selection or long-term studies. Often when working out of a specialized facility, camp, or ship, the logistics provider will maintain a system that makes satellite Internet connections limited to their own server. Acquire the websites or temporary e-mail addresses used by the research site's service and distribute this information to your contact personnel at your institution, as well as your family or friends wishing to contact you. With technologies and budgets improving and flexing over time, each field season may offer new conveniences or unforeseen setbacks; just remember to get the most current and precise details before you travel.

Regarding local computer and radio communications, decide what channels of information sharing your fieldwork will depend on. Will you have multiple teams out in the field at different locations, all at once? Do you plan to maintain an open, ongoing database? Is there an individual or group that is responsible for coordinating data or field crew outings? Intranet systems might exist in community facilities and ships, but are unlikely to be offered at temporary ice camps, so file sharing or database coordinating may need to be improvised. Recognize that once field crews are out on the ice, radio communications are the only means to maintain contact; develop a thorough protocol for field trips and radio communications.

Organizing a field party for a day trip should involve you coordinating and submitting a trip plan with your base camp. Discuss the goals, transport methods, participants, contingency plans, and the timeline of the day. File the plan with someone who understands your logistics and will monitor your movements and return. Often handheld radios are assigned to party members in the interest of maintaining communications with the base camp or for requesting aid in the case of an emergency. Because factors such as poor connections, wind and other background noise, operator error, and emotional stress can affect the quality of a radio conversation, users should be trained in the proper operation and maintenance of radios. Be certain to delegate the responsibility of radio communications to a member of your party who understands the importance of the task.

### 3.18.4.8 Equipment Maintenance

Taking the time to inventory, clean, and keep up personal gear and field equipment can at times be tedious, but it is a discipline that pays dividends beyond the relatively minimal time commitment. Keeping clothing dry and clean of salt from exposure to sea spray and seawater will prove a lasting investment as proper care will optimize the functionality of the material. Moving parts on field equipment and metal parts can be jammed or corroded by ocean salts. Likewise, snow and ice can temporarily cause malfunctions.

Practical field experience will lead to a refined routine on the ice that can limit exposure to destructive elements. One simple precaution is to make certain that parts and pieces are not casually left standing in the snow or in melt ponds. The best care for field equipment is to thoroughly rinse all pieces and cases with freshwater after each day trip into the field. However, take the time to read equipment manuals or contact the manufacture regarding maintenance in the field. Some submersible electronic equipment is designed to operate accurately without rinsing between each use.

Some equipment or parts may have been modified for a particular use or because the piece was old or damaged. Be certain to familiarize yourself and your crew members to the equipment and spare parts. Knowing how to improvise a fix in the field is a valuable skill to develop. Carry extras for parts that are small or easily lost in operation. Discuss the sensitivity, cost, method of repair, or any other pertinent details that will engender proper handling and care of the instruments.

It is all too common to take for granted the functionality of equipment, a sentiment that is sorely regretted when vital instruments are lost or damaged by poor training or oversight.

#### 3.18.4.9 Instruction

Another important issue regarding field equipment is the responsibility of educating users of the proper operation. Although the temptation may be to expect the user to know all there is to know about an instrument, consider any piece of field equipment to be as unfamiliar as an automobile is to a new driver: At first sight, the student likely knows that it is a vehicle that runs on gas, has tires and an engine, and can be utilized for transport or recreation, but only through proper instruction does he or she understand how it was intended to operate, safety concerns, limitations, and regulations. Think of an ice corer, microscope, or snowmachine in the same terms.

All of the gear that you and your research group rely on is critical to safe and successful field campaigns. Take the time to discuss all of the equipment with your group and determine who is experienced with and capable of operating each item. Provide instruction and hands-on testing opportunities to those who need it before they assume responsibility for a task or piece of gear. Consider that even if someone has vast experience on sea ice or with a certain instrument, they may not want the responsibility of training another field partner. Make sure that someone with the right preparation and attitude is available to provide instruction. As an instructor, regularly inquire if your message is clear and ask for it to be repeated back or have the student try to execute the job.

From the other perspective, knowing how to be open to instruction is equally important. Respect that equipment belonging to other groups may be similar to something you have worked with, but may function differently. Be sure to ask questions if you feel confused or too inexperienced to operate equipment or perform a task. Take a moment to demonstrate your new skills so that everyone involved can feel confident that there is universal understanding of how the work will be completed.

#### 3.18.4.10 Daypack

Those embarking on trips onto sea ice should carry gear to protect themselves from the elements; provide energy, nutrition, and hydration; compensate for lost or damaged clothing; and collect and record data observations. Ideally, a few precious items will always be worn on the person, but most will be efficiently packed into a daypack that is easy to carry while traveling and hauling other gear.

It is important to remember that field workers should not leave their daypack out of reach or in the care of someone else. The role of the daypack is to be the source of all essential and backup supplies. For example, if you travel ahead and expect your pack to be delivered later, you may not be able to begin your work if some instruments are in it. Also, imagine that the pack never makes it out to you and the weather changes or you get wet and then you find yourself without a proper change of clothes.

The following list highlights suggested items for your daypack that make a trip out on sea ice more comfortable and productive. As in other sections of this chapter, an open-ended list (Box 3.18.5) is provided at the end of the chapter to provide you with the basics and room to personalize your packing list.

***Extra clothing layers***: Pack multiples of hats, gloves, socks, and thermal underwear. To ensure the best condition, store in waterproof bags.

***Extra sunglasses or goggles***: Have backup eye protection.

***Sunscreen and lip balm***: Pack in easy-to-reach pocket of daypack.

***Snacks***: Pack a ration that will be more than you expect to eat. Choose high-calorie foods that won't freeze. Suggested energy intake while working in the cold is 5,000 calories per day. Chocolate, nuts, and candy are popular options.

***Water bottle***: Always bring water and keep in a warm or insulated place to avoid freezing. Hot water for tea or cocoa kept in a thermos is a morale booster.

***Camera***: For research or personal use. Bundle in soft material or case. Hand warmers sometimes assist in keeping batteries and moving parts from becoming too cold.

***Notebook and paper***: Water-resistant notebooks and loose sheet paper in different sizes are useful for collecting field data and for noting ideas for future fieldwork. Delegate a "secretary" from your group to record observations/minutes as they occur.

***Pencils***: Preferred over ink pens because they work in cold temperatures. Mechanical pencils retain writing point, yet traditional pencils have fewer parts to loose. Remember to pack extra lead or a sharpener. Multiples recommended.

***Flare guns or air horns***: Practical personal protection for bear encounters to be used in support of your designated bear guard who will likely have the firearm. Know how to operate and who in your group is carrying them. Pack in accessible location.

***Medications or first aid***: Pack prescription medicines, as well as a first-aid kit.

***Entertainment***: Books, games, and music that can be shared, traded or gifted. Not always in a daypack, but nice to have if out for many days or in remote camps.

## 3.18.5 RESEARCH PLATFORMS

How you design and conduct your sea ice research depends on how you will access the field site, what amenities will be present, who will assist, and which season is ideal for your work. For example, certain research methods are cost prohibitive if working from remote platforms, while others may be possible at more accessible sites that offer established logistics support or neighboring research institutions. If working with a new crew or committing time to train members, quantity of data may be reduced in favor of higher-quality data collection. Visiting a research site in the winter will pose a different set of issues regarding daylight hours, weather, and ice stability than if conducting research in the summer. All of these examples demonstrate the variables in the types of research platforms available. Consider how each of the platforms discussed below compare in the matters of travel, infrastructure, social dynamics, and seasonal use.

### 3.18.5.1 Landfast Ice

The easiest and most economical for daily trips, research sites located on coastal ice are just a short hike or snowmachine ride away. Helicopters or small boats may also be options, depending on the local facilities. The greatest challenge with travel associated with this platform would be your initial transportation to it. Coastal ice sites in the Arctic or Antarctic likely require flights with multiple connections, as not many road systems access polar coasts. However, the sites are typically near established camps or communities which rely on sea ice as a resource, so local interest and infrastructure will be available to support the needs of most research. Be sure to learn about any permitting or access issues that need to be addressed before assuming right-of-way onto land or ice.

The type of accommodations available will vary. A camp may have temporary structures, whereas a permanent community may offer residential or institutional lodging and workspace. Each setting will be different, with some experiences providing opportunities for research groups to work with the local community, while others result in isolation. Depending on the access and relationship with the neighboring community, there may be regular occasion for cultural exchange and resupply. Another consideration of coastal ice sites is the presence of migratory and resident wildlife. Prepare to work and live in the habitat of polar bears, seals, birds, and other polar species.

Because most coastal study sites will experience seasonal melt and open leads, they are typically accessed for research from winter (freezeup) through spring and early summer. Each end of the annual cycle possesses unique environmental hazards that will affect travel near or on ice, as well as the appropriate type of research design.

### 3.18.5.2 Icebreaker

Some ships are designed to provide icebreaking service in the interest of maintaining an ice-free transport route. Others are committed wholly to scientific purposes. You may be invited on to either, but to participate in a research cruise, you will likely be required to possess a passport and travel internationally to reach the ports from which you will begin and end your trip. Additionally, a ship route may travel over multiple claims, so proper documentation is required of each participant in the event that the ship needs to dock unexpectedly. Airfare to and from the cruise will be costly, as the ports utilized by the ship are typically remote. Reservations should be made to allow for delays or cancellations of connecting flights en route to the destination or complications in the ship's schedule.

Depending on the ship policies and the captain's discretion, access onto the ice may be possible via helicopter, small boat, the man-basket, or gangway. You may be able to discuss your preferential approach prior to boarding the ship, but factor flexibility into your field plan, too. Ultimately, the captain and the officers are entitled to the decision of how, where, and when anyone is allowed on the sea ice, so be respectful of the policies and take advantage of the opportunities to nurture the relationship between the research and logistics community. Provide the captain, officers, and lead scientist with a detailed research plan that includes your goals, methods, equipment, crew, as well as additional personnel required, and contingency plans so that they can understand your needs and appreciate your level of preparedness.

On board, space will be designated for lab activities and storage. As reviewed above (see Section 3.18.4) try to confirm space before the cruise, and once the ship is underway, make an effort to work with others to share and maintain the space. Cabins, or sleeping berths, may be single or dormitory style with private shower and restroom included, or shared facilities located down the hall. The mess, or dining area, will offer scheduled meals and perhaps an all-hours self-serve snack bar. A small store on board will likely sell additional snacks, basic toiletries, and souvenirs.

The ship's crew functions to ensure the continued operation of the vessel. Scientists are passengers who pose additional responsibilities and special requests for the crew. Be patient with the system and take time to explain your research and equipment to anyone assigned to work with you or others who show general interest. Participate in the safety drills and other mandatory assemblies or trainings. Consider the ship and its crew to be an extension of your field team; get to know the personalities, skills, and services available and offer your background and experience. As mutual understanding grows, respect and efficient field time will be the reward. Access to the bridge to monitor navigation or other ship equipment may be restricted, so be certain to gain proper clearance before entering.

In addition to researchers, it is common for ships to host a combination of other guests, such as vacationers, educators and students, or reporters and production crews. This brings diversity to the remote and contained community found on board. Recognize that the unique culture and social life that will inevitably evolve is part of the cruise experience. Strive to create a healthy balance of productive research and involvement with others. Exercise for physical and mental fitness. Volunteer to assist a ship crew, learn new skills working with another science group, or share your work in a casual presentation.

The obvious advantage to using icebreakers in research is that they help access sea ice cover throughout the year, particularly in summer or the melt season in higher latitudes. Buoy and mooring programs rely on the ships to be able to return to exact locations in order to deploy or recover instruments. Often the cruises are planned in order to optimize coordination with multiple research disciplines and to monitor seasonal conditions.

### 3.18.5.3 Ice Camp

Historically, camps were settled for the purpose of studying ice over the winter season and therefore utilized either large multiyear floes or tabular shelf icebergs, known as "ice islands." Some long-term camps consist of an icebreaker frozen in the ice pack for approximately a year. The most common camp will be temporary, designed to accommodate a specific discipline or to coordinate multiple projects. The chosen site will host an ice type and location suitable to the goals of the camp—for example, if the project will use scientific divers or automated underwater vehicles (AUVs), perform strength testing of the floe, collect biological samples, or be in an area of dynamic ice to observe deformation processes.

Travel to the camp may prove costly, likely requiring a long flight or ship trip to reach it. Accounting for transport of supplies and personnel will likely put the most stress on the budget. However, some camps may be accessed by snowmachines over landfast ice and therefore incur less cost.

Because of the expertise required to create camps, they are often established by a logistics provider who selects site, clears the runway, builds infrastructure, and manages the general services and schedule of the camp. Situated far from most other populations, ice camps need to be largely self-sufficient, so are typically staffed with experienced research and logistics crews. Even participants with decades of experience recognize that the conditions are formidable, and therefore a cautious, humble attitude leads to the safest and most successful field campaign. To optimize your role as a member of a field team, it is highly encouraged to continue to acquire relevant skills useful for the remote polar environment.

Some ice camps are briefly established for a few weeks during the spring to observe the dynamic breakup season. Others are built up to sustain longer field campaigns and, like the historical examples mentioned above, are located on thick multiyear floes, tabular icebergs, or fast ice and may operate for months, or serve seasonally as a semipermanent camp.

## 3.18.6. QUICK REFERENCE LISTS

This section is a summation of the chapter presented in the form of lists that are meant to provide you with practical references before heading into the field. Remember that the chapter is intended not as an exhaustive survey of personal field logistics, but as a guide to prompt awareness and instigate proactive organization among field party members. The lists are designed to start the process, with suggested items appearing in no particular order of importance and blank space provided at the end of each list to be filled with entries specific to your field research needs. Remember that lists are used by even the most experienced researcher, who recognizes their worth not just as an inventory of priorities but as a reminder that some things are too critical to overlook.

**TRAINING PROGRAMS**

- CPR
- First aid
- Wilderness first responder
- Cold weather survival
- Winter camping
- Bear awareness
- Firearms safety and awareness
- Helicopter safety and awareness
- VHF radio communications and etiquette
- ______
- ______
- ______
- ______
- ______
- ______
- ______
- ______

**Box 3.18.1.** List of optional training

**COURSEWORK**

- Field techniques in interdisciplinary sea ice research
- Arctic skills
- Snowmachine repair and maintenance
- Orienteering
- Ski mountaineering
- Ice climbing
- Outboard motor repair and maintenance
- Scientific diving certification
- ______
- ______
- ______
- ______
- ______
- ______
- ______

**Box 3.18.2.** List of suggested courses

**PERSONAL GEAR**

- Hat and/or balaclava
- Neck gaiter and/or face mask
- Sunglasses and/or ski goggles
- Sunscreen, lip balm, and lotion
- Waterproof gloves or overmitts
- Gloves and mittens
- Socks
- Base layers
- Mid-layers
- Jacket/parka and windproof layer
- Wind- and waterproof pant layer
- Overalls
- Boots
- ______________________
- ______________________

**Box 3.18.3.** Clothing basics

**DON'T LEAVE HOME WITHOUT IT**

- Sunglasses
- Sunscreen and lip balm
- High-calorie snacks
- Water bottle
- Waterproof gloves or overmitts
- Extra gloves
- Extra socks
- Warm hat
- ______________________
- ______________________
- ______________________
- ______________________
- ______________________
- ______________________
- ______________________

**Box 3.18.4.** Absolute necessities

**DAYPACK**

- Extra clothing layers
- Extra sunglasses or goggles
- Sunscreen and lip balm
- Snacks
- Water bottle
- Camera
- Notebook and paper
- Pencils
- Flare guns or air horns
- Medications/first aid
- ______________________
- ______________________
- ______________________
- ______________________
- ______________________
- ______________________
- ______________________

**Box 3.18.5.** Items for a daypack

**COMMON ERRORS IN THE FIELD**

- Underestimate dehydration
- Failure to eat enough
- Sweating (hypothermia)
- Wind and sunburn
- Missing or broken equipment
- Inadequate fuel for tools and travel
- Improper clothing
- Corroded equipment
- Poor communication
- Reliance on GPS waypoints
- Attitude affected by weather
- ______________________
- ______________________
- ______________________
- ______________________

**Box 3.18.6.** All-too-common mistakes

**LESSONS LEARNED**

- Hydrate before, during, and after fieldwork. Carry an insulated water bottle.
- Eat well and often. Carry additional snacks in pockets and daypack.
- Never sweat. Monitor your pace and dress in layers to avoid hypothermia.
- Keep skin covered. Exposure to wind, sunlight, and low temperatures is intense on ice.
- Use a checklist while packing gear. Double-check before going out.
- Track shipments. Carry extra parts for equipment and know how to do field repairs.
- Test equipment. Fill tanks and carry extra fuel.
- Know the weather forecast, yet pack additional clothing fit for a sudden change.
- Rinse anything that comes in contact with saltwater. Maintain equipment regularly.
- Engender open communication by discussing data needs and field roles in advance.
- Sea ice moves; no GPS points are static and wholly reliable for navigation.
- Counter weather-induced stupors by preparing for and expecting the worst conditions.
- ______________________________
- ______________________________

**Box 3.18.7.** Hard-won pearls of wisdom for the field

**HELPFUL WEB SITES**

- http://www.arcticscience.org/logisticsPermitting.php
  Barrow Arctic Science Consortium (BASC), Logistics & Permitting
- http://www.nsf.gov/od/opp/arctic/res_log_sup.jsp
  National Science Foundation Office of Polar Programs (NSF-OPP), Arctic Research Support and Logistics
- http://www.arcus.org/logistics/index.html
  Arctic Research Consortium of the U.S. (ARCUS), Arctic Logistics
- http://www.arcus.org/alias/
  ARCUS, Arctic Logistics Information and Support (ALIAS)
- http://www.polar.ch2m.com/
  CH2M HILL Polar Services (CPS), home
- http://psc.apl.washington.edu/home.php
  Polar Science Center, Applied Physics Laboratory, Univ. of Washington, home
- http://www.usap.gov/travelAndDeployment/documents/USAPFieldManual.pdf
  U.S. Antarctic Program (USAP), field manual
- ______________________________

**Box 3.18.8.** Relevant websites from agencies, logistics providers, and organizations

### 3.18.7 CONCLUSION

This chapter has attempted to draw your attention to some of the many facets of organization and readiness that are required to succeed in sea ice research fieldwork. On a personal level, the discussion began with an introspective look into your own knowledge and training, and then expanded to consider the talents and offerings of your team members and other collaborators. On a practical level, details regarding the environment, equipment, and infrastructure were outlined. As you prepare to learn, instruct, observe, or investigate sea ice, take time to examine your goals and evaluate what types of logistics support you will need. With each new venture into the field, your experience will become more diverse and you will gain greater confidence and therefore have more to offer as a member of the research community. Develop good habits for planning and communicating and remain open to new methods and coordination.

## Chapter 4

# Concluding Remarks: Integration of Sea Ice Field Research into Polar System Science

*R. Gradinger and D. K. Perovich*

Sea ice is of growing importance to the global system because of the many services it provides. It can be a roadway for travel, a platform for activities, a barrier for ship transport, a habitat for a rich ecosystem, and an indicator of climate change. Measuring the properties of sea ice is important for these reasons and many others. This book is designed to show how to make these measurements. In Part 3, most chapters focus on a particular property of the sea ice cover. However, this narrow focus is only a pedagogical tool to facilitate learning the various methodologies. An interdisciplinary approach is needed to understand the properties and processes governing sea ice in their entirety. More important, an integrated synthetic approach is needed to fully comprehend the services that sea ice provides to the larger system.

The very broad range of field techniques presented in the different chapters is a reflection of the diversity of interests from a disciplinary perspective. Two key observations can be made based on this information that are of immediate relevance for the development of sea ice research over the next two decades. First, it is critical that a standard methodology be applied when measuring sea ice properties. This facilitates the intercomparison of results and fosters synergy between different data sets. Many aspects of sea ice research have at least some level of standardization for the analysis of basic bulk parameters; there are some, however, such as sea ice algal chlorophyll *a* or inorganic nutrient concentrations that do not. While this book does not provide the ultimate solution for this issue, it should be considered as the first step towards the consolidation of approaches and development of best practices. This is specifically relevant when aiming for pan-arctic and antarctic synthesis efforts like the Fourth International Polar Year 2007–2009. Here the science community needs to reach consensus through a series of dedicated workshops targeting each of the listed parameters and variables, which can help inform updated editions of this book or lead to alternative or accompanying guidelines to facilitate development of best practices that may eventually lead to standardized, fully intercomparable approaches.

Second, it is very evident that sea ice is a major component of polar systems with relevance well beyond the sea ice itself or beyond a single discipline. This can be easily demonstrated when considering human use of the polar seas. Three examples of the relevance of sea ice for humans in polar seas demonstrate that the degree of refinement needed to address these issues will vary depending on the questions asked. The future of arctic seas as a shipping corridor (Figure 4.1a) might be the easiest case to tackle, as the presence, thickness, and strength of the sea ice are of most importance and can be readily assessed. The other extreme might be an analysis of the climate relevance of sea ice (Figure 4.1b), requiring a profound understanding of not only the sea ice physical properties but also the biological cycles in the water column and ice, which are again controlled by sea ice physics and chemistry. Various feedback mechanisms complicate these scenarios and highlight the need for interdisciplinary approaches when aiming at understanding the role of sea ice for polar climate systems. From the subsistence perspective (Figure 4.1c), again a smaller subset of physical and biological estimates might be sufficient to provide necessary information to coastal communities depending on marine life for food and sustenance and to other decision makers. In all of these examples, the spatial and temporal scale at which information is collected and processed is of key importance as well and may vary from case to case.

We hope this book helps stimulate interdisciplinary sea ice research. Such work is critical as we strive to understand current change, predict future changes, and determine how to respond to change. It is currently not precisely known what the sea ice cover will be like in 2050; the only certainty is that it will be very different from the historical record, and that such changes will impact ocean biodiversity, the carbon cycle, and human use as outlined above. Research in the field will remain a necessity to help develop and ground-truth model studies and discover new linkages between disciplines and ice and its environment.

Our three examples demonstrate that the list of required information will vary from question to question. The current edition of the book does not aim at being a complete field guide to sea ice research, but focuses on approaches deemed most relevant, with some constraints provided by the origins of the book out of a field course, as outlined in the Preface to this volume. Other aspects, such as sediment trap deployment or emerging new technologies such as autonomous or remotely operated underwater vehicles, are not covered in this edition. Technological advances will likely help further transform sea ice research and make it fascinating to follow the development of the field over the next decade and work on collaborative approaches to update information contained in this book. Not only will new science emerge, but the societal relevance of sea ice research will also evolve, whether it relates to shipping in the Arctic, the risk of invasive species, the

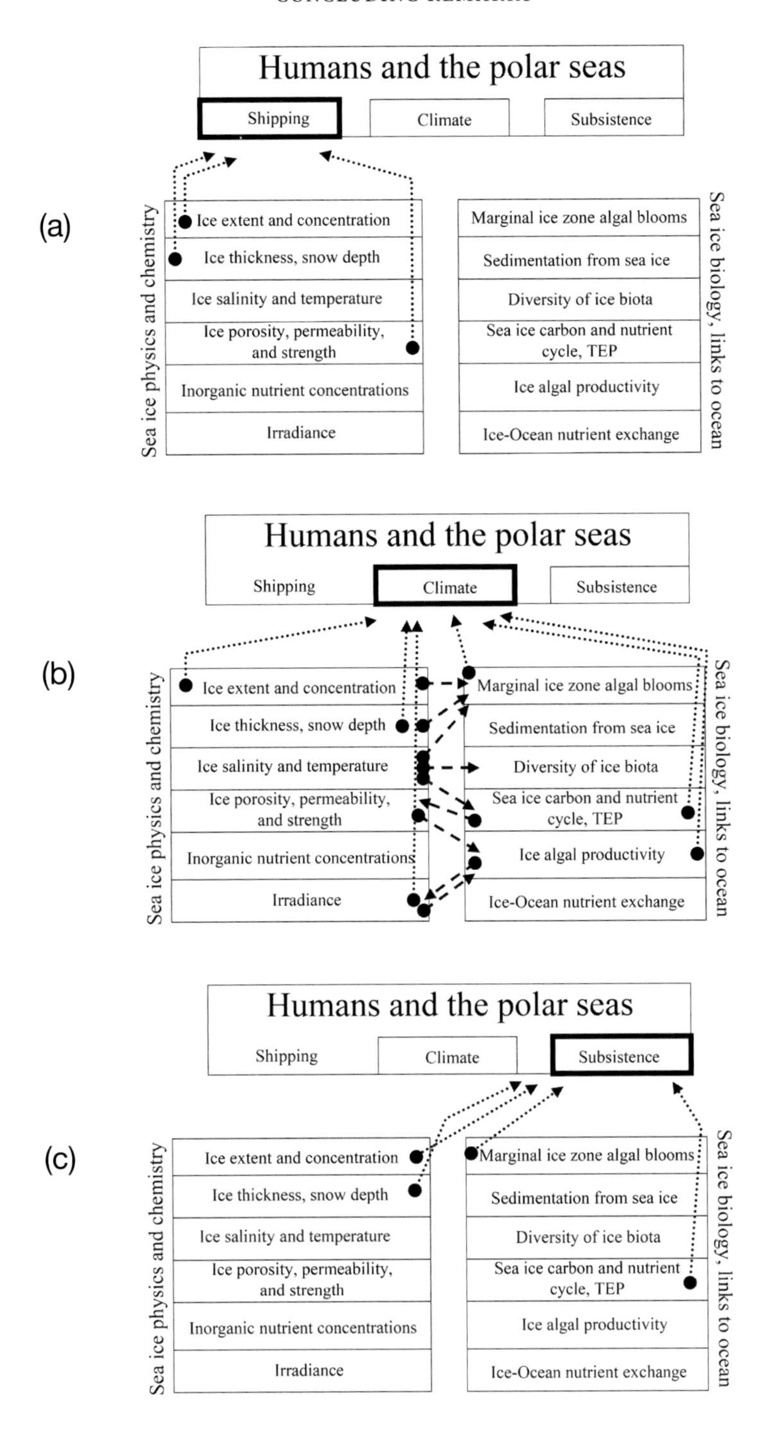


**Figure 4.1.** Three schematic examples for system science approaches needed to understand the relevance of sea ice for humans in polar seas. Shown are linkages between human use or activities: (a) shipping, (b) climate impacts, (c) subsistence hunting and relevant ice variables as well as connections between ice properties highlighting the need for interdisciplinary approaches with different refinement levels (TEP: transparent exopolymeric substance; see Chapter 3.8 for details).

development of new fisheries, or changing subsistence hunting practices. Increased human use of polar seas, in particular in the Arctic, make observing and understanding sea ice processes and their contribution to the environment and human activities a relevant endeavor for the global community.

# About the Multimedia DVD

The multimedia DVD is formatted to play in every DVD region on a standard DVD player in NTSC format. Most computers should open the DVD automatically with DVD player software such as Windows Media Player, Cyberlink PowerDVD, or DVDplayer. The chapters of the DVD are formatted to coincide with those found in the textbook. Please note that not every chapter in the textbook has a corresponding video component. In addition, some DVD menu titles have been abbreviated from their textbook form due to layout constraints. Click on the images to access the video. The following is a list of the contents available on the DVD:

**Main Menu Options**

- About this DVD: Provides information and production credits
- Animations: Direct link to all of the video animations featured in the various chapters (for interactive animations, please see the "SupplementalMaterials" directory located on the DVD-ROM section of this DVD).
- Section 1: Introduction to Sea Ice Field Research
    - Provides an overview of Sea Ice Field Research in Barrow, Alaska
- Section 2: Sea Ice System Services Framework: Development and Application
    - Interview with Amy Lauren Lovecraft
- Section 3: Sea Ice Research Techniques
    - Sea Ice Research Techniques: Chapter Menu
        - Chapter 3.1: Field Techniques for Snow Observations on Sea Ice
        - Chapter 3.2: Ice Thickness and Roughness Measurements
        - Chapter 3.3: Ice Sampling and Basic Sea Ice Core Analysis
            - Chapter 3.3a: Ice Sampling and Processing in the Field
            - Chapter 3.3b: Processing Ice Cores
            - Chapter 3.3c: Sea Ice Microstructure—Preparing Sea Ice Thin Sections

- Chapter 3.4: Thermal, Electrical, and Hydraulic Properties of Sea Ice
- Chapter 3.5: Ice Strength—In Situ Measurement
- Chapter 3.6: Sea Ice Optics Measurements
- Chapter 3.7: Measurements and Modeling of the Ice–Ocean Interaction
- Chapter 3.8: Biogeochemical Properties of Sea Ice
- Chapter 3.9: Assessment of the Abundance and Diversity of Sea Ice Biota
- Chapter 3.10: Studying Seals in Their Sea Ice Habitat: Application of Traditional and Scientific Methods
- Chapter 3.11: Community-Based Observation Programs and Indigenous and Local Sea Ice Knowledge (Indigenous Sea Ice Knowledge)
- Chapter 3.17: Integrated Sea Ice Observation Programs (Coastal Sea Ice Observatory and Landfast Ice Break-out)
- Chapter 3.18: Visual History of the Sea Ice: An Exploration in Archival and Modern Film

### Operating the DVD-ROM

To access the supplemental materials, you will need to play this disk in a DVD-ROM drive of a Windows, Macintosh, Linux, or Unix computer. Open the DVD drive to view the contents of the supplemental files found on this DVD-ROM in the directory "SupplementalMaterials."

## TECHNICAL INFORMATION

### Downloadable Documents

A number of documents are available on the Sea Ice Handbook DVD-ROM in Portable Document Format (PDF). These files can be viewed using Adobe Acrobat Reader. A free copy of Adobe Acrobat Reader can be downloaded from <http://www.adobe.com/products/acrobat/readstep2.html>.

### Contents of Multimedia DVD: Supplemental Materials

This multimedia DVD contains supplemental materials that can be accessed through the DVD's directory structure, with all materials contained in the directory "SupplementalMaterials."

The directory "SupplementalMaterials" contains the following subdirectories with further resources:

1. Animations: This file contains interactive Flash video animation files (.swf). To view SWF animation files: open the file with your web browser (Firefox, Internet Explorer, Safari, etc.). Your web browser will need the Flash Player plug-in enabled to view these files properly. You can also use the Adobe Flash Player to open the SWF files. Free copies of these applications can be downloaded: Firefox from <http://www.mozilla.com/en-US/products/>; Internet Explorer from <http://www.microsoft.com/windows/Internet-explorer/default.aspx>; Safari from <http://www.apple.com/safari/download/>; or Flash Player from <http://get.adobe.com/flashplayer/>.

2. FieldCourseReports: This directory contains an overview of the 2008 Barrow Field Course on Interdisciplinary Techniques in Sea Ice Field Research, including the instructions and materials for the course components (FieldCourseOverview.pdf). The document "FarrellFieldCourseReport Summary.pdf" is a comprehensive summary of the reports from course participants, compiled and edited by Sinéad Farrell.

3. ShipBasedIceObs: This directory contains materials and resources relevant for the chapter of the handbook describing ship-based ice observations:

   (a) ArcticSeaIceObsSoftware: This subdirectory contains program files and resources to install a sea ice observation database that runs under MS Windows, Mac OS X, and Linux (developed by James Halliday, University of Alaska Fairbanks, redesigning an earlier version written by Steve Roberts, University Corporation for Atmospheric Research) and implements the ice observation protocols developed for the Arctic as described in the corresponding handbook chapter.

   (b) MANICE: This subdirectory contains PDF files of the Canadian Ice Service's complete MANICE Manual of Standard Procedures for Observing and Reporting Ice Conditions (reproduced with permission of the Canadian Ice Service, Environment Canada).

   (c) Observing_Antarctic_Sea_Ice: This subdirectory contains a PDF-based manual and resources for completing ship-based ice observations in the Southern Ocean, released by the Scientific Committee on Antarctic Research's ASPeCt Program. The manuals and resources require the programs Adobe Acrobat and Apple Quicktime and instructions on how to install these programs (if not installed already) are provided.

(d) UAAObserversIceGuide.pdf: This is a PDF file of a Sea Ice Observer's Guide developed at the University of Alaska Anchorage.

(e) WMO_IceNomenclature: This subdirectory contains PDF files of the World Meteorological Organization's Sea Ice Nomenclature, a set of proposed revisions, and a document listing sea ice information services worldwide.

4. SnowClassification: This directory contains a PDF file of the International Classification for Seasonal Snow on the Ground, issued by the International Commission on Snow and Ice of the International Association of Scientific Hydrology and the International Glaciological Society.

**Acknowledgments**

Funding for this project was provided in part by the U.S. National Science Foundation and the Experimental Program to Stimulate Research. This DVD was created in collaboration with Film students at the University of Alaska Fairbanks, who produced much of the footage featured in this DVD. Special thanks to Theatre UAF at the University of Alaska Fairbanks. All source material for this DVD is available to the public at the University of Alaska Rasmuson Library.

**Credits**

Directed and produced by Maya Salganek. Co-produced by Hajo Eicken. Associate producer Samuel K. German. Editors-in-chief: Tyson Hansen and Maya Salganek. Editors: Samuel K. German, Lacie Stiewing, and Kimberly Maher. Animation by Miho Aoki. Visual anthropology and scientific documentary digital video students, May 2008: Archana Bali, Paula Daabach, Samuel German, Anna Edwardson, Tyson Hansen, Kimberly Maher, Lacie Stiewing, and Molly Wilson.

# List of Contributors and Affiliations

**Bluhm, Bodil**
School of Fisheries and Ocean Sciences
Institute of Marine Science
P.O. Box 757220
University of Alaska Fairbanks
Fairbanks, Alaska 99775-7220
USA
E-mail: bluhm@ims.uaf.edu

**Druckenmiller, Matthew**
Geophysical Institute
P.O. Box 757320
University of Alaska Fairbanks
Fairbanks, Alaska 99775-7320
USA
E-mail: mldruckenmiller@alaska.edu

**Eicken, Hajo**
Geophysical Institute
PO Box 757320
University of Alaska Fairbanks
Fairbanks, Alaska 99775-7320
USA
E-mail: hajo.eicken@gi.alaska.edu

**Florence Fetterer**
WDC for Glaciology, Boulder/
National Snow and Ice Data Center
1540 30th St. University of Colorado
Boulder, CO 80309-0449
USA
E-mail: fetterer@nsidc.org

**Gearheard, Shari**
National Snow and Ice Data Center
Cooperative Institute for Research in Environmental Sciences
University of Colorado at Boulder
P.O. Box 241
Clyde River, Nunavut, X0A 0E0
CANADA
E-mail: shari.gearheard@nsidc.org

**Gradinger, Rolf**
School of Fisheries and Ocean Sciences
University of Alaska Fairbanks
Fairbanks, Alaska 99775-7220
USA
E-mail: rgradinger@ims.uaf.edu

**Haas, Christian**
Dept. of Earth & Atmospheric Sciences
University of Alberta
Edmonton, Alberta
CANADA T6G 2E3
E-mail: Christian.Haas@ualberta.ca

**Huntington, Henry P.**
23834 The Clearing Drive
Eagle River, Alaska 99577
USA
E-mail: hph@alaska.net

**Hutchings, Jennifer K.**
International Arctic Research Center
930 N. Koyukuk Drive, P.O. Box 757340
University of Alaska Fairbanks
Fairbanks, Alaska 99775-7340
USA
E-mail: jenny@iarc.uaf.edu

**Ingham, Malcolm**
School of Chemical and Physical Sciences
Victoria University of Wellington
NEW ZEALAND
E-mail: Malcolm.Ingham@vuw.ac.nz

**Kelly, Brendan P.**
Office of Polar Programs
National Science Foundation
4201 Wilson Boulevard
Arlington, Virgina 22230
Present address:
National Marine Mammal Laboratory, NOAA
AFSC/TSMRI Auke Bay Laboratories
17109 Pt. Lena Loop Road
Juneau, Alaska 99801
E-mail: brendan.kelly@noaa.gov

**Leppäranta, Matti**
Division of Geophysics
Department of Physical Sciences
University of Helsinki
P.O. Box 64 (Gustaf Hällströminkatu 2)
Fi-00014 Helsinki
FINLAND
E-mail: Matti.lepparanta@helsinki.fi

**Lindsay, Ron**
Polar Science Center
Applied Physics Laboratory
University of Washington
1013 NE 40th Street
Seattle, Washington 98105
USA
E-mail: lindsay@apl.washington.edu

**Lovecraft, Amy Lauren**
Northern Studies
and
Political Science Department
University of Alaska Fairbanks
Fairbanks, Alaska 99775
USA
E-mail: allovecraft@alaska.edu

**Mahoney, Andy**
Department of Physics
University of Otago
730 Cumberland Street
Dunedin, 9016
NEW ZEALAND

**Massom, Robert A.**
Australian Antarctic Division and
Antarctic Climate & Ecosystems Coop. Research Center
Private Bag 80, c/o University of Tasmania
Sandy Bay, Tasmania 7005
AUSTRALIA

**Masterson, D. M.**
112 Silvercreek Crescent N.W.
Calgary AB
CANADA T3B 4H7
E-mail: dmasterson@shaw.ca

**McMinn, Andrew**
Institute of Antarctic & Southern Ocean Studies
Private Bag 77
University of Tasmania
Hobart TAS 7001
AUSTRALIA
E-mail: Andrew.mcminn@utas.edu.au

**Nomura, Daiki**
Institute of Low Temperature Science
Hokkaido University
Kita-19, Nishi-8, Kita-ku
Sapporo 060-0819
JAPAN
E-mail: daiki@lowtem.hokudai.ac.jp

**Orlich, Alice**
International Arctic Research Center
930 N. Koyukuk Drive, P.O. Box 757340
University of Alaska Fairbanks
Fairbanks, Alaska 99775-7340
USA
E-mail: aorlich@iarc.uaf.edu

**Perovich, Don**
Cold Region Research & Engineering Lab
72 Lyme Road
Hanover, New Hampshire 03755
USA
E-mail: donald.k.perovich@usace.army.mil

**Petrich, Chris**
Geophysical Institute
P.O. Box 757320
University of Alaska Fairbanks
Fairbanks, Alaska 99775-7320
USA
E-mail: Chris.petrich@gi.alaska.edu

**Pringle, Daniel**
Arctic Region Supercomputing Center
105 West Ridge Building, P.O. Box 756020
University of Alaska Fairbanks
Fairbanks, Alaska 99775-6020
USA

**Roberts, Andrew**
International Arctic Research Center
930 N. Koyukuk Drive, P.O. Box 757340
University of Alaska Fairbanks
Fairbanks, Alaska 99775-7340
USA
E-mail: aroberts@iarc.uaf.edu

**Rozell, Ned**
Geophysical Institute
903 N. Koyukuk Drive, P.O. Box 757320
University of Alaska Fairbanks
Fairbanks, Alaska 99775-7320
USA
E-mail: nrozell@gi.alaska.edu

**Shirasawa, Kunio**
Pan-Okhotsk Research Center
Institute of Low Temperature Science
Hokkaido University
Kita-19, Nishi-8, Kita-Ku
Sapporo 060-0819
JAPAN
E-mail: kunio@lowtem.hokudai.ac.jp

**Sturm, Matthew**
U.S. Army Cold Regions Research and Engineering Laboratory
Ft. Wainwright, Alaska 99703-0170
USA
E-mail: msturm@crrel.usace.army.mil

**Worby, Anthony P.**
Australian Antarctic Division and ACE CRC
University of Tasmania
Private Bag 80
Hobart, 7001
AUSTRALIA
E-mail: A.Worby@utas.edu.au

**Zhang, Jinlun**
Polar Science Center
Applied Physics Laboratory
University of Washington
1013 NE 40th Street
Seattle, Washington 98105
USA
E-mail: zhang@apl.washington.edu

# Index